THIS BOOK DOES
NOT CIRCULATE

KENT STATE UNIVERSITY LIBRARY, KENT, OHIO

THE MAPPING OF AFRICA

Onderzoeksprogramma Explokart

Research programme Explokart

Utrechtse Historisch-Cartografische Studies

Utrecht Studies in the History of Cartography

7

redactie / editorial board

Peter van der Krogt
Paula van Gestel-van het Schip
Günter Schilder

with the collaboration of Penelope Wiborg Betz

Published

1. D. Blonk & J. Blonk-van der Wijst, *Hollandia Comitatus: Een kartobibliografie van Holland.*

2. H. Deys, M. Franssen, V. van Hezik, F. te Raa en E. Walsmit, *Guicciardini Illustratus: De kaarten en prenten in Lodovico Guicciardini's Beschrijving van de Nederlanden.*

3. J. Smits, *Petermann's Maps: carto-bibliography of the maps in Petermanns Geographische Mitteilungen, 1855-1945.*

4. P.J. de Rijke, *Frisia Dominium: Kaarten van de provincie Friesland tot 1850: Geschiedenis en cartobibliografie.*

5. P.W.A. Broeders, *Gijsbert Franco baron von Derfelden van Hinderstein (1783-1857): Leven en werk van 'eene ware specialiteit' in kaart gebracht.*

6. Paula van Gestel-van het Schip en Peter van der Krogt (ed.), *Mappæ Antiquæ: Liber Amicorum Günter Schilder.*

7. Richard L. Betz, *The Mapping of Africa: A cartobibliography of printed maps of the African continent to 1700.*

Richard L. Betz

THE MAPPING OF AFRICA

A Cartobibliography of Printed Maps of the African Continent to 1700

Hes & De Graaf Publishers BV

This publication was made possible through the support of

VSB fonds, Utrecht
Faculteit Geowetenschappen, Universiteit Utrecht
K.F. Hein Fonds
Stichting Dr Hendrik Muller's Vaderlandsch Fonds

Graphic-design and lay-out: Fred van den Berg, Vreeland
Lithography and printing: Drukkerij WC den Ouden BV, Amsterdam
Binding: Boekbinderij Van Waarden b.v., Zaandam
Illustration title page: Wenzel Hollar, small map of Africa, see page 338.

HES & DE GRAAF Publishers BV
Tuurdijk 16
3997 MS 't Goy-Houten
The Netherlands

ISBN 978 9061944898

© 2007 HES & DE GRAAF Publishers BV, 't Goy-Houten

All rights reserved

Table of Contents

Foreword	10
Preface	11
Acknowledgements	12
The Mapping of Africa	
The Era before the Portuguese	15
Indigenous Mapmaking	
Early Understanding of Africa by the Egyptians, Greeks, and Romans	
Arab Cartography of Africa	
Chinese Cartography of Africa	
Medieval European View of Africa	
Portolan Charts of Africa	
Early European Renaissance Mapping of Africa	
European Exploration of Africa	28
The Starting Point	
The Portuguese Voyages of Exploration	
The Portuguese in Africa	
The Arrival of Other Europeans	
Ptolemaic Maps: The Earliest Printed Maps of Africa	37
Significant World Maps that Show the Continent of Africa	40
Henricus Martellus, Manuscript World Map, c.1489-1492	
Francesco Rosselli, Printed World Map, Florence, c.1492-1493	
Martin Behaim, Manuscript Terrestrial Globe, 1492	
Juan de La Cosa, Manuscript Planisphere, 1500	
Alberto Cantino, Manuscript Planisphere, 1502	
Nicolo de Caveri, Manuscript Planisphere, c.1504-1505	
Giovanni Contarini–Francesco Rosselli, Printed World Map, Venice or Florence, 1506	
Martin Waldseemüller, Printed World Map, Strasbourg, 1507	

Overview of Printed Maps of Africa from 1505 to 1700 53
1. African Cartography from the Anonymous Woodcut of c.1505 to Sebastian Münster, 1540
2. The Landmark Maps by Gastaldi and Ortelius
3. The Late Sixteenth Century – The Developing Understanding of Africa and the Mercator Map
4. The Dominance of the Dutch and the Blaeu Model
5. The Scientific View of Africa Starting with the Nicolas Sanson 1650 Map
6. The Evolution of the Mapping of Africa: The Influence of Jaillot and Duval
7. Delisle Model of 1700

Cartobibliography
Notes on the Use of the Cartobibliography 75

1	1508 Antonio Francanzano (Fracan) da Montalbodo, Milan	78
2	1513 Pliny the Elder – Melchior Sessa, Venice	81
3	1540 Sebastian Münster, Basel	83
4	1554 Giovanni Battista Ramusio – Giacomo Gastaldi, Venice	95
5	1556 Leo Africanus – Jean Temporal, Lyon	98
6	1562 Paulo Forlani, Venice	101
7	1563 Giovanni Battista Ramusio – Giacomo Gastaldi, Venice	103
8	1563 Giovanni Francesco Camocio – Paulo Forlani, Venice	105
9	1564 Giacomo Gastaldi – Fabio Licinio, Venice	108
10	1564[65] Ferando Bertelli, Venice	114
11	c. 1570-1575 Giacomo Gastaldi – Giovanni Francesco Camocio, Venice	116
12	1570 Abraham Ortelius, Antwerp	118
13	c. 1573 Giacomo Gastaldi – Donato Bertelli, Venice	126
14	c. 1573 Gerard de Jode, Antwerp	129
15	1573 Giovanni Lorenzo D'Anania, Naples	131
16	1575 François de Belleforest, Paris	132
17	1575 André Thevet, Paris	133
18	1577 Abraham Ortelius – Filips Galle (1), Antwerp	135
19	1579 Claudio Duchetti, Rome	137
20	1582 Giovanni Lorenzo D'Anania, Venice	140
21	1588 Sebastian Münster – Sebastian Petri, Basel	143
22	1588 Livio Sanuto, Venice	145
23	1589 Jodocus Hondius – Jean Leclerc, Paris	147
24	1589 Heinrich Bünting, Magdeburg	149
25	c. 1590 Giovanni Battista Mazza, Venice	152
26	1592 Heinrich Bünting – Daniel Adam z Weleslavina, Prague	153
27	1593 Cornelis de Jode, Antwerp	154
28	1593 Abraham Ortelius – Filips Galle (2), Antwerp	156
29	1594 Giuseppe Rosaccio, Ferrara	158
30	1594 Giuseppe Rosaccio, Florence	159
31	1595 Gerard Mercator II, Duisburg	162
32	1595 Giovanni Botero, Venice	166
33	1596 Cornelis de Jode, Antwerp	168
34	1596 Giovanni Antonio Magini, Venice	170
35	1597 Petrus Keschedt, Cologne	172
36	1597 Fausto Rughesi Rome	174
37	1597 Cornelis Claesz.– Barent Langenes, Middelburg	176
38	1597 Johannes Matalius Metellus, Cologne	179
39	1598 Abraham Ortelius – Pietro Maria Marchetti, Brescia	181
40	1598 Zacharias Heyns, Amsterdam	183
41	1598 Jodocus Hondius, Amsterdam	185
42	1598 Hernando de Solis, Valladolid	187
43	1600 Leo Africanus – John Pory, London	189
44	1600 Arnoldo di Arnoldi, Siena	191
45	c. 1600 Luis Teixeira–Joannes van Doetecum, Amsterdam	193
46	1600 Matthias Quad, Cologne	195
47	1601 Abraham Ortelius – Johannes van Keerbergen, Antwerp	197
48	1602 Levinus Hulsius, Nürnberg	199
49	1602/09-1617 Cornelis Claesz. – Johannes Janssonius, Amsterdam	202
50	1605 Heinrich Bünting – Jan Jansz., Arnhem	204
51	1605 Giuseppe Rosaccio, Venice	206
52	1606 Jodocus Hondius, Amsterdam	208
53	1607 Jodocus Hondius, Amsterdam	211
54	1608 Willem Janszoon (Blaeu), Amsterdam	214
55	1614 Pieter van den Keere, Amsterdam	220

56	1616 Petrus Bertius – Jodocus Hondius Jr., Amsterdam	223
57	1617 Willem Janszoon (Blaeu), Amsterdam	225
58	1619 Jodocus Hondius Jr., Amsterdam	229
59	c. 1620-1635 Godefridus de Scaicki, Rome	233
60	1623 Jodocus Hondius Jr., Amsterdam	235
61	1624 Petrus Bertius, Paris	238
62	1626 John Speed, London	241
63	1627 Petrus Bertius – Melchior Tavernier, Paris	244
64	1628 Johannes Janssonius, Amsterdam	246
65	1630 Jan Evertsz. Cloppenburch, Amsterdam	249
66	c.1636 Jean Boisseau, Paris	251
67	1638 Matthäus Merian, Frankfurt	254
68	1639 Petrus Bertius – Melchior Tavernier (attr.), Paris	256
69	1640 Giovanni Botero – Lucatonio Giunti, Venice	258
70	1640 Jacques Honervogt, Paris	261
71	1641 Philipp Cluver – Conrad Buno, Brunswick (Braunschweig)	262
72	1643 Jean Boisseau, Paris	265
73	1644 Nicolas Picart, Paris	267
74	1645 Pierre Duval, Paris	270
75	1646 William Humble, London	272
76	1646 Stefano Mozzi Scolari, Venice	274
77	1647 Cornelis Danckerts, Paris	277
78	1650 Nicolas Sanson, Paris	279
79	1650 Nicolas Sanson, Paris	281
80	c. 1650 Pierre Mariette (attr.), Paris	283
81	1651 Nicolas Berey, Paris	285
82	c. 1651 Nicolas Picart (1), Paris	287
83	1652 Henry Seile, London	289
84	c. 1655 Jan Mathisz., Amsterdam	291
85	c. 1656 Pierre Duval, Paris	294
86	1656 Nicolas Sanson, Paris	296
87	c. 1658 Nicolaas Visscher I, Amsterdam	299
88	1658 Robert Walton, London	302
89	1658 Nicolas Berey, Paris	304
90	1658 Gabriel Bucelin, Ulm	306
91	1659 Joan Blaeu, Amsterdam	307
92	c. 1659 Nicolas Picart (2), Paris	310
93	1659 Philipp Cluver – Elzevier, Amsterdam	313
94	1660 Giovanni Battista Nicolosi, Rome	315
95	c. 1660 Pierre Duval, Paris	318
96	1660 Frederick de Wit, Amsterdam	321
97	c. 1661 Hugo Allard, Amsterdam	324
98	1661 Philipp Cluver – Johann Buno, Wolfenbüttel	326
99	1661 Dancker Danckerts, Amsterdam	328
100	1663 Anne Seile, London	331
101	1664 Pierre Duval, Paris	333
102	c. 1665 Antonio Francesco Lucini, Florence (?)	335
103	1665/1666 Thomas Jenner – Wenceslaus Hollar, London	337
104	1666 Nicolas Sanson, Paris	339
105	1666 Giovanni Giacomo de Rossi, Rome	341
106	1667 Guillaume Sanson, Paris	344
107	1668 Guillaume Sanson, Paris	346
108	1668 Jacob van Meurs, Amsterdam	348
109	1668 John Overton, London	350
110	1669 Pierre Duval, Paris	352
111	1669 Richard Blome, London	354
112	1669 Alexis-Hubert Jaillot, Paris	356
113	1670 Giovanni Giacomo de Rossi, Rome	359

114	c. 1670 Frederick de Wit, Amsterdam	361
114A	c. 1680 Frederick de Wit, Anonymous, Amsterdam	365
115	c. 1671 Gerrit Lucasz. van Schagen, Amsterdam	367
116	1672 Frederick de Wit, Amsterdam	369
117	1673 Pietro Todeschi, Bologna	372
118	1674 Alexis-Hubert Jaillot, Paris	374
119	c. 1676 John Seller, London	377
120	1677 Giovanni Giacomo de Rossi, Rome	379
121	1678 Pierre Duval – Johann Hoffmann, Nürnberg	381
122	1678 Pierre Duval, Paris	383
123	1679 Nicolas Sanson – Johann David Zunner, Frankfurt	387
124	1680 William Berry, London	389
125	1680 Robert Morden, London	391
126	1681 Jonas Moore – Herman Moll, London	393
127	c. 1682 Alexis Hubert Jaillot – Johann Hoffmann, Nürnberg	395
128	c. 1682-1685 Nicolaas Visscher II, Amsterdam	397
129	1683 Nicolas Sanson – Antoine de Winter, Amsterdam & Utrecht	401
130	1683 Johannes de Ram, Amsterdam	404
131	1683 Justus Danckerts, Amsterdam	406
132	1683 Alain Manesson Mallet, Paris	409
133	1683 Alain Manesson Mallet, Paris	410
134	c. 1684 John Seller, London	411
135	1684 Nicolas de Fer – Jacques Robbe, Paris	412
136	1685 Alain Manesson Mallet – Johann David Zunner, Frankfurt	415
137	1685 Alain Manesson Mallet – Johann David Zunner, Frankfurt	417
138	1685 John Lawrence, London	418
139	c. 1685 Justus Danckerts, Amsterdam	419
140	1686 Philipp Cluver – Johann Buno – Herman Mosting, Wolfenbüttel	421
141	1686 Philip Lea and John Overton, London	423
142	1687 Nicolas de Fer – Jacques Robbe – Antoine de Winter, Utrecht	425
143	1688 Pierre Duval – Jean Certe, Lyon	427
144	1688 Giuseppe Rosaccio – Giuseppe Moretti, Bologna	429
145	1688 Nicolas de Fer – Jacques Robbe – Thomas Amaulry, Lyon	430
146	1689 Jean Baptiste Nolin – Vincenzo Coronelli, Paris	431
147	c. 1690 Johann Stridbeck, Jr., Augsburg	434
148	c. 1690 Gerard Valk, Amsterdam	436
149	1690 A. Phérotée de la Croix, Lyon	439
150	c. 1690 Joachim Bormeester, Amsterdam	441
151	1690 Sebastián Fernández de Medrano – Jacques Peeters, Brussels	443
152	1691 James Moxon, London	446
153	1691 Vincenzo Maria Coronelli, Venice	447
154	1691 Laurence Echard, London	449
155	1692 Johann Ulrich Müller, Ulm	450
156	1692 Alexis Hubert Jaillot – Pieter Mortier, Amsterdam	452
157	1693 Nicolas de Fer, Paris	455
158	1694 Alexis Hubert Jaillot – Pieter Mortier, Amsterdam	457
159	1695 Herman Moll, London	460
160	169[5] Alexis Hubert Jaillot, Paris	462
161	c. 1696 Gerard Valk, Amsterdam	465
162	c. 1696 Carel Allard, Amsterdam	468
163	1696-1698 Nicolas de Fer, Paris	470
164	1697 Abraham Ortelius – Domenico Lovisa, Venice	474
165	c. 1697 Jacob von Sandrart, Nürnberg	476
166	1697 Philipp Cluver – Johann Wolters, Amsterdam	478
167	1697 A. Phérotée de la Croix – Johann L. Gleditsch, Leipzig	480
168	c. 1699-1700 Justus Danckerts, Amsterdam	482
169	1699 Heinrich Scherer, Augsburg	485
170	1700 Nicolas de Fer, Paris	487

171	1700 Nicolas de Fer, Paris	489
172	1700 Paolo Petrini, Naples	491
173	c. 1700 Paolo Petrini, Naples	493
174	1700 Guillaume Delisle, Paris	495

Appendices
List of Lost Maps that Show Africa	500
List of Some 'Firsts' on Printed Maps of the Continent of Africa	502
Some Important Placenames on the Map of Africa	504
Chronology of Historical Events	507

Bibliography 510

Libraries Cited in this Cartobibliography 522

Indexes
Alphabetical Index of Maps by Title	524
Names of Persons	529

Photo credits 539

Foreword

The *Utrecht Studies in the History of Cartography* is a series which aims to publish the results of historical research on maps executed under the supervision of, or in close co-operation with, the staff of the URU-Explokart research group in the history of cartography. An analytical cartobibliography of a certain group of maps must be included in the research. The series was started to publish the research by the so-called Explokart working groups. Members of these working groups are generally well-educated, elderly men (only a few women), pensioners who spent their working life as high-ranking managers in trade and industry and as medical specialists. The similarity between them is that they like maps. Under the supervision of the Explokart staff, the working groups research several subjects in the history of cartography, generally cartobibliographies. The aim of this research is not to keep the volunteers off the street, but to let them publish on the subject. The cartobibliographies of the provinces of Holland (volume 1 of the series) and Friesland (vol. 4) and that of the maps and plans in Guicciardini editions (vol. 2) are examples of this work. However, the series is also open to individual volunteer researchers, who in an early stage of their research contact the Explokart staff. Their work is often discussed with the staff, but they do not necessarily research under the supervision of the staff. The research by Jan Smits on the maps in the journal *Petermanns Geographische Mitteilungen* (vol. 3) is the first of this kind of research to be published in the series. Also dissertations, such as the fifth volume on the cartographer Von Derfelden van Hinderstein, and future research by Explokart staff members will find a place in this series.

The present seventh volume of the series was compiled by an independent researcher. Richard Betz and his wife Penny lived for twelve years in Southern Africa, where he engaged in economic development work. His doctoral dissertation focused on factors that influence the successful development of rural enterprise in Africa. During this time he acquired a passion for old maps of the continent. He started to collect those maps, and, as often occurs with map collectors, started to trade maps.

I met Richard Betz for the first time at the Society for the History of Discoveries conference in Washington, DC, in 2000. A few months later at the Miami Map Fair in February 2001, he showed me a Blaeu map of Aethiopia Superior which was not included in my brand new Blaeu volume of *Atlantes Neerlandici*. So he made a good start to become a volunteer Explokart researcher! In Miami he told me about his plans to write a cartobibliography of Africa. There was a need for such a cartobibliography since the only works on maps of Africa were Norwich's catalogue of his collection and Tooley's research. The first "official" contact between Richard Betz and Explokart was at the June 2001 IMCoS Map Fair in London. Sitting on the curb of the sidewalk outside the fair venue, Richard and I discussed the conditions a cartobibliography of the African continent had to meet to be published in our series. During the following five years Richard and the Explokart staff had several meetings in Utrecht, in New Hampshire, and at conferences and map fairs, and a lot of e-mail contact. The final meeting to discuss the draft version of his cartobibliography was at the June 2006 Map Fair in London.

The editors of the *Utrecht Studies in the History of Cartography* are very glad to welcome Richard Betz among their authors. His work is a major step forward in the unravelling of the cartography of the "dark continent" with all its hypotheses on the interior, which only became known to the outside world in the nineteenth century.

Peter van der Krogt,
head of the URU-Explokart research program,
Utrecht, May 2007

Preface

'We will stifle ourselves if we dare nothing and do nothing for fear of not covering all the ground all the time'.
Anonymous.

The genesis of this cartobibliography began in the late 1990s during an unsuccessful attempt to date and describe several loose maps of Africa that I had in my possession. This experience planted the seed of trying to categorize and describe all printed maps to 1700 that showed the entire African continent. What began as a haphazard gathering of information on maps of Africa has evolved into this cartobibliography on the mapping of Africa.

Thanks to the structured format and valuable advice provided by the staff of the Explokart Research Group of Utrecht University, I have reviewed the literature, conducted an investigation in many major libraries and private collections, analyzed these findings, and compiled information on the 174 distinctly separate maps contained in this book. If I knew then, what I know now about the enormous nature of this task, I wonder if I would have started. Now, I am glad that it has been done. I have learned a lot.

Like many authors of similar books, there is a continual dread of 'missing' a map, or some piece of relevant information about a map. That has been and will continue to be a concern. As Philip Burden, the noted author and antiquarian book, print, and map dealer, wisely advised me, 'This will inevitably happen'. No one person can be expected to have seen all the maps and all the atlases that were produced during the period represented in this book. Philip told me that the important thing is to publish and to get the information contained in this book into as many hands as possible. Only then can some of these 'missing' maps be brought to light.

Some people have asked me why I ended this cartobibliography in 1700 with the Delisle map of Africa. Besides the standard answer of finishing the book with this first map which does not show the two Ptolemaic lakes in central Africa, the date of 1700 provided me with a hope to complete this book within a reasonable period of time. Once I began the research, I quickly realized that the number of distinct maps of Africa dramatically increased in the later half of the seventeenth century. This increase in maps continued to accelerate in the eighteenth century.

I did not include maps of regions of Africa or maps that showed Africa with other continents. Some of these maps are mentioned in the introduction and I hope that the information contained in the book's individual map entries can help the reader to identify regional maps of Africa.

Enjoy the maps presented in this cartobibliography. Maps are a wonderful medium. They are aesthetically pleasing to behold and they provide us a window into our own past.

Richard L. Betz
Stoddard, New Hampshire, USA
February 2007

Acknowledgements

This cartobibliography would not exist without the extensive assistance of many people. I want to thank all of the following people, and others whom I may have inadvertently omitted.

First and foremost, I want to thank my partner in this work, my wife, Penelope Wiborg Betz for her untiring support, encouragement, and good humor throughout what at times seemed to be an overwhelming task. Without her many long hours editing the text, her keen eye for observing differences between maps, and her patience on numerous library trips with their false leads and dead ends, this book would never have been completed.

I am also deeply indebted to the following people who read all or portions of the text, and gave suggestions and comments:

Günter Schilder, emeritus Professor on the History of Cartography, former head of the Explokart Research Program, Faculty of Geosciences, Universiteit Utrecht, The Netherlands.
Peter van der Krogt, head of the Explokart Research Program, Faculty of Geosciences, Universiteit Utrecht, The Netherlands.
Paula van Gestel–van het Schip, researcher and editor of the Explokart Research Program, Faculty of Geosciences, Universiteit Utrecht, The Netherlands.
Jason Hubbard, map researcher and author, France.
Wulf Bodenstein, map researcher and author, Brussels, Belgium.
Philip Burden, author and antiquarian book, print, and map dealer, Clive A. Burden Ltd., Rickmansworth, Herts, UK.
Rodney Shirley, map researcher and author, Buckingham, UK.
Douglas Sims, map researcher and author, Brooklyn, NY, USA.
Joseph Q. Walker, antiquarian map collector and researcher, UK.
Elri Liebenberg, emeritus Professor of Geography, the University of South Africa, Pretoria, South Africa.
Gerald J. Rizzo, Executive Director, Afriterra Free Library, Boston MA, USA.
Paula Flemming, Stoddard, NH, USA.

The following people, among others, provided me with numerous map images, and with assistance on individual maps:

Paul and Stephan Haas, Antiquariat Gebr. Haas, Bedburg-Hau, Germany.
Roger S. Baskes, antique book collector, Chicago, Illinois, USA.
Beatrice Loeb-Larocque and Pierre Joppen, Paulus Swaen Internet Auction and Gallery, Paris, France.
Bernard Shapero Rare Books, London, UK.
Martayan Lan Fine Antique Mapsm New York, USA.
Peter Meurer, researcher and author, Heinsberg, Germany.
Altea Gallery, London, UK.
Richard B. Arkway Antique Maps, New York, USA.
Alexandre Antique Prints, Maps & Books, Toronto, Canada.
Filip Devroe, Antiquariaat Sanderus, Gent, Belgium.
Nicolas Struck, Nicolas Struck Book and Map Dealer, Berlin, Germany.
Fritz Hellwig, Bonn, Germany.
Klaus Stopp, Mainz, Germany.
Ashley Baynton-Williams, map researcher and author, UK.
Jan Werner, Map Curator, Universiteitsbibliotheek Amsterdam, The Netherlands.
Tom Hall, Chicago, IL, USA.
Marcel van den Broecke, author and director, Cartographica Neerlandica, Bilthoven, The Netherlands.
W. Graham Arader III Galleries, New York, USA.
Leen Helmink Antique Maps & Prints, Amersfoort, The Netherlands.
Barry Lawrence Ruderman, Antique Maps La Jolla, California, USA.
Markus Heinz, Map Curator, Staatsbibliothek zu Berlin, Berlin, Germany.
Reiss & Sohn Buch- und Kunstantiquariat Auktionen, Königstein, Germany.

The following provided advice, assistance with specific library questions, and general moral support:

Paul van den Brink, researcher of the Explokart Research Program, Faculty of Geosciences, Universiteit Utrecht, The Netherlands.
Francis Herbert, map librarian, Royal Geographical Society, London, UK.
Alice Hudson, map librarian, New York Public Library, USA.
Peter Barber and the staff, especially Debbie Hall, Map Room, British Library, London, UK.
The Staff of the Département des Cartes et Plans, Bibliothèque Nationale de France, and especially its Director, Hélène Richard, and senior staff including Catherine Hofmann and Madeleine Barnoud, Paris, France.
John Hebert, Head, and Pam van Ee and the rest of the staff, Geography and Map Division, Library of Congress, Washington, D.C., USA.
Göran Bäärnhielm, map curator, Kungliga Biblioteket, Stockholm, Sweden.
Robert Karrow, curator of Special Collections and curator of maps, The Newberry Library, Chicago, Illinois, USA.
David Cobb, curator of the Harvard Map Collection, and his staff, Harvard College Library, Cambridge, MA, USA.
Annalisa Battini, Biblioteca Estense Universitaria, Modena, Italy.
György Danku, map curator, National Széchényi Library, Budapest, Hungary.
Sjoerd de Meer and Ron Brand, Maritiem Museum, Rotterdam, The Netherlands.
Tony Campbell, retired head of the Map Room, The British Library, London, UK.
Henrik Dupont, Det Kongelige Bibliotek, Copenhagen, Denmark.
Piero Falchetta, Ufficio Carte Geografiche, Biblioteca Nazionale Marciana, Venezia, Italy.
Gillian Hutchinson, curator of the History of Cartography, National Maritime Museum, Greenwich, UK.
Roberto Trujillo, head Dept. of Special Collections, Stanford University Library, Stanford, CA, USA.
Charles van den Heuvel, former head of the Collectie Bodel Nijenhuis, Universiteitsbibliotheek Leiden, The Netherlands.
Marco van Egmond, map curator, Map Library, Faculty of Geosciences, Universiteitsbibliotheek Utrecht, The Netherlands.
Theodor Bauer and the staff of the Bayerische Staatsbibliothek, München, Germany.
George Carhart, Osher Map Library, Portland, Maine, USA.
Ron Grim, curator, Norman B. Leventhal Map Center, Boston Public Library, USA.
Frits Muller, Frederik Muller Rare Books, Bergum, The Netherlands.

The Mapping of Africa

The Era before the Portuguese

Indigenous Mapmaking

Many examples of what may be called 'map form' rock art (engravings and paintings) with depictions of people, animals, and surroundings occur on the continent of Africa. The earliest rock art, from more than ten thousand years ago, is generally rough drawings of people and animals, and is focused on hunting. By ten thousand years BCE, the emergence of agricultural-based societies resulted in rock art depicting homesteads, cattle, and land and their relationships to one another.

Figures drawn on rock and cave walls, which suggest geographic relationships, exist in the Atlas and Tissoukat Mountains of Morocco and southern Algeria and throughout southern Africa (South Africa, Lesotho, Swaziland, Namibia and Zimbabwe). Short (2003: 29) states that symbols painted in caves in Mali have been interpreted as maps showing directions. The rock art of southern Africa from around 500 years ago depicts cattle, homesteads, pathways and settlements in the form of plans of the artists' surroundings.

The oldest known map representing a part of Africa, and likely one of the earliest maps still extant in the world showing easily recognizable geographical information, is a papyrus map housed at the Museo Egizio in Turin, Italy. In total, the map is about 210-280 cm long by 41 cm wide. It depicts a 15 km stretch of Wādī al-Hammāmāt ('Valley of Many Baths') in the central part of Egypt's Eastern Desert to the east of Luxor. The map contains numerous descriptive notations. It clearly shows Wādī al-Hammāmāt's long valley, other routes with information on where these routes lead, the surrounding hills (in conical forms with wavy sides), the important quarries of decorative stone and gold (identified with red), and the mining settlement at Bir Umm Fawākhir. Made about 1150 BCE, it was prepared for one of the quarrying expeditions sent to this region by King Ramesses IV (1156-1150). (Harrell 2007. Also Harrell's extensive bibliography on this subject at the end of his website article. Shore in Harley and Woodward, 1987: 117-129. Bricker 1989: 147).

Non-permanent maps were often used when giving directions by drawing maps in the sand or through the verbal recitation of geographic information that had been handed down from generation to generation. Scenes of sub-Saharan Africans drawing maps on the ground are a recurring theme in the literature related to the exploration of Africa by Europeans. Unfortunately, these maps were ephemeral. Interestingly, since their first arrival, European explorers readily understood and incorporated indigenous 'maps' into their view of Africa. This was usually accomplished through interviews with African traders, pilgrims, and other well-traveled people, probably supplemented by actual map sketches, literally in the sand, and other visual means of communication. It can be assumed that Arabs on the east coast as well as those in North Africa, such as Leo Africanus in the early sixteenth century, who ventured south of the Sahara Desert, must have obtained much of their geographic information from African sources.

One of the best known permanent maps attributed to African sources is a map that was given to the explorer, Hugh Clapperton, in 1824 by Sultan Mohammed Bello of Sokoto in West Africa (the location of this manuscript map is unknown). For a fuller discussion of indigenous mapmaking, please refer to the section on traditional cartography in Africa by Maggs and Bassett in Woodward and Lewis 1998: Vol. 2: Book 3: 13-48.

Early Understanding of Africa by the Egyptians, Greeks, and Romans

Europeans had a cartographic concept of Africa, particularly that part of Africa along the Mediterranean Sea, from classical times. According to Relaño (2002: 1), the Greeks and Romans considered North Africa as part of the *ecumene* with the result that no independent idea, i.e. no cartographic concept, of Africa arose in Classical Antiquity, nor did it appear in the Middle Ages. (From earliest times, Africa was called Libya, particularly that area known to Europeans. The term 'Africa', which came into increasing usage from the late medieval period, is derived from the allegorical representation for the southwestern wind named Africus or Afer, for a descendent of Abraham).

THE MAPPING OF AFRICA

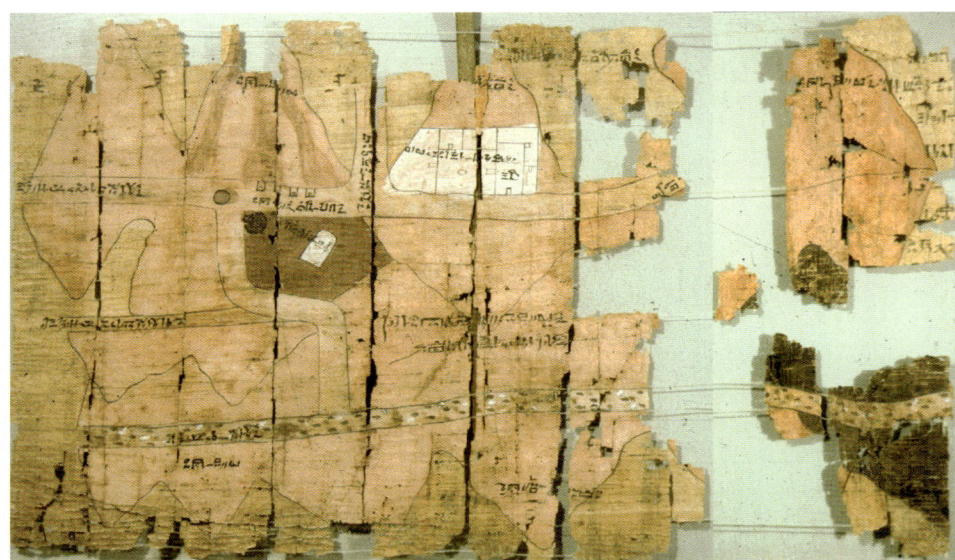

The Turin Egyptian papyrus map. South at top showing two parallel roads with an intersecting road (Museo Egizio, Torino; courtesy of James A. Harrell, University of Toledo, Ohio, USA, see Harrell, 2007).

Much of classical geography was based on an integration of geography with mathematics and the symmetry of an all-encompassing cosmography rather than on empirical discovery. Aristotle (384-322 BCE), building on Plato's concept of a global earth, developed the concept of climatic zones based on latitudes encircling the globe to determine habitability. Aware of the intense heat south of coastal North Africa, Aristotle concluded that parts of the land mass closest to the Equator, the torrid zone, were uninhabitable. Although Aristotle reasoned that there must be a southern temperate zone to balance the northern temperate zone, he believed no one could reach it because of the intervening torrid zone.

The Greek, Herodotus (489-425? BCE), was one of the first writers to describe the interior of Africa through actual travel and discovery. He visited Egypt, as far south as the Nile cataracts at Elephantine, and Libya, traveling at least as far west as Cyrene, in the middle of the fifth century BCE and collected an enormous amount of geographical, historical, and anecdotal information on Africa (or Libya as Africa was called by him). It was Herodotus who greatly underestimated the longitudinal width of Africa, an error that was perpetuated well into the sixteenth century. Herodotus also skeptically recorded an account of the supposed circumnavigation of Africa at the orders of the Egyptian Pharaoh, Necho, by Phoenician sailors around 600 BCE (Klemp 1968: 9). His view of the African continent with a peninsular outline in the south linking the western and eastern oceans was accepted by Eratosthenes in the third century BCE, by Polybius in the second century BCE, by Strabo in the first century BCE, and later by Mela, Pliny, and others. In addition, Herodotus favored a source for the Nile River far to the west in sub-Saharan West Africa. Using a classical sense of symmetry, Herodotus balanced his Nile River with a northern river in Europe, the Istar (Danube). This theory of a West African river as a source for the Nile was carried forward to become a major river running latitudinally across West Africa. Classical writers divided the interior of the African continent into Libya Interior (the Sahara) and Ethiopia Inferior (the greater Sudan).

Although the Romans were not noted African explorers, their military expeditions did probe the interior. In the first century BCE, a Roman expedition under Cornelius Balbus (Hallett 1965: 46) conquered a number of interior towns in North Africa, which included Garama, the capital of Garamantes. The Roman writer, Pliny the Elder (23-79 CE), recorded this military campaign. Pliny drew on a number of sources for his writings on Africa. From Juba, a Berber King and contemporary of Emperor Augustus, Pliny developed a belief that the Niger and the Nile were one and the same river. According to him, the Niger (and Nile) originated in lower Mauritania (southern Morocco) flowed into a lake, ran underground until it emerged in another lake in another part of Mauritania (southern Algeria), resumed its underground course across the desert country and finally reappeared joining the Nile near Meroe.

Nearly a century later, two other Roman expeditions crossed the central Sahara, one led by Julius Maternus and the other by Septimus Flaccus, to assist the Garamantes in a war against the Ethiopians. These were reported by Claudius Ptolemy, the famed second century geographer from Alexandria, who

was repeating the earlier account of Marinus of Tyre. (Hallett 1965: 47). Marinus, a Phoenician, who lived about the turn of the first century CE, extended the southern reach of the inhabited world. Earlier, Strabo and Eratosthenes had calculated that the inhabited world reached only to the cinnamon-producing country (Cinamomifera) in present day Ethiopia. According to Marinus, the southern limit was at the parallel (12 degrees south) that extended through the country of Ethiopia, then named Agisymba (a generic name for central Africa below the Equator), and the Promontory of Prasum (somewhere near Zanzibar on the coast) (Harley 1987: Vol. I: 179).

By the middle of the second century CE, Claudius Ptolemy (90-160 CE), working from the famed Library of Alexandria in Egypt, was the heir to a long classical cartographic tradition. Ptolemy, however, put forth his own view of Africa, deriving his information from the Phoenician geographer, Marinus of Tyre, and, it is assumed, from traders and military officers who had traveled into the interior and along the Red Sea and East African coast. Ptolemy abandoned the peninsular shape of Africa in favor of an eastward expansion of the southern part of the continent to form a southern shore for a completely enclosed Indian Ocean. Ptolemy's other assertion had an even longer life. It was Ptolemy who showed the Nile River rising from two lakes south of the Equator which were fed by the streams of the Mountains of the Moon further to the south. It appears that Ptolemy relied on Marinus' account, written in about 120 CE, of the Greek trader, Diogenes, who supposedly landed on the East African coast at Rhapta (in Tanzania) in about 50 CE. Diogenes learned of the origin of the Nile in the twin East African lakes, after traveling inland along existing trade routes for some 25 days to 'two great lakes and the snowy range of mountains where the Nile draws its source' (Lane-Poole 1950: 4). The lakes were originally known as Paludes Nili, but later as both Zaire (northern portion) and Zembere (southern portion) for the western lake, and Zaflan for the eastern lake, with both emerging from the Montes Lunae (Mountains of the Moon). In West Africa, Ptolemy shows Herodotus' West African river. Although this is slightly confusing based on the text, Ptolemy seems to depict the river in a 'bar-bell' pattern with a lake, the Nigritis Palus and the Libyæ Palus, at either end of the river. He does imply that the river flows from west to east, ending in a lake in West Africa, the Libyæ Palus (see Rizzo, *The Patterns and Meaning of a Great Lake in West Africa* in *Imago Mundi*. 58, I: 80-89, for a detailed discussion of this point).

Arab Cartography of Africa

The end of the classical period in geography was marked by enormous political and religious upheaval during the first millennium. With the ascendancy of Islam across much of North and East Africa with its accompanying commercial and political expansion, Arabs played a major role in a better understanding and mapping of Africa, particularly in North and West Africa, Egypt, and East Africa.

From about 750 CE, Arab traders began moving down the East African coast in search of gold, ivory, slaves, animal skins and horns, scented woods, ambergris, and other goods. As the trade developed, Arab settlements began. These were often on coastal islands which provided protection from indigenous Africans as well as other traders. Arab traders mostly stayed along the coast, usurping African villages

Example of Claudius Ptolemy's map of Africa from the Rome edition of 1478, 1490, 1507, and 1508 of his Geographia. *(Author's collection).*

and building new fortified settlements. From there, they let the Africans bring goods to them for trading purposes. These early settlements, from north to south, included: Mogadishu, Malinda (Malindi), Mombasa, Pemba Island, Zanzibar, Quiloa (Kilwa), Mozambique Island, and Sofala.

The Arab advance down the East African coast seems to have mainly stopped just beyond Sofala, hindered by the strong Agulhas current running north with force up the east coast from Natal in southern Africa. Hall (1998: 33) cites evidence that Arabs did venture into the waters of southern Africa. A settlement was established at Chibuene, an island several days' sail south beyond Sofala. From there, Arabs traded inland along the Sabi and Limpopo river valleys. An eighth century Islamic burial site has been found at Chibuene and Hall says that the town may have been founded in pre-Islamic times. Furthermore, there is evidence from Chinese sources (Hall 1998: 92) that Chinese explorers, as part of Admiral Zheng He's fleet, followed the African coastline past Sofala, but were stopped by storms from going beyond 'Habuer or Ha-pu-erh' (Kuei-sheng Chang 1970: 14: 26), an unidentified island south of Sofala. In any event, Sofala was the focal point of the southeastern coast because of its location as an entry port to the important gold mining regions of the interior around Great Zimbabwe.

Not all of the traders on the East African coast were Arabs. Numerous Indian traders sailed directly across the Indian Ocean on the northeast to southwest winter monsoons. They came from the Gujarat port of Cambray and other centers along the Malabar Coast of India. There may have been trade between the Chinese, Arabs, and Africans on the East African coast during this period. It is more likely that this trade developed through Arab or Indian intermediaries who facilitated the trade between China and Africa, based on the archaeological information available. Chinese commercial artifacts from about 800 to 1400 CE, in the form of coins and porcelain, have been found near all major Arab settlements on the coast (Snow 1988: 5-8).

THE ERA BEFORE THE PORTUGUESE

One of the greatest Arab scientists and geographers of his time, Abu Rayhan al-Biruni had a significant influence on future mapmakers of Africa. A Persian born in 973, al-Biruni worked in Afghanistan and the Punjab where he wrote a book of geography which, in part, provided plans for enabling mosque builders to precisely align their buildings with Mecca. In his further writings, al-Biruni calculated the world's circumference to within 70 miles of its actual circumference (Hall 1998: 37). Working from Ptolemy's text, he criticized Ptolemy and provided his own view of Africa's shape and size, stating that Africa extended far to the south beyond the source of the Nile into an unknown region. Though he felt the southeast coast beyond Sofala was impossible to navigate and noted that no one had returned from beyond this area, al-Biruni further stated that Ptolemy was wrong. He asserted that, at Africa's southern end, the Atlantic and Indian Oceans were connected and that the southern end of Africa did not adjoin a great southern continent thus enclosing the Indian Ocean. Al-Biruni's view was not widely adopted, as Al-Idrisi, who is described below, still showed Africa almost joined to the great southern landmass enclosing the Indian Ocean, largely using a Ptolemaic representation of Africa. (Harley 1992: Vol. II: Book 1: 160-1. Hall 1998: 38. Also see Tibbetts in Harley 1987: Vol. I: 137-155).

World map of Al-idrisi of 1154. Africa at top (Bodleian Library, Oxford University, England)

By the twelfth century, Arab cosmographers were producing vast volumes of knowledge of the world. Probably the most important of the medieval Arab cartographers was Mohammed al-Idrisi (1100-1165), the 'Nubian Geographer'. Al-Idrisi traveled extensively in Northern Africa and, together with his work at the University of Cordoba studying all geographic and cartographical sources of his time, produced a systematic account of the world and of Africa stretching from the Red Sea to the Atlantic. This account was known as the *Nuzhat al-mushtÇq...* or the *Book of Roger* after the Norman King of Sicily, Roger II, under whose patronage Al-Idrisi worked. Al-Idrisi drew on the work of Ptolemy, but also on the work of the Balkhi school of geographers, especially Ibn Hawqal. These geographers produced maps of the Islamic Empire and divided the Empire into 20 to 22 regions, with south at the top on their world maps (see Ahmad in Harley 1987: Vol. I: 156-173).

To accompany his book, in 1154, Al-Idrisi produced a world map on a silver plate, no longer extant, for the Norman King. Manuscript copies of this map do exist. The Al-Idrisi map, with south at the top, shows Africa surrounded by the traditional, medieval depiction of an 'encircling ocean' around the world. Considerable parts of his work were based on Herodotus and the Greeks with a division of the world into climatic zones including the Torrid Zone for much of sub-Saharan Africa. Al-Idrisi's map does depart from Herodotus with a western Nile River (Al-Idrisi's 'Nile of the Negroes') originating from the same lake as the Nile River of Egypt, then flowing across North Africa and exiting into the Atlantic. Of interest, he shows a third lake in Africa into which the two lower lakes flow. This depiction would appear in late sixteenth century maps of Africa such as those by Cornelis Claesz.-Barent Langenes in 1597. Al-Idrisi also shows a series of rivers originating in mountains in Southeast Africa (to the left of the Nile lakes on the above map), known to Arabs from their contacts along the eastern coast.

The *Book of Roger* first appeared in Rome, in Arabic as *Kitāb nuzhat al-mushtāq...*, in 1592; eight years later it was translated into Italian, but not published (Relaño 2002: 108). Finally in 1619, Al-Idrisi's book was published in Latin in Paris as the *Geographia Nubiensis* and was used by Nicolas Sanson, among others, as a source for some of his maps. Though the exact positioning of placenames is imprecise in his book, from the seventeenth century onward, European cartographers freely placed the numerous cities and towns mentioned by Al-Idrisi on their maps.

Other Arab writers of Africa of this period were Al-Bakri, who lived in Spain in the eleventh century and who gathered information on the western Sudan; Ibn Khalduna, a Berber historian of the fourteenth century; and Ibn Battuta, a Berber lawyer from Tangier (1325-1368). Ibn Battuta, the famous Arab traveler, ventured up the Nile as far as Aswan, and made two trips to sub-Saharan Africa to visit Mali, Timbuktu, and other areas where Muslims had settled. According to Hall (1998: 75), Ibn Battuta was astonished by the riches of West Africa, at that time the world's greatest gold producer. Later, he went as far south as Kilwa on the East African coast. Ibn Battuta's description of the geography of Africa generally conformed to the view of Al-Idrisi.

Chinese Cartography of Africa

By the fourteenth century, and as a result of growing commercial and intellectual contacts with Persians and Arabs, Chinese awareness of Africa was incorporated into their conception of a Sino-centric, single continent, world view that produced a number of world maps showing Africa. The Chinese clearly knew far more about Africa than did Europeans. A number of Chinese maps show a triangular, southward-pointing shape of Africa, rather than the eastward sloping continent of European and some Arab cartographers. The Chinese maps show rivers flowing north through the continent and a great lake in the center of Africa.

One map of primary importance was the *Yü T'u* (Terrestrial Atlas) by Chu Ssu-Pen (or as he is also known, Zhu Siben) in 1320. Though this map is lost, the *Yü T'u* was enlarged, dissected and revised by the Ming scholar, Lo Hung-hsien (or Luo Hongxian) in 1541 and published in the form of an atlas in about 1555, under the title *Guang Yü T'u* (Enlarged Terrestrial Atlas) [London BL, 15261.e.2]. This is the earliest derivative copy of Chu Ssu-Pen's work to have survived, though it is difficult to know to what degree Lo Hung-hsien deviated from the original. (see Harley and Woodward 1994: 241-246).

Among the map sheets of Lo Hung-hsien's atlas, one is entitled *The Countries in the South-western Sea* and covers a considerable portion of the Indian Ocean and a part of Africa. The atlas by Lo Hung-hsien, however, has Africa pointing south, and other evidence shows that Chu Ssu-Pen must have originally

THE ERA BEFORE THE PORTUGUESE

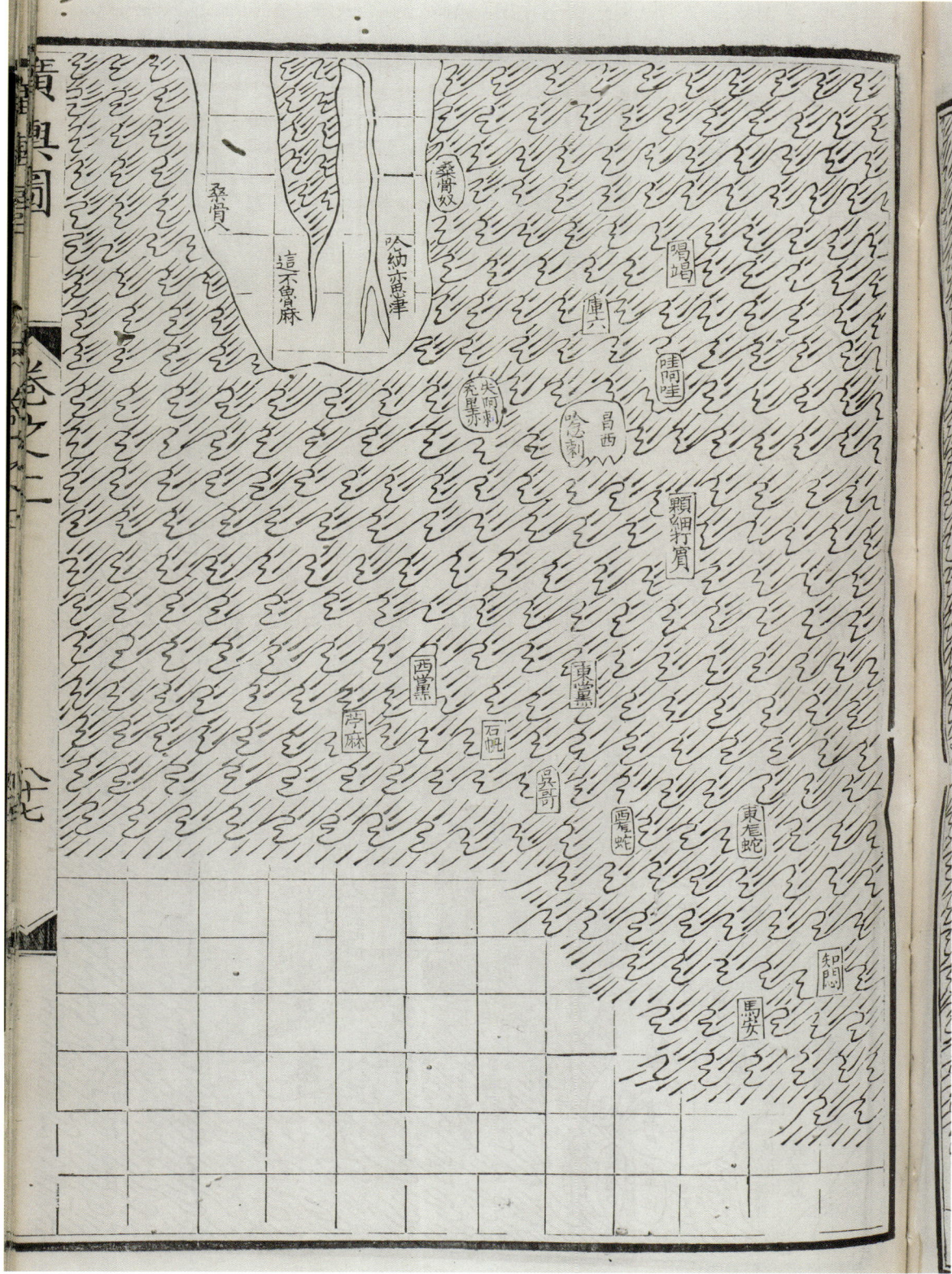

The Africa map section from Guang Yü T'u *by Lo Hung-hsien (London BL, 15261.e.2).*

drawn Africa in this way. Furthermore, in the interior of the continent, two rivers are shown flowing north, one emptying into a large body of water and the other leading further north, but terminating at the margin of the map. The name of the latter river was rendered as *Ha-na-i-ssu-chin*, which is a possible corruption of the Arabic words *Al-Nil-Azrak*, meaning the Blue Nile. The island off the east coast is called *San-pa Nu*, apparently designating the source of the Zanzibar slaves. (Kuei-sheng Chang 1970: 24: 23). On the upper left corner of the map, the coastline turns westward, possibly suggesting what might be the Guinea coast. Between the west coast and the inland water body, one sees an area named *Sang-ku*, a Chinese transliteration of the Arabic term Zangue, or the Black People, possibly the Congo region. Below the inland water area and to the southwest of the river discharging into the lake is a name

| 21

pronounced as *Che-pu-lu-ma*. The first three syllables combined are recognizable as a corruption of the Arabic word djebel, meaning 'mountains'. Kuei-sheng Chang states that it might refer to an elevated area which the Arabs called the Ma Mountains, corresponding closely to the South African Highveld with its Drakensberg Mountains.

Another important world map is the *Kangnido* or *Honil kangni yoktae kukto chi to* (map of integrated lands and regions of historical countries and capitals). In size, it is 164 x 171.8 cm. It was produced by Kwon Kun (or as he is also known, Chuan Chin) and Yi Hoe (or Yi Hewi) in 1402, using a source map of about 1330 and, in part, Islamic sources (Harley 1994: Vol. II, Book 2: 245-6; and also Tooley 1999-2004: A-D: 266). The original map is believed to be lost. A copy of this map from c. 1470, painted in color onto silk and in excellent condition, does exist in the Ryukoku University Library, Kyoto, Japan. According to Harley and Woodward (Harley 1994: Vol. II: Book 2), this is the oldest surviving world map from East Asia. There are two other versions of the *Kangnido* in Japan which are more recent than Ryukoku's *Kangnido*.

The Kangnido *map. Africa is at the lower left (Ryukoku University Library, Kyoto, Japan).*

There is a facsimile of a Kangnido-type map hanging in the Parliament Building in Cape Town, South Africa. It is based on a map existing in China. This map is known as the *Da Ming Hun Yi Tu* or Map of the Great Amalgamated Ming Empire. The Chinese government, who presented the South African Parliament with the facsimile, claims that the original was compiled in the 22nd year of the reign of the founder of the Ming Empire, Emperor Hong Wu, i.e. 1389. If this is true it would make the map older than the Ryukoku *Kangnido*. It is an enormous map (3.86 x 4.56 meters) done on silk and draped like a tapestry. Woodward mentions that the original is currently in the Palace Museum in Beijing whereas the South African Parliament gives its location as the First Historical Archive of China, Beijing.

The *Kangnido* is the culmination of many years of Chinese geographical knowledge and is based on a fifteenth century world map by Kwon Kun (or as he is also known, Chuan Chin), according to Kuei-sheng Chang. The depiction of Africa with its peninsular shape is similar to the *Yü T'u* world map. However, the *Kangnido* shows increased geographical details of Africa particularly within Southern Africa. The map shows a river flowing toward the southwest which may be the Orange River.

As Kuei-sheng Chang states, by the beginning of the 1400s, China had a good understanding of the east African coast, based primarily on trade contacts through Arab middlemen. Between 1405 and 1433, the Chinese undertook a series of enormous commercial and political ventures under the command of Admiral Zheng He, the Three Jeweled Eunuch. He led seven vast fleets from China across the Indian Ocean (Kuei-sheng Chang 1970: 14: 25-6). These expeditions, using the most up-to-date navigational technology such as the compass and stern-post rudder, varied in size from about 40 to more than 100 large junks, plus accompanying support ships with as many as 30,000 soldiers, functionaries, artisans, and commercial representatives of the Emperor. Referred to in the literature as Star Rafts, the huge treasure ships, carrying the imperial envoys, were five times the size of the Portuguese caravels that would appear in the Indian Ocean some 100 years later. For comparison, the treasure ships were about 400 feet long, while Vasco da Gama's lead ship was only 74 feet long.

The early voyages went no further than the Southern Indian coasts of Malabar and Coromandel. The fourth voyage in 1413-15 went to Hormuz and areas in the Indian Ocean. Later voyages, including the fifth voyage in 1417-19, visited the 'western regions', including Hormuz, Aden, and Mogadishu on the northeast coast of Africa, and possibly as far south as Malindi. Here, they found giraffes, lions, leopards, camels, and ostriches (Diffie 1977: 66-67). The sixth voyage (1421-1422) also went to the East African coast, while the seventh voyage (1431-1433) supposedly reached as far south as Mombasa and Malindi. These incursions into the Indian Ocean abruptly ended with the last voyage of 1431-1433 and China returned to its long isolation from the rest of the world. Yet, despite having overwhelming forces, it is interesting to note that the Chinese did not come as conquerors, but simply to trade and to extend their influence. Unlike the Portuguese, who came to dominate the continent 70 years later, the Chinese simply withdrew back to China.

Much that occurred in these Chinese voyages is documented in the *Yingyai Shenglan* (*Triumphant Vision of the Ocean's Shores*) by Ma Huan, a chronicler who went with Zheng He on several of the expeditions. Without this book, much of the history of these voyages would have been based only on fragmentary writings, as most of the records were lost (Hall 1998: 84-93). Another chronicler who described these voyages was Fei Xin, who wrote a memoir, *Triumphant Tour of the Star Raft*, about his experiences. Though he did not actually travel to the East African coast, he obtained information from other Chinese voyagers who did (Snow 1988: 26).

Medieval European View of Africa
While the Arabs and Chinese were developing a deeper understanding of Africa, European knowledge of the continent languished during the Middle Ages. Most European scholars of the time believed that the earth was in the shape of a globe. However, their maps often were portrayed symbolically, using Christian theology, in various forms, from the simple diagrams as expressed in the numerous 'T-O' maps of the period to the more elaborate mappæmundi.
David Woodward developed a useful, detailed classification system for early medieval mappæmundi dividing these maps into four main groups: tripartite (the T-O map), quadripartite, zonal, and transitional (see Campbell 1987: 2-4; Woodward 1985: 75: 510-21).

The tripartite map or the T-O map, as it is more commonly known because it showed the seas bisecting the lands in the shape of a 'T' within an encircling 'O' ocean, appeared in numerous manuscripts from the early Middle Ages and then in early printed books. The world in this type of map was generally portrayed as circular in shape with east at the top, signifying the supposed site of the Garden of Eden. Jerusalem was at the center of the world. Asia covered the top half of the circle, Europe the bottom left quarter, and Africa the other quarter with these two quarters divided by the Mediterranean Sea. Cartographic detail was generally lacking. A variation on the tripartite map was the quadripartite map, showing a T-O design with added land at the southern extreme, the Antipode. The zonal or climatic maps often appeared as the first printed maps illustrating the works of d'Ailly, Macrobius, and others. These maps reflected a Greek belief that the world was divided into five climatic zones: frigid zones at the top and bottom, a torrid zone at the equator, and habitable temperate zones in between.

T-O Map from a printed edition of Isidore's Etymologiae *of 1472. East at top (London BL, IB.5441)*

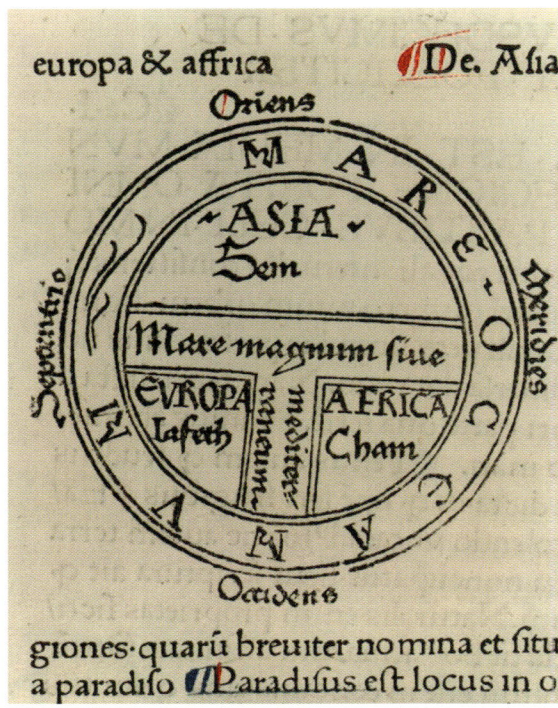

More elaborate, transitional mappæmundi were also circular in design with Jerusalem at the center and east at the top. Unlike the more basic T-O maps, many of these late Middle Age mappæmundi had a basis in contemporary travel, as well as in orthodox Christian beliefs. These mappæmundi combined factual accounts of travelers such as Marco Polo, Christian monks, and others, with legendary fables of monsters and other wonders, and with fictitious tales of travel by people such as Sir John Mandeville.

Marco Polo (in the service of the Khan from 1275-1292) never visited Africa, but he did write about the continent. He begins his description of Africa by accurately describing the Socotra Islands at the entrance to the Red Sea, though he was unclear about their position south of the Arabian Peninsula. He gives some details about Madagascar, naming it for the first time. This information was likely obtained from Arab or Indian seamen on his return from China, although he states that Madagascar was one of the biggest islands in the world with a circumference of about 4,000 miles, 2,000 miles larger than it actually is. He adds further information on the East African coast including Zanzibar and Mogadishu on the Horn of Africa. His description of Zanzibar Island seems confusing, as it groups the island with the entire region of Zanj (the Arab name for the East African coast). His account ends with a description of Abyssinia (or Middle India as it was known to Europeans of the time) and the battles of the Abyssinian Christians with the Moslems (Hall 1998: 51).

Portolan Charts of Africa

The European renaissance of scientific mapmaking began with the portolan charts, first developed in about 1200, though this date is open to continuing discussion. (See Campbell in Harley 1987: Vol. I: 380-384, for a thorough discussion of this point). The first documented mention of the existence of a portolan chart onboard a ship was a reference to the use of a sea chart on a Genoese ship carrying Saint Louis from Aigues-Mortes to Tunis in 1290. (De Jourdin et alia. 1984: 198).

These sea charts were created to meet the practical needs of sailors in the Mediterranean Sea. The portolans were based on actual, first-hand knowledge resulting from sea voyages and showed a wealth of information on placenames and other geographical data. While the place of origin of portolans is still open to some conjecture, it is generally recognized that the Catalonian island of Majorca, with its central location in the Western Mediterranean and its active shipping, was a principal focus for production during the late Middle Ages and early Renaissance (Relaño 2002: 93).

Portolan charts had a number of distinctive characteristics. The names of ports were usually inscribed perpendicularly along the coast. No geographical information appeared inland from the coast. The charts were utilitarian with little or no decorative elements. Rhumb lines and windroses were employed as guides to sail between ports and other points. As the portolans evolved over time, increasing decorative elements were added. While something of a generalization, with numerous exceptions, Relaño (2002: 92) mentions that the more austere of the utilitarian sea charts are commonly known as the Italian style of chart making, whereas the charts with more decorative elements are generally attributed to the Catalan style.

The earliest of these portolans to survive is the *Carte Pisane* sea chart of the late 1200s, now in the Bibliothèque Nationale de France (*Paris BNF*, Rés. Ge.B.1118) (Campbell in Harley 1987: Vol. I: 404). This chart represents the coast of the Mediterranean, the Black Sea, Western Europe, and Northwest Africa. The Pisane portolan shows West Africa from about 33 degrees north to the area of *Zamor* (Azemmour) (Campbell in Harley 1987: Vol. I: 412). Although the Portuguese prepared numerous portolans of their explorations, the oldest surviving Portuguese sea chart is possibly one by Reinel compiled as late as c.1483.

Majorcan mapmakers from the Balearic Islands in the Mediterranean Sea played a leading role in the development of reliable sea charts, which coincided with their role as important seafarers from this period. Most portolans that have survived are Majorcan or Italian in origin. Besides the aforementioned Pisane sea chart, notable among these earlier chartmakers, who produced charts depicting Northwest Africa and some of the Atlantic Islands, were Vesconte (1313, c.1321), Angelino Dulcert (1339, with a reference to *Rio de Oro*), and Guillelmus Soler (c.1385) (De Jourdin et alia. 1984).

Early European Renaissance Mapping of Africa

Significant among the later, transitional mappæmundi, that made use of the new African discoveries reflected in sea charts largely produced by Majorcan and Italian seafarers, are the *Catalan Atlas* of c.1375 (Bibliothèque Nationale de France), the *Atlante Mediceo* or *Medici Atlas* of c.1430 (*Firenze BML*, MS Gaddi 9, fol. 2v-3), the Catalan-Estense world map of c.1450-1460 (*Modena Estense*, C.G.A.1) and the Fra Mauro world map of 1459 (*Venezia BNM*). Though these maps may have had errors, when compared to prior mappæmundi that were only based on classical sources, these newer maps represent a great leap forward to a better understanding of the world.

The *Catalan Atlas* (c.1375) of the Majorcan, Cresques Abraham, is actually a set of vellum leaves mounted on boards comprising a world map. It depicts Africa as far south as *Buyetder* (Cape Bojador) with the great east-west river in West Africa. This is the first appearance of Cape Bojador on a chart (Relaño 2002: 95). Bojador was the southern limit of fourteenth century knowledge of the west coast and it would not be rounded until 1434 by the Portuguese, Gil Eanes. The interior of the map is filled with numerous medieval stories sprinkled with elements of fact. One notable scene shows Arabs on camels trading with African kings.

It is useful to mention the *Atlante Mediceo* or *Medici Atlas*, dated about 1430 (*Firenze BML*, MS Gaddi 9, fol. 2v-3). This atlas contains a manuscript map showing the African continent along with Europe and part of Asia. The map leaf is 40 x 55 cm. (Note: although this atlas is referred to as the *Medici Atlas*, it has no relation with that family and was actually owned by the Gaddi family, also from Florence). The map is generally devoid of geographical information on the interior of Africa. However, what makes this map of interest is its depiction of the shape of Africa. The anonymous maker of the world map in the *Medici Atlas* shows Africa as being somewhat triangular in shape. Of paramount interest is also the fact that whereas previous maps depict a 'world landmass' surrounded by a narrow encircling ocean, this map shows a wide area of water in the south surrounding Africa with a clear interlinking of the Atlantic and Indian Oceans. As Relaño states (2002: 124), it would be misleading to think that the 'portrayal of Africa displayed in the *Medici Atlas* was just the product of pure fantasy'. Earlier maps from the Chinese (discussed above) show a similar depiction of Africa. There was clearly a flow of information facilitated by Arab traders between China and the Arab world. Agreeing with Relaño, it does not seem unreasonable that information on Africa could have been relayed to Italy through trade contacts in the late Middle Ages.

The unknown author of the *Catalan-Estense* world map of c.1450-1460 (*Modena Estense*, C.G.A.1) produced a map that shows the African part of the world in greater detail than other parts. The northern half of Africa past Cape Verde and Cape Bojador to *C. Rosso* (Cape Roxo) contains considerable detail. It is filled with legends and comments and with an image of the coastline clearly based on charts of that period, particularly the chart produced by Andrea Bianco in 1448. This map and the Bianco chart are the earliest cartographic records of the landfalls of Dinis Dias and Alvaro Fernandes (Relaño 2002: 103). Past Cape Roxo, the coastline continues in a southeasterly direction creating a huge gulf in West Africa which may have been an anticipation of the actual Gulf of Guinea. (See Relaño 2002: 102-109 for a thorough discussion of the *Catalan-Estense* World Map).

Fra Mauro map with Africa upside down at the top right (Biblioteca Nazionale Marciana, Venice).

Probably the most notable map from this period is the Fra Mauro world map. Fra Mauro, a Camaldulian monk from the island of Murano near Venice, was engaged by King Afonso V of Portugal to produce a world map to include the latest discoveries. He worked on this map from 1457 to 1459 with the assistance of Andrea Bianco, another Venetian cartographer. The original map was sent to Portugal and was lost. The existing map is a duplicate of the original made in Venice at about the time of the original by Bianco and the illuminator, Francesco de Cherso (Mauro died in 1460). (Relaño 2002: 132).

What makes this large, circular map important is the wealth of geographical information that was used in the production of it. Also, Fra Mauro was willing to correct the misconceptions of Ptolemy when Portuguese or other portolans, or the reliable accounts of travelers, proved Ptolemy wrong (Bricker 1989: 54). It is apparent that Fra Mauro had access to extensive cartographic information. Venice, along with the other Italian city republics of its day, had particularly close trade and political contacts with much of the non-Western world to the east and south. In addition, King Afonso V sent him the latest details on the Portuguese discoveries along the west coast of Africa. Together with this information and by interviewing numerous people from, or who had traveled to, these regions, Fra Mauro produced a world map that could not be easily surpassed in this late medieval/early Renaissance period.

The coast of Africa is generally well presented on the Fra Mauro map, based on contemporary knowledge, as far south as *Cavo Rosso* (Cape Roxo). However, past this point to *Cavo Boiedor* (Cape Bojador), there are fewer placenames, than on other maps from this period such as the Bianco chart of 1448 (Relaño 2002: 134; Cortesão 1971: 176). This is unusual since Bianco was involved in the development of this map. Just beyond Cape Roxo, Fra Mauro places a large gulf, *Sinus Ethyopicus*, which does not reflect reality. Still, the general orientation of the west coast is fairly accurate.

Probably the most important change depicted on the map regarding Africa is a representation of the southern extreme of Africa as peninsular in shape, albeit eastward slanting, and not as joining onto the great southern continent of Ptolemy to form a landlocked Indian Ocean (Cortesão 1971: 174-175; Klemp 1968: 20). Even though by this period, 1457-1459, the Portuguese had not explored to a significant extent along the west coast, it is worth speculating that this representation was not purely conjecture on Fra Mauro's part. Though Fra Mauro was often deferential to Ptolemy, he tended to prefer a 'modern' view of Africa, based on knowledge obtained over time, that is, empirical knowledge, rather than relying on tradition alone (Relaño 2002: 133). He seems to have had access to extensive Arab sources, identifying in East Africa, for example, *Maabase* for the Arab settlement at Mombasa, *Sofala*, and *Xegiba* (Zanzibar).

Fra Mauro's use of Arab sources is supported by a legend that he places on his map. In the Indian Ocean, to the right of the southern tip of Africa (Africa is depicted upside down with south at the top on the map), Fra Mauro inserts a detailed legend in Italian with the wording 'Around 1420 one ship or junk from India [note: possibly China which was often referred to as 'Upper India'] navigated through the Indian Ocean towards the Islands of Men and Women [possibly the Seychelles] offshore the Cape of Diab [probably Madagascar] in between the Green Islands[?] and the darkness for 40 days, finding nothing but water and air, and they navigated for 2,000 miles and his fortune declined and they returned to the Cape of Diab in 70 days. On the shore there, the sailors saw a bird called 'Chroco'...'. (translated from Italian by Piero Falchetta, Curator, Biblioteca Nazionale Marciana, Venice, from his research on Fra Mauro to be published). Klemp believes the area where the crew from this ship went ashore was likely Southeast Africa (Klemp 1968: 20). Regarding the navigability around Africa to Asia, Fra Mauro concludes that 'it can therefore be asserted, without any doubt whatever, that this austral and southwestern part is navigable, and that the Indian Sea is an Ocean, not a lake'. (Relaño 2002: 133, based on a translation into English by Armando Cortesão 1960). Also, Fra Mauro does not show a torrid zone of intense heat blocking southern exploration as did prior mappæmundi.

For the geography of Abyssinia, Fra Mauro likely had the reports, as well as possible interviews with the various Coptic Church delegations from Abyssinia who were traveling to Italy and to the Pope in Rome. From these, he is able to construct a fairly accurate view of this region. He shows the Blue Nile leaving Lake Tana with its characteristic southern, then western, and then northern turns, for the first time on a map (Klemp 1968: 20). Unfortunately, he also shows it as the headwaters for the Egyptian White Nile River. The two Ptolemaic lakes along with Lake Tana appear as three small lakes in the Abyssinian Highlands.

European Exploration of Africa

The Starting Point

The first stage of European expansion is generally accepted as beginning with the capture of Ceuta in North Africa by the Portuguese in 1415. This signaled an expansion beyond traditional boundaries. This first phase of exploration culminated with the circumnavigation of the world by the Spanish ship, *Victoria*, in 1519-1522.

The Portuguese had their predecessors in the conquest of the Atlantic, but those explorers left little record of their voyages and did not have a lasting effect on world history. It is generally known that Italian and Catalan ships from the Mediterranean had sailed out into the Atlantic in the late thirteenth and early fourteenth centuries. What they sought is not known, and what they found is equally uncertain, though they probably sighted the Azores, possibly Madeira, and other islands off the African mainland. One early attempt to sail out into the Atlantic and down the West African coast was a voyage in two galleys commanded by the Genoese brothers, Ugolino and Vadino Vivaldi, in May 1291. From available records on the voyage, their goal was to reach India via the ocean. They supposedly reached a place called Gozora, but nothing further was heard from them (Relaño 2002: 94). Cortesão (1960: Vol. I: 298) identifies this place as Cape Non. There were a number of voyages following this one by Majorcan and Italian sailors. One further voyage of note was by Jaume Ferrer to the River de l'Or (Senegal River) in August 1346. It is not known if Ferrer did in fact reach this river, but his voyage is recorded some years later in the *Catalan Atlas* of Creques Abraham (c.1375) (Relaño 2002: 95).

The main motivation for this 'Age of Discovery' initiated by the Portuguese was likely a combination of religious, economic, strategic, and political factors. An important contributing factor is that Portugal was a united country for the entire fifteenth century unlike most of Europe which was going through either war or internal upheaval. C.R. Boxer (1961: 6) states that what inspired the Portuguese were, in chronological order, the motives of: (1) crusading, missionizing Christian zeal; (2) the desire for gold from Guinea to the south of the Sahara Desert; (3) the quest for Prester John; and (4) the search for spices.

This crusading spirit to extend Christianity was strengthened with a search for the legendary Prester John. In the late Middle Ages, a letter was sent to the Byzantine Patriarch from someone pretending to be an important Christian king named Prester John located in the East. For centuries, from the receipt of this fictitious letter, Europeans searched for the Kingdom of Prester John.

To Europeans, Prester John lived in the 'Indies', a somewhat vague concept that at different times was said to include upper Asia (the East), middle Asia (India), and finally lower India (Abyssinia). Implicit in this search was a hope that by finding and establishing ties with Prester John, he could become an ally and his vast army would join with Europeans in the protection of the Holy Lands from the Islamic Saracens and in the protection of all Europe from the Mongols. In the late 1400s and early 1500s, the Portuguese searched vainly for the Kingdom of Prester John in East Africa, though they did eventually find a Christian kingdom in Abyssinia.

From 1402 onwards, occasional Abyssinian Christian Coptic monks and envoys reached Europe, most notably Rome, and at least one of them made the trip to Lisbon in 1452. However, Europeans continued to have a rather vague understanding of Abyssinia.

The earlier seizure of Ceuta gave the Portuguese information on the lands of the Upper Niger and Senegal to the south, since Ceuta was the northern point for the trans-Sahara gold trade. This information provided the Portuguese with an impetus to attempt to by-pass the Barbary intermediaries in the trade with the peoples south of the Sahara by sailing down the coast.

As the Portuguese voyages of discovery slowly advanced down the West African coast in search of gold and trade goods, the acquisition of slaves took on an increased importance. Finally, the search for spices seemed only to become important after the death of Prince Henry in 1460.

The Portuguese Voyages of Exploration

Although the Portuguese voyages of exploration began in about 1420 with the European discovery (or possibly re-discovery) of Madeira, they began in earnest in 1434. In that year, a Portuguese barcha, a fishing boat with one large sail and a square sail requiring a crew of about 14, which was captained by Gil Eanes, first rounded Cape Bojador in West Africa. His was the fifteenth Portuguese attempt over a twelve-year period, due to strong winds and currents. Up to that point, the known world on the west coast of Africa was Cape Non (200 miles north of Cape Bojador) and the Canary Islands. Cape Bojador was something of a psychological barrier; the Mar Tenebroso (Sea of Darkness) to the south was assumed to be too difficult for navigation. Surmounting this barrier marked a transition point for a world previously constrained by the limited possibilities of a Ptolemaic world view to a new world of heretofore unimagined future possibilities.

At first, south of Cape Bojador, the Portuguese saw only a desolate coast. This served to reinforce the ancient geographical notion of an uninhabitable torrid zone to the south. However, by the time Dinis Dias reached Cape Verde in 1444, and then Nuno Tristão discovered the lands south of the Senegal River, the *Terra dos Negros* (Guinea), the Portuguese knew they had discovered a world new to them. As the Venetian, Alvise da Cà da Mosto, said, in service to the Portuguese king in 1456, 'Indeed if compared with that of ours, the places which I have seen and known, another world could well be said'. (Relaño 2002: 151).

An active trade ensued with estimates of some 25 caravels a year involved in this enterprise. From 1441, caravels were primarily used for these voyages, replacing the smaller barcha and barinel. The caravels were possibly based on similar ships designed by the Arabs. They had two or three masts with lateen sails, a castle at the stern, and a slender shape, useful for sailing the waters of coastal Africa. The caravel was used into the sixteenth century, but by this time larger vessels increasingly took their place in the transoceanic voyages (Diffie 1977: 118-9).

The first fortified trading outpost or feitoria (factory) was established on the island of Arguim, south of Cape Blanco in West Africa, in 1445. Here, European textiles, hardware and other goods were exchanged for gold dust, slaves, malagueta (a pepper-like spice), ivory, animal skins, and other products. This became the prototypical trading outpost and the Portuguese, and then later the Dutch, French, English and others, established similar factories on both coasts of Africa. Probably the most notable factory and accompanying fort on the West African coast was established on the Guinea Gold Coast, near Cape Coast under the command of Diogo de Azambuja. This fort, named *SÇo Jorge da Mina* (or El Mina), was to long serve Portuguese commercial interests in those areas of the Guinea coast called Malagueta, the Ivory Coast, the Gold Coast, and the Slave Coast. Among the participants in the Azambuja voyage was Christopher Columbus who had arrived in Portugal in 1476 (Diffie 1977: 154).

These successes seemed to encourage further Portuguese expansion and numerous expeditions continued down the coast. In 1460, Pero de Sintra discovered *Serra Leoa*, the southernmost point reached by the Portuguese during the life of Prince Henry. Following a short period of inactivity and indecision, a contract was signed between the Portuguese crown and a rich Lisbon merchant, Fernão Gomes in 1469. Gomes was granted a trading monopoly in Guinea with the stipulation that his captains had to advance exploration down the coast at a rate of 100 leagues per year for a five- and then six-year period. By this time, the Portuguese were convinced that the sea route to Asia lay just ahead. This private enterprise resulted in exploration down to Cape St. Catherine at 1°55' S, just south of the Equator.

A series of voyages of discovery followed, assigned by King João II to Diogo Cão, which ventured past the Congo to Southwest Africa. On the first voyage of 1482-1484, Cão reached as far south as Cape Santa Maria at *Monte Negro*, at about 13° S, and, planting a padrão (a carved stone column containing the particulars of each voyage that the Portuguese expeditions carried on board their vessels and which they erected as proof of their furthest position down the African coast) in August 1483, he named the place, *Santo Agostinho*. (Diffie 1977: 155). During this first voyage, Cão discovered the Congo River. João de Barros, in *Décadas da Asia*, wrote that Cão found a very large river flowing into the ocean with fresh water some twenty leagues out to sea (Diffie 1977: 155). Cão named the river, *Rio Poderoso*, known later as the Zaire and then as the Congo River.

By January 1486, on his second voyage, Cão had apparently reached *Ponta dos Faralhões* (Walfish Bay) at 22° 10' S on the southwest coast just above the Tropic of Capricorn (Diffie 1977: 156), but the sea route to the East still lay mysteriously ahead. This place is identified on the Martellus world map with an interesting notation that ends with the wording 'hic moritur'; while this could refer to Cão's death, it likely refers to the end of his voyage.

These discoveries by Cão are shown in the *Cornaro Atlas* of c.1489 (*London BL*, MS 73) containing 37 charts generally attributed to Christoforo Soligo (Relaño 2002:155. Cortesão 1960: 198-200). As was continually the case, information concerning the Portuguese discoveries rapidly found their way principally to Italy from Portugal.

By this time, the Portuguese were already convinced that the way to the East around Africa lay open to them. In fact, in December 1485 in a speech delivered to the Pope by the Portuguese envoy, Vasco Fernandes de Lucena, it was stated that the opening of the sea route to India was a certainty and that in the near future Portuguese ships were due to enter the Indian Ocean sailing past *Promontory Prasso* on the southeast coast of Africa (Boxer 1978: 35).

Almost immediately after Cão's return, preparation began on a new expedition further to the south to find this sea route. The person charged with this task was Bartolomeu Dias, who left the Tagus in August 1487 in command of two caravels and one supply ship. Going past Cão's farthest point, Dias left his supply ship at *Angra das Voltas* (Luderitz Bay in Namibia) in late 1487. In early 1488, on his way further south into uncharted waters, Dias' ships were blown south and then eastward by hurricane force winds, missing the Cape of Good Hope entirely and finally sighting land to the east at a river that they named *dos Vaqueiros*, from the cattle they observed in the river. Unable to land, the ships sailed on eastward landing at *São Bras*, what is now known as Mossel Bay in South Africa, on February 3, 1488. Dias, realizing that the land spread before him in an easterly direction, continued on to *Baia da Lagoa* (Algoa Bay near present day Port Elizabeth) at which point a timber cross was planted. He sailed for several more days in a more northeasterly direction to a point named *Rio de Infante* after João Infante, the captain of the ship Pantaleão and the first man to step ashore (this point was possibly the Great Fish River of today). Axelson calls it the Keiskama River (See Axelson 1973: 15, for a discussion of this topic). From here, Dias sailed on for a few days, tantalizingly reaching the entrance to the Indian Ocean, but was forced to turn back by his mutinous crew who were likely sick and short on provisions.

At the same time as the Dias voyage, as part of the strategy to find information on the sea route to the Indies and sailing conditions in the Indian Ocean, and to obtain a better understanding of India and the East coast of Africa including the Kingdom of Prester John, João II sent Afonso de Paiva and Pero da Covilhã overland via Cairo on May 7, 1487. After agreeing to meet again in Cairo, Paiva and Covilhã separated with Paiva traveling to Abyssinia, where he died on route, and Covilhã traveling to India. In late 1489, Covilhã was on the East African coast, traveling as far south as Sofala. Covilhã returned to Cairo where he met with two representatives of the Portuguese king. After Covilhã gave them a written account of his travels, he was ordered to finish Paiva's trip to Abyssinia. Upon reaching Abyssinia in 1491, Covilhã was welcomed, but for some unknown reason was not allowed to leave the country. He eventually died in Abyssinia.

At this point there was a seeming delay of almost nine years in pursuing the sea route, from 1488 when Dias returned to Portugal, until 1497 when Vasco da Gama departed. Some have stated that this was due to multiple reasons: distracting events in Ceuta and North Africa; the death of King João's heir and then King João; the tremendous burden placed on the small economy and population of Portugal by the voyages of exploration; and the disruption caused by Columbus' first voyage and the subsequent Treaty of Tordesillas (1494), effectively dividing the new discoveries into separate spheres of influence. All of these reasons have merit. Boxer (1978: 36) puts forth another interesting theory. It is known that Dias brought back some information on the prevailing wind patterns of the west coast of Africa: he found clockwise, circular wind patterns with a southward wind down the coast for the region north of the Equator, but counter-clockwise wind patterns northward up the coast for the region south of the Equator. For shorter sailing distances, primarily north of the Equator, this was not a problem. For the long-distance ocean voyage of Dias and the anticipated next one finally to India, a better strategy was needed than the extremely time-consuming method of simply beating against the trade winds down the coast of Africa. Thus, a plausible additional reason for the delay in Da Gama's voyage is that, during this intervening period, the Portuguese sent unrecorded ships into the South Atlantic to better understand sailing conditions in this part of the world. This would explain why Da Gama sailed with the wind down the coast as far as Sierra Leone, and then headed westward out into the ocean, almost reaching as far west as Brazil, before heading southeasterly to catch the prevailing trade winds. After leaving the Cape Verde Islands some 96 days before, Da Gama finally attained landfall approximately 80 miles northwest of the Cape of Good Hope. At the time, this was the longest voyage out of sight of land made by a European ship, far surpassing Columbus' 1492 voyage across the Atlantic to the Caribbean.

Much has been written about Da Gama's voyage to India from 1497 until his return to Portugal in 1499. In summary, after entering the Indian Ocean, he called at Quilimane, Mozambique Island, and other East African, Arab-Swahili ports, long controlled by Moslems traders, who were not receptive to what they saw as Portuguese and Christian interference in their established patterns of trade. Finally, in Malindi he was able to acquire the services of an Arab pilot, Ahmad ibn Madjid, who guided him across the Indian Ocean to Calicut in India, the major entrepot for the spice trade on the Malabar Coast. By sheer good fortune, Da Gama arrived in East Africa during March and thus was able to take advantage of the favorable monsoon winds which blow in a northeasterly direction toward India during the spring and summer.

Due to the absence of a strong Asian naval power, the Portuguese were able to quickly establish their control over the Indian Ocean, except for the Red Sea and a few other coastal areas. This power culminated in Francisco de Almeida's victory over an Egyptian fleet off Diu in 1509 and another victory off Malacca in 1511. The key strongpoints of Goa (from 1510), Hormuz (1515), and Malacca (1511) were soon supplemented by a large number of fortified trading posts or feitorias (factories) extending from points along the West African coast to Sofala, Mozambique Island, Mombasa, and Malindi on the East African coast. This Portuguese power over sub-Saharan Africa and over many of the lands surrounding the Indian Ocean was not seriously challenged until the arrival of the Dutch, English, and French in Africa and the Indian Ocean nearly a century later.

The Portuguese in Africa

During the fifteenth, sixteenth, and seventeenth centuries, the Portuguese penetrated inland on the West African coast up the Senegal River to Timbuktu (in 1487) and into the Congo and Angola. To a greater extent, they explored inland from the East African coast to Abyssinia and from Sofala to the gold-producing country of Monomotapa.

The first Portuguese group to reach the Kingdom of Prester John in Abyssinia and return to Portugal was led by Rodrigo de Lima and Jorge de Abreu and included Father Francisco Alvares. They finally reached the Kingdom of Prester John in 1520. There they met the emperor of Abyssinia, Lebna Dengel, or King David II as he was known to the Europeans. Interestingly, they also met Pero da Covilhã, who had been sent to the East by João II in 1487 and who was stranded in Abyssinia since 1491. Alvares took extensive notes during his six-year stay in the kingdom. In 1527, Alvares and other members of his party arrived back in Portugal. A complete account of his travels was published as *Ho Preste Joam das indias...* in Lisbon in 1540. This book revealed the Kingdom of Prester John to be less in grandeur, but it took more than another 100 years to put this well-entrenched myth to rest. Although they began to question the wealth of the kingdom, Europeans considered the land of Abyssinia to extend far into the interior of the continent and far to the south. (Drawing on the earliest designation for the area of Africa south of North Africa and Guinea, Europeans labeled much of Africa as Ethiopia Superior [upper or northern] and Ethiopia Inferior [lower or southern], though they sometimes used the term Abyssinia interchangeably with Ethiopia, at least initially). This accounts for the placement of numerous Abyssinian placenames in Southern Africa on the maps of the period and on those maps from later in the century by Gastaldi, Ortelius, and others.

Further visits followed to the Kingdom of Prester John. In 1541, in response to an invasion by Somali Moslems, the Portuguese king sent a force under Cristóvão da Gama, younger son of Vasco, to assist the Kingdom of Prester John. Miguel de Castanhoso accompanied the campaign and wrote extensively on everything he saw. Numerous soldiers remained in Abyssinia and from about 1555 to 1633 the Portuguese maintained a missionary presence there. By 1633, the Portuguese were expelled and all Catholic missionary activity was prohibited.

The myth of the extent of Prester John's empire was largely finished by the seventeenth century with the arrival of the Jesuits, Pero Paez, Manuel de Almeida, and Jerome Lobo, who documented their visits.

Paez, in 1618, was the first European to describe the source of the Blue Nile as Lake Tana. Almeida, who was in Ethiopia from 1624 to 1633, wrote on his observations, *Historia de Ethiopia a alta*, in 1645 (Relaño 2002: 65). This manuscript contained a map. While the original map is lost, some copies have remained (Cortesão 1960: Vol. V: 109-112). Lobo also left an account of his travels of 1623.

Father Baltazar Tellez used these various narratives in his *Historia Genral de Ethiopia a Alta*, published in Portugal by the Society of Jesuits in 1660. He used the Almeida map of Ethiopia. Through this means, Almeida's map was disseminated throughout Europe and numerous derivatives developed such as Hiob Ludolf's 1683 map of Ethiopia.

Hiob Ludolf's 1683 map of Ethiopia (Author's collection).

Madagascar, the world's fourth largest island after Greenland, New Guinea and Borneo, experienced a number of outside invaders from the African continent, India and elsewhere in the Indian Ocean, likely including Australia. The written history of Madagascar began in the seventh century, when Arabs established trading posts along the northwest coast.

Diogo Dias, a ship captain, was the first Portuguese to view Madagascar on August 10, 1500 after his ship became separated from the rest of Pedro Álvares Cabral's fleet in a storm during their voyage to India (There is some dispute with this, as some believe that Afonso de Albuquerque was the first Portuguese to view Madagascar) (see Axelson 1973: 29). Interestingly, the Cabral fleet was the first to sight the coast of Brazil after strong winds had blown them too far to the west. Diogo Dias had also earlier discovered some of the Cape Verde islands with António Noli. Dias named the island St. Laurentis because he first sighted it on St. Lawrence's feast day. This island later came to be known as Madagascar, a reference to its much earlier name from the writings of Marco Polo. The Portuguese were unsuccessful at establishing a permanent base on Madagascar. In 1638, the French took possession of Diego Suarez on the northern tip of the island and settled in the Bay of Antongil in 1642. They founded Fort Dauphin on the extreme southern coast in 1643. In 1664, the French posts were turned over to the Société des Indes Orientales and the island was called La France Orientale and Ile Dauphin. In 1672, Fort Dauphin was destroyed by the Malagasy; the French did not return to the area until the eighteenth century when they established a base on the small island of St. Marie off the East coast.

On the East coast and particularly in Southeast Africa, the Portuguese found a chain of Arab-Swahili coastal settlements which had been active on the coast for almost 1000 years. Although still proud of their Islamic heritage, these Arabs retained much of the indigenous culture and traditions. These settlements traded with the African tribes in the interior for gold, ivory, and other goods, in exchange for cloth, beads, and other merchandise mostly originating in India. The Portuguese wanted to monopolize this trade by circumventing the Arab traders with the establishment of their own fortified factories. At the same time, they wanted to secure re-supply points on the coast for those ships from Portugal making the long trip to Goa and further east. In 1505, they built a fortified factory at Sofala and established factories and subsidiary stations elsewhere up the coast at Quelimane near the mouth of the Cuama (Zambezi) River, Mozambique Island (in 1507), Kilwa to a lesser degree as a port of call, Mombasa, and Malindi (in 1509).

While they were able to obtain gold in Southeast Africa, the amount was never as much as on the Guinea coast from São Jorge da Mina. As a result, the burden on Portugal to maintain settlements on the East coast was immense. The arrival of the Portuguese in Southeast Africa coincided with the disintegration of the Makalanga confederation of tribes ruled by Mutapa, the paramount chief of the area. The Portuguese grandly called this area around present-day Zimbabwe the Empire of Monomotapa (See Randles 1981: 5-6; and Axelson 1973: 25-26). Over long periods the interior was in a state of strife and civil war, and conditions were such that trade was almost impossible. Another constant difficulty that the Portuguese faced was the deadly diseases which were prevalent in these tropical regions, particularly along the coast and in other low-lying areas.

There is some record of the Portuguese exploration of the Zambezi River valley which occurred when they first arrived on the coast, but much of the further exploration was conducted by the numerous, undocumented traders up the Zambezi and its tributaries in search of gold and other trade goods. (For a detailed discussion see the two books by Eric Axelson, *Portuguese in South-East Africa 1488-1600*; and *Portuguese in South-East Africa 1600-1700*). Probably the first notable explorer into the interior was Antonio Fernandes, a degredado (a criminal sentenced to overseas exile). From 1511, he was sent into the interior by the Portuguese authorities on a series of expeditions up the Cuama River to find provisions and to determine where gold might be located. Some scant records of these trips prepared by the clerk in Sofala survive (Axelson 1973: 80). Axelson calls him the first explorer of southern central Africa.

Much of what was known in Europe of Southeast Africa was obtained from João de Barros' *Décadas da Asia,* published in Lisbon in 1552. This volume included detailed information on Sofala and the Kingdom of Monomotapa which he described as being between the Cuama River and what was called the Rio da Lagoa (renamed the Espirito Santo by Lourenço Marques as a result of his 1545 exploration into the region). Among the tributaries of the Cuama were six rivers named the Ruenya, Nyaderi, Mazoe, Ruia, Hunyani, and Luangwa which indicated the extent of Portuguese penetration into the interior (Axelson 1973: 141). De Barros also referred to the gold mines in the district of Toroa, known as the Kingdom of Buta, a vassal of Monomotapa, which would appear on maps of this period. De Barros provides a reference to buildings called Symbaoe, or a court where the Monomotapa might reside. Based on the detailed description provided by De Barros, Axelson believes that this is an early reference to the present-day Great Zimbabwe ruins (Axelson 1973: 142).

King Sebastião, upon his ascension to the throne of Portugal in 1568, decided that he would act decisively to finally control the gold fields of Monomotapa. An expedition was launched in 1569 headed by Francisco Barreto. In Sofala, he was joined by Vasco Fernandes Homem. Finally in 1571, the expedition headed up the Cuama River and reached Sena, the main center for the ten Portuguese traders who worked the Zambezi area (Axelson 1973: 158). Here, beside the village of Inhaparapara, the members of the expedition established themselves, protected by a small fort, which they dedicated to São Marçal. From this point with a force of over 650, they began a march inland from the river but disease and African attacks whittled their numbers and the force retreated to Sena where Barreto died leaving Homem in charge of a greatly reduced force of 180. Next, the Portuguese, by this time reinforced, marched inland fighting their way through the Kingdom of Quiteve finally reaching Manica, a gold-producing area. With Africans resisting his further advances, Homem again retreated to Sena, where he reached the fort of São Marçal in late October 1575. Further sorties into the interior from the river proved equally futile. As Axelson states (1973: 163), the Barreto-Homen expedition was a protracted, dismal and expensive failure. The ascension to the throne of Portugal in 1580 by the Spanish king virtually stopped any further attempts at exploration of the interior, save for continued activities of the traders along the Cuama and its tributaries. In spite of these military disasters, the Portuguese had well-established trading bases on the Zambezi at Sena and Tete. A Dominican friar, João dos Santos, who was stationed at Sofala from 1586 to 1590, traveled up the Zambezi to visit Sena and Tete in 1590. Here, he commented on the forts of stone and lime containing several artillery pieces (Axelson 1973: 173). Tete had seven or eight pieces of artillery and 600 Christians, of whom 40 were Portuguese. Santos described his travels in *Ethiopia Oriental*, published in Evora in 1609.

One mystery on sixteenth century maps of Southern Africa is the appearance of *Cast. Portugal* (Castle Portugal), located almost dead-center in the middle of Southern Africa (southeast of *Zachaf Lacus* and slightly north of the Tropic of Capricorn). The first appearance of this 'Portuguese Castle' was in Gerard Mercator's world map of 1569 where it is shown to the south of where the Zambere River splits into the Cuama and the Spirito Santo Rivers. Numerous later maps show *Cast. Portugal* (Van Linschoten, 1596), or variations of this name well into the eighteenth century (De Fer lists it as *Chateau de Portegal*). Randles (1956: 13: 84) states that the origin of this feature is untraceable and theorizes that the reference may be to a fort actually constructed in Abyssinia by the expedition of Cristóvão da Gama in 1542, badly misplaced into Southern Africa as are so many of the placenames of Abyssinia. However, there is also the possibility that the castle may be a reference to either Sena or Tete which were established as fortified, trading outposts by this time, though hardly as forts. What is important to remember is that many, if not most of the names that appeared on maps showing Southern Africa during this period had a basis in reality, even though the placenames were misplaced.

Diogo Cão was the first European to have contact in 1482 with the Kingdom of the Congo. The center of this kingdom was located in what is now northern Angola, around the city of Mbanza Kongo, which was later renamed São Salvador. It was bounded on the north by the Congo River and on the south by the Loje River. The rulers, or Mani, of this kingdom soon adopted Christianity and one, Nzinga Nvemba or Dom Afonso I, 1506-1545, established strong relations with the Portuguese, welcoming numerous missions to the Congo which included skilled workers and artisans (Boxer 1969: 28-29). However, these relationships broke down over time due to the increasing demands for slaves, initially taken from lands outside of the Kingdom of the Congo, for plantations on São Tome and elsewhere. For their part, the Portuguese did not try to impose their will on the country or try to conquer it by force. This decision not to impose Portuguese will on the Congo was a reaction to the more pressing concerns in Guinea and on the East African coast. In fact, the kings of the Congo continued to request further assistance which was generally ignored. Apart from sending a force to repel a Jaga invasion in 1568-1573, little was done by the Portuguese authorities in this area.

Just to the south in present day Angola, the Portuguese had a colonial presence from 1520 until late in the twentieth century. In 1520, King Manuel requested that two envoys travel to the Kingdom of Angola to meet with Africans in the area. Much of the initial interest was as a potential center for slaves, but also as a missionary site. In February 1575, the Portuguese began the settlement of Angola at São Paulo de Loanda.

From 1578 to 1584, a Portuguese explorer, Duarte Lopes, journeyed through the Kingdom of the Congo recording detailed information on his travels. At the conclusion of his travels, the ruler of the Congo sent Lopes back to Europe on a mission to report to the King, Phillip II, and to the Pope on the affairs of the Congo. During his time in Rome from 1588-1589, Lopes came into contact with Filippo Pigafetta (c.1533-1604). Based on Lopes' travels, Pigafetta wrote a book entitled *Relatione del reame di Congo et delle circon-*

vicine... published in Rome by Bartolomeo Grassi in 1591. The book contained a map of the Congo Region and a map showing most of Africa, excepting West Africa, engraved by Natale Bonifacio.

Concerning the cartography of Africa, there are two reasons why this book and its accompanying map of Africa are significant. Firstly, Pigafetta rejects the long-held Ptolemaic belief of the Mountains of the Moon as the source for the Nile River, and secondly, he rejects the Nile as flowing north from two side-by-side lakes in central Africa. In his depiction, there are two lakes in central Africa, one above the other. The lower lake feeds the *Rio de Manhice* (Zambezi) and the *Lorenzo Marches* (Limpopo), and the northern one feeds the Nile, with a river connecting the two lakes. Also, the northern lake is placed above the Equator; Ptolemy placed the lakes further to the south. Pigafetta does locate another lake to the west as a source for the Congo and other rivers flowing east, a theme first introduced by Waldseemüller. (See Relaño 2002: 209-211 for a discussion of the Pigafetta map). This work became an important source on the Congo and was frequently used by later writers and mapmakers. The earliest translation from Italian was into Dutch in 1597 by Cornelis Claesz. The De Bry brothers published a German translation in 1597 with two similar maps as a first part of their series, *Petits Voyages* (Schilder 2003: Vol. VII: 188).

What is probably of most significance with this map is that Pigafetta, unlike the other cartographers of this era, consciously chose to ignore the primacy of Ptolemy when depicting the interior of central Africa. In other maps of Africa in atlases of this time, notably Ortelius' Africa map in his *Theatrum Orbis Terrarum*, Africa continued to be depicted relatively accurately along the coasts with some correct information for limited sections of the interior which were open to exploration (Southern Africa, Southeast Africa, the Congo, Abyssinia, and parts of West Africa). The majority of cartographers mapping the interior continued to follow Ptolemy. The next to seriously challenge Ptolemy's authority were the French school of geographers of the latter half of the seventeenth century. This culminated in Guillaume Delisle's landmark map of Africa of 1700 which was devoid of most Ptolemaic information in the interior.

The Arrival of Other Europeans

The wealth generated by the Portuguese overseas possessions in Africa encouraged the Dutch, English, French, and others to try to imitate them. The Spanish and Portuguese thrones were united in 1580 (which lasted until 1640) following the deaths of the Portuguese King Sebastian in North Africa and his successor. The Dutch, who had been contesting their independence from Spanish rule since 1568, now turned their wrath on the Portuguese as well and in particular on the Portuguese overseas possessions using their superior sea power supported by greater economic resources and manpower.

By the middle of the sixteenth century, English, French, and Dutch trading ships began to visit the coasts of Africa, especially in West Africa. Whereas the Portuguese were in Africa in the service of their King, the others, particularly the Dutch and English, sailed as agents of commercial concerns and, from 1600, of the famed East India Companies (The English East India Company was founded in 1600 and the Dutch East India Company in 1602).

In a series of rapid actions over the next 50 years, the Dutch expanded into Africa at the expense of the Portuguese. In 1612, the Dutch established a settlement on the Guinea coast at Mouri (Fort Nassau). In 1621, they took the trading posts of Arguim, south of Cape Blanco in West Africa, and Gorée, off the Cape Verde peninsula in Senegambia, from the Portuguese. In 1638, the Dutch seized the fort at São Jorge da Mina (El Mina) and built numerous other forts on the Guinea Coast. They tried, but failed, to take Mozambique Island which, in part, forced the Dutch to found a colony at the Cape of Good Hope to supply the ships enroute to their new possessions in the East. This occurred in 1652 under Jan van Riebeeck. In 1666, the Dutch built a fort at Cape Town and soon after began their first explorations and expansion into the interior. By the middle of the 1600s, the Dutch were, with little dispute, the greatest trading nation on earth with their far-reaching operations.

In 1626, the French established a settlement named St. Louis at the mouth of the Senegal River. Later, from 1637, they explored the Senegal River for some 100 miles, establishing various trading posts. In 1677, the French seized the Dutch trading posts in Senegal, completing their conquest of the Senegal region, and advanced further up the river to Mambuk.

THE MAPPING OF AFRICA

The De Bry example of the Pigafetta map (Author's collection).

Though later than the Dutch and French, the English also established their presence in Africa by building a fort at the mouth of the Gambia River on James Island in 1662, as well as trading elsewhere in Africa. To a lesser degree, other European countries, also established a presence in Africa. The Swedes built a fort at Christiansborg near Accra in West Africa in 1652 which, five years later, was captured by the Danes who held it for almost 200 years.

On the east coast of Africa, in the late 1600s, the Portuguese were forced from their remaining posts by Omani Arabs. They remained in the region of Mozambique which they retained into the late twentieth century.

Ptolemaic Maps: The Earliest Printed Maps of Africa

Greek culture and knowledge were preserved in Byzantium during the early Middle Ages and then by Islamic scholars as the old Byzantine Empire crumbled. Greek geography, including the work of Ptolemy, influenced numerous early Arab scholars. In the case of Claudius Ptolemy's *Geographia,* the complete translation from Greek or Syriac into Arabic was concluded in the ninth century (Tibbets in Harley 1992: Vol. II: Book 1: 99).

As the Byzantine Empire continued to disintegrate, a large number of Greek manuscripts came to Italy in the fourteenth century, including copies of Ptolemy's *Geographia.* Given its detailed description of the known world, this book must have been a revelation, expanding the world view of most Europeans. Instead of a provincial, restricted way of viewing the world, many now perceived it in a vast, new way. This utter sense of wonderment and the resulting complete adoption of Ptolemy's world view were both a blessing and a curse. The arrival of the book in the West, coinciding with the development of the printing press in the middle of the fifteenth century, enabled it to be widely distributed, thus entrenching Ptolemy's position of supremacy in Europe. In fact, so powerful was the book that its influence was retained, in spite of the rash of new discoveries, well into the seventeenth century. This proved unfortunate as it tended to retard the promulgation of other, more suitable maps. The last vestiges of Ptolemy's influence did not entirely disappear until 1700, when Delisle produced his landmark map of Africa. Even then, some maps continued the Ptolemaic traditions when portraying the interior of Africa into the eighteenth and early nineteenth centuries.

The translation of the *Geographia* into Latin from Greek was begun by the Byzantine scholar Emanuel Chrysoloras (? – 1415) and was finished in 1406 by his pupil, Jacopo d'Angiolo (Jacobus Angelus) in Italy. The Latin translation appeared to have been widely distributed in manuscript, first without maps, and then later with the maps prepared, using the over 8,000 coordinates for placenames described in the text.

There exist two versions of the *Geographia* with maps: an 'A Recension' version with a world map and 26 maps of parts of the world including four of Africa; and a 'B Recension' version with a world map and 64 maps of parts of the world, including eight of Africa in Book IV. The Biblioteca Medicea Laurenziana, Florence, contains the earliest known Greek manuscript edition of the *Geographia* from the 'B Recension' (*Firenze BML*, Plut.28.49). It is dated from the early fourteenth century.

The first printed edition of the *Geographia* with maps was published in Bologna in 1477 (Campbell 1987: 129). (There was an earlier printed edition, without maps, published in Venice in 1475). The Bologna 1477 edition was followed by editions of the *Geographia* in Rome in 1478, 1490, 1507, and 1508; in Florence (the distinctly separate Berlinghieri edition) in 1482; and, north of the Alps in Ulm in 1482 and 1486. Numerous further editions of the *Geographia* by Sylvanus (1511), Waldseemüller (1513), Fries (1522), Münster (1540), and others followed.

The text of the *Geographia* contained a section on Africa as Book IV describing geographical information and map coordinates to enable the skilled reader to prepare maps of Africa. Book IV was composed of eight chapters: '1. Mauritania Tingitana, 2. Mauritania Caesariensis, 3. Numidia Africa,
4. Cyrenaica, 5. Marmarica, which is properly called Libya. Entire Egypt both Lower and Upper, 6. Libya Interior, 7. Ethiopia which is below Egypt, and 8. Ethiopia which is in the interior below this' (Stevenson 1991: 93-109.).

As has been noted above, the 'A Recension' versions of Ptolemy's *Geographia* contain four maps of Africa. These are Tabula Prima Africae (Morocco and Northwest Africa), Tabula Secunda Africae (central North Africa), Tabula Tertia Africae (Egypt), and Tabula Quarta Africae (all of known Africa). The first three maps of coastal North Africa contained numerous placenames and geographical features, particularly as this was an area readily known to Ptolemy and his contemporaries.

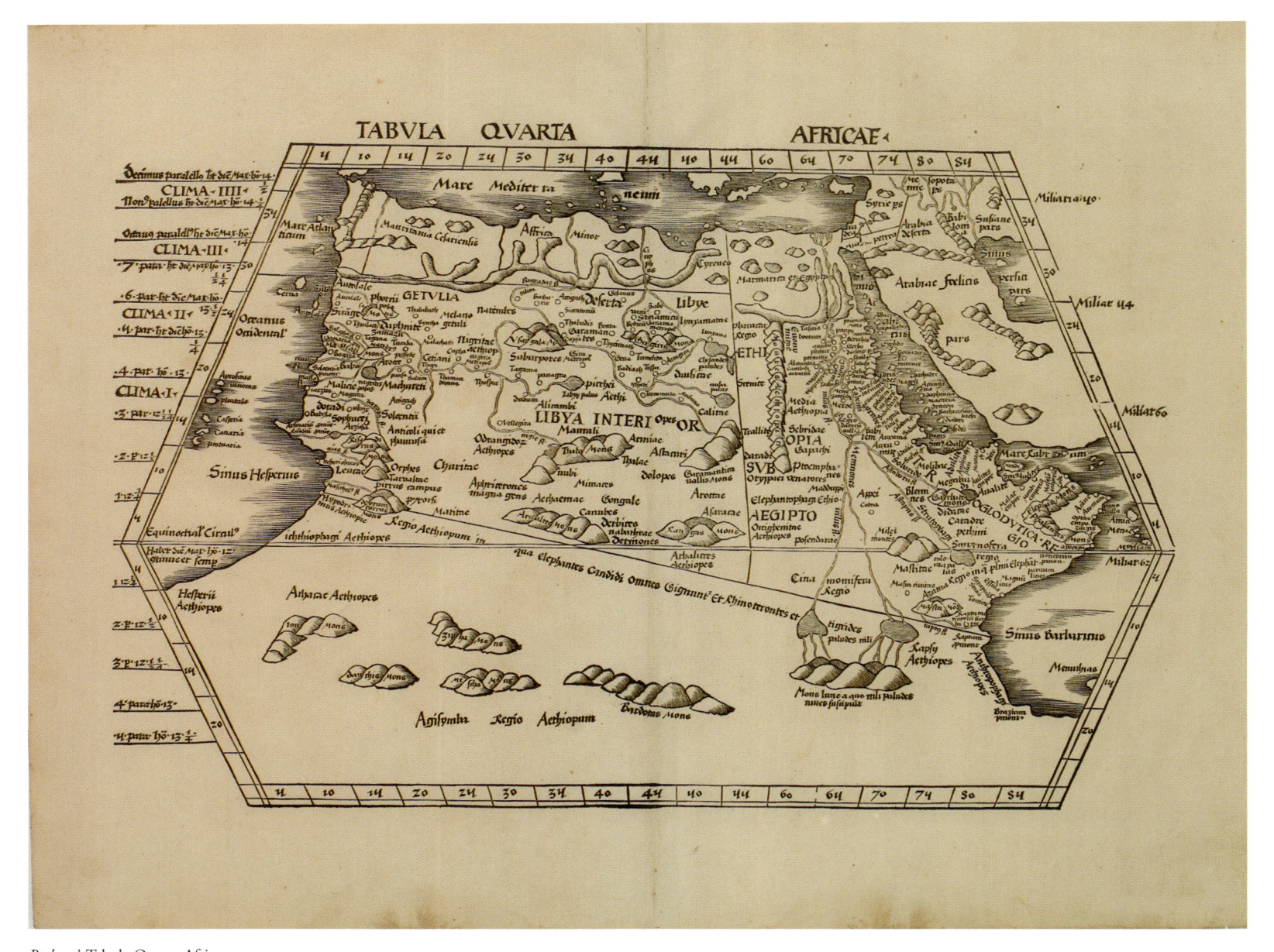

Ptolemy's Tabula Quarta Africae, *from Waldseemüller's* Geographia *of 1513 (Author's collection)*

On Tabula Quarta Africae, Ptolemy extends Africa far to the south past his line for the Equator to *Agisymba Regio Aethiopum*, his name for the most southern region on this map. A line divides the inhabited lands of Egypt, Libya Interior, Libya, etc., from the unknown area of Aethiopia Interior.

On the east coast, the line ends near *Raptum Promontory*. The farthest geographical point shown on this coast is *Prasum Promontory* (possibly Cape Delgado), to the south of which the coastline takes an easterly direction. (See Dilke in Harley 1992: Vol. II: Book 1: 179). The Indian Ocean off of East Africa is referred to as *Sinus Barbaricus* (another name that is apparent on numerous maps of the sixteenth and seventeenth centuries). Working from Alexandria in Egypt, Ptolemy had better access to direct information about the East coast from traders up the Nile River, on the Red Sea, and along the northeast coast of the Indian Ocean.

The farthest point shown to the south on the west coast is *Hesperii* (western) *Aethopia*. *Canaria* is identified as one of the islands shown off the coast of West Africa. There are numerous references to the existence of the Canary Islands (*Insulae Fortunatae* on Ptolemy's world map) in classical literature, but they had long been unvisited until their first documented 're-discovery' in about 1336 and possibly as early as 1312 by Lanceroto or Lanzarotto Malocello, a Genoese sailor (Cortesão 1971: vol. II: 68-73; Diffie and Winius 1977: 27-29).

Ptolemy places two twin lakes, *Paludes Nili*, as the source for the Nile River, to the south of the Equator. A separate lake is shown in Abyssinia as the source for the Blue Nile River which joins the White Nile by the Kingdom of Meroe. Ptolemy's line for the Equator is placed too far to the north, reflecting a latitudinal reduction in the known world. Also, Ptolemy stretches the Mediterranean longitudinally by about 20 degrees, giving the distance between Tangiers and Alexandria as 54° (the actual distance is 35°39'). He is thought to have done this to conform to earlier Greek calculations on the size of the world, thus making the inhabited part of the world reach the total distance of 180°. Arab geographers corrected the length of the Mediterranean, but this correction did not affect maps produced by Europeans until much later. (Tibbetts in Harley 1992: Vol. II: Book 1: 101).

Ptolemy's *Geographia* did not contain a printed map showing information from the new Portuguese explorations until the Rome edition of 1508. (Berlinghieri had begun the practice of adding Tabulae Novae [modern maps] to the *Geographia* for his edition in Florence of 1482, but without a modern map of Africa.) In the 1508 Rome edition, there is a map of the world prepared by Johannes Ruysch. It shows the extent of Africa, including the sea route to India, recently discovered by Da Gama. Though it was not as precise as its manuscript counterparts from this period, the Ruysch world map did depict Africa in a relatively accurate way. The first printed maps to appear in an edition of the *Geographia* that focused on Africa and utilized information from the new discoveries were Martin Waldseemüller's maps of North and South Africa (Tabula Moderna Prime Partis Aphricae, and Tabula Moderna Secunde Porcionis Aphrice) which appeared in Waldseemüller and Matthias Ringman's edition of Ptolemy's *Geographia* of 1513. However, it was not until Sebastian Münster's edition of the *Geographia* in 1540 that a modern map of the entire continent of Africa appeared.

Significant World Maps that Show the Continent of Africa

The mapping of Africa during the fifteenth and early sixteenth centuries was largely based on the record of Portuguese voyages of exploration. From the onset of Portuguese expansion after the seizure of Ceuta on the North African coast, the Portuguese crown conscientiously gathered information from voyages by requiring their caravel pilots to continuously add to charts and, where possible, determine precise latitudes by taking frequent surveys from land. The result was a generally pragmatic view of the world compiled from the experiences of sailors and not solely based on ancient sources such as Ptolemy.

All of this new information on Africa was entered into the *PadrÇo Real* or *PadrÇo Oficial*, the royal standard cartographic prototype, which was held in the Armazém da Guiné e das Indias (Armoury of Guinea and the Indies). These charts were constantly kept up-to-date with new discoveries; these secrets supposedly could not be divulged under penalty of death.

If there was, in fact, a policy of maintaining secrecy under penalty of death, it was totally ineffective. As will be shown, time and time again, the latest information on the new Portuguese discoveries was often known outside of Portugal by the Italians and others within a few years, if not within a few months, of the actual discoveries. It seems logical that the Italians saw the Portuguese expansion around Africa in search of Asian spices and other trade goods as a direct threat to Italy's almost virtual monopoly on trade with the East. For numerous years, the Italians, and especially the Venetians, had maintained a monopoly on purchasing spices from the Moslem merchants of the Mameluke Empire in Egypt and Syria, who, in turn, obtained the spices further to the east. As a result, the Italians would do anything to obtain up-to-date information on the Portuguese discoveries and their future plans for exploration. Bribery and even thefts were common means of obtaining information on this cartographic record using the many Italian diplomatic and trade agents in Portugal.

Another mystery related to the dissemination of Portuguese cartographic information is why so few of the maps and charts directly made by the Portuguese are still in existence. Today, the only evidence of Portuguese cartography from this period is the work of Pedro Reinel, showing an area of present day Zaire (the undated chart is likely c.1484-1492), and a chart by Jorge de Aguiar of 1492. While an in-depth discussion of this subject is beyond the scope of the present book, Cortesão and Da Mota among others have written extensively on this subject (Cortesão 1960). Evidently, in 1504, the Lisbon government ordered the total destruction of certain categories of maps and prohibited all exports of cartographic instruments and related materials outside the kingdom. The Royal Edict of November 13, 1504 forbade the preparation of navigational charts showing any detail south of the Congo River and ordered a recall of extant maps to allow for deletion of that specific information. This destruction was compounded by a severe earthquake that destroyed much of Lisbon in the eighteenth century and, with it, supposedly much of the remaining Portuguese cartographic record. In any event, most of the cartographic record from this period has survived due to Italian copies of Portuguese source material, or, in the case of Martin Waldseemüller's 1507 world map, copies of Portuguese maps that came via Italy.

The following maps are of importance as they demonstrate an evolving understanding of Africa and serve as sequential prototypes for later maps of the continent. Traditionally and from ancient times, the east coast of Africa was better mapped than the west coast, owing to information indirectly received in Europe from Arab and Islamic sources. This gradually changed as further information from the European voyages of exploration lead to a new view of Africa.

Henricus Martellus, Manuscript World Map, c.1489-1492

The Martellus world map is extant in four copies: one within Martellus' codex *Insularium Illustratum* at the British Library (Add.MS 15760, fol. 68v-69r); one at the Universiteitsbibliotheek Lieden (Vossianus lat., fol. 23); one within Buondelmonti's *Liber insularum* at the Biblioteca Medicea Laurenziana, Florence (Plut. 29. 25, fol. 66v-67); and one as a much larger, separate version at the Yale

Martellus manuscript world map (London BL, Add.MS 15760, fol. 68v-69r).

University Library, New Haven (See Relaño 2002: 165, for information on the locations of these copies).

Henricus Martellus, a German cartographer working in Italy, prepared a world map in manuscript that is important because of its representation of the continent of Africa. For the first time, Martellus gave Europeans a view of Africa with a definite and more accurate shape, albeit with a more southeastward sloping orientation. Although Martellus placed the Cape of Good Hope too far south at more than 40° S, he truly captured a new image of the north-south extent of the African continent, except for the still unexplored area along the southeast coast. Martellus likely based his own world map on an earlier one by Nicolaus Germanus, a cartographer working in Florence, for an edition of Ptolemy's *Geographia*. Altering and improving Germanus' world map, Martellus made use of the up-to-date geographical information then flowing between Portugal and Italy.

As part of Martellus' new depiction of Africa, this is the earliest known manuscript map to show the rounding and discovery of the Cape of Good Hope by Bartolomeu Dias in early 1488, the year before this map was probably prepared. It includes 30 new names (Relaño 2002: 166) up to *ilha de fonti*, seen by Dias just north of the farthest point he reached before entering the Indian Ocean. Next to this point, there is a legend which reads 'up to here, next to *ilha de fonti,* arrived the last Portuguese expedition in the year of the Lord 1489'. On the west coast of Africa, many of the numerous new placenames are based on the Diogo Cão and Dias discoveries.

The east coast, particularly the uncharted southeast coast, is based on non-modern sources. There is no indication that Martellus knew of the accounts of Pêro da Covilhã, sent by the king of Portugal overland to the east coast of Africa. The placenames on the east coast, *Raptu Promontorium* and *Prassum Promontorium*, are of Ptolemaic origin. The interior of Africa was also influenced by Ptolemaic conventions, among which were a Nile River system emanating in the twin lakes and the Mountains of the Moon.

Rosselli printed world map (Firenze BN, VMM P/27).

Francesco Rosselli, Printed World Map, Florence, C.1492-1493

The Rosselli world map is extant in one example at the Bibliotheca Nazionale Centrale, Firenze (Florence) (*Firenze BN*, VMM P/27). It is a copperplate engraved map; 37.5 x 52.5 cm.

Francesco Rosselli was one of the earliest makers and sellers of printed maps with his known cartographic output extending from 1482 to 1508. His 1492-1493 world map was discovered in Florence with some regional maps, and it is believed that they were intended for an unpublished edition of Ptolemy's *Geographia*.

At first glance, it appears that this map is a close copy of the Martellus manuscript world map. This is not entirely the case. Though Rosselli likely had access to the Martellus map, possibly in Florence, there are significant differences (based on the Martellus example at the British Library). Of the names below *Monte Negro* on the Rosselli map, just half are common to both the Martellus and Rosselli maps, and each has a number of placenames not found on the other's map (Campbell 1987: 72).

Probably the most prominent feature in Rosselli's world map is his depiction of Africa. Extensive cartographic detail is shown along the entire west coast of Africa to the Cape of Good Hope and just beyond. Rosselli places the image of a padrão at the point which signifies the easterly extent of Dias' voyage. As such, this is the earliest printed map to show the rounding and discovery of the Cape of Good Hope by Bartolomeu Dias in 1488 and the potential sea route to the Indies. The landmass between this padrão and the Arab cities in Southeast Africa is blank. This map distorts the southeasterly projection of the African landmass, still honoring a Ptolemaic conception since earliest times of a landmass that trended eastward. This configuration of Southern Africa places the longitude for the Cape of Good Hope much too far to the south and east.

Rosselli's placenames for East Africa follow Ptolemy, and not more modern sources. He places a *Prassum promoroum* and a *Raptum promoroum* on the two capes on the East African coast, along with a reference to the peoples of that region as *Anthropophagi Ethiopes*. Also of significance, Madagascar is shown on the map off the Southeast African coast with Zanzibar just below it. This placement was clearly based on medieval understanding, including information on Madagascar which Marco Polo had heard and mentioned in his writings. A more precise location for these two islands was not known to Europeans until the sighting of Madagascar by Cabral's fleet and of Zanzibar by Da Gama's voyage.

While there are differences, it is important to note that there are also similarities between the world maps by Martellus and Rosselli, and the globe by Martin Behaim (see below), thus signifying and capturing an explosion of information based on the newly explored coasts of Africa and particularly the beginning of the sea route to the Indies. As Relaño states (2002: 167), it is known that Behaim used a printed map in the construction of his globe; he may well have used aspects of this printed world map by Rosselli to prepare his globe.

Behaim Globe, section showing Africa (Nürnberg, Germanisches Nationalmuseum).

Martin Behaim, Manuscript Terrestrial Globe, 1492
The Behaim globe is located at the Germanisches Nationalmuseum, Nürnberg.

This is the oldest terrestrial sphere made in Europe that is extant today, although there were references to various globes since antiquity. As has been noted above, this globe has numerous similarities to the Martellus and Rosselli maps. Behaim likely had access to the printed Rosselli map as a resource, but unlike Martellus and Rosselli, he actually lived in Lisbon and interacted with scholars advising the king from 1484 to some time before he returned to Nuremberg, probably in the spring of 1490, to prepare and complete this globe. In Lisbon, he had direct access to information on Portuguese discoveries along the African coast from various maps and charts. In addition to these, Behaim probably had access to the reports of contemporary explorers. He was still in Lisbon when Dias returned from his famous voyage that had rounded the Cape of Good Hope going as far as the entrance to the Indian Ocean. Besides his access to charts and firsthand accounts, Behaim asserted that King João II sent him to Guinea in 1484-1485, a statement that has not been proven.

Much of the significance of Behaim's globe is with its spherical shape and with the skills of the actual globemakers, Glockengiesser and Kalperger. While the globe portrays the outline of Africa in general accordance with the Martellus and Rosselli maps, Behaim does not faithfully use the geographical knowledge exhibited on the previous two maps. In fact, his outline for Southern Africa is quite different from the earlier two maps with an exaggerated extension to the southeastern portion of the continent.

Along the entire west coast of Africa, Behaim provides numerous placenames, including two not on the other two maps. These are *Rio Behemo* and *Tiera da Peneto*, which might be a reference to his supposed visit to the Guinea coast, especially as they appear on no other maps (Relaño 2002: 169). As noted by Relaño, Behaim places the padrão of São Jorge incorrectly, although its correct placement was well-known by that time. He also provides inaccurate information on the Atlantic islands off the West African coast, incorrectly stating that sugar production occurred on the Azores rather than on Madeira, and includes some mythical islands in this area, such as *Insula de sant bandan* (The Irish Saint Brendan?) and *Insula Antillia*.

At the southern end of Africa, while Behaim seems to refer to the Dias expedition through a legend, the toponym does not reflect this expedition and the present-day Cape of Good Hope is not even mentioned. On the east coast, he, like Martellus and Rosselli, primarily follows Ptolemy. Here, Behaim adds an assortment of fanciful medieval traditions which is in part a reliance on the writings of Marco Polo. Behaim does include the Rosselli depiction of Madagascar and Zanzibar. The interior of Africa is based on a Ptolemaic understanding of the river systems and peoples of the interior. Also, like other cartographers, Behaim extended Abyssinia far into Southern Africa and filled unknown space with any information that was available from classical sources.

Juan de la Cosa planisphere (Madrid, Museo Naval).

Juan de la Cosa, Manuscript Planisphere, 1500

The De la Cosa Manuscript Planisphere is located at the Museo Naval, Madrid. It is on two vellum leaves in the form of an illuminated manuscript assembled into a chart; 95.5 x 177 cm.

Juan de la Cosa prepared this map in 1500, in the port of Santa Maria, near Cadiz. De la Cosa was the owner and pilot for the ship, *Santa Maria*, and accompanied Columbus to the New World on his first two voyages.

This map is most noted for its depiction of the newly discovered lands to the west. However, important information on Africa is also represented on this map. It is the first known manuscript map to show information from Vasco da Gama's voyage to India (the first to show the Da Gama voyage on a printed world map was the one prepared by Giovanni Matteo Contarini and engraved by Francesco Rosselli in 1506. The first map to show the sea route to India was a small printed woodcut map of 1505 on a Nuremberg newsletter).

The west coast of Africa is faithfully and accurately based on the preceding maps as far as Bartolomeu Dias' landfall in Southern Africa at which point De la Cosa states, 'up to here was discovered by The Exalted King John of Portugal'. Another annotation in Asia, near *Rio Yndo*, confirms that De la Cosa was also aware of the Da Gama voyage with the insertion of wording that this was 'land discovered by King Manuel of Portugal' (Relaño 2002: 178). De la Cosa also uses the placement of ships around Africa to signify the sea route to the Indies.

Significantly, especially when compared with the earlier maps, De la Cosa corrected the shape of Africa toward a much more accurate orientation. By changing it from the southeastward slanting direction found on the previous maps, he removed the last vestiges of Ptolemaic influence on the outline of the continent. De la Cosa also correctly placed for the first time the Equatorial line across Africa. Along with these improvements were some notable problems (Relaño 2002: 179). Instead of placing the Cape of Good Hope too far south as on the previous maps, he moved it about 6° north of its correct latitude, thus slightly flattening the shape of Africa at its bottom. This resulted in a shorter distance between Guinea and *C. de Boa Esperança* (Cape of Good Hope). Also, to the east of *R. do Ynfante* (the Great Fish River) on the southeast coast, the toponomy is hard to recognize. De la Cosa used some placenames of Portuguese origin; though nothing of Da Gama's landfalls in Southeast Africa is shown. He also apparently added placenames from local languages and traditional sources, as well as those he had invented. The islands of Zanzibar and Madagascar are placed far off the African coast in the Indian Ocean.

While De la Cosa had some basic knowledge of the Portuguese discoveries, including the Da Gama voyage, he did not have adequate information, or at least did not adequately show this information in his view of Africa. These deficiencies were to be corrected in the Cantino planisphere of 1502.

Alberto Cantino, Manuscript Planisphere, 1502

The Cantino Planisphere is located at the Biblioteca Estense Universitaria, Modena (*Modena Estense*, C.G.A.2). It is on three vellum leaves, illuminated and assembled into a chart; 22 x 105 cm.

This anonymous chart was obtained by Alberto Cantino in Lisbon where he was working as a diplomatic agent for the Duke of Ferrera, Ercole d'Este. Cantino likely obtained the chart clandestinely from an unknown Portuguese cartographer in the summer of 1502 (the precise date based on existing records) and smuggled it out of Portugal. The Cantino chart was copied from the *PadrÇo Real* or *PadrÇo Oficial*, the royal standard cartographic prototype, held in the Armazém da Guineé das Indias.

The Cantino planisphere is the first to provide a modern depiction of the African outline. The unknown author delineated with precision the Equator and the latitude lines for the Tropics of Cancer and Capricorn. As part of their orders to bring back detailed sailing information, Portuguese pilots had specific instructions to determine precise latitudes by taking frequent surveys from land. The coastlines of Africa had been surveyed from 1484 to 1499 during the voyages of exploration, principally of Diogo Cão, Bartolomeu Dias, and Vasco da Gama. This work produced a chart of Africa with heretofore unknown accuracy for the coastlines, along with the proper placement of corresponding toponyms for the ports of call on the way to the East. The chart is up-to-date with information on Portuguese discoveries prior to the summer of 1502. It records 68 placenames from the Zaire River to the Cape of Good Hope, twice as many as on the Martellus map and about 20 more than on the De la Cosa planisphere (Relaño 2002: 180). The Cape of Good Hope,

THE MAPPING OF AFRICA

Cantino manuscript planisphere. Close-up showing Africa (Modena Biblioteca Estense Universitaria).

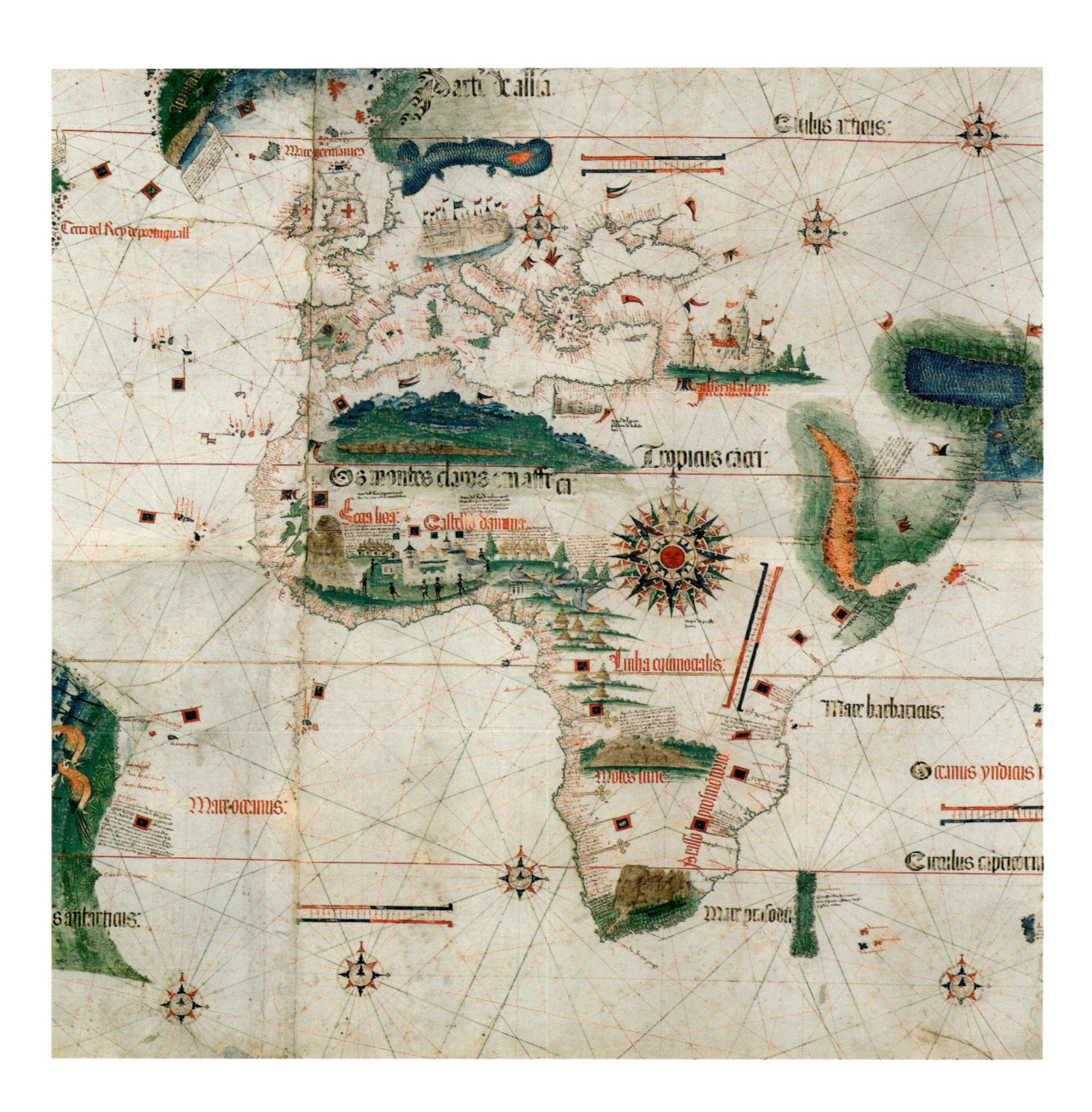

which was misplaced by Martellus too far to the south and by Juan de la Cosa too far to the north latitudinally, is shown accurately on this planisphere.

This is the earliest map to show precisely the recent explorations of Da Gama and Cabral, particularly along the southeast coast of Africa. The Portuguese standards and crosses representing the padrões are placed to continually mark the major advances of the Portuguese on their way to the Indies. At the entrance to the Red Sea, there is a notation that states 'Up to here was discovered by the King of Portugal', proving that the cartographer for this chart was aware of the Diogo Dias ship which had become separated from Alvares Cabral (whose fleet sailed to India from 1500-1502) at the Gulf of Aden (Relaño 2002: 179).

Besides detailed toponymical information along the coasts, there are a number of legends placed in the interior which record information on the Portuguese activities with Africans. For example, at *S. Jorge da Mina* (the gold mines at Mina), information is provided on the Portuguese trade with Guinea, as well as minerals that were available to the Portuguese. At the Kingdom of the Congo, an account is provided on the missionary efforts of Diogo Cão in 1491. The Kingdom of Prester John in East Africa is shown for the first time on this planisphere (Relaño 2002: 62). Below the compass rose in the center of Africa is a legend that identifies the Land of Prester John. Not knowing the extent of Ethiopia or the size of the Land of Prester John, the original cartographer probably assumed that this Kingdom extended across most of central and even Southern Africa. Confusion over the exaggerated extent of Ethiopia and the Land of Prester John was to continue well into the seventeenth century. This resulted in the on-going appearance of names which should have been associated with East Africa being placed too far to the south.

Even along the east coast, further precision is provided, thus correcting the inaccuracies of the De la Cosa map. Six padrões are shown on the eastern coast of the Cantino Planisphere. The nomenclature (63 place-names are used on the east coast) corresponds to cities and regions visited by Da Gama and Cabral: *Caffalla, Mocambique, Quilluá, Mobacá, Melindé, Mogadoxo,* and *Barbara* (Relaño 2002: 180). Only three place-names of Ptolemaic origin are found on the east side of Africa: *Mare Prasodu* (first seen by Bartolomeu Dias as he entered the Indian Ocean), *Mare Barbaricus* (the sea near the entrance to the Red Sea), and *Prasso Prosmōtorio* (a cape in East Africa, corresponding to Cape Delago in present day Mozambique, and, on this chart, a vast region).

The vertical line through Brazil on the left side of the Cantino chart is the first reference to the Treaty of Tordesillas on a map (Molllat du Jourdin 1984: 214). This treaty, which Castile and Portugal signed in June 1494, effectively divided the world, 370 leagues west of the Cape Verde Islands, into Spanish and Portuguese spheres of influence.

The representation of the outline of Africa as depicted on the Cantino planisphere and its successors, the chart of Maggiolo (1504), the planispheres of De Caveri (1504-5) and King Hamy (c.1504), and the world map by Waldseemüller (1507), firmly provided a newer, more modern view of Africa in the eyes of Europeans. For almost 50 years, apart from increasing information on placenames and on the interior, only small corrections in details would be added to the outline of Africa. These improvements were those made by Jorge Reinel (1510) with adjustments in the shape of Madagascar, by Diogo Ribeiro (1525) who corrected the orientation of the Mediterranean, and by João de Castro in his Roterio do Mar Roxo (1541), which presented a much more accurate orientation and depiction of the Red Sea (Cortesão 1960: Vol. I: 92-3).

De Caveri manuscript planisphere (Paris BN, Cartes et Plans, Ge SH Arch 1).

Nicolo de Caveri, Manuscript Planisphere, C.1504-1505

The De Caveri Planisphere is located at the Bibliothèque Nationale de France in Paris. It is on ten vellum leaves, illuminated and assembled into a chart; 115 x 225 cm.

This large manuscript seachart of ten sheets of vellum was made by Nicolo de Caveri of Genoa and represents the entire world as known in 1502-1504 (no discovery made after 1504 is shown on the map, thus the date of c.1504-1505). In the center of the chart is a symbolic mappæmundi placed in Africa from which rhumb lines emanate and connect with compass roses that surround Africa. Along the left margin of the chart, an important new device is introduced: a latitude scale from 55º S to 70º N, which, according to Du Jourdin, appears for the first time on this chart (Molllat du Jourdin 1984: 216). This device enabled ships to sail with much greater efficiency and precision far from the coast. Since voyages across the oceans had become a possibility, determining latitude accurately became essential to record compass directions. With this information navigators could return directly to their home port and later revisit the newly discovered lands. Dead reckoning and a few latitude readings were helpful for the early Portuguese explorers as they sailed closely along the western African coast, but for repeated and quicker oceanic travel more scientific methods were required.

The fact that the chart's configuration and nomenclature is nearly identical to those of the Cantino Planisphere, which was made in Lisbon in 1502, means that De Caveri probably had access to an unknown common Portuguese prototype. The De Caveri Planisphere is undated, but signed as follows: *Opus Nicolay de Caveri Januensis* (The work of Nicolo de Caveri, Genoese). That is, this chart was copied by De Caveri, possibly in Portugal. Some have theorized that if De Caveri had executed his work in Italy, there would have been no reason for inscribing the legends in the Portuguese language, and he would have had time to translate the Portuguese words into Italian (Mollat du Jourdin 1984: 216).

Literally hundreds of placenames line the coasts along with symbolic representations of where the Portuguese placed their stone columns or padrões. As on the Cantino Planisphere, the coastlines of Africa are remarkably accurate. However, unlike Cantino, De Caveri's chart still depicts the Red Sea, which the Portuguese had not yet explored, with a medieval west-to-east orientation widening the overall shape of Africa.

De Caveri's planisphere was used as a source for a continuous series of maps over the next 25 years, principally the 1507 twelve-sheet printed world map of Martin Waldseemüller. De Caveri's nomenclature and outlines are apparent in the Waldseemüller 1507 map, and even more so in Waldseemüller's 1516 Carta Marina. It is assumed that Duke René, the patron for a group of cartographers working at St. Die in Lorraine in France, obtained a copy of the De Caveri chart, likely through a diplomatic agent of his in Italy, and passed it on to Waldseemüller.

These maps served to present the image of Africa to Europeans until news of further explorations of the Portuguese, particularly on the East African coast, corrected and helped to complete the coastal cartography of Africa.

Giovanni Contarini–Francesco Rosselli, Printed World Map, Venice or Florence, 1506

The only example of the Contarini-Rosselli world map is located at the British Library (*London BL*, Maps C.2.cc.4). It is a copper engraved map; 42 x 63 cm.

Designed by Giovanni Matteo Contarini and engraved by Francesco Rosselli, this map was separately published and the only example known was re-discovered in 1922. The map has a fan-shaped coniform projection, a modification of Ptolemy's first conic-like projection. The date of 1506 and the names of the designer and engraver are on a legend to the east of the Cape of Good Hope. The map has no title.

Shirley (1993: 23) and Randles (2002: 70) state that there appears to be no single source for this printed map. It is assumed that several different maps were used in its design. The Contarini-Rosselli map contains aspects of the De la Cosa, Cantino, and De Caveri maps or one of their prototypes. As has already been stated, information and numerous manuscript copies of maps were flowing easily between Portugal and Italy during this period. Contarini likely had access to this information and used examples of these maps in the preparation of his own map.

SIGNIFICANT WORLD MAPS THAT SHOW THE CONTINENT OF AFRICA

*Contarini-Rosselli world map
(London BL, Maps C.2.cc.4).*

This map portrays an Africa similar in shape and orientation to the Cantino and De Caveri planispheres. It is the first printed world map to show the discoveries of Da Gama and Cabral, particularly along the southeast coast of Africa. Although Contarini uses new information for the outline of Africa, he, as so many others from this period, still relied on Ptolemy's *Geographia* for his presentation of the interior. The two Nile River source lakes are traditionally shown side by side with multiple streams running into them from the Mountains of the Moon further to the south. Contarini adds two further lakes to the west and southwest of the Mountains of the Moon which eventually connect to the Nile. The Niger River in West Africa is linked by lakes on its east and west sides. Generally, his portrayal of the hydrographical system of the interior is confusing and does not seem to be based on prior sources. Contarini shows three rivers running towards the Southeast African coast originating from mountains (with no source lakes) and one which runs toward the southwest African coast (unlike on previous maps). According to Randles (1956: 13: 70), this southwest river, *S. Tiago*, might be a reference to the River Samtiagua near where Da Gama landed by St. Helena's Bay. Of the three rivers running towards the southeast, he also states that the northernmost river is the *Rio dos Bons Sinais*, so named by Da Gama and not known today. The other two rivers are not known, and, as Da Gama does not mention them, Randles speculates that information on them may have been obtained from Arab sources. Between the two rivers in Southern Africa, there is the town, *Vigict Magna*. Randles believes that *Vigict* is possibly a corruption of the river Wanshit in Abyssinia, again demonstrating the tremendously exaggerated size that Europeans attributed to the area of Ethiopia, extending it well into Southern Africa. Further up the east coast, Contarini shows *Mylinde* (Malindi), the name of the city kingdom and an important Portuguese ally.

The islands of Madagascar and Zanzibar are well off the eastern coast of Africa in the Indian Ocean, reflecting a presentation similar to the De la Cosa world map, and unlike the Cantino and De Caveri planispheres, which show Madagascar as too far south, but closer to the African coast. The Mediterranean Sea still has too great an east-to-west width.

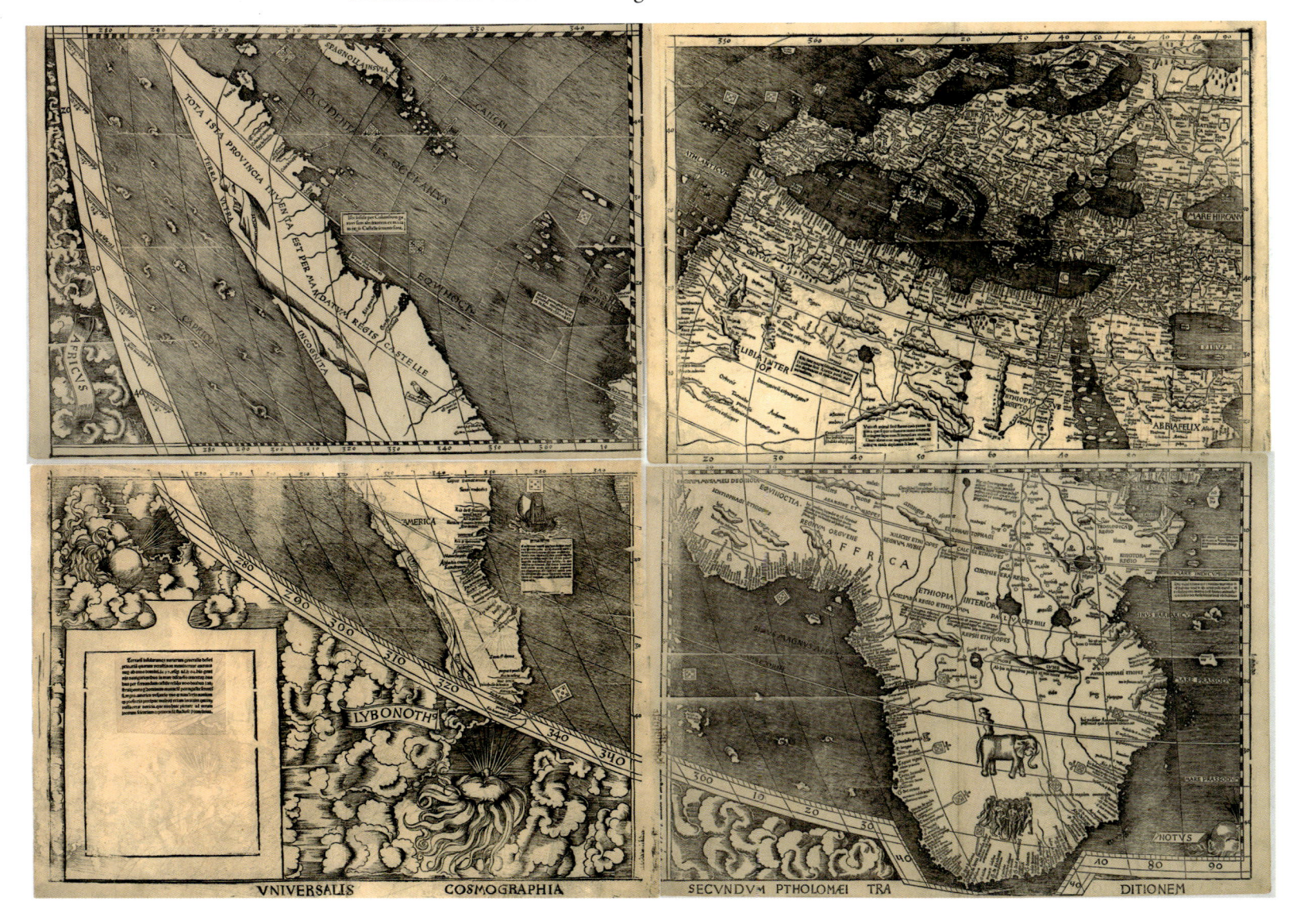

Waldseemüller's printed world map, section showing Africa (Washington LC).

Martin Waldseemüller, Printed World Map, Strasbourg, 1507
The only example known of the Waldseemüller world map is at the Library of Congress, Washington, DC. It is on 12 woodcut printed sheets; 132 x 236 cm. The map's title is: *Universalis Cosmographia Secundum Ptholomaei Traditionem Et Americi Vespucü Aliorûque Lustrationes.*

Martin Waldseemüller (1470-1518) was probably the most influential cartographer of the first half of the sixteenth century. His wall map, which survives in only one known example, seems to have been a great success in its day based on existing evidence. This map was quickly followed by his edition of Ptolemy's *Geographia* of 1513 which contained modern maps of Northern and Southern Africa. Both the wall map and the two modern maps from the *Geographia* seem to have a common source in the De Caveri- or a Cantino-De Caveri-type model and were likely prepared around the same time.

On the wall map, Waldseemüller demonstrates a thorough knowledge of recent discoveries in the world. Support was given to him and the other members of his academic circle in St. Die, near Strasbourg in present-day France, by their patron, René, Duke of Lorraine. René passed on numerous source materials to Waldseemüller obtained from his agents who likely were based in Italy. Besides a copy of the De Caveri map mentioned above, copies of letters of various explorers and travelers were also sent to Waldseemüller.

From the De Caveri planisphere, Waldseemüller borrowed the shape, the coastal placenames, and the legends for his depiction of Africa. The two maps agree on many minor details including the same placement of the padrões (a carved stone column containing the particulars of each voyage that the Portuguese expeditions carried on board their vessels and which they erected as proof of their furthest position down the African coast), the elephant in Southern Africa (also used by Sebastian Münster for the Africa map in his *Geographia* of 1540), and the Portuguese banners on the island groups off the western coast of Africa. So close are the similarities that Fischer (1903: 21) speculates that the map Waldseemüller used was, in fact, the De Caveri itself now in the Bibliothèque Nationale.

Though Waldseemüller used the De Caveri model for his map, his drawing of the outline of Africa is rough, in comparison to the earlier De Caveri and Cantino planispheres. The exception is the region of Southern Africa, especially where the map actually 'breaks' the gridline at the bottom of the map. Waldseemüller was also working with wood which often had less precision than copperplate. In addition, he made corrections and ad-hoc adjustments, while the woodcut was being prepared, which further resulted in some rough engraving.

While Waldseemüller used current information available to him from the De Caveri planisphere, he did not altogether abandon his faith in Ptolemy. He continues the errors in longitude of Ptolemy, though some of this was corrected in his Carta Marina of 1516. For example, he includes Ptolemy's *Prassum Promontor* just north of the padrão at Mozambique. His depiction of the interior is often based on the classical understanding of Africa contained in Ptolemy's *Geographia*. He does not seem to have used Arab sources in his depictions of the interior and of the coast of East Africa. At this point in time, the Portuguese had yet to make a significant impact along the coasts of Africa and in the interior, particularly in the Congo, Southeast Africa, and Abyssinia.

Egyptus Novelo, North is at left (*Paris BN, Manuscrits occidentaux, Latin 4802 Folio, 130v-131*).

One primary source for the interior that Waldseemüller did use was the Egyptus Novelo, a 'modern' map of greater Ethiopia, drawn by Pietro del Massaio. Information for this map was provided to Del

Massaio by Abyssinian monks who were in Florence for the Church Council of 1441 (Randles 1956: 13: 73-4). The Egyptus Novelo was added to a manuscript edition of Ptolemy's *Geographia* by Jacopo Angiolo (*Paris BN*, MS Lat. 4802, 130v-131). Two other copies are in the Vatican Library (Fischer 1903: 20, Relaño 2002: 201 and Randles 1956: 13: 120-122). As there are some differences in placenames between the Waldseemüller 1507 map and the Egyptus Novelo, Randles theorizes that Waldseemüller likely had access to a similar map, again probably supplied to him by the Duke of Lorraine.

It is important to note that the Egyptus Novelo map was not drawn to scale and that the details of Abyssinia provided by the monks were assumed to include the greater portion of central and Southern Africa. Thus, the geography and placenames for Abyssinia, derived from Abyssinian sources, filled much of the vast unknown central and southern interior of Africa. This depiction of an enormous Abyssinia far to the interior and to the south, resulting in the misplacement of numerous placenames, was to be perpetuated well into the seventeenth century.

The twin Nile River source lakes, identified as *Paludes Nili*, are conventionally placed just south of the Equator with the Mountains of the Moon below. These mountains also show two rivers flowing from them to the southeast coast, the *Rio S. Vincenco* and the *Rio de Lago*. Ravenstein (Randles 1956: 13: 70) identifies the *Rio S. Vincenco* as the Pungwe River and the *Rio de Lago* as the Umbelasi River. In the Abyssinian highlands, Waldseemüller places another lake at the Equator connecting to the Nile River. This lake was intended to be Lake Tana, the actual source for the Blue Nile River. It was not until later in the seventeenth century that Lake Tana was properly placed north of the Equator. Waldseemüller lists numerous placenames on and around the headwaters of the Nile as well as a general name for the region, *Ethiopia Interior*, stretching far to the south.

Another feature on the 1507 map borrowed from the Egyptus Novelo is a lake to the west of the *Mons Lune* (Mountains of the Moon) with a river flowing north and ending in another lake in the Sahara. This lake is labeled *Sachaf Lacus* and it was to remain a common feature on maps depicting the interior well into the seventeenth century.

Overview of Printed Maps of Africa from 1505 to 1700

The maps represented in this cartobibliography are grouped into six general categories or cartographic models, based on key characteristics on the maps. These characteristics include the depiction of the shape of the continent, information on the hydrography (the lakes and river systems), uses of mountain ranges, and the appearance of regional names, placenames, and other text on the map.

The process used to develop the cartographic models included a close examination of all the maps presented in this cartobibliography, the identification of the key characteristics of each map, the development of generalizations leading to cartographic models, and finally the association of each map with one of the six models. There will be some exceptions to the general models, and some unintentional omissions of data (missing lakes, etc.) by the mapmakers or map engravers. However, the author believes that this use of cartographic models is a valuable tool to better understand the development of the mapping of Africa during the critical period from 1500 to 1700.

1. African Cartography from the Anonymous Woodcut of c.1505 to Sebastian Münster, 1540
The first printed maps to primarily show the African continent were crude woodcuts. In c.1505 an anonymous woodcut appeared on the frontispiece of a newsletter, *Den rechten weg ausz zu faren von Liz / bona gen Kallakuth. vo meyl zu meyl….* (The Right Way to go from Lisbon to Calicut, mile by mile…). This is the first printed map to show the Cape of Good Hope, apart from the Rosselli world map of c.1492-1493, and the first to publicly disseminate the Portuguese sea route to India around Africa. It shows an eastward-sloping orientation for the continent with a wide and flat Southern Africa, extending eastward to almost the 100th meridian.

Anonymous 1505 woodcut showing the Portuguese sea route to India (Washington LC).

The publication of the newsletter, although not stated, is attributed to Nuremberg, as a symbol (N) for that city has been placed on the map, an unlikely occurrence unless it had importance to the publisher. In 1505, there was significant German and specifically Nuremberg involvement in the India trade, through their support to the Portuguese king. The Welser, Fugger, and Imhof banking firms invested in Lisbon's spice fleets and maintained permanent representatives in Lisbon (Parker 1956). In all likelihood, agents for one or all of these firms were responsible for the relatively speedy transfer of information from Lisbon to Nuremberg, particularly on the Da Gama and the forthcoming Da Albuquerque and Da Cunha voyages to India. From the appearance of the wording, lettering and numbering on the map, the newsletter was printed in haste reflecting a desire to disseminate this information quickly and further promote German investment in the East India trade. The author has noted two different variants of this map. One example at the Bell Library of the University of Minnesota (*Minneapolis Bell*, 1505 Re) has 'clouds' in the upper left and right corners of the map. One example at the Library of Congress (*Washington LC*, DS411.7 .R4) is 'without clouds' in the upper left and right corners.

The first map to show only the continent of Africa was published in 1508 by Antonio Francanzano (Fracan) da Montalbodo (See Map # 1 for further information). It appears on the title page of the book, *Itinerarium Portugallensium e Lusitania in Indiam & indem occidentem & demum ad aquilonem*. The Da Montalbodo map corrects the severe eastward-sloping orientation of Africa in the anonymous 1505 map, but still represents Africa as too wide longitudinally, with the Cape of Good Hope at least 15 degrees to the east of its actual position. Many placenames on the map are based upon Portuguese discoveries: *C. Biancho* (Cape Blanco) and *C. Verdo* (Cape Verde) on the west coast; *Monte Negro* with a padrão cross (an engraved stone marking the limits of exploration) along the southwest coast; and *C de Spaza* (for Cabo de Boa Esperança or the Cape of Good Hope) in the south. The African-Arab port of *Melido* (Malindi) on the east coast is also shown. Although the rivers are not identified, they appear to be the Niger in the west with its confusing hydrography, the Congo/Zaire River in the central region, the Blue Nile in the Abyssinian Highlands, and the White Nile originating in the Mountains of the Moon and the two Ptolemaic lakes. Interestingly, a third distinct river which also flows into the Nile River is placed to the west of the two Ptolemaic Lakes originating far into Southern Africa. The Zambezi and Limpopo in the southwest originate, not in the Mountains of the Moon, but in separate mountains in Southern Africa. Further to the east, the Malabar peninsula in India is shown along with the important trade port of Calicut.

An example of the 1508 edition of *Itinerarium Portugallensium e Lusitania in Indiam & indem occidentem & demum ad aquilonem* at the Österreichische Nationalbibliothek (*Wien ÖBN*, 394092-C.K) contains a unique set of six maps showing Africa and parts of the coast of Africa from Lisbon to West Africa. This is the only known example of the book containing this set of maps. These maps are intended to illustrate the voyages of Cadamosto and Da Sintra to Senegal. In style, the regional maps are very similar to the maps prepared by Martin Waldseemüller for the 1513 Strasbourg edition of Ptolemy's *Geographia*. In size, these woodcut maps are 15.5 x 22.5 cm.

Another early woodcut map showing Africa was on the title page of a book, *Der welt kugel Beschrybung der welt … mit vil seltzaman dingen*. ('Globe of the World…'), published by Johann (Joannes) Grüniger in Strasbourg in 1509. This map is hemispherical in shape and shows the entire continent of Africa with a portion of Asia, America, and most of Europe. Africa is depicted in full, but with no information, other than the latitudes for the Equator and the Tropics of Cancer and Capricorn. The book is a geographical tract describing the world. The title of the book conveys the impression that the book was meant to accompany a real globe with the text, 'Exposition or description of the world, and of the terrestrial sphere constructed as a round globe similar to a solid sphere…'. No coastal landfalls or river systems are shown. The shape of Africa follows a somewhat more conventional representation without the eastward slant to the southern part. The southern tip of the continent is flat and wide, along the lines of the anonymous 1505 map of Africa, described above. The cartographer and the author of the work are not absolutely known. There is some speculation by Church (1951: 74-75) and others that the map and the text are the work of Martin Waldseemüller. There is a reference to Amerigo Vespucci on the title page, but not to Columbus, and the name America is employed later in the text to designate the New World. Waldseemüller was an early promoter of the work of Vespucci and is credited with providing the name America to the New World. Furthermore, Waldseemüller's *Cosmographiae Introductio* was published in the same year in Strasbourg by Grüniger.

Grüniger map of 1509 (New York, PL).

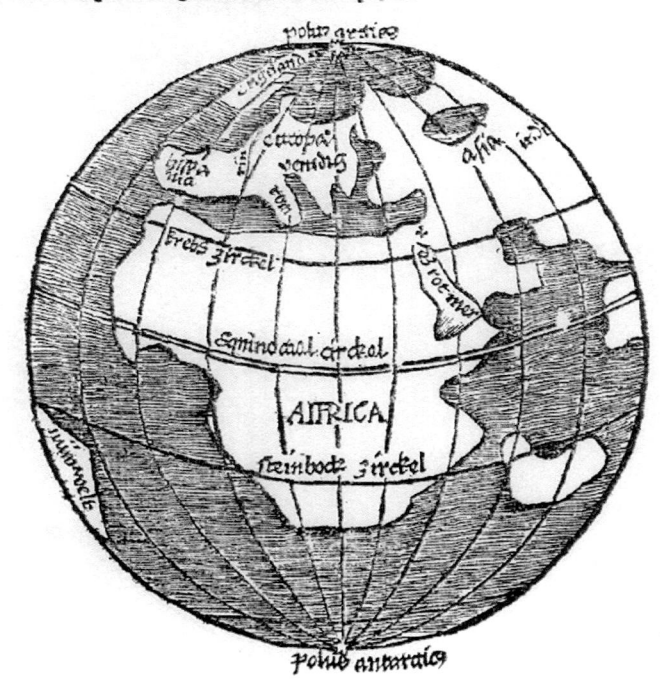

The next two most noteworthy maps of Africa were the modern maps of North Africa (TABVLA MODERNA PRIME PARTIS AFRICAE) and South Africa (TABVLA MODERNA PORCIONIS APHRICE) produced by Martin Waldseemüller for editions of Claudius Ptolemy's *Geographia* in 1513 and 1520. Directly modeled after the Africa section of Waldseemüller's 1507 wall map of the world, these maps provide detailed placenames along the west and southern African coasts. Laurent Fries reproduced these maps in a slightly reduced format for editions of Ptolemy's *Geographia* in 1522, 1525, 1535, and 1541. Other important maps of North and South Africa which appeared in an edition of Claudius Ptolemy's *Geographia* of 1548 were prepared by Giacomo Gastaldi.

There were other maps from this period that also depict parts of Africa. In 1542, Johannes Honter published a map showing the northern half of Africa in an edition of *Rudimentorum cosmographiae*.

Probably the most noteworthy map of Africa in the middle of the sixteenth century was prepared by Sebastian Münster in 1540 (Map # 3). While primitive in its depiction of Africa, the map's importance lies in the fact that it was included in what were probably the two most widely circulated and read books of the mid-sixteenth century – Sebastian Münster's *Geographia* and *Cosmographia*. Münster's map of Africa is the earliest, readily available, printed map to show the entire continent of Africa. Like his predecessors, Münster places the southern tip of the continent more than 10 degrees too far to the east, thus creating a long, sloping southwestern coast of Africa. This was not to be corrected until Gastaldi's eight-sheet wall map of 1564. Münster remains true to the Ptolemaic view of the interior of Africa. His source for the White Nile River rises in the unnamed twin lakes, which in turn are fed by the Mountains of the Moon, though these are not labeled on his map. He does show a separate lake in the Abyssinian Highlands near *REGNVM Melinde* as the source for the Blue Nile, though this placement is too far south to be an accurate representation of the real source, Lake Tana. Interestingly, Münster does not include a westward extension of the Nile River across West Africa, as later cartographers did by

THE MAPPING OF AFRICA

Laurent Fries' 1535 reduced version of the Waldseemüller maps of Northern and Southern Africa (Author's collection).

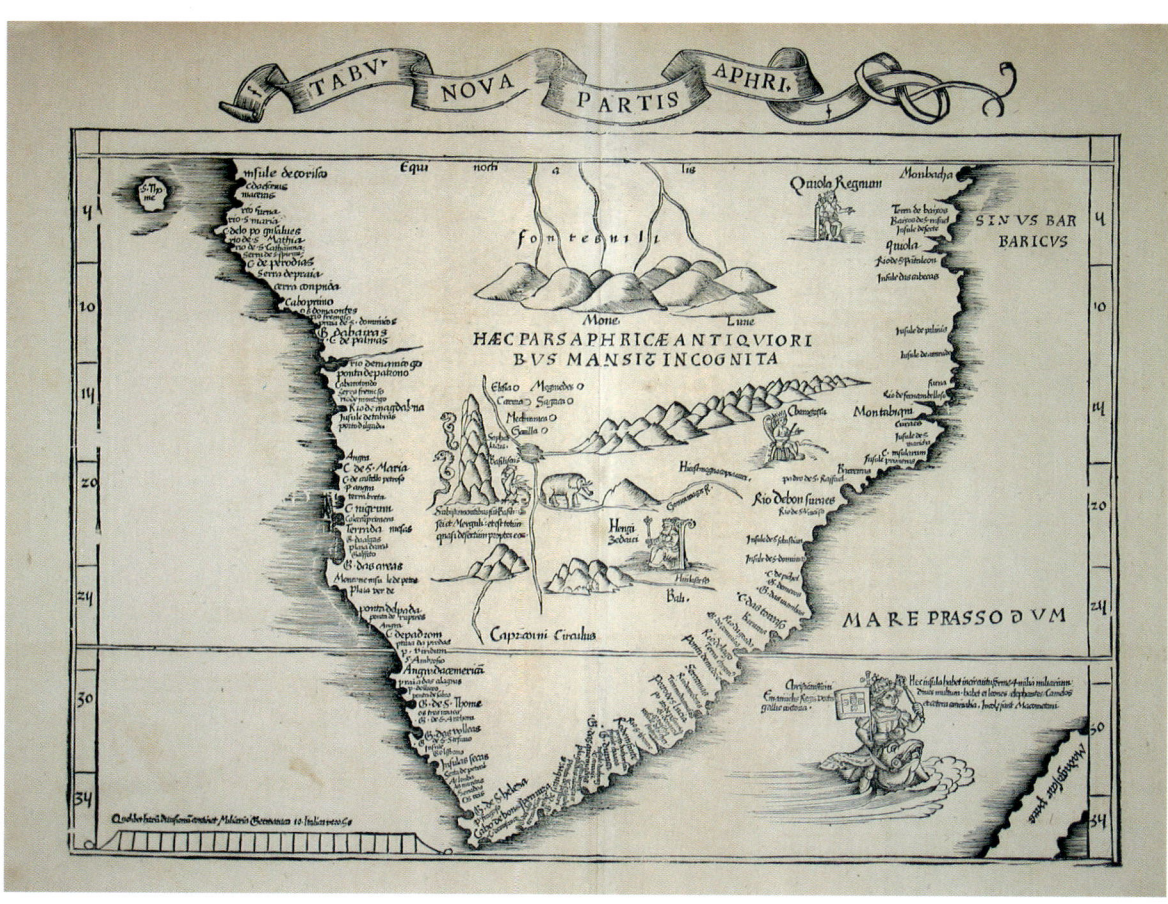

adapting the much earlier tradition of the Greek, Herodotus. Münster does not show the island of Madagascar off the southeast coast of Africa, which is unusual considering it had been reported in Europe since the time of Marco Polo. He does show the island of *Zaphala Aurifodina* in the Indian Ocean from which King Solomon, according to legend, obtained silver and gold.

Another map from this early period is a simple outline map of Africa by Pliny-Sessa of 1513. Concluding this period, there are four maps of importance which are all depicted with south at the top of the map: the Ramusio-Gastaldi maps of 1554 and 1563, the Africanus-Temporal of 1556, and the Bertelli of 1565. The common geography for these four maps is the work of Giacomo Gastaldi, and serves as a bridge to Gastaldi's landmark map of 1564.

Early Maps of Africa

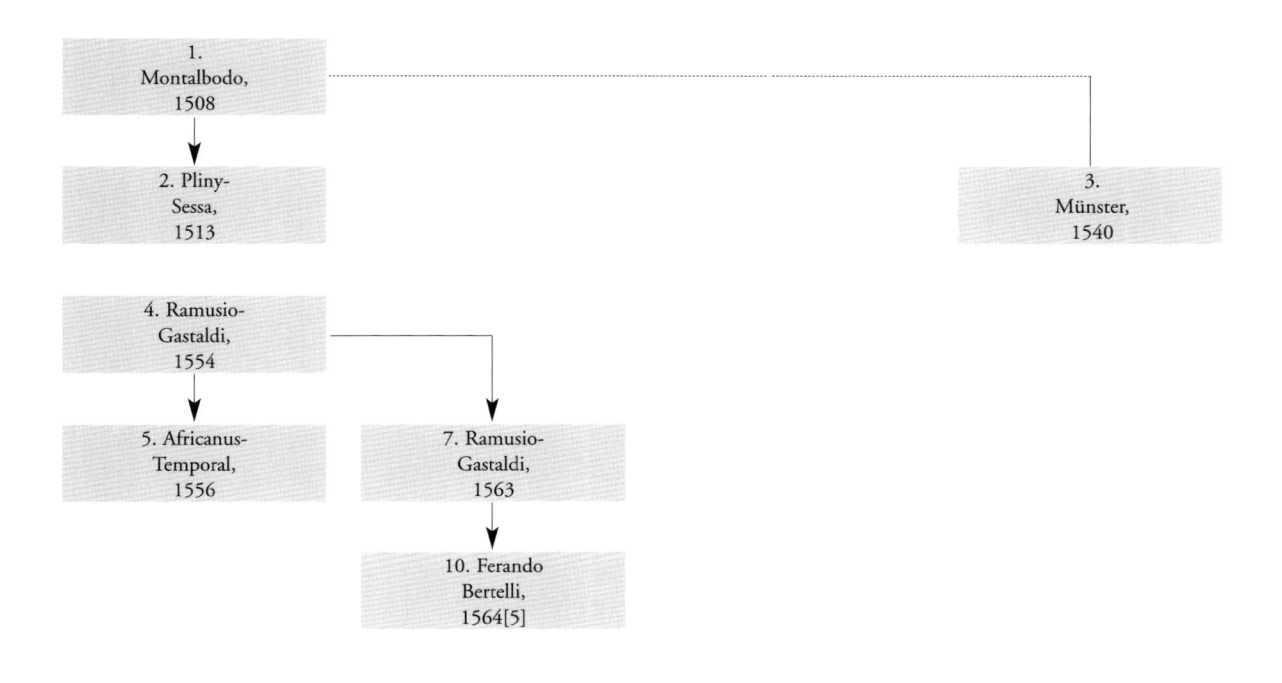

1. Montalbodo, 1508
2. Pliny-Sessa, 1513
3. Münster, 1540
4. Ramusio-Gastaldi, 1554
5. Africanus-Temporal, 1556
7. Ramusio-Gastaldi, 1563
10. Ferando Bertelli, 1564[5]

Map sources not apparent, some due to small size:

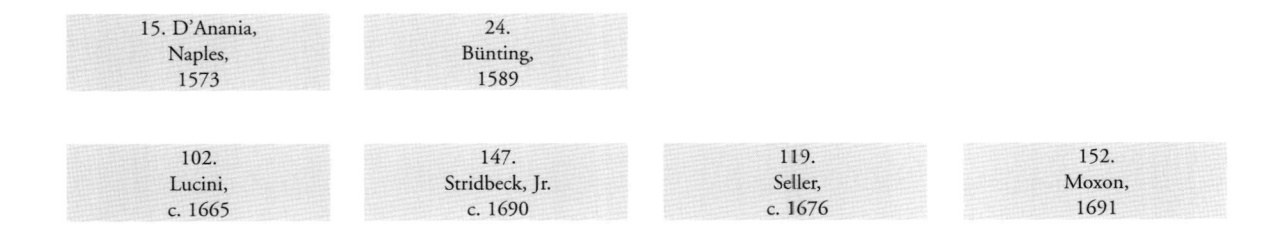

15. D'Anania, Naples, 1573
24. Bünting, 1589
102. Lucini, c. 1665
147. Stridbeck, Jr. c. 1690
119. Seller, c. 1676
152. Moxon, 1691

2. The Landmark Maps by Gastaldi and Ortelius

The next major advance in the mapping of Africa was made by Giacomo Gastaldi working from Venice. Gastaldi prepared detailed maps of North and South Africa in 1548, and a map of Africa in an edition of Giovanni Ramusio's *Delle Navigationi et Viaggi* in 1554, with up-to-date geographic information.

These smaller maps culminated in Gastaldi's eight-sheet wall map of Africa of 1564 (Map # 9). This map is a testament to Italian cartography. Without much dispute, it is probably the single most important and influential map of Africa of the sixteenth century. This map represents a quantum leap from Sebastian Münster's Africa map of 1540, which had no direct influence on future mapmakers.

Following his smaller 1554 map that appeared in Ramusio's book, Gastaldi corrected the long, sloping southwest coast of Africa from the typical presentation of Münster and others. Working with the large-format size afforded to him with eight sheets, Gastaldi was able to add significant detail to his map, including numerous new placenames. Gastaldi shows a southbound river, the unnamed Zembere, flowing out of the western Ptolemaic Lake (named *Zembere* and *Zaire*), passing through the unnamed Mountains of the Moon before splitting into the *Cuama* (Zambezi) and *Spirito Santo* (Limpopo) Rivers flowing to the southeast coast of Africa. The ancient city of Great Zimbabwe (*Simbaoe*) is depicted in the interior of Southern Africa with a drawing of a large city. The West African Niger River is shown now to begin in a *lacus Niger* (Lake Niger) in central Africa.

Gastaldi's wall map spawned a number of imitators both in Italy (Gastaldi-Camocio c.1570-1575, Gastaldi-Bertelli c.1573) and in the Netherlands (Gerard and Cornelis de Jode). There were also a number of folio-sized maps that were modeled after the Gastaldi wall map by Sanuto, Forlani, Duchetti, and Gerard and Cornelis de Jode.

Abraham Ortelius, from Antwerp, used the Gastaldi wall map as the model for his own folio-size map of Africa that appeared in his *Theatrum Orbis Terrarum* starting in 1570 (Map # 12). Like Münster some thirty years before, the *Theatrum* was widely dispersed and was printed in numerous editions in a variety of languages into the early seventeenth century. The tremendous commercial success of the *Theatrum* helped to cement the Gastaldi image of Africa in the minds of Europeans.

Geographically, the Ortelius map closely follows Gastaldi's wall map of 1564. This is especially evident in the choice of placenames, although there are misspellings and sometimes widely differing placements of towns.

Ortelius introduced two important changes to the shape of the continent not shown on the Gastaldi map; the shape of the southern subcontinent became more pointed, and the eastward extension of the continent was significantly reduced. The distance from Ceuta to Cairo was also reduced. Both of Ortelius' distances are in fact close to the actual measurements today: 6,800 km from east to west at its widest point, and 3,450 km from Ceuta to Cairo.

Ortelius also had access to Gerard Mercator's 1569 wall map of the world and probably Paulo Forlani's 1562 map of Africa. However, Ortelius disregarded Mercator's classical sources for the connection of the Niger River as a western branch of the Nile River and, instead, uses Gastaldi's representation of the river and lake hydrology of the African interior. A curious omission on Ortelius' map is the lack of the Mountains of the Moon, a prominent feature in most prior maps of Africa. It appears that Ortelius closely followed Gastaldi who, in turn, did not specifically include the Mountains of the Moon on his 1564 wall map. In the earliest Ptolemaic maps and texts, the Mountains of the Moon are depicted as a continuous mountain chain at 12 degrees south, separating Africa known to the ancients from terra incognita in the south. As more became known of the southern extent of the continent, the lake source for the Nile was progressively moved southward in Africa so that the Mountains of the Moon appeared to many mapmakers as a source for the Spirito Santo (the present day Limpopo) and the Cuama (Zambezi) river basins to the southeast.

The Gastaldi and Ortelius maps were used as the basic models for numerous maps of Africa by other mapmakers in Germany, Belgium and the Low Countries, France, and Italy well into the seventeenth century.

Gastaldi-Ortelius Model

Some Major Elements
- North-south correction in the shape of Africa
- Southern African river systems are depicted
- West African Niger River begins in Lacus Niger in central Africa
- Increased number of placenames
- The shape of the southern subcontinent becomes more pointed
- Eastward extension of the continent is significantly reduced (Ortelius)

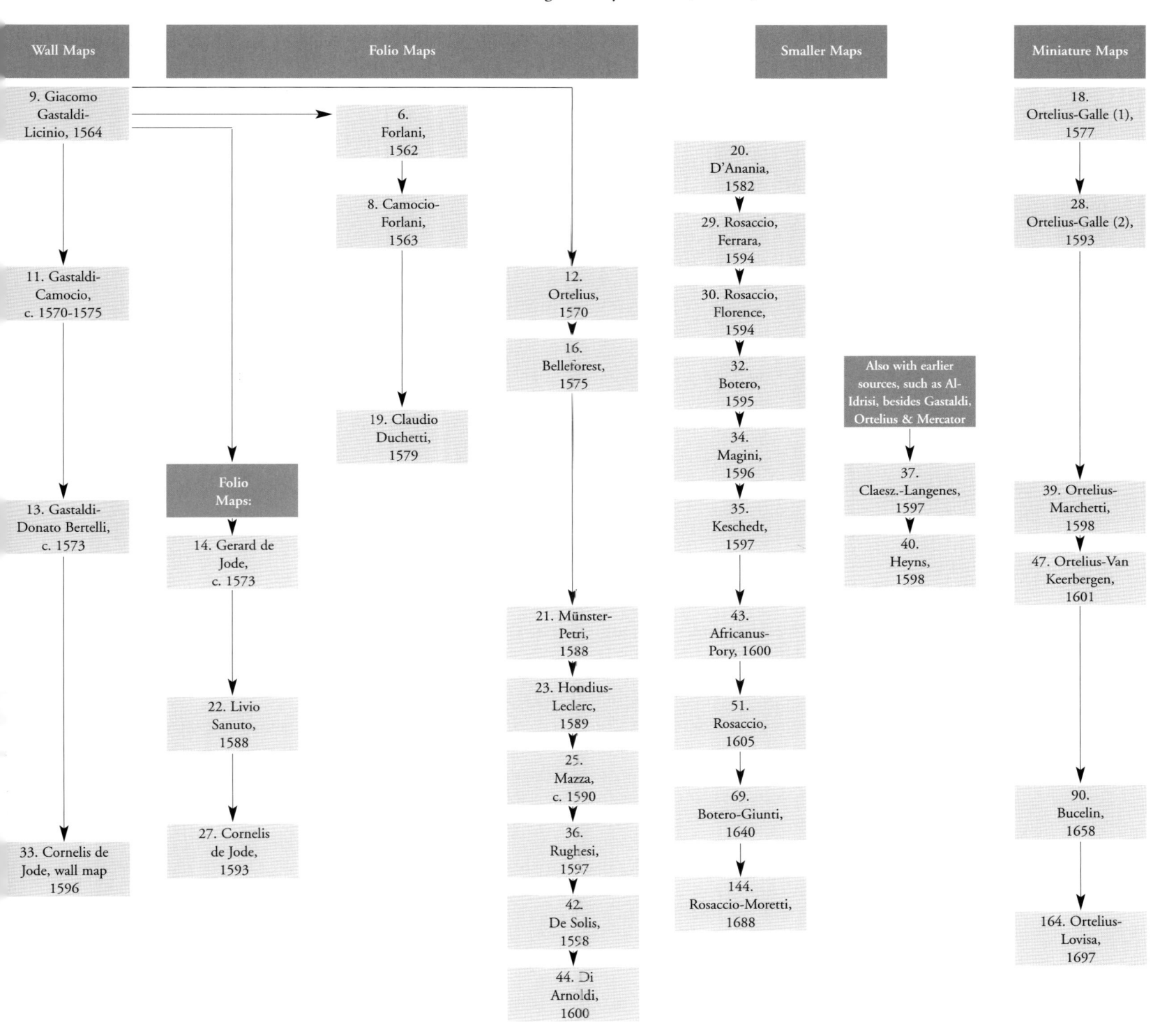

Wall Maps	Folio Maps	Smaller Maps	Miniature Maps
9. Giacomo Gastaldi-Licinio, 1564	6. Forlani, 1562	20. D'Anania, 1582	18. Ortelius-Galle (1), 1577
11. Gastaldi-Camocio, c. 1570-1575	8. Camocio-Forlani, 1563	29. Rosaccio, Ferrara, 1594	28. Ortelius-Galle (2), 1593
13. Gastaldi-Donato Bertelli, c. 1573	12. Ortelius, 1570	30. Rosaccio, Florence, 1594	
	16. Belleforest, 1575	32. Botero, 1595	
	19. Claudio Duchetti, 1579	34. Magini, 1596	
	Folio Maps:	35. Keschedt, 1597	
	14. Gerard de Jode, c. 1573	Also with earlier sources, such as Al-Idrisi, besides Gastaldi, Ortelius & Mercator	
	22. Livio Sanuto, 1588	37. Claesz.-Langenes, 1597	39. Ortelius-Marchetti, 1598
	27. Cornelis de Jode, 1593	40. Heyns, 1598	47. Ortelius-Van Keerbergen, 1601
	21. Münster-Petri, 1588	43. Africanus-Pory, 1600	
	23. Hondius-Leclerc, 1589	51. Rosaccio, 1605	
	25. Mazza, c. 1590	69. Botero-Giunti, 1640	90. Bucelin, 1658
	36. Rughesi, 1597	144. Rosaccio-Moretti, 1688	
	42. De Solis, 1598		164. Ortelius-Lovisa, 1697
	44. Di Arnoldi, 1600		
33. Cornelis de Jode, wall map 1596			

Africa section of Mercator's wall map of the world of 1569 (Rotterdam MM).

3. The Late Sixteenth Century – The Developing Understanding of Africa and the Mercator Map

The next important printed map of Africa was actually the Africa section of Gerard Mercator's famous wall map of the world of 1569. To prepare his world map, Mercator appears to have relied on Waldseemüller's modern maps of North and South Africa of 1513, the Gastaldi wall map of 1564, and the other maps produced by Gastaldi (the 1548 maps of North and South Africa and the Ramusio-Gastaldi map of 1554) for the general shape and coastal features. Mercator's map has a fair degree of accuracy especially along the coastline that was unsurpassed by most sixteenth century cartographers, though he portrays Madagascar as slightly too elongated to the northeast.

Gerard Mercator, like many of his predecessors, used the writings of Ramusio, Leo Africanus, and others for additional details. He also used classical sources for his depiction of the interior of Africa. These include the work of Herodotus to show the river systems of West Africa with the more northern river (*Giras* and then *Nubia flu*) flowing eastward into the Nile River, and the more southern river, which he labels as the *Nil* (actually the Niger River), flowing due westward into the Atlantic Ocean.

Mercator continues the evolving Ptolemaic tradition with his placement of the Mountains of the Moon as the headwaters of the two Nile lakes in the far southern area of Africa. Ignoring the Gastaldi-Ortelius model as evidenced in Gastaldi's 1554 map as well as Ortelius' 1570 map, Mercator portrays the source of the Cuama River (Zambezi) and the unnamed Spirito Santo (Limpopo) River as being a common river, the *Zambere*, originating in a *Sac[haf] Lac* and not the western Ptolemaic lake of Zembere. He places this third lake, *Sac.haf lac*, to the west of the Mountains of the Moon. Interestingly, this third lake also is a source for the Nile River as well as the Zaire (Congo) River in the west. Partly owing to the importance of the Mercator name, Sachaf Lacus was to appear in maps of Africa well into the late seventeenth century. Importantly, Mercator also places a river to the north above the Cuama, the *Zuama flu.*, which was to become a more accurately situated Zambezi River in late seventeenth century maps of Africa, particularly on those by the French (Ortelius had also shown a *Zuama fl.* on his map, but this river appears to be the Cuama River placed further to the south.).

Mercator seems to have had access to Portuguese texts as well, including one by João de Barros that describes European advances into the interior. For example, Mercator shows Portuguese exploration up the *Cuama* (Zambezi) River into the interior of south central Africa in the region of Monomotapa or *Benamataxa*, as he names it on his map. *Ca. Portogal* (the Portuguese Fort) is placed on the map within the junction of the Spirito Santo and Cuama rivers.

Andre Thevet's map of Africa in 1575 was the first map to be modeled on the Africa section of the Mercator wall map of the world of 1569. The importance of this map is often overlooked in the literature.

Gerard Mercator II prepared a folio map of Africa for his family's publication of an atlas in 1595. Mercator II was the grandson of the famous cosmographer, Gerard Mercator, who had died in the previous year. This map of Africa by the younger Mercator is a reduction of the Africa section of his grandfather Gerard Mercator's famous world map of 1569.

The 1595 Mercator map of Africa (Map # 31) was used as a model for maps of Africa by a number of mapmakers through the middle of the seventeenth century, owing in part to the importance of the Mercator name.

Mercator Model

(From Mercator's 1569 world map; some maps with elements from Blaeu, Texeira & the Portuguese, etc.)

Some Major Elements
- Further refinement of the shape of Africa
- Origin of Southern African rivers in Sachaf Lacus
- Additional placenames from the Portuguese, etc.
- Placement of a river above the Niger River flowing eastward into the Nile River

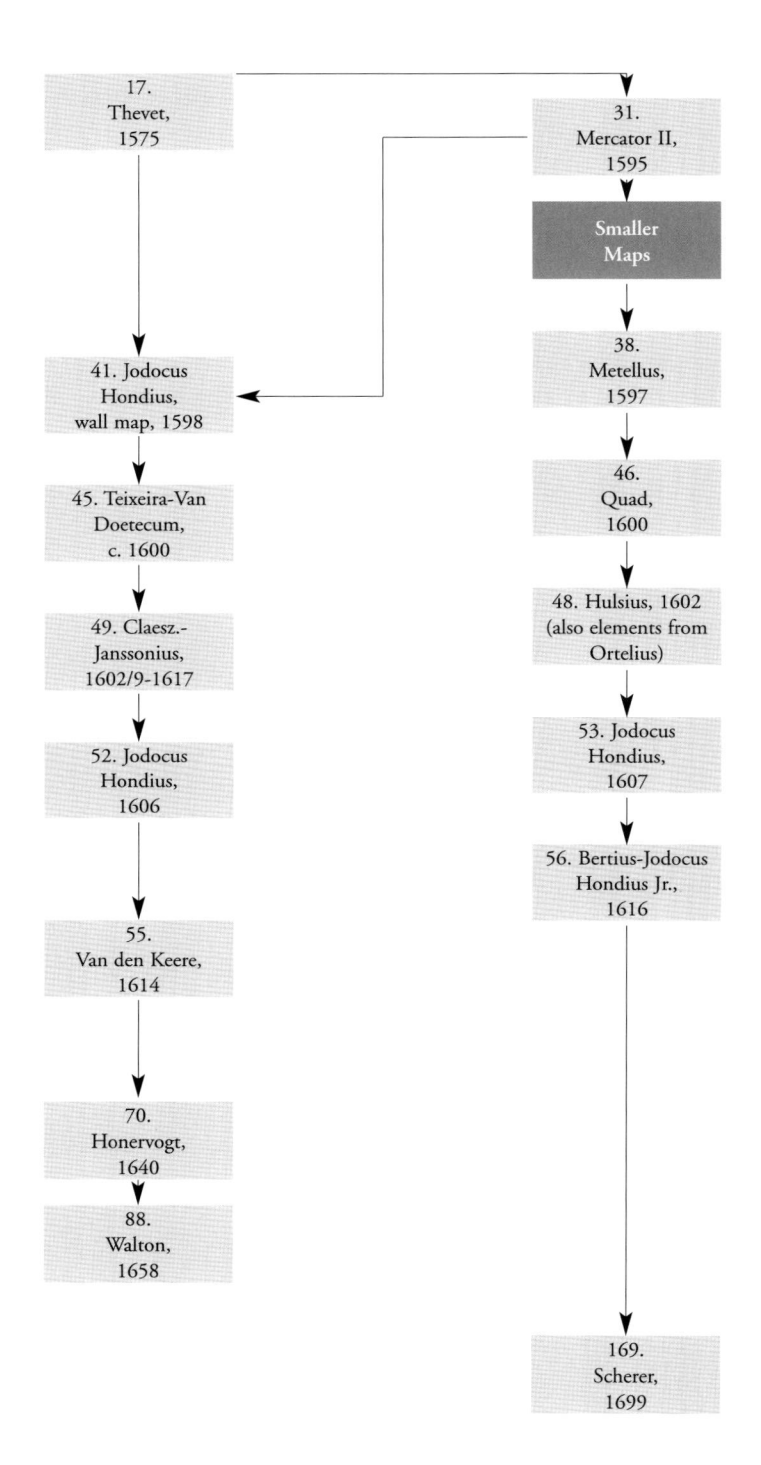

```
17.                              31.
Thevet,  ─────────────────►     Mercator II,
1575                            1595
                                  │
                                  ▼
                                Smaller
                                Maps
                                  │
  ┌───────────────────────────────┤
  │                               ▼
  ▼                              38.
41. Jodocus   ◄──────────────   Metellus,
Hondius,                        1597
wall map, 1598                    │
  │                               ▼
  ▼                              46.
45. Teixeira-Van                 Quad,
Doetecum,                        1600
c. 1600                           │
  │                               ▼
  ▼                              48. Hulsius, 1602
49. Claesz.-                    (also elements from
Janssonius,                      Ortelius)
1602/9-1617                       │
  │                               ▼
  ▼                              53. Jodocus
52. Jodocus                      Hondius,
Hondius,                         1607
1606                              │
  │                               ▼
  ▼                              56. Bertius-Jodocus
55.                              Hondius Jr.,
Van den Keere,                   1616
1614                              │
  │                               │
  ▼                               │
70.                               │
Honervogt,                        │
1640                              │
  │                               │
  ▼                               ▼
88.                              169.
Walton,                          Scherer,
1658                             1699
```

4. The Dominance of the Dutch and the Blaeu Model

The next major development in the mapping of Africa was the Blaeu model as evidenced by Willem Blaeu's wall map of 1608. Some major elements of the Blaeu model are the addition of placenames from the more recent discoveries (e.g. Dutch naval expansion and presence around Africa); an inclusion of elements from both the Gastaldi-Ortelius and Mercator models; and, in Blaeu's folio map of 1617, a further refinement to the river systems in southern Africa.

The geography for the Africa wall map of 1608 (Map # 54) is similar to Blaeu's 1605 double-hemisphere wall map of the world, which is not surprising given the close proximity in dates. By this time, there was a thriving exchange of ideas and cartographic information between Northern Europe, Italy, and to some degree Portugal. This exchange was probably facilitated, in part, by the regular Frankfurt Book Fair, which sat astride most north-south travel routes; the exchange also occurred through direct commercial contacts. Blaeu likely had access to Gastaldi's 1564 detailed wall map of Africa and those of the Northern European imitators such as De Jode. He also had access to the books of exploration of Africanus, Pigafetta (based on the travel account of Duarte Lopes), Ramusio, De Barros, and others. These were supplemented by the more recent Dutch sources such as Van Linschoten's *Itinerario* of 1596 (with charts based on earlier Portuguese sources) and the accounts of the subsequent voyage to the East under Admiral De Houtman.

The geographical divisions within Africa are similar to those of the earlier sixteenth century maps of Ortelius, Gastaldi and others (*Barbaria, Biledulgerid, Libya Interior*, etc.). In West Africa, *Niger fluvius* (the Niger River) combines with *R. Senega* (the Senegal River) to make a vast river system and flows directly westward into the Atlantic Ocean. The source continues to be a lake, *Niger lacus*, in central Africa. North of this river is *Tombotu* (Timbuktu), a major trading center linking the minerals and other trade goods of sub-Saharan West Africa with North Africa. The depiction of the Gold Coast in Guinea is based on a drawing from a trip to El Mina which was offered to Luis Teixeira in Lisbon. (Schilder 1996: 155, Cortesão 1960: 67-90). This Teixeira map found its way to Amsterdam where it was used for a 1602 map drawn by Pieter de Marees for Cornelis Claesz. and was the source for this as well as other maps of the region.

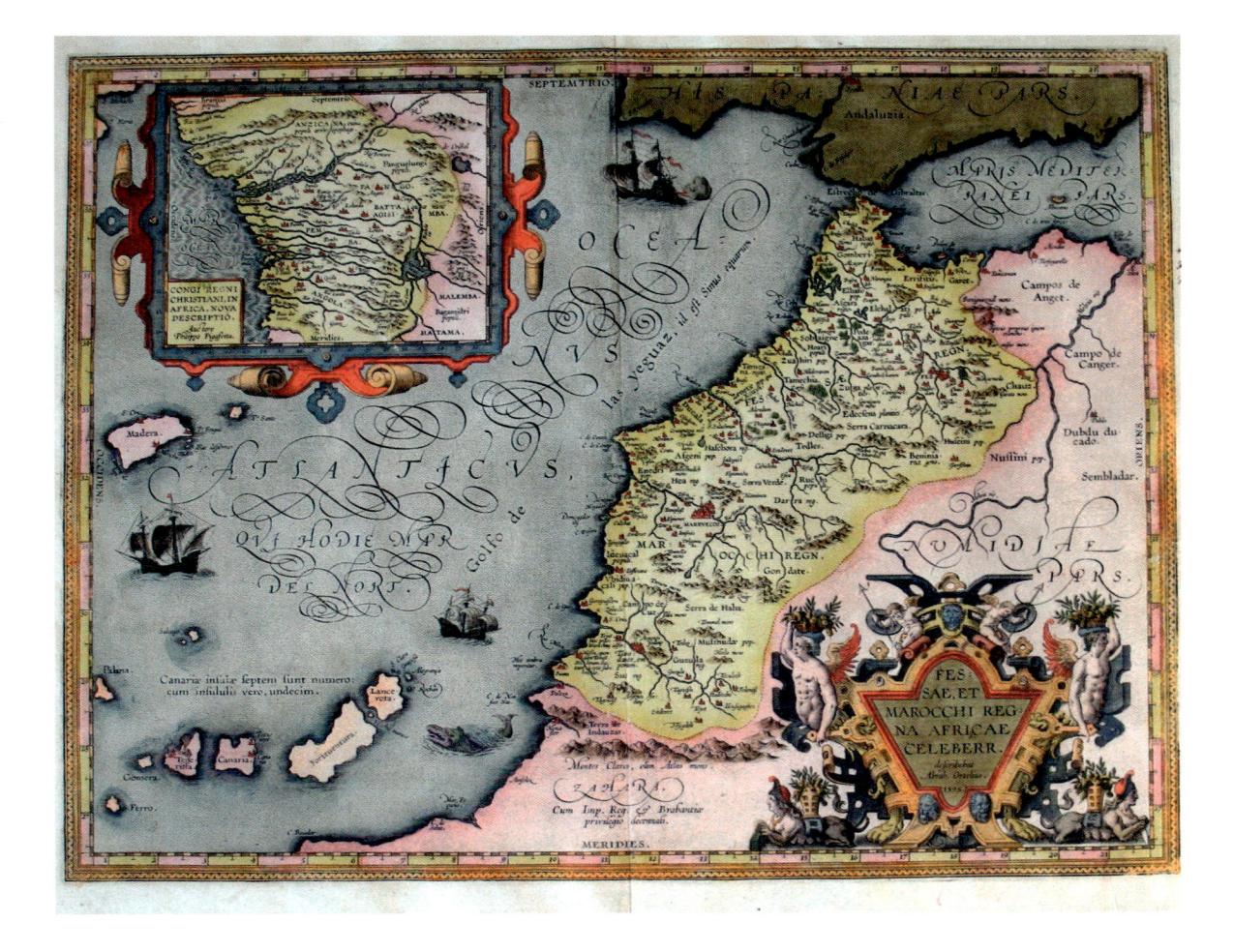

Abraham Ortelius' map of Northwest Africa with the inset map of the Congo (Author's collection).

In central Africa, the geography of the Congo Region is mainly from a map that was inserted into Ortelius' *Theatrum* from 1595 (Fessae et Marocchi... with the Congo inset), which was in turn derived from the travel accounts of Duarte Lopes, as published by Pigafetta in 1591.

Along the southern African coast, the more recent voyage to the east by De Houtman is reflected by references to Dutch names: *Vlejis baij, Vis baij,* and *Mossel baij*. The interior of southern Africa uses placenames from Portuguese explorations up the river systems from their bases at Mozambique Island and *Cefala* (Sofala), the most notable being *Cast Portogal* (Castle Portugal) and various African kingdoms known to Europeans, such as *Butua* and *Monomotapa*. The sources for the Nile River are faithful to Ptolemy with the western lake in central Africa of *Zaire lacus/Zembre lacus* and the eastern lake of *Zaflan lacus*. The *Lunae Montes* (Mountains of the Moon) serve as a divider between the Nile Lakes and the major southern river systems of *R de Spirito* and *Cuama Fluvi*. Both of these two latter rivers have a common source in a lake south of the Mountains of the Moon, *Sachaf lacus,* and an initially common river in the *Zambere flu*. This follows Mercator, who, in his 1569 world map, placed a large *Sachaf lacus* as the source for the southern rivers, based initially on the writings of the Portuguese, João de Barros, in 1552.

Abraham Ortelius' map of East and Central Africa (Author's collection).

In East Africa (Abyssinia), Blaeu based his work on Ortelius' map of the region (the Prester John map: Presbiteri Iohannis Sive Abissinorum...) and numerous texts of Portuguese exploration along the coast and into the interior. As with most commentators of that day, the Christian Kingdom of Prester John was greatly exaggerated in size.

Another Blaeu map that was important, again owing to its influence on numerous mapmakers almost to the end of the seventeenth century, was Willem Blaeu's folio-size map, initially published in 1617 (Map # 57). This map is essentially a reduction of Blaeu's famous and influential wall map of 1608. One notable difference on this 1617 map, when compared with Blaeu's 1608 wall map, is that this map breaks the long-standing tradition of the single southern river, the Zambere, which originates in Sachaf

Lacus, and then divides into the Cuama and Spirito Santo Rivers. On Blaeu's 1617 folio map the Cuama River is shown as originating in the highlands to the south of the Mountains of the Moon and is not connected to the Zambere River.

The Blaeu 1608 wall map and 1617 folio map of Africa were followed by numerous maps directly modeled after Blaeu. These include the imitations of Blaeu's wall map in Italy by De Scaicki, Scolari, De Rossi, and Todeschi, and the French copy by Jaillot. There were derivatives as well by Cornelis Danckerts in 1647, Jan Mathisz. in c.1655, Joan Blaeu, Justus Danckerts, and Gerard Valk as late as c.1690. Numerous folio-size maps thorough the seventeenth century were also modeled after either the Blaeu wall map of 1608 or the variation in the Blaeu folio-size map of 1617. A notable copy of the Blaeu folio map was a map by Jodocus Hondius, Jr. in 1619, which spawned a number of imitators, with and without decorative borders, both in the Netherlands and in France. In fact, Hondius' 1619 map in its later map states appeared in numerous editions of Hondius-Janssonius atlases to about 1680.

Visscher's finely detailed and engraved map of c.1658, which was modeled after the 1608 Blaeu wall map of Africa, generated numerous imitators ending with Justus Danckert's 1699-1700 map. The Blaeu wall map also served as a model for smaller maps of Africa that appeared beginning with Janssonius' *Atlas Minor* of 1628 and continuing with a series of close copies by Boisseau in c.1643, Berey in 1651, and two maps by Picart in c.1651, and c.1659.

Blaeu Model

Some Major Elements
- Additional placenames from the more recent discoveries (e.g. Dutch in West and South Africa)
- Elements from both Gastaldi-Ortelius and Mercator models are present
- Folio map modifies river systems in Southern Africa

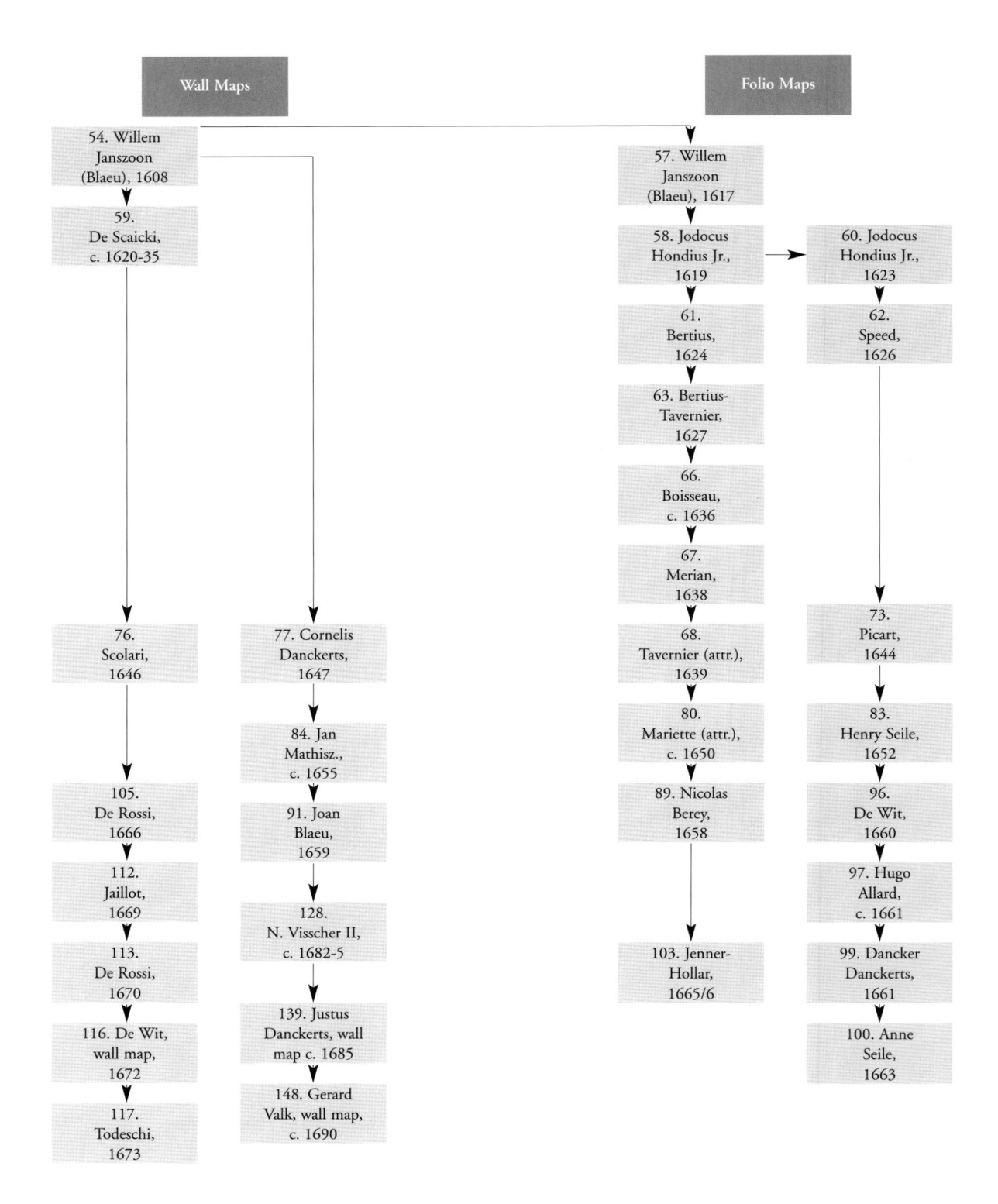

Wall Maps

54. Willem Janszoon (Blaeu), 1608
↓
59. De Scaicki, c. 1620-35
↓
76. Scolari, 1646
↓
105. De Rossi, 1666
↓
112. Jaillot, 1669
↓
113. De Rossi, 1670
↓
116. De Wit, wall map, 1672
↓
117. Todeschi, 1673

77. Cornelis Danckerts, 1647
↓
84. Jan Mathisz., c. 1655
↓
91. Joan Blaeu, 1659
↓
128. N. Visscher II, c. 1682-5
↓
139. Justus Danckerts, wall map c. 1685
↓
148. Gerard Valk, wall map, c. 1690

Folio Maps

57. Willem Janszoon (Blaeu), 1617
↓
58. Jodocus Hondius Jr., 1619 → 60. Jodocus Hondius Jr., 1623
↓   ↓
61. Bertius, 1624   62. Speed, 1626
↓
63. Bertius-Tavernier, 1627
↓
66. Boisseau, c. 1636
↓
67. Merian, 1638
↓
68. Tavernier (attr.), 1639   73. Picart, 1644
↓   ↓
80. Mariette (attr.), c. 1650   83. Henry Seile, 1652
↓   ↓
89. Nicolas Berey, 1658   96. De Wit, 1660
↓   ↓
103. Jenner-Hollar, 1665/6   97. Hugo Allard, c. 1661
                                    ↓
                                   99. Dancker Danckerts, 1661
                                    ↓
                                   100. Anne Seile, 1663

Blaeu Model
Continued

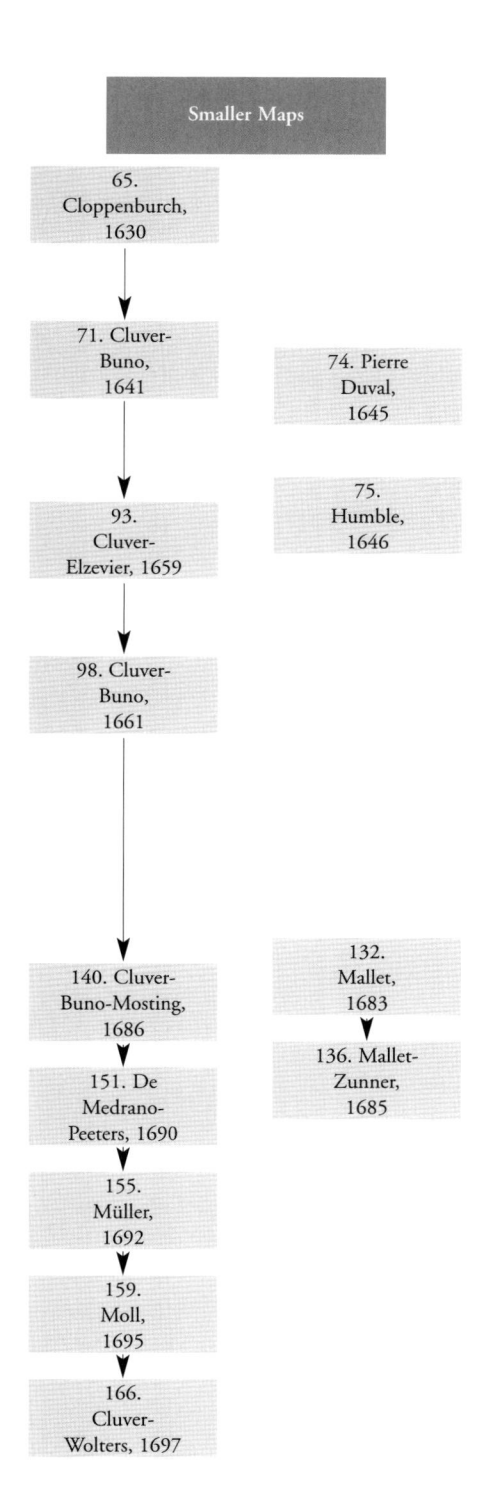

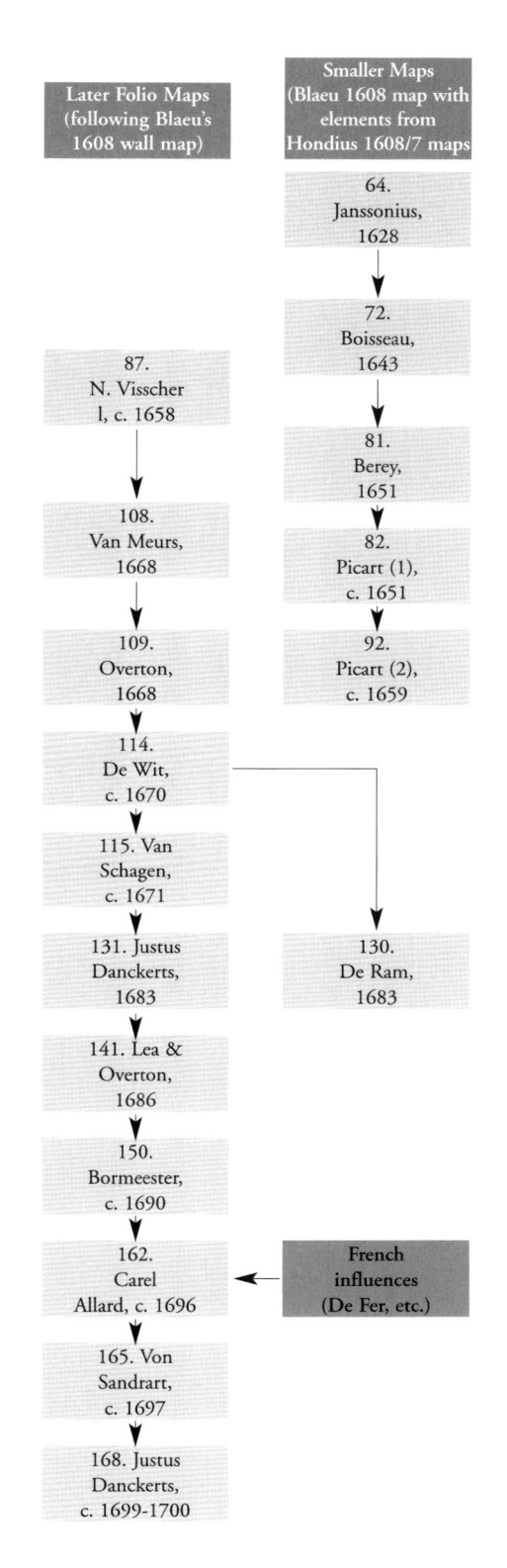

Smaller Maps
- 65. Cloppenburch, 1630
- 71. Cluver-Buno, 1641
- 93. Cluver-Elzevier, 1659
- 98. Cluver-Buno, 1661
- 140. Cluver-Buno-Mosting, 1686
- 151. De Medrano-Peeters, 1690
- 155. Müller, 1692
- 159. Moll, 1695
- 166. Cluver-Wolters, 1697

- 74. Pierre Duval, 1645
- 75. Humble, 1646
- 132. Mallet, 1683
- 136. Mallet-Zunner, 1685

Later Folio Maps (following Blaeu's 1608 wall map)
- 87. N. Visscher I, c. 1658
- 108. Van Meurs, 1668
- 109. Overton, 1668
- 114. De Wit, c. 1670
- 115. Van Schagen, c. 1671
- 131. Justus Danckerts, 1683
- 141. Lea & Overton, 1686
- 150. Bormeester, c. 1690
- 162. Carel Allard, c. 1696 ← **French influences (De Fer, etc.)**
- 165. Von Sandrart, c. 1697
- 168. Justus Danckerts, c. 1699-1700

Smaller Maps (Blaeu 1608 map with elements from Hondius 1608/7 maps)
- 64. Janssonius, 1628
- 72. Boisseau, 1643
- 81. Berey, 1651
- 82. Picart (1), c. 1651
- 92. Picart (2), c. 1659
- 130. De Ram, 1683

5. The Scientific View of Africa Starting with the Nicolas Sanson 1650 Map

Austere in comparison to the Dutch folio-size maps of this period, Sanson's 1650 map of Africa (Map # 78) is precise and scientific in its approach to presenting information. Sanson, known as the father of French cartography, is sparse in his information, tending to omit placenames, one may assume, because they were based only on hearsay.

More information, however, is provided on those areas that were coming under increasing French control in Madagascar and in West Africa. The *R da Volta* (Volta River) is more precisely placed on the map and *Accara* (Accra) appears near its mouth on the Guinea coast in present-day Ghana.

For his depiction of much of the interior, Sanson relies on the still dominant Blaeu wall map of 1608. The river systems generally follow Blaeu, though Sanson reverts to the 1608 depiction of the *Cuama* and *R. de Spirito Santo* Rivers as both flowing from the unnamed Zambere which originates in the unnamed Sachaf lacus. (In Blaeu's 1617 folio map the source of the Cuama changed to the southern side of the Mountains of the Moon).

For his major divisions within Africa, Sanson tends to use even older conventions. He has the land of *Ethiopie* stretching far into southern Africa, below the Nile source lakes. *Libye* dominates most of West Africa, and above that is *Afrique*, an early reference to that part of Africa along the North African coast. He does provide geographic subdivisions that are more in line with the conventions of his day.

The Sanson 1650 map was used as a model by his son, Guillaume, and, at least initially, by Duval, his nephew, in his 1664 folio map and 1660 atlas minor map. Outside of France, Sanson influenced the Italian mapmakers Nicolosi, De Rossi (1677), and Petrini. In England, Blome (1669) used his map. Sanson continued to influence later mapmakers to 1700 through his association with Jaillot and, to a lesser degree, Duval.

OVERVIEW OF PRINTED MAPS OF AFRICA FROM 1505 TO 1700

Sanson 1650 Model

Some Major Elements
- Additional placenames in areas under French influence
- Increased precision of the delineation of the Volta River
- Austere and scientific with omission of conjectural placenames

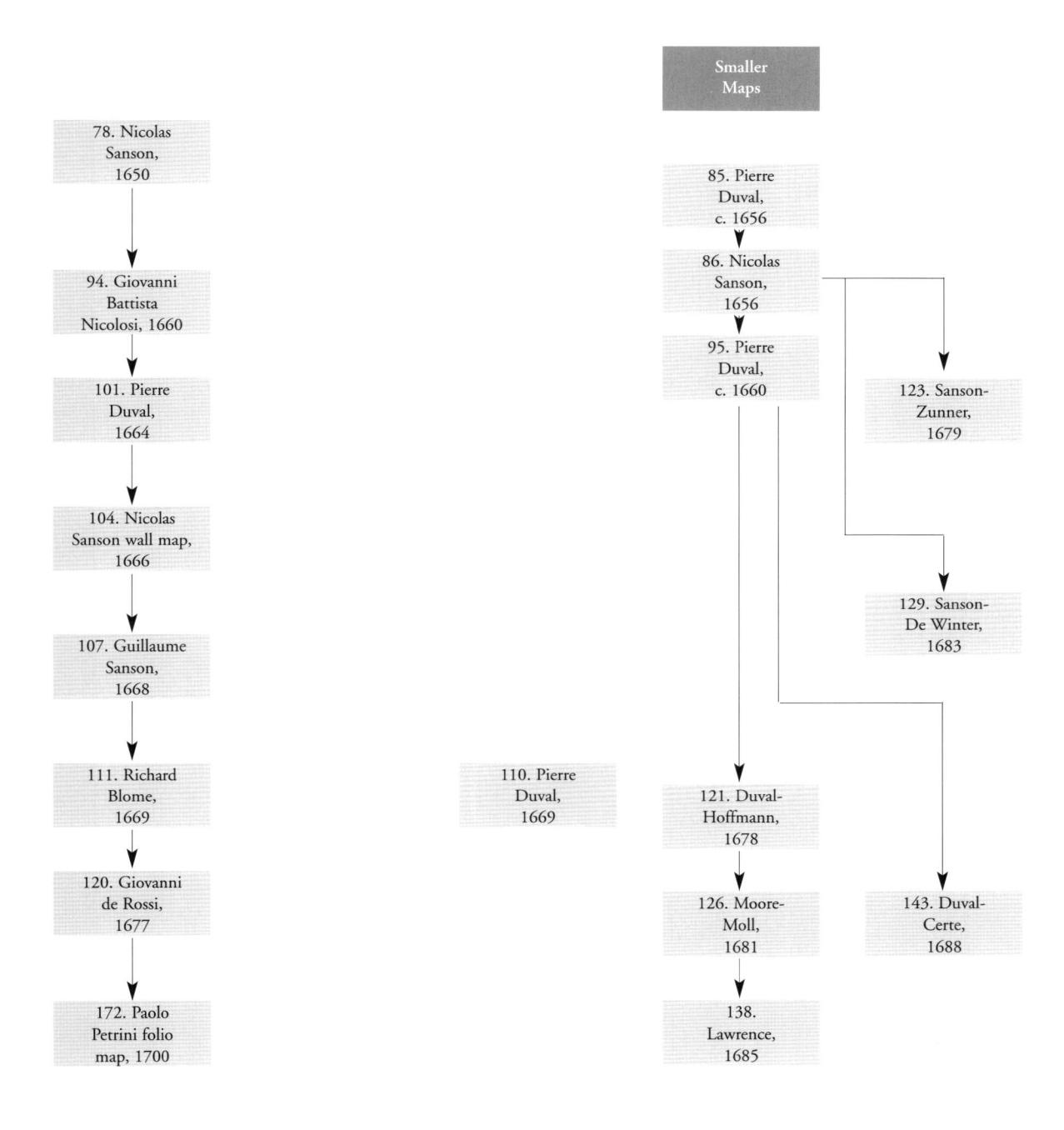

```
78. Nicolas Sanson, 1650
    ↓
94. Giovanni Battista Nicolosi, 1660
    ↓
101. Pierre Duval, 1664
    ↓
104. Nicolas Sanson wall map, 1666
    ↓
107. Guillaume Sanson, 1668
    ↓
111. Richard Blome, 1669
    ↓
120. Giovanni de Rossi, 1677
    ↓
172. Paolo Petrini folio map, 1700
```

Smaller Maps

```
85. Pierre Duval, c. 1656
    ↓
86. Nicolas Sanson, 1656 ─────────┐
    ↓                              ↓
95. Pierre Duval, c. 1660      123. Sanson-Zunner, 1679
                                   ↓
                               129. Sanson-De Winter, 1683

110. Pierre Duval, 1669

121. Duval-Hoffmann, 1678
    ↓
126. Moore-Moll, 1681          143. Duval-Certe, 1688
    ↓
138. Lawrence, 1685
```

| 69

6. The Evolution of the Mapping of Africa: The Influence of Jaillot and Duval

As the influence of the Dutch on the mapping of Africa began to wane in the later part of the seventeenth century, the French mapmakers came into increased prominence. To a large degree, this was due to the influence of Louis XIV through his active promotion of a scientific academy, the Académie Royale des Sciences, but also to a rise in the influence of a centralized and strong France in the world.

Some major elements of the Jaillot-Duval model were a new depiction of the river system of southern Africa (from Jaillot 1674); a removal of the connection between the Nile River and the two central African lakes (from Duval 1678); and a correction of the Blue Nile River in Abyssinia and its source (from Duval 1678).

The Alexis Hubert Jaillot 1674 map (Map # 118) has a number of departures from Sanson's earlier work. Of most significance, this map introduces a new depiction of the rivers south of the Ptolemaic lakes. Attempting to correct the misplaced modern-day Zambezi River, which appeared too far south on the earlier maps, Jaillot places a new major river, the *Zambeze*, above the *Zambere R.* and *Rio de Spiritu Santo*, which exits into the Indian Ocean by the trade port of Quelimane. This new alignment correctly places Sofala to the south of the Zambezi. The new Zambeze River originates in a new, unnamed lake to the east of Zachaf Lac, the traditional source for the Zambere/Cuama and Spirito Santo Rivers.

Following the 1650 and 1668 Sanson maps of Africa, the Jaillot map still shows long-held, incorrect interpretations. The *Empire des Abissins* (Abyssinia) extends far to the south of the Equator. The fictitious island of 'New' St. Helena in the South Atlantic is on the map. However, the coastline, especially south of the Equator, takes on a more modern appearance in this map when compared to the earlier Sanson maps. Yet, the Jaillot map shows little new information on the South African coast, which is somewhat unusual, since the Dutch by this time had established a permanent supply base at the Cape of Good Hope.

The Pierre Duval 1678 map (Map # 122) is an important transitional map of Africa. For the first time on a map of Africa, the connection between the Nile River and the two Ptolemaic central African lakes is purposely removed and they are not shown as a source for the Nile. The map also clearly shows a correct orientation of the Blue Nile in the Abyssinian highlands and its source as Lake Tana. While the Ptolemaic lakes do not connect to the Nile, the western lake of *Zaire* is the primary source for the *Zaire* (previously Congo) River. Many other aspects of this map follow the Jaillot map of 1674. The third southern river, named *Indires R. al Zambere* (Zambezi) by Duval, exiting into the Indian Ocean at Quelimane and to the north of the Cuama River is shown, though without its separate source lake. In West Africa, the Niger River still shows its much earlier representation with an east-to-west direction entering into the Atlantic through the Senegal and Gambia Rivers. Above the Niger River, there remains the even older representation of a river (*Nubie*) flowing eastward into the Nile.

These Jaillot-Duval maps were followed with various degrees of refinement by a number of mapmakers: in France with De Fer (1684) and the De Fer derivatives, Coronelli-Nolin (1689), and De la Croix (1690); in England with Berry, Morden, and Seller; and to a lesser degree in the Netherlands with the Mortier and Valk copies.

Jaillot - Duval Model

Some Major Elements
- New depiction of the rivers systems of Southern Africa (Jaillot 1674)
- Removal of the connection between the Nile River and the two central African lakes (Duval 1678)
- Blue Nile River originates in Lake Tana in Abyssinia (Duval 1678)
- Correction of the Nile River in Abyssinia and its source (Duval 1678)

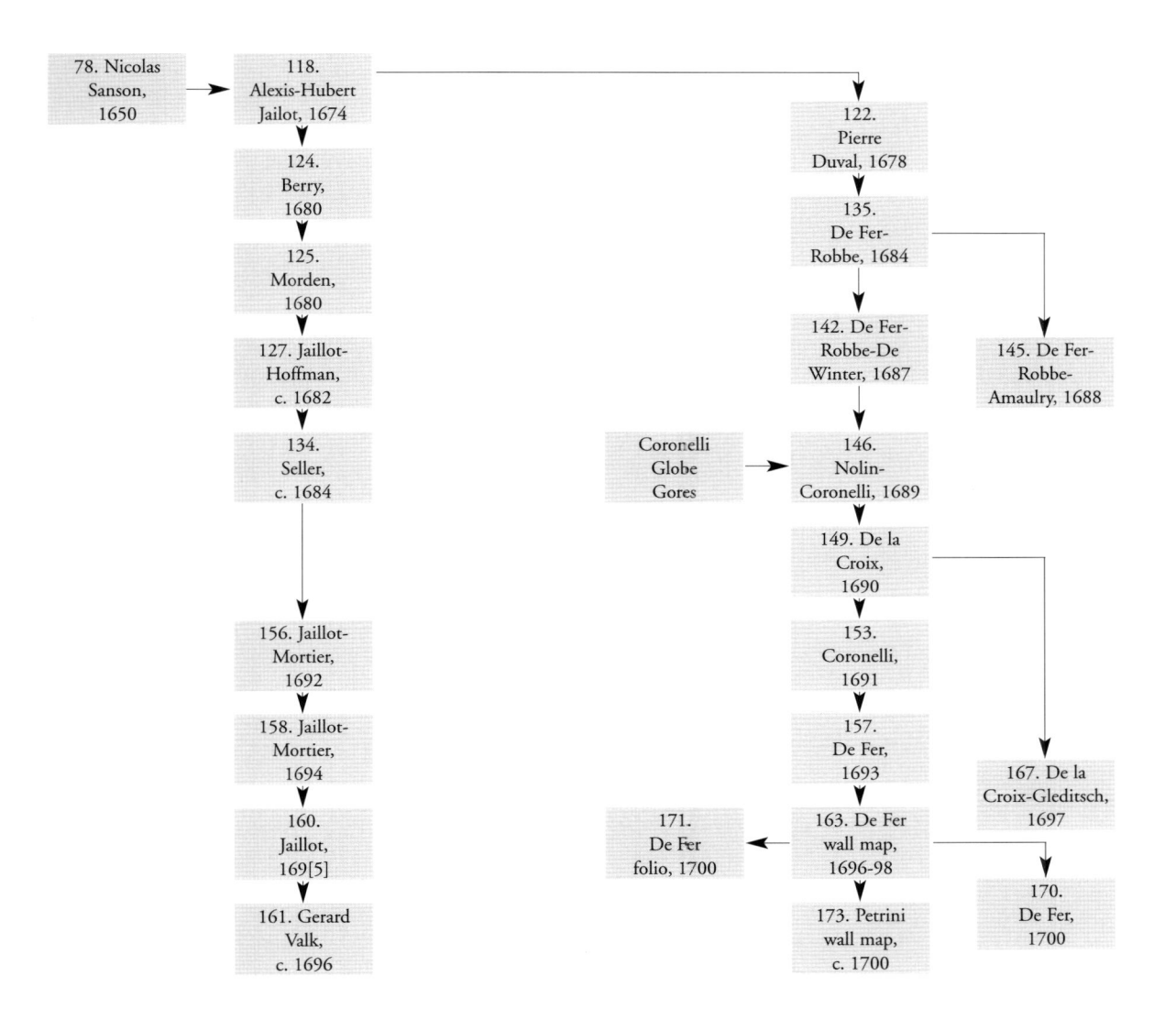

```
78. Nicolas        118.
Sanson,    →   Alexis-Hubert ─────────────────┐
1650           Jailot, 1674                   ↓
                    ↓                     122.
                124. Berry,               Pierre
                1680                      Duval, 1678
                    ↓                         ↓
                125. Morden,              135. De Fer-
                1680                      Robbe, 1684 ──────┐
                    ↓                         ↓              │
                127. Jaillot-             142. De Fer-        │
                Hoffman,                  Robbe-De       145. De Fer-
                c. 1682                   Winter, 1687   Robbe-
                    ↓                                    Amaulry, 1688
                134. Seller,    Coronelli    146.
                c. 1684         Globe    →  Nolin-
                                Gores       Coronelli, 1689
                    │                         ↓
                    │                     149. De la
                    │                     Croix,
                    │                     1690 ──────────┐
                    ↓                         ↓          │
                156. Jaillot-             153.           │
                Mortier,                  Coronelli,     │
                1692                      1691          │
                    ↓                         ↓          │
                158. Jaillot-             157.           │
                Mortier,                  De Fer,        │
                1694                      1693      167. De la
                    ↓                         ↓     Croix-Gleditsch,
                160.            171.      163. De Fer  1697
                Jaillot,        De Fer  ← wall map,     ↓
                169[5]          folio, 1700  1696-98   170.
                    ↓                         ↓        De Fer,
                161. Gerard               173. Petrini  1700
                Valk,                     wall map,
                c. 1696                   c. 1700
```

| 71

7. Delisle Model of 1700

This is a landmark map of Africa. The 1700 Delisle (Map # 174) is the first map to show Africa without the two Ptolemaic-based, Nile River source lakes. Delisle also gives the correct longitude for the Mediterranean Sea of 42°, thus correcting the width of the northern shape of Africa.

The interior of Africa has the main regions identified with large, dark capital letters for *BARBARIE, NIGRITIE, HAUTE GUINÉE, ABISSINIE, and PAYS DES CAFRES*. Lines divide all of Africa into numerous sub-regions labeled Royaumes, Estats, etc. Within the regions, considerable detail and additional placenames are shown, particularly around those areas settled or explored by Europeans in West Africa, Abyssinia, the Congo, Southeast Africa on the Zambezi River, and South Africa. At the Cape of Good Hope, he shows the Dutch settlements at *Fort des Hollandois* (the Dutch Fort at the Cape of Good Hope) and at *Hellenbok* (Stellenbosch).

In West Africa, Delisle orients the Niger River in an east-west axis, but its previous source lake, Lake Niger is omitted on this map. The Niger seems to flow westward into *Lake de Guarde* where it terminates; exiting *Lake de Guarde* in the west is the Senegal River which flows into the Atlantic. A faint line connects the Niger River with the Nile River, which was Delisle's way to indicate that, although some cartographers in the past depicted the Niger as connected with the Nile, he did not believe this to be true. The *Nil fl* (Nile River) is shown originating in Lake Tana in the Abyssinian highlands. The Nubia River, also shown in previous maps flowing eastward into the Nile, ceases to exist on this map.

Delisle does show a vaguely engraved and unidentified lake, at about 5° S, not far inland from Zanzibar in East Africa with no rivers entering or exiting it. This lake is probably based on reports by traders of the inland lakes that are, in fact, found in this part of Africa. It could represent any of the lakes in this region, or it could be an amalgam of Lakes Victoria, Malawi, and Tanganyika.

In southern Africa, Delisle removes the middle (the Cuama) of the three traditional, major southern rivers and replaces it with the smaller *Sabia* River, by Sofala. Also, the Infante River is removed. He does show the *Magnice ou R. de S. Esprit* (Spirito Santo) River (the modern Limpopo). On the *Zambeze* River, both of the Portuguese interior trading towns of Tete and now Sena are shown.

The 1700 Delisle map of Africa served as the model for European mapmakers and was much copied throughout the eighteenth century. The map is noteworthy for its attention to a scientific approach in the preparation of the map. Delisle exercised care, wherever possible, in the inclusion only of verifiable information. His map was constantly being updated, which means that there were a number of later states quickly following on his original 1700 publication of this map.

Delisle Model

Some Major Elements
- Removal of the Central African source lakes for the Nile River, based on Ptolemy
- East-to-west flowing Niger River remains but without Lacus Niger
- Refinements to river systems in Southern Africa
- Removal of Cuama River and correction of Zambezi River
- Additional details in those areas under European control
- Correction of the width of the northern part of Africa

Sanson-Jaillot-Duval maps → 174. Delisle, 1700

Cartobibliography

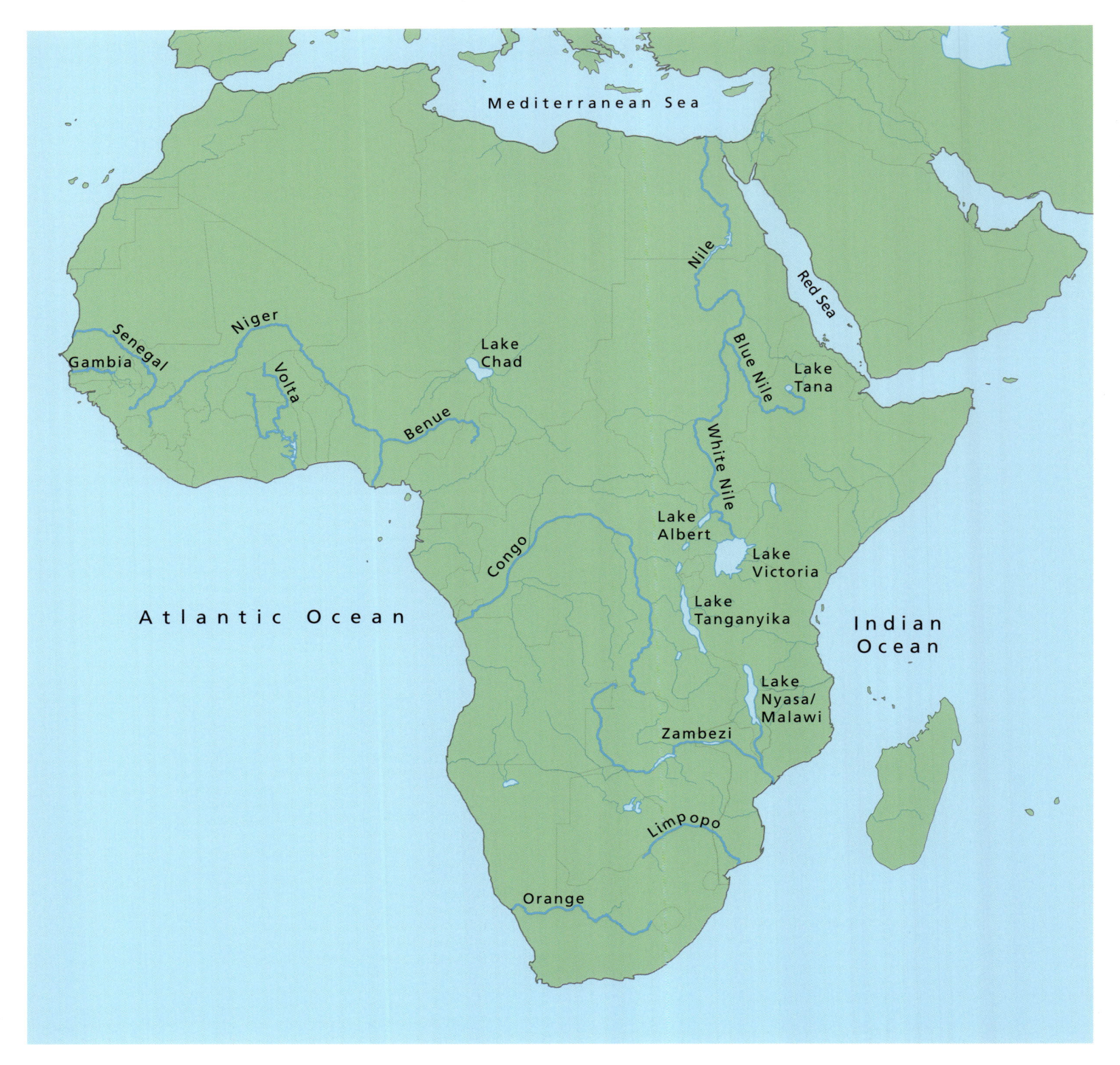

Map of Africa today showing major lakes and rivers.

Notes on the Use of the Cartobibliography

This cartobibliography is intended to describe all printed maps up to 1700 that depict the continent or most of the continent of Africa. As such, maps that depict Africa along with much of Europe and/or Asia, or maps that only depict regions of Africa are not described.

The maps in this book are arranged chronologically, starting with the earliest known printed map of the continent of Africa. The book ends with the first map to depict the continent of Africa generally not using ancient sources; that is, the Delisle map of 1700.

The explanatory parts for each map described in this book are as follows:

Map Number: A chronological number starting from 1.

Date: The date of the printing of the first issue of the map (and its book or atlas, if not separately issued). When a map can be precisely dated, that date is used. Where the map date is approximate, then c. (circa) is used.

Mapmaker: The name of the cartographer, mapmaker, or author to whom the map is generally attributed. A publisher is sometimes included to denote a specific issue of a map.

Map Title: The title is given verbatim, generally in full. If the title is too long, sufficient wording is used to easily identify the map. The use of ' ... ' indicates that a portion of the title is omitted. The exact wording, spelling, and capitalization are used. The use of [ ] shows where missing letters or words are inserted into the title. The use of ' | ' signifies when a new line of text begins in the map title.

Map Image: An image of each map is included. Sometimes, for important maps, other states are also shown.

Scale bar: The verbatim wording and measurements for the scale(s) are used.

Location of the publication of the map, with publisher and the first date: Where the map was first printed along with the name of the publisher and the date of the first appearance of the map. Brackets [ ] are used when this information does not appear on the map, but only within the book or atlas.

Title for the Book: An abbreviated title of the atlas or book in which the map appeared is used with sufficient wording to easily identify the book. If the map was first issued separately, that is, not as part of an atlas or book, that is noted.

Printing Process: Maps of this period were printed using either a woodblock or a copperplate.

Dimensions: The height is listed first and then width. The dimensions are provided in centimeters, as the use of millimeters would be superfluous due to distortions caused by normal environmental and printing factors. If the map is made from more than one sheet, this is noted as well.

Coordinates: Description of the latitude and longitude coordinates used on the map, if any.

Map Orientation: A compass direction for the top of the map is provided to orient the reader; this is useful if the map is unusually placed on the page. If compass roses and windroses are found on the map, these are noted.

Location of the Title: Where the map title is located on the map, if one is present.

Notable Elements or Characteristics on the Map: These may be ships, waves, animals, location of cartouches, coats-of-arms, etc. They are intended to help identify the map.

Description of the Map: This includes the history of the map and its significance, the map's purpose, uses of the map, the geography represented on the map, sources for the geography, etc. Toponyms that appear on the map are in italics.

Publication Information: This is information related to the publication of the map. It includes a description of the different states or variants of a map, if there are differences.

For the purposes of this cartobibliography, a new 'state' is generally acknowledged if there are purposeful, physical changes to the woodcut or copperplate that was used during the first print-run to produce the map. These visible differences in the woodblock or copperplate may be a change in the map title, map date, placenames, gridline, decorative elements, engraved page number or notation, etc.

A map 'variant' is a variation in a map, particularly, for example, a wall map that may or may not have a separate title strip above the map or text panels on the borders, or changes in those map elements. A variant may also result particularly with woodcut maps that use insertable type. These variations may be inadvertent type-setting errors, changes in lettering style and size, 'missing' placenames, etc.

If the map is in different states, then a numbering sub-set is used; for example, 1.1, 1.2, etc. with a date for each state. Whenever possible, the locations of the different states of a map are noted along with their shelfmarks or call numbers. This is particularly useful for a scarce map or an unusual state of a map. When a map is not scarce, several locations may be noted or the term, various, may be used. If the map is in different variants, then a numbering sub-set is generally used.

In those cases when the map is from an atlas or a book, it is important to note that editions of atlases and books from this period can contain maps prepared for a previous edition. Unlike today, where there is uniformity of pages within all books from a certain edition, publishers from this period often inserted unused pages and maps from previous editions into an edition being printed, rather than waste precious paper stock. Thus, it is not always possible to declare with absolute certainty the particular map or its state that appears in a specific edition. To address this problem and where possible, multiple examples of the same edition of an atlas or book were examined to ascertain the particular map that was used.

References: Sources that were consulted in the course of research for the map description and more detailed background information on the map and mapmaker are listed by author, date, and page number. Full bibliographical citations are provided in the Bibliography.

Glossary
A few terms used in this book are explained below.

Folio - The format for the book or atlas. The term is also used to describe the size of a book. For a folio, the paper is folded once before binding it into the book. A **quarto** format is folded twice before binding into the book; another fold is called an **octavo**.
Gridline - Outer-most numerical line surrounding the map, usually containing numbers for latitude and longitude.
Imprint - This refers to the printed information on the map for the publisher, printer, mapmaker, engraver, and location of publication.
Neatline - This is the outer printed line surrounding a map.
Platemark - The impression made by the copperplate into the paper during the copperplate printing process just outside of the neatline (the outer printed line).
Recto - The side of the printed sheet of paper showing the map impression. It can also refer to the right side of an open book.
Signature - A printer's mark on the page, usually a letter or number in order that enables the binder to place the pages, especially the pages containing the maps, in their proper order in the book.
State - Purposeful, physical changes to the woodcut or copperplate that was used during the first print-run to produce the map.
Variant - A variation in a map, particularly, for example, a wall map that may or may not have a separate title strip above the map or text panels on the borders, or a woodcut map which uses insertable type. These variations in a woodcut map may be inadvertent type-setting errors, changes in lettering

style and size, 'missing' placenames, etc.
Verso - The back of the printed sheet showing the map, often containing text. The term may refer to the left side of an open book.

Abbreviations
The following abbreviations are used when describing the cartographic details for the map entries in this book.

attr. = attribution, when the author of a map is not absolutely known.
c. = circa, indicating an approximate date for a map.
cm = centimeters, the measurements for a map's dimensions.
no., nos. or # = number or numbers.
Vol. = Volume.

Latin Terms used on Maps
Author of a Map = If listed on the map, the author may be identified with the wording: *auctore, delineavit, descripsit, inventit*, or contractions of these words.

Engraver of a Map = If listed on a map, the engraver may be identified with the wording: *sculpsit, fecit, caelavit, incidet*, or a contraction of one of these words.

Publisher of a Map = If listed on a map, the publisher may be identified with the wording: *excudit, formis, sumtibus, apud, ex officina*, or a contraction of one of these words.

Uses of Latin Placenames in Books and Maps
As Latin was the language used by the educated class in Europe from the fifteenth century into the seventeenth century, some commonly-used Latin placenames that appear on maps from this period are listed below, along with their current English equivalents.

Agrippina Colonia = Köln or Cologne.
Argentorum or Argentina = Strasbourg.
Batavia = Passau in Germany, or, more recently by the Dutch, Jakarta in Indonesia.
Brunsvicum = Brunswick; now Braunschweig.
Guelpherbytum = Wolfenbüttel.
Lugdunum = Lyon.
Lugdunum Batavorum = Leiden.
Lutetia = Paris.
Mediolanum = Milano or Milan
Melita = Malta.
Monasterium = Münster.
Trajectum ad Rhenum = Utrecht.
Trajectum ad Mosam = Maastricht.
Ulyssipons = Lisbon.

CARTOBIBLIOGRAPHY

1   1508 Antonio Francanzano (Fracan) da Montalbodo, Milan

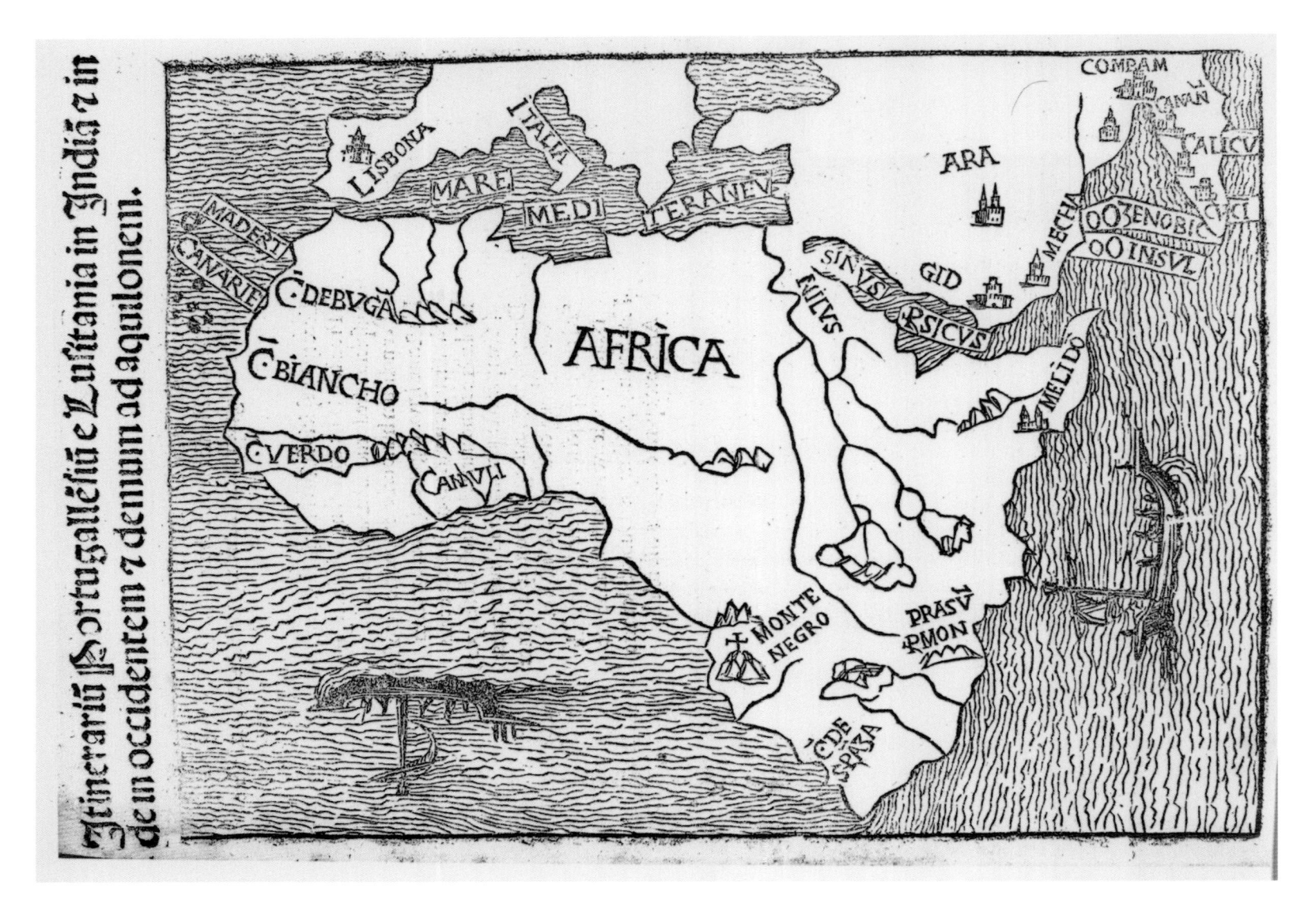

1.1 First State 1508

[Map of Africa].
- No scale bar.
- [Milan : Antonio Francanzano (Fracan) da Montalbodo, 1508].
- From: *Itinerarium Portugallensium e Lusitania in Indiam & indem occidentem & demum ad aquilonem.*
- 1 map : woodcut ; 22.5 x 16.5 cm.
- No latitude and longitude coordinates.
- West at the top; no compass rose.
- At the top of the page as title to the whole book: *Itinerarium Portugallensium e Lusitania in Indiam & indem occidentem & demum ad aquilonem (*Travel route of the Portuguese from Lusitania [Portugal] to India, as well the western as the eastern).
- On the top half the wave lines are running vertically and there is one ship (vertical) in the sea; on the bottom half of the map, the wave lines are running horizontally and there is one ship (horizontal) in the sea.
- The Red Sea is labeled *SINVS P\SICVS.*

Example: *Washington LC*, E101 .F88 (map on title page of book; the verso has 18 lines of text beginning *Presbyteri Francisci Tantii Cornijeri...*).

1.2 Second State 1508

The Red Sea is now labeled *SINUS ARABICVS*. The verso has 18 lines of text beginning *Presbyteri Francisci Cornijeri…*

Examples: *London BL*, G 20.e.17 and G.6988 within books; *Providence JCB*.

Detail of Map 1.1 showing SINVS P\SICVS.

Detail of Map 1.2 showing SINVS ARABICVS.

Description

This map of Africa appeared on the title page with the book's title above the map. It is the earliest known printed map of the continent of Africa in which the continent is represented alone as surrounded by the ocean. The source for this map is not known, but it likely was an Italian in the employ of the Portuguese king or an Italian employed by one of the many Italian enterprises in Portugal at that time.

In 1508 Antonio da Montalbodo published a Latin translation of *Paesi Nouamente retrouati Novo Muodo…* by Archangelo Madrignano, a Milanese Cistercian monk. The *Paesi* is one of the earliest printed collections of voyages and is sometimes called the collection of Càdamosto. The *Paesi* was first published in Italian in Vicenza in 1507, without the map of Africa. There were later editions in Latin, German, and French (ending in 1528). The map of Africa only appeared in the Latin edition of 1508 published in Milan. This 1508 edition of the book is of 'excessive rarity' (Sabin 1868: 299/50058).

The *Paesi* begins with the first published description of the Venetian explorer Cadamosto's (Alvise da Cà da Mosto) two voyages in the service of the Portuguese to Cape Verde and Senegal. Further descriptions include: Cadamosto's and Da Sintra's voyages along the west coast of Africa; the voyage of Vasco da Gama around Africa to India; and several explorers' voyages to Brazil, including Pedro Cabral, Columbus, and Vespucci.

The map of the continent of Africa is oriented with West Africa at the top of the page to best fit the shape of the book. The Montalbodo map corrects the severe eastward-sloping orientation of Africa in the anonymous 1505 printed map, described in the introduction, but still represents Africa as too wide longitudinally, with the Cape of Good Hope at least 15 degrees to the east of its actual position. Many placenames on the map are based upon Portuguese discoveries: *C. BIANCHO* (Cape Blanco) and *C. VERDO* (Cape Verde) on the west coast; *MONTE NEGRO* with a padrão cross (an engraved stone marking the limits of exploration) along the southwest coast; and *C DE SPAZA* (Cape Boa Esperança or the Cape of Good Hope) in the south. The African-Arab port of *MELIDO* (Malindi) on the east coast is also shown.

Although the rivers are not identified, they appear to be the Niger in the west with its confusing hydrography, the Congo (Zaire) River in the southwest region, the Blue Nile in the Abyssinian Highlands, and the White Nile originating in the unnamed Mountains of the Moon and the two Ptolemaic lakes. Interestingly, a third, distinct river which also flows into the Nile River is placed to the west of the two Ptolemaic lakes originating far into Southern Africa. The Zambezi and Limpopo in the southwest originate not in the Mountains of the Moon, but in separate mountains in Southern

Africa. Further to the east the Malabar peninsula in India is shown along with the important trade port of Calicut.

An example of the 1508 edition of *Itinerarium Portugallensium e Lusitania in Indiam & indem occidentem & demum ad aquilonem* at the Österreichische Nationalbibliothek (*Wien ÖNB*, 394092-C.K) contains a unique set of six maps showing Africa and parts of the coast of Africa from Lisbon to West Africa. This is the only known example of the book containing this set of maps. These maps are intended to illustrate the voyages of Cadamosto and Da Sintra to Senegal. In style, the regional maps are very similar to the maps prepared by Martin Waldseemüller for the 1513 Strasbourg edition of Ptolemy's *Geographia*. In size, these woodcut maps are 15.5 x 22.5 cm.

Publication Information
This map is known in two states, both from 1508, in Latin with different wording for the Red Sea. It is not clear as to why there are two different states of the map as they both appeared only in the 1508 Latin edition. Some have speculated that the woodblock was possibly badly damaged. When a new woodblock had to be cut, the Latin name for the Red Sea may have been inadvertently changed. The author does not believe the woodblock was re-cut as no other changes were noted to the block. It is likely that the map's author decided to change the name from the Persian (*P\SICVS*) to the Arabian (*ARABICVS*) Sea. Upon closer examination of this section of the map, it appears that the wording of '*P\SI*' was cut from the block, leaving the original '*CVS*', causing an enlargement of this space on the map. After this was done, a new piece of wood with the wording '*ARABI*' was affixed to this space.

References
Tooley 1966: No. 29: 14; Church 1951: 65-67 (with the image of the second state on p. 66); Maggs Brothers Catalogue 1929: 24 (with image of the first state); Sabin 1868: entry # 50058.

2   1513 Pliny the Elder – Melchior Sessa, Venice

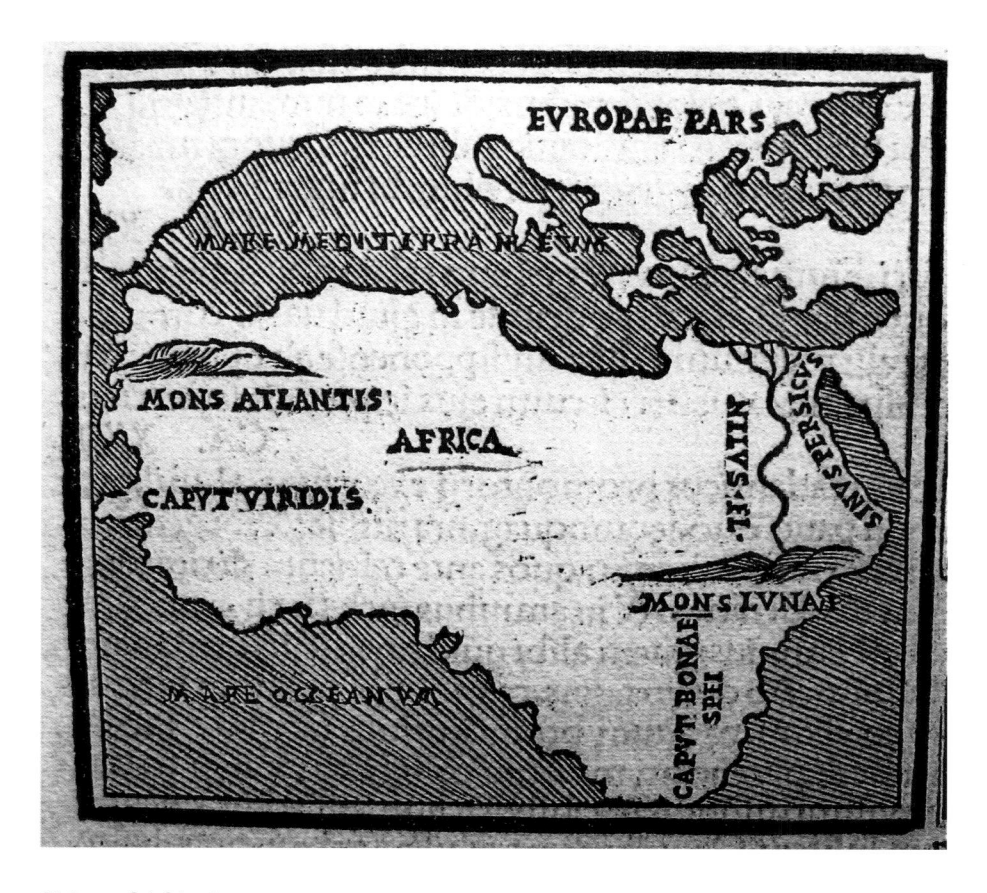

[Map of Africa].
- No scale bar.
- [Venice : Melchior Sessa, 1513].
- From: *C. Plinij Secundi Veronensis Historiae naturalis libri xxxvij, aptissimis figuris exculti ; ab Alexandro Benedicto Ve. physico emendationes redditi.*
- 1 map : woodcut ; 7.3 x 8.0 cm.
- No latitude and longitude coordinates.
- North at the top; no compass rose.

Examples: *Bologna BU*, A.IV.G.II.29; *New York PL,* *KB+ 1513 Pliny. C. Plinij Secundo c.3; *Pisa UB*, S.r.3.14.

Description

This is a rare map prepared by Melchior Sessa for an edition of the Roman writer Caius Plinius Secundus' (or as he is better known, Pliny the Elder), multi-volume encyclopedia, *Historiae naturalis* (Natural History), rediscovered and widely read during the Renaissance. The book covers the characteristics of the physical universe, geography of the known world, anthropology, zoology, botany, mineralogy, medicinal properties of various substances, and a history of the fine arts including the uses of artists' pigments. Pliny's chapter on mankind describes various mythical creatures inhabiting the non-European world such as the people called the Arimaspi with one eye in the middle of their forehead. Much of this information was taken from Herodotus. Depictions of these various creatures were to appear on maps by numerous mapmakers into the seventeenth century. Of note, Pliny (23-79) died during the volcanic eruption of Mt. Vesuvius near Pompeii.

The map is devoid of cartographic detail. The wording *AFRICA* is in the northern center of the map and is meant to represent all of the continent, not just that portion known to early classical writers. The *MONS ATLANTIS* (Atlas Mountains) are in the northwest. Further down the coast is *CAPVT*

VIRIDIS (Cape Verde) though why it has greater prominence over Cape Bojador or other capes of the West African coast is not known. At the southern tip, *CAPUT BONAE SPEI* (Cape of Good Hope) is shown. The *NILVS FL.* (Nile River) and its source as the *MONS LVNAE* (Mountains of the Moon) are placed in the area of Abyssinia, though without the Ptolemaic source lakes. This is more in line with early classical writers who saw the Nile source as being at the edge of the known world in Northeastern Africa.

Geographically, the map generally follows the Montalbodo map of 1508. Africa continues to be depicted slanting in its southern portion excessively to the east.

Publication Information
This map is on a page of text in a book of 1513 titled *C. Plinij Secundi Veronensis Historiae...* . The map appears on the verso of folio XXII. Melchior Sessa was the publisher for the book, and both the book and the map were likely printed by Agostino de Zanni.

The book and the map were re-published in 1516 in Venice by Sessa and Pietro di Ravani. There were numerous editions of this book that do not contain this map. The New York Public Library has a Paris edition of 1516 (*New York PL*, *KB +1516 Pliny); a Venice edition, by the heirs of Aldi, dated 1535-6 (*New York PL*, *KB 1535); and a Basel edition of 1554 (*New York PL*, *KB +1554), all without this map.

References
The author is indebted to Douglas Sims for bringing this map to his attention. Nicolardi 1984.

3 1540 Sebastian Münster, Basel

Map 3 (Variant 1)
from Geographia *1540.*

AFRICA XVIII. NOVA TABVLA.
- No scale bar.
- [Basel : Heinrich Petri, 1540].
- From: *Geographia Universalis vetus et nova….Claudii Ptolemae... .*
- 1 map : woodcut with insertable type ; 25.5 x 34.5 cm.
- North at the top; no compass rose.
- Title across the top above the map (there is a black leaf motif before the title).
- At the bottom left of the map there is a text box within a scrollwork cartouche.
- A large ship is in the Atlantic Ocean between the cartouche and the southwestern coast of Africa.

Description

This is the earliest, readily available, printed map to show the entire continent of Africa. It is famous for the medieval depiction of the Cyclopses, called the *Monoculi* on the map, and the parrots and elephant in Southern Africa. Crowns and scepters identify various African kingdoms throughout

CARTOBIBLIOGRAPHY

Map 3 (Variant 2) from Geographia *1542.*

the continent of Africa, including *Hamarich*, the capital of the famous fictional Christian king, Prester John.

While primitive in its depiction of Africa, the map's importance lies in the fact that it was included in what were probably the two most widely circulated and read books of the mid-sixteenth century – Sebastian Münster's *Geographia* and *Cosmographia*. This map first appeared in Münster's edition of Claudius Ptolemy's *Geographia* in 1540 and again in 1542, 1545, and 1552. Like his contemporaries, Münster saw the mastery and embellishment of the ancient geography as reflected in Ptolemy, Solinus, Pomponius Mela, and others as a necessary precursor to a better understanding of modern geography. The map also appeared in the numerous editions of Münster's *Cosmographia* from 1544 to 1578.

While Münster does not specifically indicate his sources or prototypes for this map, Skelton (1966: xviii) states that the basic form of Africa for this map is taken from the delineation of Africa on Münster's world map of 1532. On this map, West Africa has a vertical orientation and the southeast coast of Africa slopes uniformly to the northeast. Like his predecessors, Münster places the southern tip of Africa more than 10 degrees too far to the east, thus creating a long, sloping southwestern coast of Africa. This was not to be corrected until Gastaldi's eight-sheet wall map of 1564.

Münster's sources for his map of Africa varied from classical writers to his contemporaries, reflecting his great intellectual curiosity and attempt at 'scientific cartography'. Much information on West Africa was available from the voyages of Cadamosto (Oehme, 1968: ix). A source for the interior of North Africa may have been the *Descrittione dell 'Africa* written in manuscript by al-Hassan (Leo Africanus) in 1526, though this did not appear in published form until Giovanni Ramusio's *Navigationi et Viaggi* of 1550. From Africanus, Münster could obtain useful information on those areas of Africa under Arab influence (see the entry for the Africanus-Temporal map [Map # 5] for further information on Leo Africanus).

The majority of Münster's geographical information for this map comes from the Portuguese, whose travels, by this time, had been widely documented and disseminated throughout Europe. Even though Münster relies heavily on Portuguese sources, he does not depict the cartographic details known at that time from extensive Portuguese exploration along the Zambezi River in Southeast Africa (the Zambezi is not even shown). In West Africa, Münster ignores the Portuguese commercial operations on the Gold Coast.

Münster remains true to the Ptolemaic view of the interior of Africa. His source for the White Nile River rises in the unnamed twin lakes, which, in turn, are fed by the Mountains of the Moon which are not labeled on this map. He does show a separate lake in the Abyssinian Highlands near *REGNVM Melindae* as the source for the Blue Nile, though this placement is too far south to be an accurate representation of the real source, Lake Tana.

Münster does not include a westward extension of the Nile River across West Africa, as some later cartographers did by adapting the much earlier tradition of the Greek, Herodotus. Münster shows an unnamed Niger River (on his map, the *Senegal fl.*) rising in mountains in West Africa and then flowing westward into the Atlantic Ocean. Above this river, he shows another river, *Dara fl.*, joined in the east by *Lacus Libyæ* and in the west by *Nigir palus*, but without an exit to the ocean. His *Dara fl* is a modification of the Ptolemaic presentation of twin west African lakes joined by a river, but without Ptolemy's source for the river in mountains further to the southwest (for a fuller description of the West African rivers, see Rizzo 2006: 80-89).

Münster also does not show the island of Madagascar off the southeast coast of Africa, which is unusual considering that this island had been reported in Europe since the time of Marco Polo. He does show an island named *Zaphala Aurifodina* in the Indian Ocean from which King Solomon, according to legend, obtained silver and gold. Here, Münster could be confusing *Zaphala* with Sofala or Cefala, the port on the east coast most closely associated with trade with the gold mining region of Monomotapa in the interior.

There is a text box set within a strapwork cartouche at the lower left of the map. The text within the box describes the discovery of Africa by the Europeans and the movement down the west African coast to the Cape of Good Hope, around the Cape, up the East African coast, and across to Calicut in India in search of spices.

Publication Information
Sebastian Münster (1489-1552) was one of the three great cartographers who dominated the sixteenth century, along with Mercator and Ortelius, 'and of these three, Münster probably had the widest influence in spreading geographical knowledge throughout Europe in the middle years of the century' (Moreland & Bannister 1989: 78). '[Münster's] *Cosmographia...* contained not only the latest maps and views of many well-known cities, but included an encyclopedic amount of detail about the known - and unknown - world and undoubtedly must have been one of the most widely read books of its time.'.

Münster published four Latin editions of the *Geographia* in 1540, 1542, 1545 and 1552. There were 29 editions of the *Cosmographia* published from 1544 to 1578: 5 in Latin, 15 in German, 6 in French, and 3 in Italian.

The same woodcut was used in all editions of the *Geographia* and *Cosmographia*. It is not known who cut this map, though the woodcut borders surrounding the type on the verso of some of the maps are largely attributed to Hans Holbein, since some of these borders appeared in earlier book illustrations that he is known to have created.

Progressive cracks occurred over the course of the woodblock's long printing history. Around 1544 with the first *Cosmographia*, a faint crack appeared in the woodblock. This vertical crack extends from just above the island of Sardinia to the bottom of the map. In the 1545 *Geographia*, the crack begins to become more noticeable. A diagonal crack is also apparent below the text box in the bottom left of the map, entering the text box to the left of the bottom strapwork. In 1552 a second crack begins to appear to the left of the longer one which now extends through Sardinia. Over the life of the *Cosmographia*, these cracks became progressively more noticeable.

Münster was one of the first to regularly use insertable metal stereotype for placenames by creating spaces in his woodblocks into which metal type could be affixed. Over the course of the many editions of the *Geographia* and *Cosmographia*, words were constantly being inserted and removed, and, in some cases during the actual printing process, words fell out of their slots and were even replaced upside down. The two images below are from Variant 8.A of the 1552 Latin edition of the *Cosmographia* showing an upside down *Regnū* and *QVIOLA*.

Details from Map 3, variant 8.A of the 1552 Latin edition of the Cosmographia showing an upside down Regnū *and* QVIOLA.

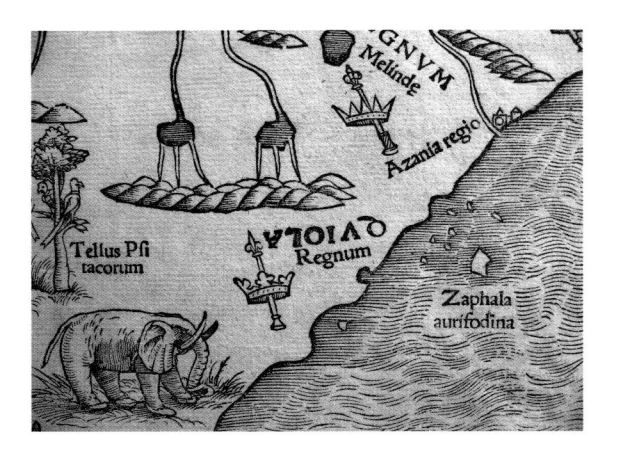

The text box in the lower left corner of the map was obviously set with moveable type. The author noticed slight differences in the text box in each of the Latin, German, and French editions of the *Geographia* and *Cosmographia*, particularly in the use of dashes to show breaks in words, slight spelling changes, and even some changes in word usage. It is possible this was due to changes in the conventions of language, spelling, and printing over time.

Examples seen from the three Italian editions of 1558, 1571, and 1575 appear to be identical. No differences were noted in the text boxes, the maps, or the versos. Although unlikely, as these Italian editions were published in the geographically separate cities of Basel, Venice, and Cologne, it is possible that all of the Italian maps were printed at the same time and later inserted into subsequent editions.

The author noted 15 variants of the map of Africa (please see the chart below). Although other changes were noticed on the map of Africa, these additional changes were disregarded as superfluous because they did not alter the total number of variants. The use of information on variants along with information on changes in the text on the verso of each map (as presented later), will determine the specific date for any loose map.

To identify loose maps as one of fifteen variants, based on the changes in placenames and typeface that occurred over time, follow these directions carefully:

1. Look carefully at the Münster 1540 map, Variant 1, on page 83.
2. Look at the chart on page 88. The first row going across is from Variant 1 and describes names on the map that experience changes over time. As you scan down the columns (A to J), changes are described and are given a new number (e.g. A1 changes to A2).
3. The letters and numbers (e.g. A1) also correspond to the examples of changes in each placename based on the images from actual maps (see "Changes on Münster's Map of Africa" on page 89). Please use this page as a check to be sure the description (e.g. A1) on the chart matches what is seen on the undated loose map.
4. Using the 1540 map, find all of the placenames listed on the row for Variant 1. Check the undated loose map and note any differences in these placenames. Scan down the columns of the chart until the description of the change is found. Check that the corresponding number (i.e. A1) on the chart matches the image on page 89.
5. Finally, a variant can be identified when all of the placename changes from columns A to J match on the same row of the chart.
6. Next, to ascertain an exact date or possible dates for the undated map, this map can then be compared with the language and wording of the title and with the text on the verso for each edition of Münster's map.

CARTOBIBLIOGRAPHY

Chart Describing Changes in the Variants of Münster's Map of Africa

	A	B	C	D	E	F	G	H	I	J
Variant 1	A1 *AETHIOPIA* in large block capital letters across center of map	B1 *HISPANIA* in small block capital letters	C1 *AFRICA* in small block capital letters in North Africa is present	D1 *Hippa* in North Africa in small type	E1 *Satolea* in small cursive letters near Meroe by Nile River is present	F1 *Garamantes* above *LIBYA Interior* is present	G1 *Gambra fl.* is present above *Regnum MELLI* in West Africa.	H1 *Regnum MELLI* is present in West Africa	I1 *Cayrū* in Egypt reads downward	J1 *QVIOLA Regnum* is present in Southeast Africa
Variant 2	A1	B1 "	C1 "	D1 "	E1 "	F1 "	G1 "	H1 "	I2 *Cayrū* in Egypt reads upward	J1 "
Variant 3	A2 *Morland* in Gothic type replaces *AETHIOPIA* across center of map	B2 *Hispania* changed to Gothic type	C1 "	D1 "	E1 "	F1 "	G1 "	H1 "	I1 *Cayrū* in Egypt reads downward	J1 "
Variant 4	A2 ?	B2 "	C2 *AFR* missing from *AFRICA*	D1 "	E1 "	F1 "	G1 "	H1 "	I1 "	J1 "
Variant 5	A3 *ETHIOPIA* in large block capital letters replaces *Morland* across center of map	B2 "	C2 "	D1 "	E1 "	F1 "	G1 "	H1 "	I1 "	J1 "
Variant 6	A2 *Morland* in Gothic type	B2 "	C1 *AFRICA* is present	D1 "	E1 "	F1 "	G1 "	H1 "	I1 "	J2 *Regnum* missing below *QVIOLA*
Variant 7	A2 "	B2 "	C1 "	D2 *Hippa* in larger type	E1 "	F1 "	G1 "	H1 "	I1 "	J1 *QVIOLA Regnum* is present in Southeast Africa
Variant 8	A2 "	B2 "	C1 "	D2 "	E2 *Satolea* is missing	F1 "	G1 "	H1 "	I1 "	J1 "
Variant 9	A4 *AETHIOPIA* in smaller block capital letters	B2 "	C1 "	D2 "	E3 *Satolea* is present in new small block letters	F1 "	G1 "	H1 "	I1 "	J1 "
Variant 10	A4 "	B2 "	C3 Partial *C* after *AFRICA*	D2 "	E3 "	F2 *mantes* is missing from *Garamantes*	G1 "	H1 "	I1 "	J1 "
Variant 11	A4 "	B2 "	C3 "	D2 "	E3 "	F2 "	G2 *Gambra fl.* is missing	H1 "	I1 "	J1 "
Variant 12	A5 *IOPIA* is missing from *AETHIOPIA*	B2 "	C3 "	D2 "	E3 "	F2 "	G2 "	H1 "	I1 "	J1 "
Variant 13	A4 *IOPIA* is present in *AETHIOPIA*	B3 *His* missing in *Hispania*	C3 "	D2 "	E3 "	F2 "	G2 "	H1 "	I1 "	J1 "
Variant 14	A4 "	B4 *HISPANIA* in larger block letters	C3 "	D3 *Hippa* is missing	E3 "	F1 *mantes* is present in *Garamantes*	G1 *Gambra fl.* is present	H2 *'m'* in *Regnum* missing in *Regnum MELLI*	I1 "	J1 "
Variant 15	A4 "	B4 "	C3 "	D4 Both *Melma*, just above *Hippa*, and *Hippa* are missing	E2 *Satolea* is missing	F3 under *Garamantes*, *LIBYA Interior* is missing	G1 "	H3 *LI* is missing in *MELLI*, along with *m* in *Regnum*	I1 "	J1 "

MAP 3

Changes on Münster's Map of Africa

A1

A2

A3

A4

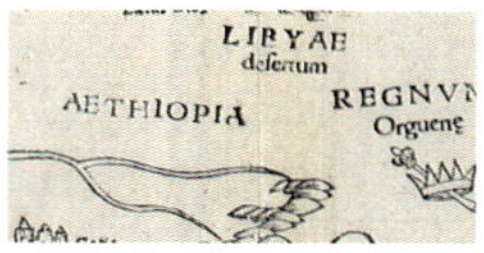

A5

B1

B2

B3

B4

C1

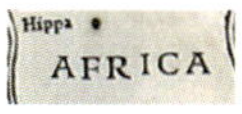

C2

C3

D1

D2

D3 & also D4 (with no *Melma*)

E1

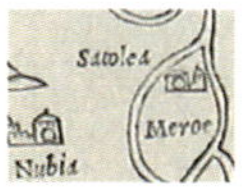

E2

E3

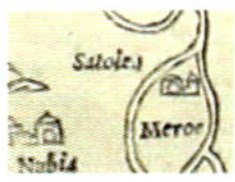

F1

F2

F3

G1

G2

H1

H2

H3

| 89

The following known editions of Münster's *Geographia* and *Cosmographia* are grouped by the language of the text. Examples of the Africa map in Münster's *Geographia* and *Cosmographia* were consulted at those libraries identified beside each edition. Along with a use of the previous chart of map variants, by using the information provided on the title and on the verso of any Africa map by Münster, more precise information on the date of a map may be obtained.

GEOGRAPHIA (published in Latin text editions only)

1540 Variant 1
Title: *AFRICA XVIII NOVA TABVLA* (black leaf motif before title).
Text Box: last two lines of text begin with 't' and 'b' respectively.
Verso: AFRICA NOVA | *Africa una ex tribus...* The second line ends with the word '*regno/* (Note the slash after the '*o*'). At bottom '46'.
Examples: *Cambridge Harv MC*, MAP-LC G1005 1540 f*; *London BL*; *New York PL*, *KB+ 1540; *Washington LC*, G1005 1540.

1542 Variant 2
Title: *AFRICA XVIII NOVA TABVLA* (black leaf motif before title).
Text Box: 't' in left margin below second to last line of text. and 'b' in left margin below last line of text.
Verso: AFRICA NOVA | *Africa una ex tribus...* The second line ends with the word '*regno*' (note <u>no</u> slash after the '*o*'). At bottom '46'.
Examples: *Cambridge Harv MC*, Mp 2.1542 Pf; *London BL*; *Washington LC*, G1005 1542.

1545 Variant 5
Title: *AFRICA XXV NOVA TABVLA* (black leaf motif before <u>and</u> after title).
Verso: AFRICA NOVA | *Africa una ex tribus...* The second line ends with the word '*regno/* (note the slash after the '*o*'). At bottom '53'.
Examples: *London BL*; *New York PL*, *KB+ 1545; *Washington LC*, G1005 1545.

1552 Variant 9
Title: *AFRICA XXV NOVA TABVLA* (black leaf motif before <u>and</u> after title).
Latitude and longitude bars outside the map.
Verso: AFRICA NOVA | *Africa una ex tribus...* At bottom '53'.
Examples: *Cambridge Harv RB*, Gp 120.24 F*; *Washington LC*, G1005 1552; *Private Collection*.
NB: only example with the numerical grid surrounding the map.

COSMOGRAPHIA

Latin Text Editions

1550 Variant 7
Title: *Totius Africae tabula & descriptio uniuersalis etiam ultra Ptolemaei limites extensa.*
Verso: AFRICAE | NOVA DESCRIPTIO, QVAE | *peninsulae instar undique feré mari cingitur,*|... At bottom '13'.
Examples: *Basel UB*, E.V.II.1; *London BL*, 566.i.14; *London RGS*, 265.e.2; *Washington LC*, G.113.M7 1550.

1552 Variant 8
Title: *Totius Africae tabula & descriptio uniuersalis etiam ultra Ptolemaei limites extensa.*
Verso: AFRICAE | NOVA DESCRIPTIO, QVAE | *peninsulae...* At bottom '13'.
Examples: *Cambridge Harv RB*, Typ 565 52.584 F; *New York PL*, *KB+ 1552.

The author has seen several examples of this variant with <u>Quiloa</u>, <u>Regnū</u> in Regnū Gambrae in West Africa, and part of <u>Garamantes</u> upside down.
Examples: *Amsterdam UB*, 1802 C 6; *Author's collection*.

1554 Variant 11
Title: *Totius Africae tabula, & descriptio uniuersalis, etiam ultra Ptolemaei limites extensa.*
Verso: *AFRICAE | NOVA DESCRIPTIO, | quae peninsulae instar undique ferè mari cingi- |*...(Note hyphen after '*cingi*'.) At bottom '13'.
Examples: *London BL*, 216.d.12; *Washington LC*, G113.M7 1554.

1559 Variant 11
Title: *Totius Africae tabula, & descriptio uniuersalis, etiam ultra Ptolemaei limites extensa.*
Verso: *AFRICAE | NOVA DESCRIPTIO, | quae peninsulae instar undique ferè mari cingi |*...(Note no hyphen after '*cingi*'.) At bottom '13'.
Examples: *New York PL*, *KB+ 1559; *Washington LC*, G113.M7 1559.

1572 Variant 14
Title: *AFRICAE TABVLA* with black leaf motif before and after title.
Verso: *AFRICA NOVA | Africa una ex tribus...* At bottom '24'.
Examples: *Amsterdam UB*, Res.1802 C 3; *Washington LC*, G113.M77 1572.

German Text Editions

1544 Variant 3
Title: *Africa/Libya/Morland/mit allen künigreichen so zů unsern zeiten Darin gefunden werden.*
Verso: *Des gantzen | lands Africe ein gemei/ | ne beschreibung.* within border of four woodcuts, top figures of horned men on seahorses, bottom figures of cherubs and left & right leaves. At bottom 'XXiii'.
Examples: *Basel UB*, E.U.I.55 1544; *London BL*, c.8.a.14; *New York PL*, *KB+ 1544.

1545 Variant 4
Title: *Africa/Libya/Morland/mit allen künigreichen so zů unsern zeiten darin gefunden werden.*
Verso: *Des gantzen | lands Africe ein gemei/ | ne beschreibung.* within border of four woodcuts, top figures of horned men on seahorses, bottom figures of cherubs and left & right leaves. At bottom 'XXVII'.
Examples: *London BL*, 1297.m.6.

1546 Variant 6
Title: *Africa/Libya/Morland/mit allen künigreichen so zů unsern zeiten darin gefunden werden.*
Verso: *Des gantzen | lands Africe ein gemei/ | ne beschreibung.* within border of four woodcuts, top figures of horned men on seahorses, bottom figures of cherubs and left & right leaves. At bottom 'XXVII'.
Example: *Basel UB*, E.U.I.62 1546.

1548 Variant 3
Title: *Africa/Libya/Morland/mit allen künigreichen so zů unsern zeiten Darin gefunden werden.*
Verso: *Des gantzen | lands Africe ein gemei- | ne beschreibung.* within border of four woodcuts, top figures of horned men on seahorses, bottom figures of cherubs and left & right leaves. At bottom 'XXVII'.
Example: *München SB*, Res/2 Geo.u.47.

1550 Variant 7
Title: *Africa/Libya/Morland/mit allen künigreichen so zů unsern zeiten darin gefunden werden.*
Verso: *Des gantzen | lands Africe ein gemei- | ne beschreibung.* within border of four woodcuts, top figures of horned men on seahorses, bottom figures of cherubs and left & right leaves. At bottom 'xiii'.
Examples: *Dresden LB*; *Washington LC*, Hauslaub Collection folio #49-34.

1553 Variant 10
Title: *Africa/Libya/Morland/mit allen künigreichen so zů unsern zeiten Darin gefunden werden.*
Verso: *Des gantzenn | lands Africe ein gemei- | ne beschreibung.* within border of four woodcuts, top figures of horned men on seahorses, bottom figures of cherubs and left & right leaves. At bottom 'xiii'.
Example: *Berkeley UC*, BANC: \xf\ G113 .M75 1553.

1556 Variant 11
Title: *Africa/Libya/Morland/mit allen künigreichen so zů unsern zeiten darin gefunden werden.*
Verso: *Des gantzenn | lands Africe ein gemei | ne beschreibung.* within border of four woodcuts, top figures of horned men on seahorses, bottom figures of cherubs and left & right leaves. At bottom 'xiii'.
Example: *London BL*, 10003.s.5.

1558 Variant 12
Title: *Africa/Libya/Morland/mit allen künigreichen so zů unsern zeiten darin gefunden werden.*
Verso: *Des gantzenn | lands Africe ein gemeine | beschreibung.* within border of four woodcuts, top figures of horned men on seahorses, bottom figures of cherubs and left & right leaves. At bottom 'xiii'.
Example: *Author's collection.*

1561 Variant 13
Title: *Africa/Libya/Morland/mit allen künigreichen so zů unsern zeiten darin gefunden werden.*
Verso: *Des gan | tzenn lands A- | frice ein gemei- | ne beschrei- | bung.* within single woodcut border of knight on horseback at bottom with other figures and cherub at top center. At bottom 'xiii'.
Examples: *Bern LB*, Aq 9486 Res K; *München SB*, Res/2 Geo.u.56.

1564 Variant 13
Title: *Africa/Libya/Morland/mit allen künigreichen so zů unsern zeiten darin gefunden werden.*
Verso: *Des gan | tzen lands A. | frice ein gemei | ne beschrei/ | bung.* within single woodcut border of knight on horseback at bottom with other figures and cherub at top center. At bottom 'xiii'.
Examples: *London BL*, cup.406.d.36; *Washington LC*, G113.M75 1564.

1567 Variant 13
Title: *Africa/Libya/Morland/mit allen künigreichen so zů unsern zeiten darin gefunden werden.*
Verso: *Des gan | tzen lands A- | frice ein gemei | ne beschrei/ | bung.* within single woodcut border of knight on horseback at bottom with other figures and cherub at top center. At bottom 'xiii'.
Example: *New York PL*, *KB+1567.

1569 Variant 13
Title: *Africa/ Libya/Morland/mit allen Künigreichen so zů unsern zeiten darin gefunden werden.*
Verso: *Des gan | tzen lands Afri- | ce ein gemei- | ne beschrei- | bung.* within woodcut border of cherubs and satyrs grasping leafy vine. At bottom 'xxv'.
Examples: *München SB*, Geo/2 Geo.u.59; *Princeton UL*, 1007.663.11q; *Stanford SC*: KB 1569\.M8\f.

1572 Variant 15
Title: *Africa/Libya/Morenland/mit allen Künigreichen so zů unsern zeiten darin gefunden werden.*
Verso: *Des gan | tzen Lands A. | frice ein gemei | ne beschrei- | bung.* within single woodcut border of knight on horseback at bottom with other figures and cherub at top center. At bottom 'xxv'.
Example: *Author's collection.*

1574 Variant 15
Title: *Africa/Lybia/Mórenland/mit allen Künigreichen/so zů unsern zeiten darin gefunden werden.*
Verso: *Des gan= | tzen Lands A= | frice / ein Ge= | meine beschrei= | bung.* within a single woodcut border of decorated framework with a cherub clasping each of side pillars. At bottom 'xxv'.
Example: *Washington LC*, G113.M75 1574.

1578 Variant 15

Title: *Africa/Lybia/Mórenland/mit allen Künigreichen/so zü unsern zeiten darin gefunden werden.*
Verso: *Des gan= | tzen Lands A= | frice / ein gemei= |ne beschrei= | bung.* within a single woodcut with a Phoenix rising from the sun at the top and a group of men at the bottom. At bottom 'xxv'.
Examples: *Basel UB*, EU I 58; *London BL*, 569.h.25.

French Text Editions

1552 Variant 10

Title: *La table & description uniuerselle de toute l'Afrique, uoire estendue outre les limites de Ptol.*
Verso: *NOVVEL | LE DESCRIPTI | ON D'AFRIQVE, LA- | quelle...* At bottom '13'.
Example: *London BL*, 568.h.5-7.

1555 Variant 10

Same Stock As 1552 (From Burden).
Title: *La table & description uniuerselle de toute l'Afrique, uoire estendue outre les limites de Ptol.*
Verso: *NOVVEL | LE DESCRIPTI | ON D'AFRIQVE, LA- | quelle...* At bottom '13'.
Example: *Paris BN*, Ge FF 3058.

1556 Variant 11

Title: *La table & description uniuerselle de toute l'Afrique, uoire estendue outre les limites de Ptol.*
Verso: *NOVVEL= | LE DESCRIPTI= | ON D'AFRIQVE, LA- | quelle...* At bottom '13'.
Example: *Washington LC*, g113.M74 1556.

1560 Variant 11

Title: *La table & description uniuerselle de toute l'Afrique, uoire estendue outre les limites de Ptol.*
Verso: *NOVVEL= | LE DESCRIPTI | ON D'AFRIQVE, LA- | quelle est presque toute enuironnée de la | l...* At bottom '13'.
Example: *Washington LC*, Stuart *KB+ 1560.

1565 Variant 13

Title: *La table & description vniuerselle de toute l'Afrique, voire estendue outre les limites de Ptol.*
Verso: *NOVVEL= | LE DESCRI- | PTION D'AFRIQVE, | laquelle est presque toute enuironnée de la | mer...* At bottom '13'.
Example: *Cambridge Harv MC*, Pusey, MA 16.65*.

1568 Variant 13

Title: *La table & description uniuerselle de toute l'Afrique, uoire estendue outre les limites de Ptol.*
Verso: *NOVVEL= | LE DESCRI= | PTION D'AFRIQVE, | laquelle est presque toute enuironnée de | la | mer...* At bottom '13'.
Example: *Providence JCB*, J568.M948c.

Italian Text Editions

1558 Variant 11

Title: *Tauola, & discrizzione uniuersale di tutta l'Africa, distesa anche piu la che i termini di Tolomeo.*
Verso: *NVOVA | DISCRIZZION | dell' Africa, la...* At bottom '13'.
Examples: *Basel UB*, Es.258; *London BL*, 10005.h.7.

1571 Variant 11 (published in Venice).

Title: *Tauola, & discrizzione uniuersale di tutta l'Africa, distesa anche piu la che i termini di Tolomeo.*
Verso: *NVOVA | DISCRIZZION | dell' Africa, la...* At bottom '13'.
Example: *Washington LC*, G113.M77 1575a (N.B.: dated 1575 in catalog but the title page is from 1571).

1575 Variant 11 (published in Cologne)
Title: *Tauola, & discrizzione uniuersale di tutta l'Africa, distesa anche piu la che i termini di Tolomeo.*
Verso: *NVOVA | DISCRIZZION | dell' Africa, la...* At bottom '13'.
Examples: *Chicago NL; London BL*, 566.i.13; *Washington LC*, G113.M77 1575.

In 1580, the map was also included in a work by Coelius Augustin Curio titled *Saracenische Geschichte und Kriegsrüstung wider das Römische Reich und die Christen : deßgleichen von der Saracenen und ihrer Könige Ursprung ...In teutsch gebracht durch Nicol. Honinger von Tauber Königshofen*. This book was published in Basel in 1580 by Sebastian Petri (1546-1627) or Sebastian Henricpetri as he was sometimes identified. This map is distinquished from the map in Münster's *Geographia* and *Cosmographia* by the inclusion to each side of the map title of *Xlij*. Also, on the verso at the top of this example, there is the heading of *Das Erste Buch* with two full pages of text in German.
Examples: *Washington LC*, Hauslab-Lichtenstein Collection, catalogued as folio #46-157 (as a separate map); *München SB*, Hbks/Hbks F 89 k#Beibd.1,S.42/43; *Basel UB*, Pa 39; *Berlin SB*, 2' Um 188.

There is a small map of Africa on the title page of the chapter describing Africa within Münster's *Cosmographia*. It is not included in this cartobibliography as it does not show the entire African continent.

References
Burden 1996: map # 12 (Americas map); Karrow 1993: 410-434; Norwich 1993: 4: map # 2 and map # 3; Moreland & Bannister 1989: 78; Klemp 1968; Oehme 1968; Skelton 1966a; Tooley 1966; Burmeister 1963; Ruland 1962.

4   1554 Giovanni Battista Ramusio - Giacomo Gastaldi, Venice

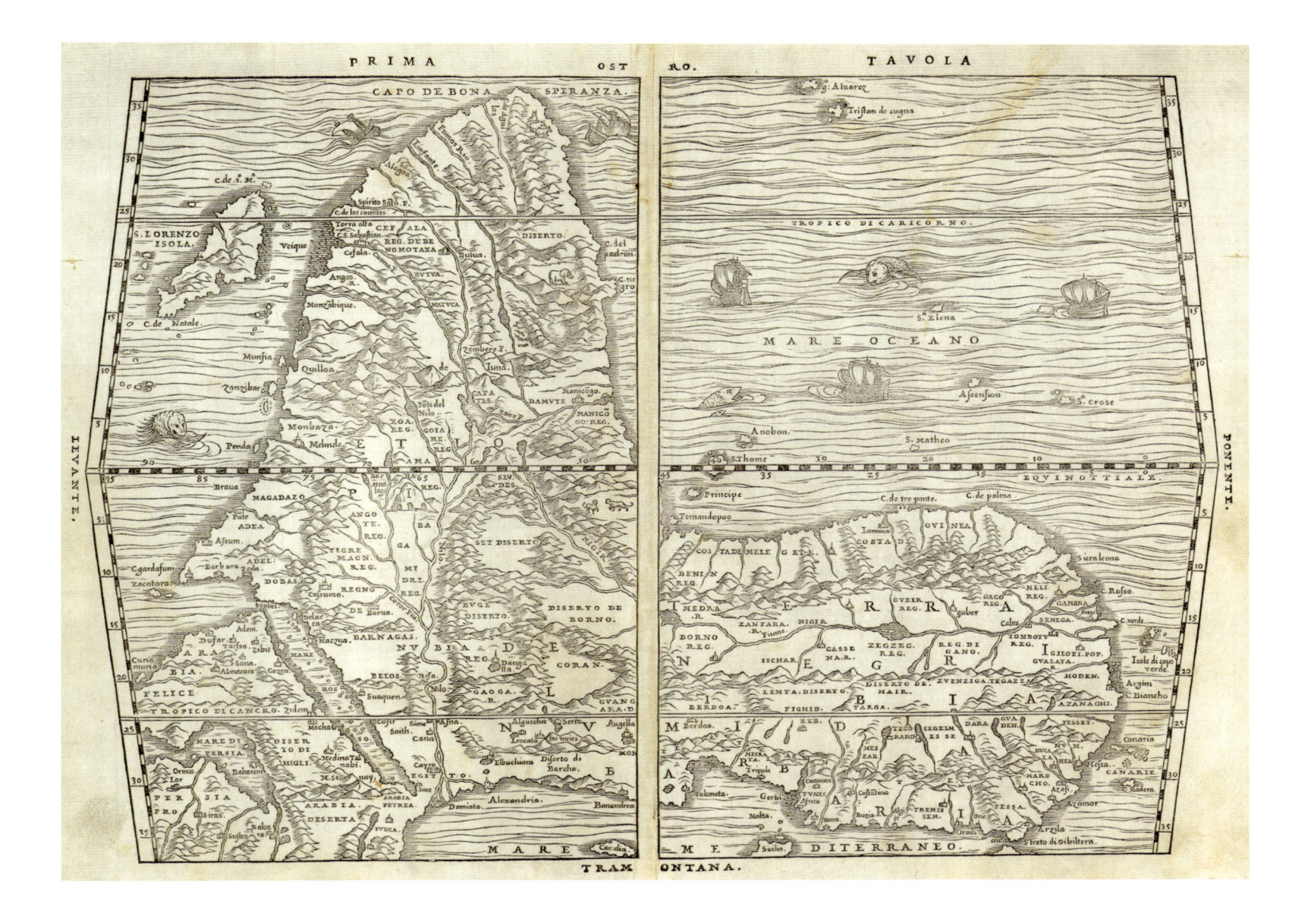

PRIMA TAVOLA.
- No scale bar.
- [Venice : Giovanni Battista Ramusio, 1554].
- From: *Delle Navigationi et Viaggi...* (*Primo Volume, Seconda Editione*).
- 1 map : woodcut ; 2 map sheets from two woodblocks ; each block 27.5 x 19.0 cm (across Equatorial Line).
- Latitude coordinates are on the left and right sides of the map. Longitude coordinates are along the Equatorial Line.
- South at the top; no compass rose.
- Title is at the top above the blocks containing the map. PRIMA TAVOLA (first map) refers to this as the first map in the book (The other two maps in the book are of India, and of China and East Asia).
- Wave lines run horizontally around the entire coast of Africa. One ship is found on each side of the southern tip of Africa. One sea monster is on the left just above the equator. On the right of the map under the TROPICO DI CARICORNO are three ships and two sea monsters. Some antelope and one elephant are near the Atlantic coast of Southern Africa. Two leopards are north of the equator in the center of the map.

Examples: *New York PL*, *KB+ 1554 (Ramusio, G. B. Primo volume, & seconda editione); *Author's collection*.

Description

Though by tradition, this map is usually referred to as the Ramusio map of Africa since it appears in Ramusio's book of travels, it rightly should be referred to as the Ramusio – Gastaldi map of Africa. Giacomo Gastaldi (c.1500-1566) actually designed and possibly cut the block that produced this map (Karrow 1993: 227). Gastaldi, from Villafranca in Piedmont, was active in Venice by 1539, initially involved with mapping the waterways around Venice. He prepared the maps, including 34 modern maps, for an important, reduced-size edition of Ptolemy's *Geographia*, published by Pietro Mattiolo in 1548. This edition contained modern maps of Northern and of Southern Africa, and considerably updated the two earlier maps of Northern and Southern Africa by Waldseemüller in 1513 and then Fries in 1522. In Venice, Gastaldi was associated with Ramusio, who, as an influential Venetian citizen, supported Gastaldi's work. Gastaldi is best known for his landmark wall map of Africa of 1564 (See Map # 9).

Giovanni Ramusio (1485-1557) was Secretary to the Council of Ten in Venice for 43 years, and his collection of reports on voyages was renowned in his own and later times. The information used to produce this map of the African continent was based on knowledge obtained from Arab geographers and Portuguese discoveries. Foremost among the Arab sources was the manuscript of the famed Arab geographer, al Hassan Ibn Muhammad al Wazzan (c.1483-1552), or as he is better known, Leo Africanus (See Map # 5 for more information on Leo Africanus).

Much of the Portuguese information for this map was taken from Montalbodo's *Paesi Nouamente retrouati Novo Muodo…* with its descriptions of the voyages of Cadamosto, Da Sintra, Da Gama, and others. This information included: a routier of the west African coast; letters written in Lisbon on July 10 and August 28, 1499, by the Italian merchant Girolamo Sernigi to report what he had learned of the Da Gama voyage on the return of the first two of the surviving ships; and a narrative of the second Portuguese voyage to India. According to Skelton (1967: I: 3-4), Ramusio is thought to have edited a manuscript version of the *Paesi* which is not known to have survived.

This map was printed from two woodcuts with south at the top. It is the first printed map of Africa in a book to show a southbound river, the *Zembere F*, flowing out of the western Ptolemaic Lake (the unnamed Zembere/Zaire), passing southward through the *Monti de Luna* (Mountains of the Moon), and then splitting into the *Cuama F.* (Zambezi) and *Spirito Santo F.* (Limpopo) rivers which flow to the southeast coast of Africa. While the ancient city of Simbaoe (Great Zimbabwe) is not shown on the map in the interior of Southern Africa, a drawing of a large city is shown with the notation beside it, *Reg. de Benomotaxa*, a reference to the Southern African kingdom of Monomotapa. The east coast shows the numerous Arab ports that in some cases were seized by the Portuguese such as *Cefala* (Sofala) near the mouth of the Cuama River in Southeast Africa. Much of this information was taken from João de Barros' *Décadas da Asia*, published in Lisbon in 1552.

This map also shows for the first time on a printed map of the entire continent, the island of Madagascar or *S. Lorenzo Isola* as it was then known. This information was based on Portuguese source material, particularly from Diogo Dias' first sighting of Madagascar for the Portuguese on August 10, 1500. The earlier maps of Southern Africa by Martin Waldseemüller and Laurent Fries placed Madagascar far to the southeast of the mainland and without much definition, and were based on the writings of Marco Polo. It was Gastaldi's map of Southern Africa which was included in his edition of Ptolemy's *Geographia* that provided the first reasonable semblance of the proper shape for Madagascar.

In West Africa, the westward flowing *Nigir* (Niger River) has its source just south of the Equator in the earliest depiction on a printed map of the unnamed Lacus Niger (Lake Niger). The ancient city of *Tombotv* (Timbuktu) is shown in one of its earliest depictions on a printed map, though placed too far to the west. A second West African river (Garamas Flu & Nubia Flu), north of the Niger River, flowing eastward into the Nile is not shown, though it would appear in later sixteenth century maps, notably those based on Mercator. The map shows evidence of the Portuguese settlements in West Africa with *Lamina* (El Mina, the Portuguese fort) on the coast of Guinea.

In this map, Gastaldi corrected the long, sloping, west coast of Africa from the typical presentation of Münster and others that showed the southern tip being 10 degrees too far to the east.

Publication Information

This map appeared for the first time in the second edition of Volume I of Ramusio's *Delle Navigationi et Viaggi* in 1554. The first edition of Volume I from 1550 only contained a woodcut of the Nile and six plans of the rock-cut churches of Lalibala in Abyssinia. This map is only found in the second edition and was replaced in the third edition of 1563 by a copperplate engraved map. The map of Africa is the first of three contiguous maps, the second and third covering India and China.

The 1554 map is rare. A fire in the Thomaso Giunti print shop in November 1557 likely destroyed the woodblocks that produced this map along with the other blocks of Volume I, shortly after the death of Ramusio earlier in 1557. As a result, few examples were printed before the destruction of the woodblocks. A copperplate replacement was prepared for the edition of 1563 and was used for the subsequent editions of 1588, 1606, and 1613 (See Map # 7 for a description of the Ramusio-Gastaldi copperplate map of 1563).

The woodcut map of 1554 contains three ships and two sea monsters off the Namibian coast under *TROPICO DI CARICORNO* in the South Atlantic. On the verso of the 1554 map, there are eleven lines of text in Italian including the heading: *A GIL STUDIOSI DI GEOGRAPHIA | NElle...* .

In the 1563 copperplate, two additional sea monsters were added in the South Atlantic above *TROPICO DI CAPRICORNO*. The verso of the 1563 map is blank.

References

Karrow 1993: 216-249; Skelton 1967-70.

5   1556 Leo Africanus - Jean Temporal, Lyon

5.1 Variant 1 1556

[Map of Africa].
- No scale bar.
- [Lyon : Jean Temporal, 1556].
- From: *Historiale description de l'Afrique, tierce partie du monde : contenant ses royaumes, regions, viles, citez, chateaux et forteresses: ...* .
- 1 map : woodcut ; 2 map sheets from 2 woodblocks ; each block 27.5 x 18.75 cm (across Equator).
- Latitude coordinates are on the left and right sides of the map. Longitude coordinates are along the Equatorial Line.
- South at the top; no compass rose.
- Wave lines run horizontally around the entire coast of Africa. One ship is found to the right of the southern tip of Africa. One sea monster is on the left side of the map in the Indian Ocean above the Equator. Off the coast of southwest Africa below the *TROPIQVE DE CAPRICORNE* are three ships and two sea monsters.
- Some antelope and one elephant are near the Atlantic coast of Southern Africa. What appear to be two leopards are north of the Equator in the center of the map.
- North (the bottom of the map) is labeled *TRAMONTANE*. City of *Manicōgo*. to south of (above) *MANICŌGO R.*

Examples: *New York HSA*, *G.490.L57.1556 (this variant is the first map in the book at the front of volume one); *Author's collection* (as a separate map).

5.2 Variant 2 1556

North (the bottom of the map) is labeled *SEPTENTRION*. City of *Manicōgo*. to south of (above) *MANICŌGO R.*

Examples: *New York HSA*, *G.490.L57.1556 (this variant is the second map in the book at the front of volume two); *Stanford SC*, Norwich Collection (See Norwich 1997: map # 7, for map image).

5.3 Variant 3 1556

North (the bottom of the map) is labeled *SEPTENTRION*. Within Arabian peninsula below (north) *TROPIQVE DE CANCER*, the letter 'T' in the word *DESERT* is much lower than in the first variant and in the second variant examples. There is no city of *Manicōgo* as in the first variant and in the second variant examples.

Example: *London BL*, 567.K.25.

Map 5, Variant 1 with the city of Manicōgo. *Map 5, Variant 3 without the city of Manicōgo.*

Description

This map of Africa, with south at the top, appeared in *Historiale description de l'Afrique*, an important and influential book on the history, geography, language, customs, and natural history of Africa by the Arab geographer, al-Hassan ibn-Mohammed al-Wezaz al-Fasi (c.1483-1552), or as he is better known, Leo Africanus. Volume one of the book describes Africa, based on Leo Africanus, and the explorations of Cadamosto, De Sintra, and others. Volume two covers further Portuguese explorations along with a detailed description of Ethiopia, from Alvares.

Born in Granada, al-Hassan spent his early years traveling throughout North Africa and the Near East. In 1518 he was captured by Christian corsairs and taken to Rome where, under the protection of Pope Leo X, he converted to Christianity and assumed the names of his benefactor, Johannes Leo de Medici. At the request of the Pope, he translated the Arabic manuscripts, maps, and sea charts, which he had been carrying with him at the time of his capture, into Italian in 1526 as *Descrittione dell'Africa*. The original manuscript does not exist. It is supposed that this text was written in rather awkward Italian. Ramusio says that he polished the text for inclusion in his first volume of *Delle Navigationi Et Viaggi*, one of the most popular 'geographic' works of the time (Skelton 1967: I: 3).

The Africanus text drew the immediate attention of readers worldwide. He revealed a world mostly unknown to Europeans. European knowledge of Northern Africa was at that time scanty, despite the Portuguese and Spanish conquests along the Barbary Coast. The interior was generally closed for Christian travelers, save some Italian traders, and the images of the area were for the most part based on the classical Greek and Latin writers, whose texts were being reprinted.

Geographically, this map is a close copy of the 1554 Ramusio-Gastaldi map, prepared for Ramusio's book. In the Leo Africanus - Jean Temporal woodcut map, the names have been translated into French, and the ships and sea monsters are engraved in a new, slightly larger style. In this map, north (the bottom of the map) is labeled *TRAMONTANE., or SEPTENTRION.*, and in the Gastaldi- Ramusio map, the woodcut map is labeled *TRAMONTANA*. In this map, the Tropic of Capricorn is labeled *TROPIQVE DE CAPRICORNE* and in the Gastaldi-Ramusio map, the woodcut is labeled *TROPICO DI CARICORNO.*

Publication Information

Africanus' work, as it appeared in the Ramusio edition, was translated into French by Jean Temporal in 1556 for Temporal's *Historiale description de l'Afrique*. Africanus' writings on Africa had a considerable influence on all later writers on Africa. The book, a detailed account of Africa, its trade routes, geography, terrain, and people was an exceptionally important source of information on the continent and is generally considered the first book written by a person of primarily African descent.

This map appeared only in the one edition of *Historiale description de l'Afrique* of 1556. The map in different variants was observed in both volumes of this work.

Africanus' information on Africa was well-received and was re-published numerous times over the next 100 years by other publishers. The following is a summary of some of those subsequent editions of the book that contain a map of Africa:

1600 John Pory, London. A new map of Africa based on Giovanni Botero.
The book, containing a new map of Africa modeled after the 1595 Giovanni Botero map, was translated into English from the Latin in 1600 by John Pory, who was supported by Richard Hakluyt. This first English edition of Africanus' work is titled *A Geographical Historie of Africa*. The book was published in London by George Bishop (Example: *London BL*, G.4258) (See Map # 43 for the Africanus-Pory map).

1665 Arnout Leers, Rotterdam. 1630 Cloppenburch map of Africa
The book, titled *Pertinente beschryvinge van Africa...*, was published in Rotterdam in Dutch in 1665 by Arnout Leers (*Washington LC*, DT7.L48). This was the first and only Dutch edition of the book. This book contains the 1630 Cloppenburg map of Africa (See Map # 65).

There were other editions of Africanus' work that contained no map of Africa. These include with details:

A French edition was published in 1556 in Antwerp by Plantin under his own imprint and partly under the imprint of his Antwerp colleague Jean Bellere (Examples: *Washington LC*, DT7.156; *London BL*, 279.B.30). There is no map of Africa in this edition seen by the author.

The book was translated from French into Latin by Johannes Florian for editions published in Antwerp in 1556 and 1558/9, with no map of Africa in the London British Library example of 1556 (published by Jan Laet as *Ioannis Leonis Africani, De totius Africae descriptione...*) or in the Library of Congress (Example: *Washington LC*, DT7.L564). The book also appeared in Zurich in 1559, in Leiden in 1632 (no map of Africa from 1632 edition at *London BL* or *Washington LC*) and 1639.

References

Rauchenberger 1999; Norwich 1997: map # 7; Mendelssohn 1993: Vol. I: 884-886; Reiss & Auvermann Buch und Kunstantiquariat Auction Catalog 1989: 68; Skelton 1967; Gay 1875: no. 258.

6 1562 Paulo Forlani, Venice

6.1 First State 1562

[Map of Africa].
[text box at the bottom left (western) map as follows]:
Al Ecc.^{mo} Philosopho, Mathematico, Medico, et Cauallier aureate, beniemerito Guardiano grande della | Scola de S. Marco: il Sig.^{or} THOMASO Rauenna… descript- | tione dell'Africa una delle principali parte del mondo con tutti ….. da Venetia il di . 9 . di maggio . M . D . LXII . | Di V^{ra} Mag.^{a} Ecc.^{ma} perpetuo seruitor | Paulo Forlani Veronese.
- No scale bar.
- [Venice] : Paulo Forlani, 9 May 1562.
- From: separately published / Italian composite atlas.
- 1 map : copperplate engraving ; 2 map sheets ; 44.0 x 59.0 cm.
- A latitude and longitude numerical grid surrounds the map on all four sides; at the top and bottom of both sheets of the map, the numbers are outside the grid. The numbers are inside the grid along the latitude grids. Longitude coordinates are below the gridline which runs along the Equator.
- North at the top; compass rose in the Indian Ocean.
- Apart from the text box at the bottom left of the map and the decorative compass rose in the

Northern Indian Ocean, the map is undecorated. There are no ships or sea monsters on the map. There is a dedication to Thomaso Rauenna within the text box.

Examples: *Chicago NL,* Novacco 4F 393; *Greenwich NMM,* BP.C.5309 and PB C 3995 (both within Lafreri Italian composite atlases); *London BL,* C.7.e.1; *Sint-Niklaas KOKW,* IATO 64 (Lafreri composite atlas, since restauration in 1994 all maps are stored separate); *Bedburg-Hau, Antiquariat Gebr. Haas; London, Bernard Shapero Rare Books.*

6.2 Second State c.1566
In this state, the left (western) map sheet has text in the text box changed to M.D.LX, with evidence of the erasure to 'II'. Geographically, there is no change to the map. The date of c.1566 is based on research by David Woodward (1990: 8).

Example: *Chicago NL,* Novacco 4F 392.

Description
This map is by Paolo Forlani, originally from Verona, who was active as an engraver and publisher from 1560 to about 1571. His cartographic production occurred in Venice, where he worked at 'al segno della colonna in merzaria' and 'in Merzaria alla libreria della Nave' (Tooley 1999-2004, E-J: 87). He was among a group of Italian publishers who issued finely prepared copperplate engraved maps in the second half of the sixteenth century. He is often associated with Giovanni Camocio, Claudio Duchetti, and Ferando Bertelli.

For this map, Forlani appears to have largely drawn upon the earlier works of Giacomo Gastaldi. These include Gastaldi's modern maps of Northern and Southern Africa that Gastaldi produced for a 1548 edition of Ptolemy's *Geographia*, and his 1554 map of Africa for Ramusio's book of travels.

Compared to the Ramusio-Gastaldi 1554 map, Forlani's map shows additional placenames, particularly in Southern Africa. Forlani also makes further refinements in the shape of Madagascar. The source lake for the Niger River (the unnamed Niger lacus) is moved from where it is placed in the Ramusio-Gastaldi map to just north of the Equator.

The *Cuama* River and the *Spo Santo f* (Spirito Santo River) divide from an unnamed river (the Zembere River) exiting *Lago Zembere* (Lake Zembere). The *Zaire* River flows to the west from Lake Zembere. A much smaller Nile source lake, *Zaflan,* is to the right of *Lago Zembere.* There are no Mountains of the Moon. The map shows considerable detail about the trading ports on the Eastern Africa coast. Important ports identified from south to north are: *Cefala* (Sofala), *Mozambique* Island, *Quiloa, Mombaza,* and *Melinde* (Malindi). *Zenzibar* Island is shown to the southeast of *MELINDE R*[egnum].

It is interesting to note that this map appears to greatly resemble Gastaldi's eight-sheet map prepared two years later in 1564. Karrow (1993: 243) states that it is possible that Forlani had access, with or without Gastaldi's approval, to the large map of Africa that Gastaldi made for the wall of the Doge's palace in Venice in 1549 to show the new discoveries. The author tends to agree with this theory.

Publication Information
This is a separately published map that often appears unjoined in some Italian composite atlases assembled during this period. These specially assembled atlases are often referred to generically as Lafreri atlases, after Antonio Lafreri (1512-1577), though others in Italy were also assembling these atlases during the later part of the sixteenth century. Lafreri was originally from Becançon in France, but, from 28 years old, worked the rest of his life in Rome. Lafreri is often considered to be the first to issue collections of maps assembled at the request of and dependent on the particular interests of clients.

References
Meurer 2004: map # 116; Sotheby's 2000; Tooley 1999-2004, E-J: 87; De Witte 1994; Karrow 1993: 243 (30/98.1); Woodward 1990: 8-9; Mickwitz et al. 1979-1995: Vol. 2: 31-33; Almagia 1944-55: Lloyd Triestino Atlas no. 114; Tooley 1939: 3: 12-21 (described as map # 67).

7 1563 Giovanni Battista Ramusio - Giacomo Gastaldi, Venice

PRIMA TAVOLA.
- No scale bar.
- [Venice : Thomaso Giunti, 1563].
- From: *Delle Navigationi et Viaggi...(Primo Volume, Tertia Editione)*.
- 1 map : copperplate engraving ; 27.5 x 39 cm (across Equatorial Line to the plate marks).
- Latitude coordinates are on the left and right sides. Longitude coordinate is along the Equator.
- South at the top; no compass rose.
- Title is at the top above the map. PRIMA TAVOLA (first map) refers to this as the first map in the book (The other two maps in the book are of India, and of China and East Asia).
- The map was printed from one plate, with a section in the middle left open where the map was stitched into the book. Wave lines run horizontally around the entire coast of Africa. One ship is found on each side of the southern tip of Africa. One sea monster is on the left just south of the Equator. Off the coast of southwest Africa under *TROPICO DI CAPRICORNO* are three ships and two sea monsters. Some antelope and one elephant are near the Atlantic coast of Southern Africa. North of the Equator, two leopards are in the center of the map. Unlike the similar woodcut map by Ramusio, this copperplate map has two additional sea monsters in the South Atlantic above *TROPICO DI CAPRICORNO*.

Examples: 1563 edition: *New York PL,* *KB+ 1563; *Private collection*; 1588 edition: *London BL,* C.79.e.4; *Helsinki AEN; New Haven Yale; Portland USM;* 1606 edition: *London BL,* 566.i.8-10; 1613 edition: *New York PL,* *KB+ 1613.

Description

This is the copperplate version of the Ramusio-Gastaldi map of Africa. It generally follows the earlier Ramusio-Gastaldi woodcut map of Africa. The differences in the two maps are noted below. This map was engraved not due to newer information, but because the 1554 woodcut map was destroyed in 1557, and a new map was needed for the subsequent editions of Ramusio's *Delle Navigationi et Viaggi*.

For a fuller description of the Ramusio-Gastaldi map, please refer to Map # 4.

Publication Information

This map appeared for the first time in the third edition of Volume I of Ramusio's *Delle Navigationi et Viaggi* which was published in 1563. A woodcut map had been used for the first time in the second edition of Volume I of Ramusio's book in 1554. The first edition of Volume I of 1550 only contained a woodcut of the Nile and six plans of the rock-cut churches of Lalibala in Abyssinia. The map of Africa is the first of three contiguous maps, the second and third covering India and China.

With the probable destruction of the woodblock in a fire in the Thomaso Giunti print shop in November 1557, this copperplate was prepared to replace the woodcut for the edition of 1563 and was used for the subsequent editions of 1588, 1606, and 1613.

The woodcut map of 1554 contains three ships and two sea monsters off the Namibian coast, on the southwest coast of Africa, under *TROPICO DI CARICORNO* in the South Atlantic. On the verso of the woodcut map, there are eleven lines of text in Italian including the heading: *A GIL STUDIOSI DI GEO-GRAPHIA | NElle...*

In the 1563 copperplate, two additional sea monsters were added in the South Atlantic above *TROPICO DI CAPRICORNO.* The verso is blank.

References

Norwich 1997: map # 6 (N.B.: describes woodcut but the image is of the copperplate version); Karrow 1993: 216-249; Skelton 1967-1970; Sabin 1868: entry 67733.

8 1563 Giovanni Francesco Camocio - Paulo Forlani, Venice

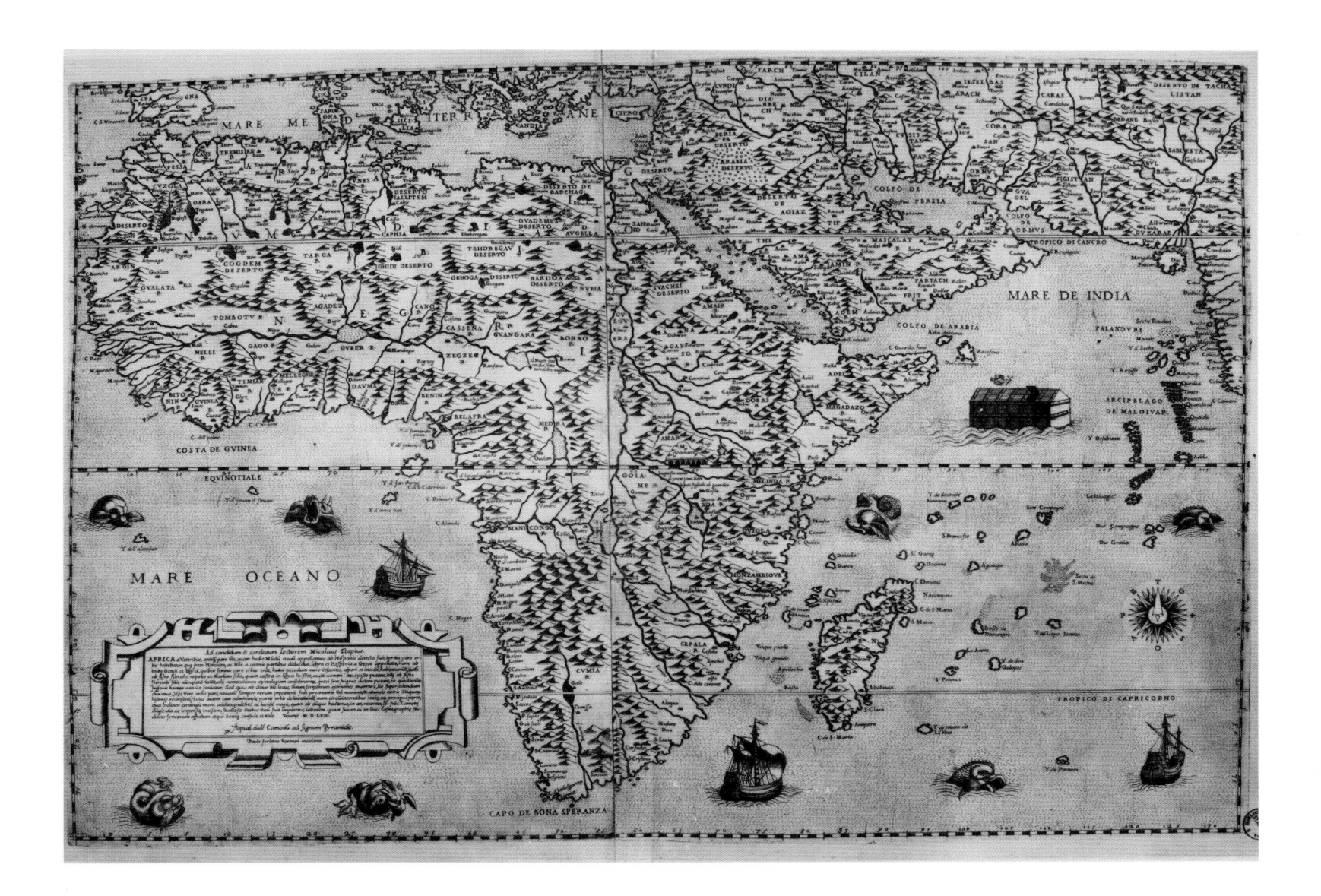

8.1 First State 1563

[Map of Africa].
[within text box cartouche on bottom of left sheet, twelve lines of text including the line with the date]:
Ad candidum et cordatum lectorem Nicolaus Stopius AFRICA a Veteribus, anteque pars illa, quam hodie Mudum Novum appellamus … Venetijs . M . D . LXIII. | Apud eúndém Camotium ad signum Pyramidis.
[below cartouche border]:
Paulo forlano Veronen incidente.
- No scale bar.
- Venice: Giovanni Francesco Camocio, 1563.
- From: separately published / Italian composite atlas.
- 1 map : copperplate engraving ; 2 map sheets ; 43.5 x 29 cm (left map sheet), and 43 x 33 cm (right map sheet).
- A latitude and longitude numerical grid surrounds the map on all four sides; the numbers are on the inside of the grid on all four sides of the map. There is also a numerical gridline on the Equatorial Line.

- North at the top; compass rose within Indian Ocean to right of Madagascar.
- In the Atlantic Ocean on the west sheet, there are two sea monsters below the text box cartouche and two sea monsters and one sailing ship above this cartouche. On the east sheet in the Indian Ocean, there are three sea monsters, two sailing ships, and Noah's Ark (with a bird returning to the Ark with a branch).

Examples: The first state of this map is only known in two examples: *Venezia BNM*, 138.c.4 (map # 64 within a composite atlas); *Wolfegg Schlossbibliothek*.

8.2 Second State 1566
Date within text box cartouche changed to M . D. LXVI.

Examples: *New York PL*, *KB+++1572 Lafrery (within composite atlas); *Paris BN*, Ge DD 1140[63].

Map 8.1 Title cartouche.

Map 8.2 Title cartouche.

Map 8.4 Title cartouche.

8.3 Third State 1566+
Erasure to copperplate of 'Apud eúndém Camotium ad signum Pyramidis.' from within text box cartouche.

It is possible that the third state of this map was prepared as late as 1575, the date of Camocio's death, given the fact that the imprint of the maker of the map, Camocio, has been removed.

Examples: *Chicago NL*, Novacco 4F 394; *Boston, Private collection*.

8.4 Fourth State 1593

Addition of 'Apud Donatum Bertellum' to the right of the evidence of the above erasure within the text box cartouche. Also the date on the cartouche has been changed to 'M. D. LXXXXIII'.

Example: *Rotterdam MM*, WAE.788.

Description

This is an expertly engraved map reflecting Forlani's excellence as an engraver. It is a close copy of Paulo Forlani's 1562 map of Africa. Unlike the Forlani 1562 map, however, the right sheet of this map is extended to show the Malabar Coast of the Indian peninsula reflecting this region's importance, particularly to the Portuguese, with its numerous trading ports.

For information on the geography of this map refer to the Forlani 1562 map (Map # 6).

Publication Information

Giovanni Francesco Camocio (active from about 1558 to c.1575) was the publisher for this map (Apud eúndém Camotium ad signum Pyramidis., below the date at the end of the text box cartouche) and Paulo Forlani was the engraver, with his imprint just below Camocio's (Paulo forlano Veronen incidente.).

The map bears a dedication to Nicolaus Stopius at the top of the bottom left text box cartouche. As a result, it is sometimes referred to as the Stopius map of Africa.

This is a separately published map that appears in some Italian composite atlases from this period. The map is known in four states, with the fourth state being un-recorded by Woodward.

References

Sotheby's 2000 auction catalog; Norwich 1997: map # 9; Woodward 1990: 12-13; Tooley 1939: 3: 12-21 (map # 70 in first state and map # 74 in second state).

9 1564 Giacomo Gastaldi – Fabio Licinio, Venice

MAP 9

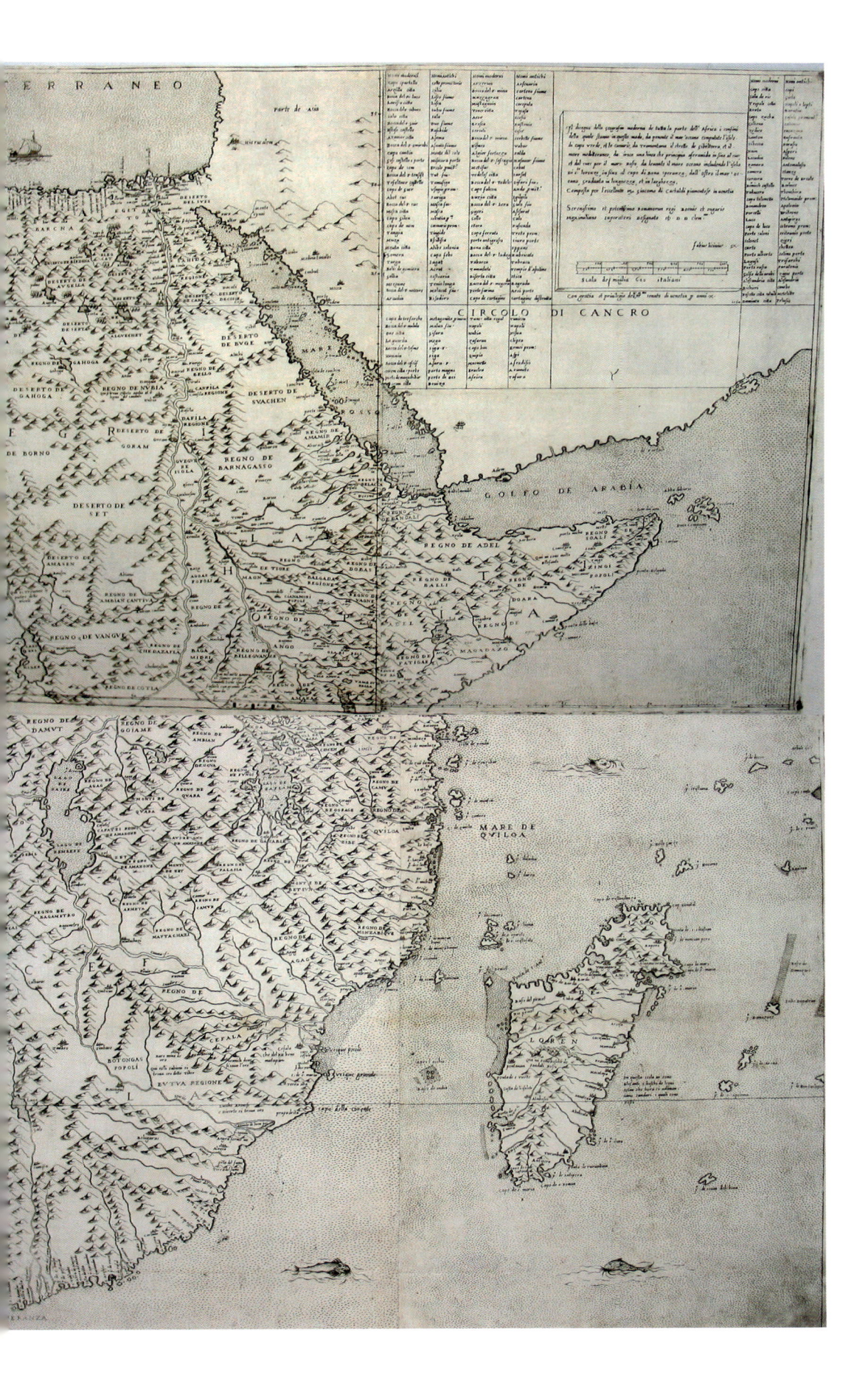

9.1 First State 1564

[Map of Africa].
[a large rectangular cartouche at upper right, sheet 4]:
Il disegno della geografia moderna de tutta la parte dell'Africa i confini | della quale stanno in questo modo, da ponente il mar' oceano computate l'isole | di capo verde, et le Canarie, da Tramontana il stretto de Gibelterra, et il | mare meditteraneo, da siroco una linea che principia aferamida in sino al sues, | et dal sues per il mare Rosso, da leuante il mare oceano includendo l'isola | Di S.^{to} lorenzo, in sino al capo di Bona Speranza; dall'ostro il mar' oc- | eano, graduata in longhezza, et in larghezza. | Composta per l'eccellente m. Giacomo di Castaldi piamontese in uenetia.
[under the text, the dedication]:
Serenissimo et potentissimo Romanorum regi Boemie et ongarie | Maximiliano inperatori Designato etc. D.D. Clem^{mo}. | Fabius licinius. ex.
[below dedication and scale bar]:
Con gratia et priuilegio dell'ill^{mo} senato di uenetia p[er] anni x. | 1564.

- Scale bar in Italian: Scala de miglia 600 italiani. 600 = 11.7 cm.
- [Venice] : Fabio Licinio, 1564.
- From: separately published.
- 1 wall map : copperplate engraving ; 8 map sheets (2 rows of map sheets with 4 sheets on each row when joined) ; 106 x 141.5 cm (overall size when joined) ; sometimes found in four quarters, each composed of two sheets as in the Basel example with sheets 1 & 2 as 53.5 x 74 cm, sheets 3 & 4 as 54 x 69 cm, sheets 5 & 6 as 53.5 x 74 cm, and sheet 7 & 8s: 54.5 x 68.5 cm).
- This map is often found closely cropped on all sides, but the latitude and longitude coordinates from the Basel example are as follows:
  Sheet 1: Latitude line on left side. Longitude line across bottom.
  Sheet 2: Latitude line on right side. Longitude line across bottom.
  Sheet 3: Latitude line on right side. Longitude line across bottom.
  Sheet 4: Longitude line across bottom only.
  Sheet 5: No latitude and longitude lines.
  Sheet 6: Latitude line on right side. Longitude line across top.
  Sheet 7: Longitude line across top only.
  Sheet 8: Longitude line across top only.
- North at the top; two compass roses in Atlantic (one in each of the two bottom left sheets).
- The title, in the form of descriptive text, appears at the upper right (in the upper part of sheet 4) in a simple frame. Under the title is the dedication. In the lower part of the title frame is the scale bar. Left (4 columns) and right (2 columns) of the title frame are lists indicating and comparing ancient and modern placenames.
- The following are decorative elements found throughout the map: one sea monster and one sailing ship in the upper left (sheet 1); one sea monster in upper left center (sheet 2); one ship in the Mediterranean at upper right center (sheet 3); nine sea monsters and two ships in the Atlantic Ocean off of the southwestern coast of Africa (sheet 5); six sea monsters and three ships directly off the southwestern coast of Africa (sheet 6); one sea monster in the Indian Ocean (sheet 7); and two sea monsters further to the east in the Indian Ocean (sheet 8).
- There are upwards of 1200 placenames on the map.

Examples: The following are the twelve examples of this map known to the author. These are found in various configurations: joined into a wall map comprising all 8 sheets; joined as an incomplete wall map with 7 sheets (missing sheet 5, i.e. the bottom left sheet); as loose maps with all 8 or 7 (sheet 5 missing) sheets; or bound into a composite atlas (with 7 or 8 sheets). Perhaps some buyers of the map felt the lower left sheet was unnecessary, since it contained nothing but an expanse of ocean, the shape of the continent rendering this unavoidable.

9.1.A. *Basel UB*, AA 110-113 (8 sheets which have been joined as pairs into 4 sheets).

9.1.B & 9.1.C. *Greenwich NMM*, BP C 5309 (bound within an untitled Italian composite atlas dated c. 1568 as 8 sheets), and *Greenwich NMM*, G290: 1/8 A-H (as 8 separate loose sheets; acquired via Maggs Brothers from the 1934 sale of the Mensing collection).

9.1.D. *Chicago NL* (7 of 8 sheets, missing sheet 5, the lower left sheet, and mounted and framed as a wall map).

9.1.E. *Paris BN*, Ge DD 5077 (8 sheets as loose maps).

9.1.F. *London BL,* Maps 189.c.1 (7 of 8 sheets, missing sheet 5, as loose maps).

9.1.G & 9.1.H *Bedburg-Hau, Antiquariat Gebr. Haas (*two examples: one example missing sheet 5 and with sheet 1 and 2 joined, sheet 3 and 4 joined, and sheet 6, 7, and 8 joined. Both examples bound into Italian composite atlases).

9.1.I. *Modena Estense*, A.49.Q.7 maps #93 and #94 (4 top sheets joined and 3 bottom sheets joined, missing sheet 5 and bound into the Alessandro composite atlas).

9.1.J. *Milwaukee AGS*, OV / 300 / A-1564 *(*7 of 8 sheets, missing sheet 5, the lower left sheet, and mounted and framed as a wall map).

9.1.K. *Stanford SC*, Norwich no. 0008 (4 top sheets joined and 4 bottom sheets joined).

Regarding the Norwich example at Stanford University Libraries Special Collection, Almàgia (1970: 65) describes an example belonging to Beans - Triestino as having *no date with the privilege*. The Triestino example was subsequently acquired by Oscar Norwich. According to Norwich (1997: 12), this example *does* have the date of 1564. This example evidently was obtained via the New York dealer, H.P. Kraus, as his Monumenta Cartographica, catalog 124, item 115 states that his example of this map was from the Lloyd Triestino collection and was incorrectly identified without the date, when, in fact, it has the date on the map.

9.1.L. *Helsinki AEN* (8 sheets). This example is illustrated in Nordenskiöld's *Periplus*.

9.2 Second State 1565

Date changed in bottom large rectangular cartouche to 1565.

This state is in three titled parts as 'La prima parte dell'Africa...' (title on bottom left of sheet 1); 'La parte seconda dell'Africa,...' (title on bottom right of sheet 4) ; and, 'La parte terza dell'Africa,...(title on bottom right of sheet 8) with the engraver's name and publication date of 1565 as 'fabius licinius ex. 1565' on each new title. Otherwise, the map appears to be identical to the Africa wall map of 1564, based on an examination by the author.

In this state, the first part is comprised of the original two top left sheets, with this new title stamped on the bottom left of sheet 1 and with a total size of 53 x 73 cm. The second part is comprised of the original two top right sheets with the newly stamped title on the bottom right of sheet 4 and with a total size 53 x 68 cm. The third part is comprised of three of the lower sheets (sheets 6, 7 and 8) with the newly stamped title on the bottom right of sheet 8 and with a total size of 53 x 102 cm. The further left lower sheet (sheet 5) is not included.

Example: *Venezia BNM*, 138.c.4 (only example known; the three parts of the map in this second state are bound into a composite atlas).

Description

This map is a testament to Italian cartography. Without much dispute, it is probably the single most important and influential map of Africa of the sixteenth century. It represents a quantum leap from Sebastian Münster's significant Africa map of 1540.

Made from eight separate sheets, this was the largest map of Africa produced up to that time. Large-scale wall maps enabled mapmakers to include a much greater amount of geographical detail than smaller maps. As a result, wall maps tended to be proto-types from which the later atlas maps were produced.

Tooley (1969: 46) locates only five examples of this map. According to Almàgia (1970: 65), there are eight or nine examples. The author has located twelve examples, referenced above. In some examples, the extreme bottom left sheet is not present, as this sheet does not show any part of the continent, but only the islands of *S*^{a.} *Elena* (St. Helena), *y*^d *della Ascencion* (Ascension), and y^d *de iouan de steuam* from south to north.

Giacomo Gastaldi (1500-1566) is known to have been at work on a large wall map as early as 1545, and he continued to improve it up to the map's publication in 1564. Biasutti (1920: 416-19 and 433) believes that the map was drawn in two stages: from 1544-45 and then again from 1554, after João de Barros' work on Africa, *Décadas da Asia,* was published in Lisbon in 1552.

For this eight-sheet map, Gastaldi drew on information used in his earlier maps: the maps of North and South Africa that he prepared for the 1548 edition of Ptolemy's *Geographia*, his mural map on the wall of the Doge's Palace of 1549, and his map of Africa in Giovanni Ramusio's book of travels, first appearing in 1554. The large size of this wall map, unlike the earlier, much smaller maps, enabled Gastaldi to adequately depict all of the geographical details that he wanted to record on his map.

A primary text source appears to be Volume I of Ramusio's *Navigationi et Viaggi* of 1550 with its accounts of the African travelers, especially Leo Africanus (Skelton 1967). Another source for Northeast Africa was Father Francisco Alvares in his *Verdadeira informacão das terras do Preste João das Indias*, published in Lisbon in 1540. Alvares visited Ethiopia as part of a mission from the Portuguese king in 1520. Gastaldi also relied on João de Barros' *Décadas da Asia* of 1552 (with subsequent volumes appearing in 1553 and 1563) for the interior lakes region, and Duarte (Odoardo) Barbosa's account of Southeast Africa. His representation of the river systems of Central Africa was based directly on Africanus, Alvares, and De Barros. According to Sims (2003), Gastaldi also relied on nautical charts, evidently from Portuguese sources.

Following his much smaller map of 1554, Gastaldi corrects the long, sloping west coast of Africa from the typical presentation of Münster and others which shows the southern tip being 10 degrees too far to the east. Gastaldi shows a southbound river, the unnamed Zembere, flowing out of the western Ptolemaic Lake (named *Zembere* and *Zaire*), passing through the unnamed Mountains of the Moon, and then splitting into the *Cuama* (Zambezi) and *Spirito Santo* (Limpopo) Rivers which flow to the Southeast coast of Africa. The ancient city of Great Zimbabwe, depicted on the map as the twin cities of *Zimbro* and *Simbaoe*, appears in the interior of Southern Africa. To the east is text describing the gold mines of this region which by this time were well known to the Portuguese.

In West Africa, the westward flowing Niger River has its source just south of the Equator in *Lago del Niger* and from there flows into *Lago de Borno* (approximately Lake Chad) and then into *Lago Guber* before finally flowing into the Atlantic Ocean.

From Sims (2003: 222), 'Gastaldi for the first time gave them (that is, wall maps of the continents) adequate size and correct form, and flushed them out with a wealth of interior detail, even if not always correct. It will not be fanciful to say that he brought the extra-European world out of hiding and into open view for the first time, and, besides the fact that it is his cartography that informs the depictions of the extra-European world in the universal mapping efforts which began in the years immediately after, including the Italian composite atlases, Ortelius's *Theatrum*, De Jode's *Speculum*, the four-continent sets, and others, one cannot doubt that Gastaldi's works also played a great inspirational role in initiating these projects. Neither those projects, nor the idea for producing them could ever have arisen without the preceding efforts of Gastaldi.'

Publication Information

The engraver of the copperplates for this wall map is Fabio Licinio. Licinio (1521-1565) was considered one of the most able engravers of maps of the sixteenth century. He is best known through his work with Gastaldi during the latter's Venetian period. Licinio was the publisher as well for this wall map with his imprint just above the scale bar within the text box: 'fabius licinius ex[cudit]'.

There is also a privilege on the map from the Senate of Venice. The dedication on the map is to Maximilian II (1527-1576), the Habsburg 'Emperor Designate' of the Holy Roman Empire. Maximilian became Emperor upon his father Ferdinand's death on July 25, 1564. As a result, it is logical that this map was finished shortly after July 1564 so as to include this dedication on the map.

The following later derivatives of this Gastaldi - Licinio map are known: Gerard de Jode, Antwerp, 1569 (no example is extant of the first state); Giovanni Francesco Camocio, Venice, c.1570-1575 (in one example); Donato Bertelli, Venice, c. 1573; and, Cornelis De Jode, Antwerp, 1596. These maps are described separately.

References
Meurer 2004: map # 117; Sims 2003; Norwich 1997: map # 8; Relaño 1995: 54; Karrow 1993: 243-245; Kraus 1970: 65; Skelton 1967-1970; Skelton 1961b; Gallo 1954: 24-26 [nos. 65-67] for State 2; Almàgia 1944-55; Tooley 1939: 3: 12-21 (map # 71, map # 68, and map # 72, the Beans-Triestino example); Biasuti 1920: Vol. IX: 327-346 and 387-436.

10   1564 [65] Ferando Bertelli, Venice

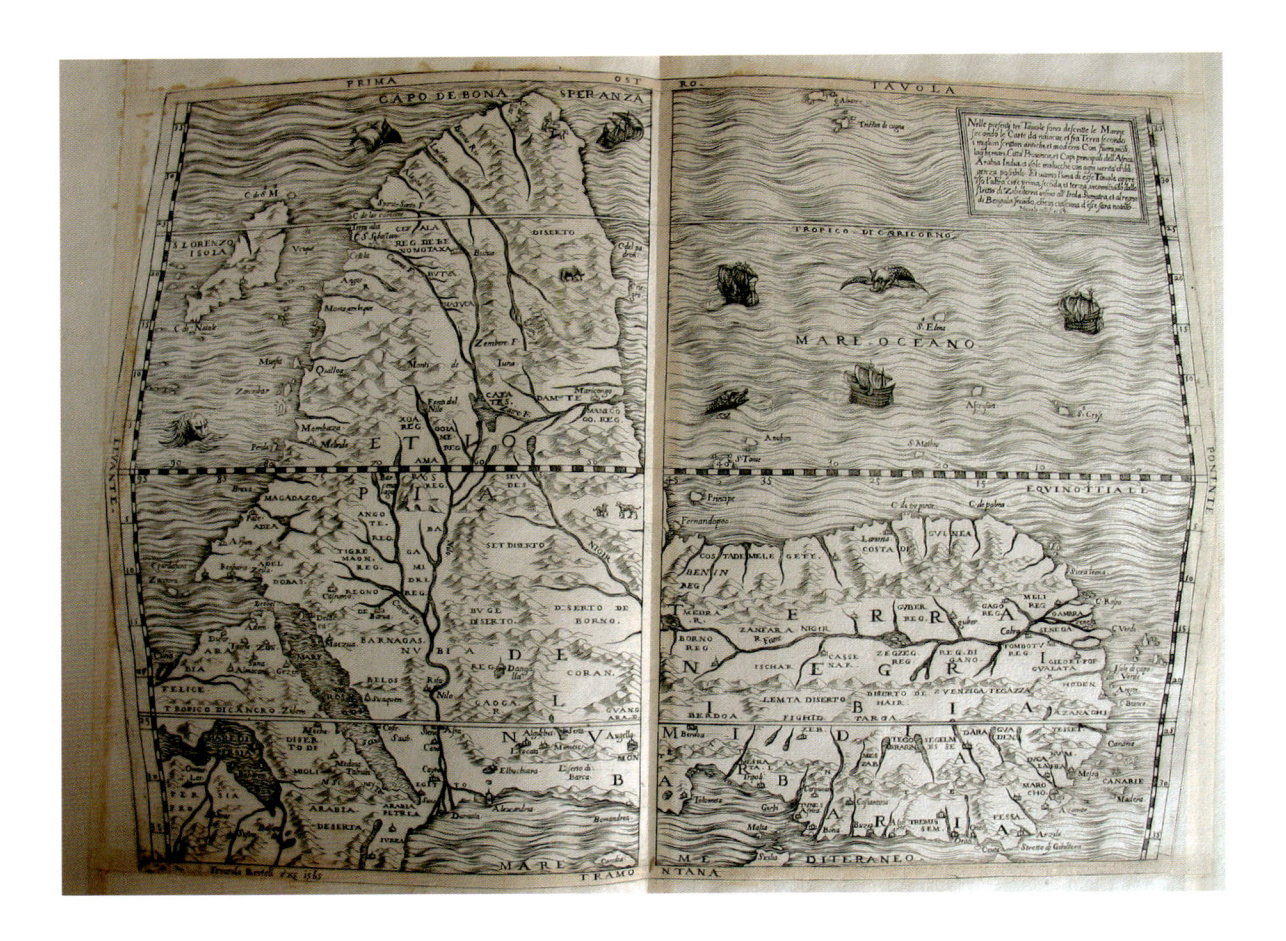

PRIMA TAVOLA.

[in text box at upper right corner]:

Nelle presenti tre Tauole sono descritte le Marine | secondo le Carte de nauicar, et fra Terra secondo | i migliori scrittori antichi, et moderni. Con fiumi, monti, | laghi, mari, Citta Prouincie, et Capi princi-pali dell' Africa, | Arabia India, et isole molucche con ogni uerita' et' dili= | genza possibile. Et uanno l'una di esse Tauole appre= | sso l'altra cioe prima, seconda, et terza, incominciando dallo | stretto di Zibelterra insino all'Isola Sumatra, et al regno di Bengala', secondo, che in ciascuna d'esse sara notato, | Niccolo nelli f. 1564.

[below map at bottom left within outer lines]:

Ferando Bertel[l]i exc 1565.

- No scale bar.
- Venice : Ferando Bertelli, 1565.
- From: separately published / Italian composite atlas.
- 1 map : copperplate engraving ; 28.5 x 39 cm.
- Latitude coordinates on both left and right borders (35N-35S). Longitude coordinates along the Equatorial Line (0-95 degrees E).
- South at the top; no compass rose.
- Title above map within outer lines.
- The right side of the map (between the Tropic of Capricorn and the Equator) contains three ships and two sea monsters. There are two floundering ships on either side of Southern Africa and one sea monster in the Indian Ocean.
- There are two animals in Central Africa, and an elephant and cattle in Southern Africa.

Examples: *Cambridge Harv RB* (as a separate map with Liechtenstein Collection); *Chicago NL*, Ayer 135 L2 1575; *Greenwich NMM*, PB C 5309 (within Italian composite atlas); *London BL,* K Top 117.2

and Maps C.7.e.1 (within an Italian composite atlas); *Paris BN,* Ge D 17757; *Bedburg-Hau, Antiquariat Gebr. Haas; London, Bernard Shapero Rare Books.*

Description

This rare map is a close copy of the Ramusio-Gastaldi copperplate engraved map of 1563. The common model for this Bertelli map and the Ramusio-Gastaldi map of 1563 is the Ramusio-Gastaldi woodcut map of 1554, as designed by Gastaldi. The Bertelli map shows extensive information on the Portuguese ports on the east coast of Africa.

The map is such a close copy that Bertelli uses the same wording of '*PRIMA TAVOLA*' above the map as in the Africa map in Ramusio's *Delle Navigationi et Viaggi...* . This is a reference to the map's place as the first of a three map series; a map of the region of India and a map of the region of China and East Asia are the other two in the set.

The map is distinguished from the Ramusio-Gastaldi map by a text box that is found in the upper right corner of the map. The text begins with '*Nelle presenti tre Tavole sono descritte le Marine..*'. and ends with '*Niccolo nelli f. 1564*'. The text box refers to the three maps in the series. In the lower margin of the map toward the left corner there is the publisher's imprint: *Ferando Bertel[l]i exc 1565*.

Publication Information

This map was published by Ferando Bertelli in 1565 (*Ferando Bertel[l]i exc 1565*) as a separate map. It also appeared in Italian composite atlases from this period. The copperplate that made this map was engraved in 1564 by Niccolò Nelli (*Niccolo nelli f. 1564*).

The map is sometimes referred to as the Niccolò Nelli map of Africa, but the author has attributed the map to its publisher, the more well known Venetian, Ferando Bertelli. Ferando Bertelli was active as a publisher of specially assembled Italian composite atlases (the so-called Lafreri-type atlases) in Venice, from about 1556 to 1572, and was associated at different times with Camocio, Forlani, and Zenoni. Ferando's relative Donato Bertelli, also from Venice, continued the trade when he assumed Ferando's business in about 1574. Little is known of the map's engraver, Niccolò Nelli, who was active in Venice from about 1562 to 1573 and who was employed by Bertelli to engrave this map.

Tooley (1966: 21) mentions only eight examples of the map. Though rare, it is likely that other examples of this map exist in various composite Italian atlases.

References

Meurer 2004: map # 113; Karrow 1993: 228 (30/73.2); Kraus 1970: 67; Tooley 1939: 3: 12-21 (map # 73).

11   c. 1570-1575 Giacomo Gastaldi –
Giovanni Francesco Camocio, Venice

[Map of Africa].
[a large, decorative cartouche at upper right, sheet 3 & 4]:
TAVOLA DE NOMI ANTICHI ET MODERNI | DELA PRESENTE CARTA.
[and just below, six columns comparing ancient and modern placenames].
- Scale bar in Italian: Scala de miglia 600 Italiani. 600 = 11.7 cm.
- [Venice : Giovanni Francesco Camocio, c.1570-1575].
- From: separately published / Camocio Atlas; formerly the Harmsworth Atlas.
- 1 wall map : copperplate engraving ; 12 map sheets ; 110 x 125 cm.
- Latitude coordinates on left and right sides of sheets 1, 2, 3 & 4 joined, 5, 6, 7 & 8 joined, 9, 10, and 11 & 12 joined. Longitude coordinates across Equatorial Line.
- North at the top; compass rose in the Atlantic on bottom left sheet.
- Numerous scrollwork text boxes, sea monsters, and ships are placed throughout the map. Animals are within Africa.

Examples: *Minneapolis Bell*, 1560 fCa (as separate sheets within the Camocio atlas; only known example of the complete map); *Bedburg-Hau, Antiquariat Gebr. Haas* (of the 12 total sheets, this example contains separate map sheets nos. 2, 7 & 8 joined, 11 & 12 joined).

Description

This beautiful map is a re-engraving of Giacomo Gastaldi's wall map of 1564 and was published by Giovanni Francesco Camocio as part of a four-continent set. See the Gastaldi-Licinio map (Map # 9) for a description of the geography. The only complete example known is located at The James Ford Bell Library of the University of Minnesota and is part of a bound atlas containing maps that comprise the four continents and a smaller world map.

The map is undated. Woodward (1997) dates this map as having been engraved in c.1570-1575. In 1575 the plague devastated Venice, and it is likely that Camocio died at this time as no further work of his is known after this date. Somewhat similarly, Caraci (1962: 59) suggests a date of 1571-74.

The example at the Bell Library is likely an earlier version of the map, as it does not bear a formal title, but only a blank cartouche in the lower left that was intended for a dedication or a title. In fact, the blank cartouche bears the wording translated from Italian as: 'Here you can put the dedicatory and explanatory note for the map'.

Publication Information

The maker of this map, Giovanni Camocio, was one of the best known and prolific of the Venetian map publishers with a shop at 'all'insegna della Piramide.' Camocio was active from c.1560 to c.1575. He probably fell victim to a plague that ravished Venice in 1574-75, as all records related to him end at that time. He came into possession of some of the plates of Fabio Licinio, Gastaldi's principal engraver, and almost all of his plates were acquired by Donato Bertelli after his death.

The wall maps in the Bell set were all designed to contain twelve sheets; that is, the nine full-sheets were arranged in three rows of three, and the three half-sheets were attached onto the right side. These maps have sometimes been referred to as nine-sheet maps, causing a little confusion. The Europe and the Asia wall maps in the set both have dedications signed by Camocio. The map of Africa does not have Camocio's imprint on it.

References
Sims 2003; Woodward 1997: Number 34; Caraci 1927: 29:178-92.

12   1570 Abraham Ortelius, Antwerp

12.1 First State 1570

AFRI= | CAE TA= | BVLA | NOVA.
[above the map title]:
AFRICA M | GRAECI | LIBYAM APP:
[below the map title in a semi-circular attachment]:
EDITA ANT= | VERPIAE | 1570.
[at the upper left corner of the map]:
Cum Privilegio.
- No scale bar.
- Antwerp : [Abraham Ortelius, Gielis Coppens van Diest], 1570.
- From: *Theatrum Orbis Terrarum.*
- 1 map : copperplate engraving ; 38 x 51 cm.
- Latitude coordinates at left and right sides. Longitude coordinates across the bottom of the map.
- North at the top; no compass rose.
- The title appears at the bottom left in a decorative strapwork frame.

- There are three sea monsters in the Atlantic Ocean and three ships engaged in a battle in the Indian Ocean. No flora or fauna are found as decoration on the map image.

Examples: *Amsterdam UB*, OF 72-30 (1570 Latin A Version; this atlas was the model for the Theatrum/Nico Israel facsimile); *Leiden UB*, COLLBN Atlas 36 (1570 Latin A Version; this atlas was the model for the Sequoia and Rinsen facsimiles); *Rotterdam MM*, WAE 50 (1570 Latin AB Version).

There were at least four different Latin editions or versions published in 1570-71. In the first state, *BI=LE=DUL= | GERID.* is missing above and below the Tropic of Cancer. At 20° and 30° South, at the right side gridline, the latitude lines do not extend to the outer decorative border.

12.2 Second State 1570

In some copies of the 1570 Latin B and C versions and in all later editions, *BI=LE=DUL= | GERID.* is now present above and below the Tropic of Cancer. At 20° and 30° South, at the right side gridline, the latitude lines still do not extend to the outer decorative border, which is the same as in the first state.

Example: *Amsterdam UB*: 1802 A 14.

12.3 Third State 1584

The latitude lines now do extend to the outer decorative border at 20° and 30° South, at the right side gridline.

Example: *Private Dutch collection*. Observed in various 1584 Latin and 1587 French editions.

Map 12 Women in title cartouche with and without extra hachuring (states 1-3 and 4 respectively).

12.4 Fourth State after 1588

The two women in the title cartouche now have extra shaded lines (additional and thicker hachuring).

Examples: *Various locations.*

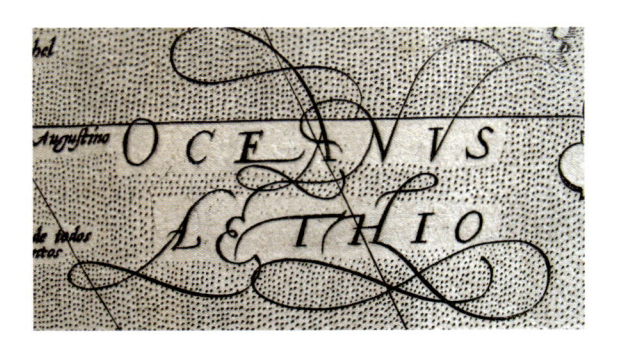

Map 12 The wording of OCEANVS | AETHIO= with and without dashes (states 1-4 and 5 respectively).

CARTOBIBLIOGRAPHY

12.5 Fifth State 1595

'=' now appears after *OCEANVS | AETHIO* in the Atlantic Ocean. Also, a crack has started to appear in the title cartouche, running down the right hand sequence of letters. This has evidently occurred during re-working of the plate.

Examples: *Various locations.*

Map 12.5

12.6 Sixth State after 1606/12 [1641]

Sometime between 1606 and 1612 the date of 1570 is removed. Some 1612 Spanish copies still have the date, and some do not.

Examples: *Various locations.*

Description

This beautiful map is one of the cornerstone maps of Africa. It generally replaced Sebastian Münster's widely circulated map of Africa of 1540. Modern in appearance compared to Münster's map and others, this 1570 map by Abraham Ortelius (1527-1598) remained the standard map of Africa well into the seventeenth century.

By 1570, the coast of Africa had been relatively well explored, primarily by the Portuguese, but also by occasional explorers from other European countries and by the Arabs on the east coast. Successive expeditions into the interior in Abyssinia, West Africa, the Congo River, and the Zambezi River basin had also yielded a significant amount of information.

In the text to accompany this map in the *Theatrum*, Ortelius freely acknowledges his sources for the preparation of his maps in his *Catalogus Auctorum*. Like all fifteenth and sixteenth century cartographers, Ortelius used the writings of Herodotus (484-425 BCE), Strabo (58 BCE-25 CE), and Ptolemy. He also relied heavily for this map of Africa on the more recent sources: primarily Ramusio in *Navigationi et Viaggi* (Venice,1550 and later editions), João de Barros in *Décadas da Asia* (Lisbon, 1552), and Leo Africanus in Jean Temporal's *Historiale description de l'Afrique* (Paris, 1556).

Geographically, the map closely follows Gastaldi's wall map of 1564. This is especially evident in the choice of toponyms, although there are misspellings and sometimes widely differing placements of towns. In one case, Ortelius shows a town symbol and a name (*Manalba* in Madagascar) which Gastaldi does not. In all but two cases (*Rio de Infante* and the *Rio de turme* in Madagascar), the remaining 31 rivers bear names on the Ortelius map that can be matched with those on the Gastaldi map. Of the coastal names, only six cannot be traced back to Gastaldi. While Gastaldi names 17 of his lakes, Ortelius only keeps three of these names. Surprisingly, Ortelius does not name the Ptolemaic lakes of Zaire/Zembere in the west and Zaflan in the east. For a more complete comparison, refer to Bodenstein's excellent analysis of placenames (1998 in *Abraham Ortelius and the First Atlas*: 185-207).

However, unlike Gastaldi, Ortelius prepared a map surprisingly free of excessive ornamentation: Ortelius has three sea monsters on his map while Gastaldi has 20; Ortelius has a sea battle with three ships, Gastaldi has seven; Ortelius only has four text entries on the map, Gastaldi has numerous ones; Ortelius only uses a standard circle with a few buildings to denote a city or town, apart from a cross on Mount Sinai, whereas Gastaldi uses six different symbols; unlike Gastaldi, Ortelius uses neither compass roses nor a scale bar.

The sea battle on this map is copied in reverse from the very decorative Diego Gutiérrez wall map of the Americas of 1562 (Van den Broecke). On early examples of this map a 'ghost sea monster' is present in the sea east of the Arab peninsula which fades away after 1584.

Ortelius introduced two important changes to the shape of the continent: The Cape of Good Hope became more pointed and the eastward extension of the continent was significantly reduced by about 1,700 kms, to 7,000 kms. The distance from Ceuta to Cairo was also reduced by 1,600 kms., to 3,500 kms. Both of Ortelius' distances are in fact close to the actual measurements today: 6,800 kms. from east to west and 3,450 kms. from Ceuta to Cairo.

As well as the aforementioned sources, Ortelius corresponded regularly with Gerard Mercator in Duisburg, and likely used information from Mercator's famous 1569 wall map of the world. Ortelius probably also had seen Forlani's 1562 map of Africa. However, Ortelius disregarded Mercator's classical sources for the connection of the Niger River as a western branch of the Nile River; instead he used Gastaldi's representation of the river and lake hydrography of the African interior. A curious omission on Ortelius' map is the Mountains of the Moon, a prominent feature on most prior maps of Africa. It appears that Ortelius closely followed Gastaldi who, in turn, did not specifically include the

Mountains of the Moon on his 1564 wall map. Bodenstein (1998: 189) postulates that Gastaldi did not include the Mountains of the Moon so as to be more faithful to Ptolemy's view of the interior. In the earliest Ptolemaic maps and texts, the Mountains of the Moon are depicted as a continuous mountain chain at 12 degrees south, separating Africa known to the Ancients from terra incognita in the south. As more became known of the extent of the continent, the lake source for the Nile was progressively moved southward in Africa so that the Mountains of the Moon appeared to many mapmakers as a source for the *Spirito Santo* (Limpopo) and the *Cuama* (Zambezi) river basins to the southeast.

Another interesting representation on the map is the placement of the name *Zanzibar* in southwest Africa. It appears that Ortelius is using the term Zanzibar a traditional way to describe a people and area of East and Southern Africa. He uses the name *Zenzibar* as the placename for the island off the East African coast that we know now as Zanzibar.

Publication Information

This map is from Ortelius' *Theatrum Orbis Terrarum*. Initially published in 1570, the *Theatrum Orbis Terrarum* was the first atlas to purposely produce a uniform series of maps. The publication of the *Theatrum Orbis Terrarum* can be considered the starting point for over 100 years of Dutch supremacy in the production of detailed and decorative maps and atlases.

The map of Africa was engraved by Frans Hogenberg who was responsible for the engraving of many of the maps in the *Theatrum*. There was only one copperplate used to produce this map over the entire publishing life of the *Theatrum Orbis Terrarum* from 1570 to 1612 [1641]. It was re-engraved on a number of occasions, most notably in 1595-1598. At this time, a crack in the copperplate began on the title, showing initially on the map at the end of the first line of the title and progressively working its way down the lines of text on the right side of the title in subsequent editions.

The *Theatrum Orbis Terrarum* was first printed by Gielis Coppens van Diest at Ortelius' own expense from 1570 to 1573. Van Diest was succeeded by his son, Anthonis, who printed French and Latin editions in 1574. Gillis van den Rade next printed an edition of the *Theatrum* in 1575. Starting with the 1579 edition, Christoffel Plantin printed the *Theatrum*. Plantin and his heirs continued printing the *Theatrum* until Ortelius' heirs sold the copperplates and the publication rights to Jan Baptist Vrients in 1601. After 1612, when Vrients died, the copperplates and rights passed to the Moretus brothers who were the successors to the Plantin publishing firm. After the death of his brother, Balthasar Moretus intended to print another edition of the *Theatrum*. After numerous delays and close to death, in 1641 Balthasar paid Jan Galle to print 25 copies of each of 138 plates, including the one of Africa, after they had been re-worked for a last Spanish edition.

Although only one copperplate was used to produce this map over its lifetime, the map is known in six states.

Of importance, it should be noted that in the reproduction of the Africa map in the 1570 *Theatrum Orbis Terrarum* Latin edition, published as a facsimile by Nico Israel in 1964, the left and right gridlines have symmetrical hachuring. It was initially assumed by the author that this was a new state of the map of Africa. In fact, after a close examination of the actual map used for this facsimile in the Universiteitsbibliotheek Amsterdam (*Amsterdam OF*, 72-30), it became apparent that the photograph of the Africa map for the facsimile was retouched along the entire left side gridline due to defects on the actual map. In the course of this retouching, the left and right gridlines were inadvertently made symmetrical.

The following table may also assist in the identification of the edition for a particular loose map. The information for the dates for the *Theatrum*, publisher, and language are from Van der Krogt (2003: 51). The information regarding the text on the verso is from Van den Broecke (Cartographica Neerlandica website, http://www.orteliusmaps.com, date consulted July 2006):

Date	Van der Krogt no.	Publisher	Lang. on verso	Page # and Text on verso
1570 A	31:001	G. Coppens van Diest	Latin	page '4' (last line, centered like six lines above it, & Hieronymi Fracastorij ; diamond-shaped full stop after title.).
1570 B	31:001	G. Coppens van Diest	Latin	page '4' (last line, centered like six lines above it, Fracastorij. ; diamond-shaped full stop after title.),
1570 C	31:001	G. Coppens van Diest	Latin	page '4' (last line, centered like six lines above it, Fracastorij. ; circle-shaped full stop after title.),
1571	31:002	G. Coppens van Diest	Latin	page '4' (last line, centered like six lines above it:, Hieronymi Fracastorij),
1571/73	31:101	G. Coppens van Diest	Dutch	page '4' (last line, centered like one line above it, in cursive script: 'tueren secreten,hier mede besich laten,endt gaen voort onse landen beschryuen.'),
1572	31:301	G. Coppens van Diest	French	page '4'.
1572	31:201	G. Coppens van Diest	German	page '4'.
1572/75	31:290	Kohler ed.	German	
1572/3	31:211	G. Coppens van Diest	German	page '4' (last line, italic like the entire text, centered like two lines above it: hiermit arbaitten lassen,vnd ghen fort,vmb vnsere Lænder zubeschreiben.), ,
1573	31:011	G. Coppens van Diest	Latin	page '4' (last line, centered like five lines above it: nymi Fracastorij.),
1574	31:012	G. Coppens van Diest	Latin	page '4' (large page number, 11 mm, similar, but not identical to 1575L; 9th text line ends: Atlanticû ver ; last line, centered like four lines above it: Rhamusij,& Hieronymi Fracastorij.),
1574	31:311	G[A.] Coppens van Diest	French	page '4'.
1575	31:013	Gillis van den Rade	Latin	page '4' (small page number, 7 mm, similar, but not identical to 1574L; 9th text line ends: Atlanticum ver ; last line, centered like four lines above it: Rhamusij,&Hieronymi Fracastorij),
1579	31:021/022	C. Plantin	Latin	page '4' (last line, left aligned: Baptistæ Rhamusij,& Hiernonymi Fracastorij. ; 14th line from the top ends: Advertedum),
1580/89	31:221/222	C. Plantin	German	page '4' (last line, centered like seven lines above it: nymi Fracastorij.),
1581	31:321	C. Plantin	French	page '4' (last line, left aligned: re, & passons outre aux autres Pays.),
1584	31:031	C. Plantin	Latin	page '4' (last line, left aligned: nis Baptistæ Rhamusi, & Hieronymi Fracastorij.),
1587	31:331	C. Plantin	French	page '4' (last line, left aligned: fons outre aux Pays.),

Date	Van der Krogt no.	Publisher	Lang. on verso	Page # and Text on verso
1588	31:431	C. Plantin	Spanish	page '4' (last line, left aligned: Baptista Rhamusi, y Hieronymo Fracastorio.),
1592	31:041	Plantin Press	Latin	page '4' (text and page number, but not typesetting identical to 1595L; 13th line from the bottom ends: apud ; last line, left aligned: ximo,habes litteras Ioannis Baptistæ Rhamusi,& Hieronymi Fracastorij.),
1595	31:051	Plantin Press	Latin	page '4' (text and page number, but not typesetting, identical to 1592L; 13th line from the bottom ends: habes ; last line, left aligned: ximo,habes litteras Ioannis Baptistæ Rhamusi,& Hieronymi Fracastorij.),
1598	31:121	[Cornelis Claesz.]	Dutch	page '3' (last line, centered like three lines above it, Gothic script: landen beschryven.),
1598	31:351	Plantin Press	French	page '4' (last line, full width: en faire la diligéte recherche aux inquisiteurs des secrets de nature,& passons outreaux autres Pays.),
1601	31:052	Plantin Press	Latin	page '4' (13th line from the bottom ends: Medorumque ; last line, left aligned: ximo,habes litteras Ioannis Baptistæ Rhamusi,& Hieronymi Fracastorij.),
1602	31:451	J.B. Vrients	Spanish	page '4' (last line, left aligned: Hieronymo Fracastorio.),
1602	31:251	J.B. Vrients	German	page '4' (last line, centered like five lines above it: Hieronymi Fracastorij.),
1603	31:053	J.B. Vrients	Latin	page '4' (text and page number, but not typesetting, is identical to 1609/1612S/L; last line, left aligned: teras Ioannis Baptistæ Rhamusi,& Hieronymi Fracastorij.),
1606	31:551	J. Norton	English	page '4' (last line, left aligned: 'Ierom Fracastorius. Of' Africa 'likewise you may reade at large in the second volume of' M. Richard Hakluyts 'English voyage.'),
1608/12	31:452-454	J.B. Vrients	Italian	page '5' (last line, left aligned: con l'innondatione del Nilo, & del monte Sinai, doue caualcò in persona.),
1609/12	31:054/055	J.B. Vrients	Latin	page '4' (text and page number, but not typesetting, is identical to 1603L; last line, left aligned: litteras Ioannis Baptistæ Rhamusi,& Hieronymi Fracastorij.
1609/1612	31:452-454	J.B. Vrients	Spanish	page '4' (last line, left aligned: tierra,tienes las cartas de Ioan Baptista Rhamusi, y Hieronymo Fracastorio.).
1612/3	31:122	Plantin Press	Dutch	page '3' (text identical to 1598D).
1641		Plantin Press	Spanish	page '4' (Same as 1609/12S).

References
Van den Broecke (Cartographica Neerlandica website, http://www.orteliusmaps.com), map 8; Van der Krogt 2003: map 8600:31; Van den Broecke 1998: 29-54; Bodenstein 1998: 185-208; Norwich 1997: map # 10; Van den Broecke 1996: map # 8; Tooley 1969: 88.

13   c. 1573 Giacomo Gastaldi – Donato Bertelli, Venice

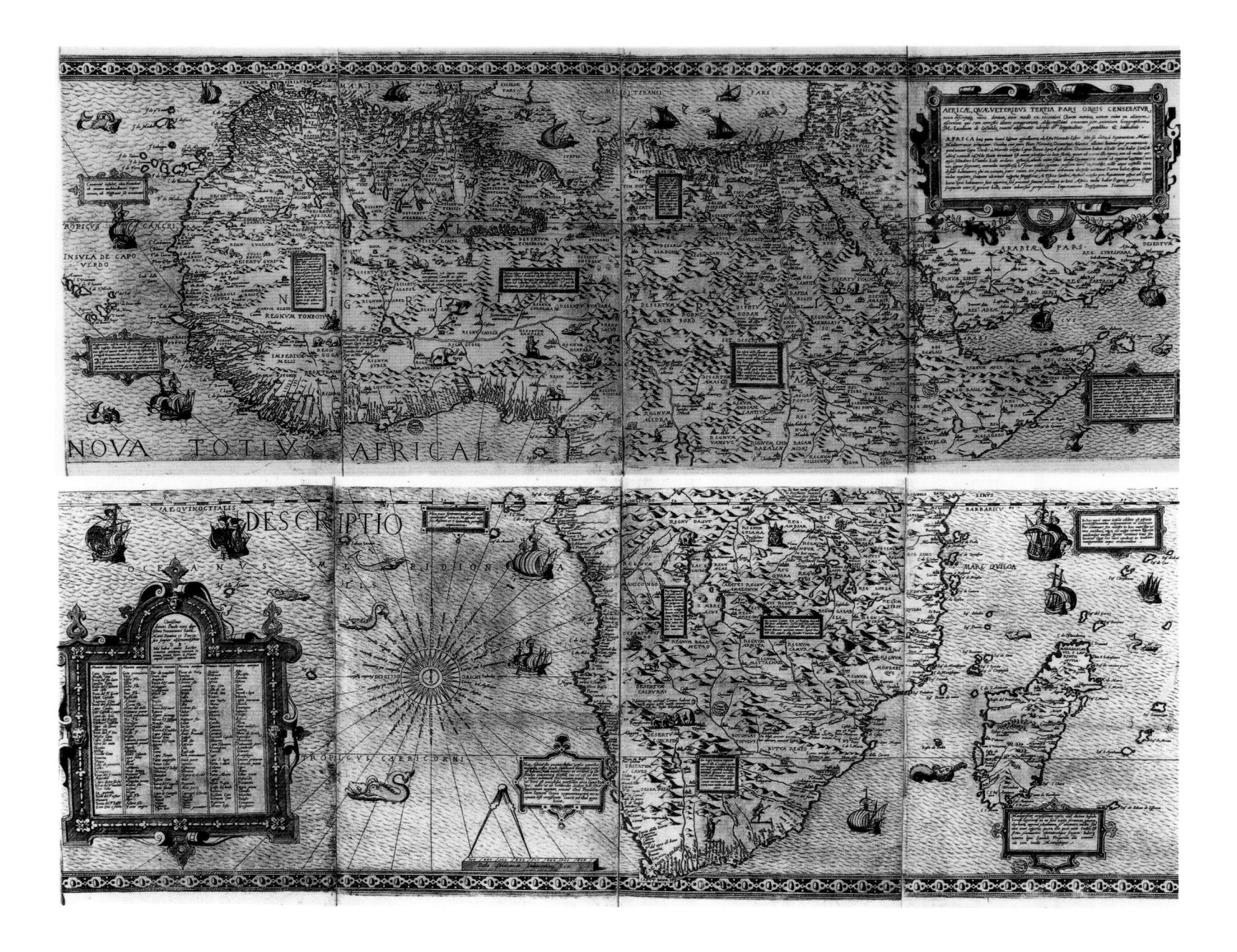

NOVA TOTIVS AFRICAE. [at the bottom of sheet 1 and 2]
DESCRIPTIO [at top of sheet 5 and 6].
[on scrollwork cartouche at upper right]:
Africae Quae Veteribus Tertia Pars Orbis Censebatur noua descriptio, nunc demum non modo ex recentiori charta marina, verum etiam ex aliorum assertione qui eam terrestri itinere peragrarunt, diligentissime exaratam per eximium Geographum M. Iacobum de Gastaldo, exacte obervatis ubique et longitudinis gradibus et latitudinis [and then ten lines of text describing Africa].
[the dedication is set within a cartouche on sheet 5]:
Clarissimo Domino Paulo Nani dignissimo Procuratori Sancti Marci Domino et Patron suo Semper observandissimo | D.B. ... [six columns of ancient and modern placenames] ...[and on bottom strapwork of cartouche] Ad signum Bibliothecae divi Marci Donatum Bertellum formis.
- Scale bar: Miliarium Italianorum 800 = 13.5 cm.
- Venice : Donato Bertelli, [c. 1573].
- From: separately published.

- 1 wall map : copperplate engraving ; 8 map sheets ; the top 4 sheets are joined together and the bottom 4 sheets are joined together. The top 4 sheets are 46.0 x 125.0 cm (trimmed at left and right sides). The bottom 4 sheets are 45.5 x 125.0 cm (again, trimmed on left and right sides). Total is 91.5 x 125.0 cm.
- Latitude coordinates are on sheet 2 (right side) and sheet 7 (left side). Longitude coordinates run across the Equatorial Line.
- North at the top; compass rose to left of Southern Africa on sheet 6.
- The title is at the bottom of sheets 1 and 2. It is not set within a cartouche.
- There are seven text boxes, ten ships, and seven sea monsters on sheets 1-4. There are seven text boxes, nine ships, and seven sea monsters on sheets 5-8.

Examples: There are two known examples of the c.1573 map as follows:
Venezia Correr, Carte geografiche a stampa del' 500 (folder no. 32).
The set in the Museo Correr is in a folder. The shelf mark for the Africa map is Cartografia cartella 32/68. In the Correr example, the top 4 sheets are joined together and the bottom 4 sheets are joined together. Most of the maps in the folder were obtained as a gift from Girolamo Ascanio Molin in the early nineteenth century, and the sheets of the four continent maps do not occur together, but at different places in the folder (Europe, no. 4/Asia, 67/Africa, 68/America, 70), though they must have originally been part of one set. Gallo (1950: 100); Gallo (1954: 6, 34, 51-52).
Firenze BML. This example was part of the Ashburner Collection, which was purchased by the Biblioteca Medicea Laurenziana, Florence (Karrow 1993). The author was not able to locate and examine the Florence example.

Variant A c.1662
This variant of the map is readily identified by the inclusion of a separately printed title strip across the top of the map: 'NOVA ET ACVRATA TOTIVS AFRICAE TABULA, auct: G.I.' and the imprint of 'Stefano Scolari forma'.

This variant of the map was published by Stefano Scolari (Sims 2003). Scolari acquired the Donato Bertelli copperplates via Andreas Bertelli and then finally from Francesco Vallegio or Valesio (Sims 2003, Borroni-Salvadori, 1980).

This variant is known in three examples as follows:
Austin HR, cartography collection 12. This example is part of a complete set of the four continents (Kraus 1970). Burden (1996) dates the America's map in the set as c.1655 as the Asia map in the set carries that date of 1655.
New York, Graham Arader Galleries. This example is also part of a set of the four continents. The Africa map is dated c.1662, based on the date of 1662 on the Asia map.
New York, Graham Arader Galleries. This example is part of a three continent set (missing the Europe map). It is dated c.1662, again based on the date of 1662 on the Asia map.

Description
This map is a faithful re-engraving of Giacomo Gastaldi's wall map of 1564 published by Licinio (Map # 9) and is part of a four-continent set. Like the Gastaldi-Licinio map, it was also produced on eight sheets. This map was published by Donato Bertelli and has a dedication to Paolo Nani. The engraver is not known.

The figure of Prester John is at the top of sheet 7, far to the south of Abyssinia between the two central Nile Lakes of Zaire/Zembere on the west and Zaflan on the east with the notation 'In hoc regno...est Regio Sedes Presbiteri Ioanis'. The *Zembere* River exits Lake *Zembere* and divides into the Cuama and Spirito Santo Rivers. Bertelli follows Gastaldi by not specifically identifying the Mountains of the Moon south of the two central African lakes. However, Bertelli does show a much larger *Mons Betsum*, south of *Lago de Zaflam*. A traditional depiction for the Niger River flowing westward to the Atlantic Ocean is shown on this map.

The map contains a dedicatory text box to Paolo Nani, procurator of St. Mark's cathedral in Venice, on sheet 5 of the map.

Publication Information

This map is not dated, but the date of c. 1573 is assigned here based on the dedication on it to Paolo Nani, who became procurator of St. Mark's Cathedral in Venice on November 22, 1573 (Sims 2003). Quoting Karrow (1993: 244), according to Caraci, the map was supposedly published sometime between 1564 and 1566, but this appears to be too early.

Bertelli was active in Venice as a publisher and mapseller from c.1558-1592 under the shop imprint of 'all'insegna di S. Marco' or 'Ad signum Bibliothecae divi Marci'. He assumed the business of his relative Ferando Bertelli by 1574 (Almagia 1948: 117-18).

References

Sims 2003; Burden 1996: map # 37 (Americas map); Karrow 1993: 244 (30/98.4); Borroni-Salvadori 1980; Kraus 1970: 26; Gallo 1954; Gallo 1950; Caraci 1927: 29:178-92.

Map 13 (detail).

14   c.1573 Gerard de Jode, Antwerp

14.1 First State c. 1573

Africæ ut terra mariq[ue], lustrata est, proprijßima | ac verè genuina descriptio, obseruatis ad unguem gra= | dibus longitudinis et latitudinis. Autore M. Iacobo | Castaldo .
[at bottom right]:
Ioannes a Deutecum | Lucas a Deutecum | fecerunt.
- No scale bar.
- [Antwerp : Gerard de Jode, c.1573].
- From: separately published / *Speculum Orbis Terrarum* (in second state).
- 1 map : copperplate engraving ; 34.0 x 47.5 cm.
- Latitude coordinates only along the left side of map. Longitude coordinates along Equatorial Line.
- North at the top; there is an elaborate, large windrose at the bottom left with inscriptions for all the names of the wind directions.
- Title is at the top right set within a strapwork cartouche.
- There are two ships and one sea monster in the Atlantic Ocean, and one ship and one sea monster in the Indian Ocean.

Example: *Rotterdam MM*, WAE 739 (as a separate map).

14.2 Second State 1578
Addition of cum privilegio in the title cartouche. Addition of Latin text on verso.

Examples: *Amsterdam UB*, 1806 A 20; *Rotterdam MM*.

Map 14.2 Title cartouche.

Description

This is an exceedingly rare, folio-size map of Africa by Gerard de Jode. It is clearly based on the famous eight-sheet Giacomo Gastaldi wall map of Africa of 1564. Within the title, De Jode recognizes Gastaldi as the author of this work (Autore M. Iacobo | Castaldo). As De Jode was represented at the important Frankfurt Book Fair where he bought and sold maps, he possibly obtained a copy of the Gastaldi wall map of Africa during one of these fairs. Other than reducing Gastaldi's heavily detailed, eight-sheet map to one folio-sized map, thus omitting numerous text passages and placenames, the basic outline for Africa and its hydrographical and topographic features are the same.

Publication Information

Gerard de Jode was born in Nijmegen in 1521 and died in Antwerp in 1591. In 1547 he was admitted to the Guild of St. Luke and became a print seller. In 1550 he was licensed as a printer. Among his output were wall maps of the continents: Africa, based on Gastaldi, in 1569 and no longer extant; America in 1576; Asia in 1577; and Europe in 1584. He also seems to have issued a number of separately published, folio-size maps without a publisher's privilege or copyright (cum privilegio).

Koeman (1967-71) theorizes that De Jode may have been inspired to issue his own atlas, based on the immediate success of Abraham Ortelius' *Theatrum Orbis Terrarum*. It is therefore likely that this map of Africa may have been prepared sometime after 1570, but before 1578, with the intention that it would eventually be included in his own atlas pending De Jode's receipt of a publisher's privilege intended to prevent unauthorized copying of his maps.

Although De Jode attempted to receive a royal privilege starting in about 1573, he was not to be granted his royal privilege until 1577. There is some belief (Denucé 1912-13: Vol. I: 169) that Ortelius used his well-connected supporters to prevent the issuance of De Jode's privilege so as to protect the publication of his own *Theatrum Orbis Terrarum* of 1570. Prior to 1578, it appears that De Jode printed a number of separate, folio-size maps without text on the verso while he waited for his privilege. According to Van Ortroy in Skelton's introduction to the facsimile edition of 1965, there are certain entries by Plantin from 1567 to 1576 that mention the sale of separate, folio-size maps by De Jode.

It is likely that the second state of De Jode's map of Africa, with the addition of the 'cum privilegio' at the bottom right within the title cartouche below the De Jode imprint, was prepared just before the publication of De Jode's *Speculum Orbis Terrarum* in 1578.

The map was engraved by Johannes and Lucas van Deutecum (with the imprint at bottom right of: *Ioannes a Deutecum | Lucas a Deutecum | fecerunt.*).

De Jode's atlas was not a financial success and no other editions were published. In total, a dozen or so examples of his atlas are said to have survived. Consequently, all maps by Gerard de Jode are quite rare.

A revised map of Africa, using a new copperplate, was issued in 1593 by his son, Cornelis. This new map was included in Cornelis De Jode's' atlas of 1593, the *Speculum Orbis Terrae*.

References

Van der Krogt 2003: map 8600:32A; Koeman 1967-71: II, Jod 1 (3); Skelton 1965b; Van Ortroy 1963; Tooley 1939: 3: 12-21 (described as map # 69); Denucé 1912-13: Vol. I:169.

15   1573 Giovanni Lorenzo D'Anania, Naples

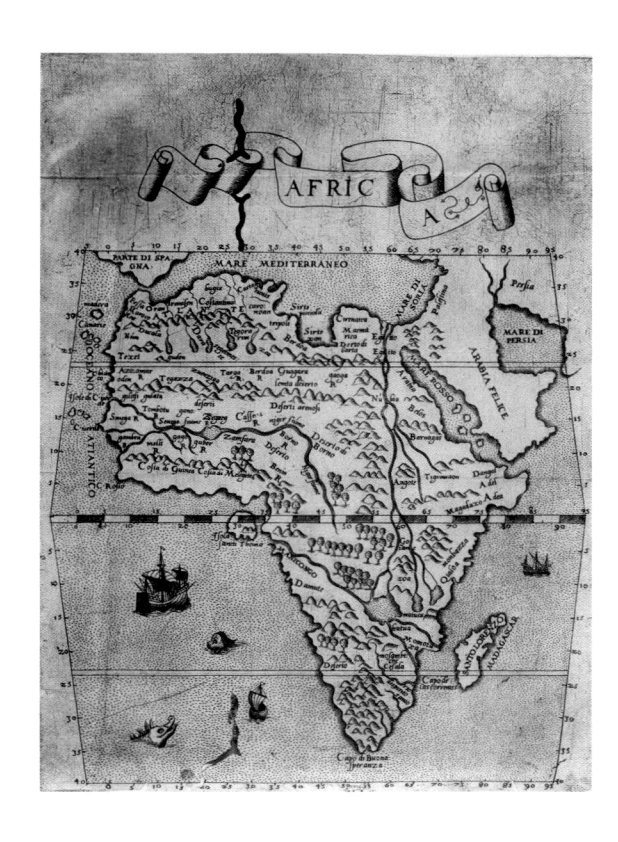

AFRIC | A.
- No scale bar.
- [Naples : Gioseppe Cacchii dell'Aquila, 1573].
- From: *La Universale Fabrica del Monde overo Cosmograpfia*.
- 1 map : copperplate engraving ; 20 x 18.5 cm.
- Latitude and longitude coordinates on all four sides of the map. Longitude coordinates along Equatorial Line.
- North at the top; no compass rose.
- Title on scroll above the map.
- There are two ships and two sea monsters in the Atlantic Ocean and one ship in the Indian Ocean.

Examples: *Madrid BM*, CCPB000007114-5; *Private German collection*. There is a copy of the 1573 edition at the New York Public Library (*New York PL*, Rare Books 02-154) that does not contain maps. There was mention in the literature of an an example of this map at the Library of the Palacio Real, Madrid, but this is not correct.

Description
This map is by Giovanni Lorenzo D'Anania (c.1525-c.1582). D'Anania was the noted author of *La Universale Fabrica del Monde...*, which made its first appearance in Naples in 1573.

The map has a trapezoidal projection which was typical of the early Ptolemaic maps of Africa. The cartographic sources for this map are not readily apparent. Elements of the Ramusio-Gastaldi 1554 map are present in the use of placenames. Unusually, this map has three lakes in Central Africa rather than the traditional two of Ptolemy. These are a larger western lake and, to its east, two smaller lakes all on the same general latitude. The far eastern lake may be a misplaced representation of the lake that appears in maps of the period by Ortelius and others on the Equator; it later was eventually moved northward to become the Blue Nile River source, Lake Tana, in Abyssinia.

Publication Information
The map only appeared in the first edition of D'Anania's *La Universale Fabrica del Monde* which was published in Naples by Gioseppe Cacchii dell'Aquila. This map of Africa is one of a set of the four continents. It is extremely rare and is known by the author in only two examples.

The second edition of *La Universale Fabrica del Monde*, published in Venice in 1576, is not known to contain any maps (Examples: *London BL*, 1295.l.9; *Chicago NL*, Ayer 7 A65 1576). Starting with the 1582 edition of *La Universale Fabrica del Monde*, published in Venice, a new map attributed to D'Anania was used.

References
Correspondence with Professor Dr. Fritz Hellwig, who kindly provided the author with further information and a photograph of this map; Burden 1996: map # 43 (Americas map); Hellwig 1994: 105-120.

16   1575 François de Belleforest, Paris

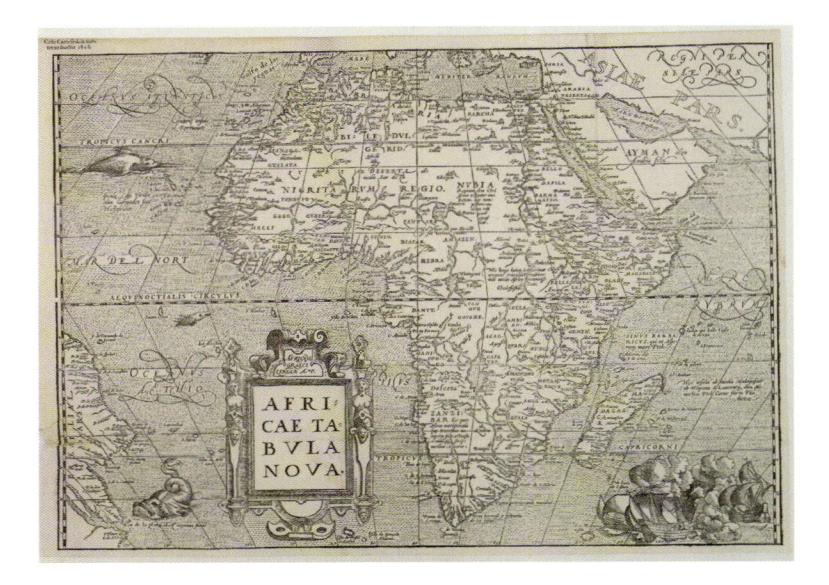

AFRI= | CAE TA= | BVLA | NOVA.
- No scale bar.
- [Paris : Michael Sonnius, 1575].
- From: *La cosmographie universelle de tout le Monde.*
- 1 map : woodcut ; 37 x 50 cm.
- No latitude and longitude coordinates.
- North at the top; no compass rose.
- Title at bottom left of the map within a decorative strapwork frame.
- The decorative elements on the map closely copy Ortelius' 1570 map: similar title cartouche, three sea monsters in the Atlantic Ocean, and three ships engaged in battle in the Indian Ocean.

Examples: *London BL*, 568.h.7 (with Sonnius imprint); *Washington LC*, G113 .M74 1575 (with Sonnius imprint); *Various other locations*. The Africa map is not in the example at the New York Public Library.

Description
This map is from a French translation in 1575 of Sebastian Münster's *Cosmographia* by François de Belleforest (1530-1583). De Belleforest did not choose to use a copy of the Münster map of Africa, but instead prepared a new map, reflecting the lessening importance of Münster by this time. The De Belleforest map, though a woodcut, is a close copy of the more popular Abraham Ortelius map of Africa and has the same title cartouche and decorative scenes in the oceans as on the Ortelius map.

Publication Information
The map only appeared in De Belleforest's 1575 *Cosmographia*. It was usually folded into the book and appeared in Book 16 on page 1868 of the second of three volumes. In the upper left corner above the map image is the bookbinder's instruction: *Ceste Carte-doit met-/tre, au fueillet 1868*. There is no text on the verso of the map.

According to Pastoureau (1984), the same map appeared in two different editions of De Belleforest's *Cosmographia* in 1575. 'Belleforest A' was published by Nicolas Chesneau and 'Belleforest B' was published by Michael Sonnius, with their respective imprints on the title pages of their books. The map was unchanged in both editions.

References
Pastoureau 1984: 55-56; Pastoureau 1980: 51-54.

17   1575 André Thevet, Paris

TABLE D'AFRIQVE.
- Scale bar (set within decorative element containing title):
  Lieues Italiques 600 = 3.8 cm.
  Lieues Françoyses 200 = 3.8 cm.
  Lieues Marines 100 = 3.0 cm.
- [Paris : Pierre l'Huilier and Guillaume Chaudiere, 1575].
- From: *La cosmographie universelle d' Andre Thevet.*
- 1 map : woodcut ; 35 x 45.5 cm (to outer black line beyond grid).
- Latitude and longitude grid surrounds the map on all four sides. At the bottom right corner of the map there is a notation for degrees of latitude (Degrez de haulteur) and longitude (Degrez de largeur).
- North at the top; no compass rose.
- Title is set within a decorative cartouche at the top right of the map. At the bottom left corner of the map is a text block within an undecorated box with the following: 'A. Thevet Cosmographe du Roy' with eight lines of text ending with 'Cosmographique'.

- There are 15 ships and eight sea monsters in the Atlantic Ocean, five ships in the Mediterranean Sea, three small ships in the Red Sea, and six ships and three sea monsters in the Indian Ocean.

Examples: Pierre l'Huilier 1575 edition: *Cambridge Harv RB,* Typ 515 75.831; *Chicago NL*; *London BL*, 568.H.3 1575; *Washington LC*: (also contains third state of Americas map, according to Burden); *Private US Collection*;
Guillaume Chaudiere 1575 edition: *New York PL*, Arents collection number 21.

Variant A 1581
There is a variant of this map in Thevet's *La cosmographie universelle*, published by Guillaume Chaudiere in 1581. In this variant, there is letterpress below the map impression as follows: 'A Paris, chez Guillaume Chaudiere, Rue S. Iaques, à l'enseigne du Temps & de l'Homme Sauuage. 1581.'

Examples: *Paris BN*, Ge D 11360; *Bedburg-Hau, Antiquariat Gebr. Haas.*

Description
The author of this map is André Thevet who was born in 1504 or 1516/17 and died in 1592 (Karrow 1993: 529). Due to numerous contradictions in Thevet's own writings of his life, as well as fanciful additions in his texts, there is some confusion over basic facts associated with his life and with his travels. It is known that Thevet traveled extensively in Europe. Unlike most cartographers of the time, he visited the Near East and North Africa, and is even reputed to have made several voyages to the Americas. From 1558, Thevet was royal cosmographer to the French kings. He is best known as the author of *La cosmographie universelle*, and the four continent maps that are found in this book.

This is an important map in the progression of the mapping of Africa. It is the first folio map to show Africa largely based on the Mercator model and not solely that of the Gastaldi-Ortelius model. It is modeled on the Africa section of Gerard Mercator's famous world map of 1569. Following Mercator, a major lake is shown to the southwest of the traditional twin Ptolemaic Nile source lakes. This third lake, the unnamed Lac Sachaf, is now shown feeding the river system to the south, notably the Cuama and Spirito Santo Rivers. This same lake is the source for the Congo River to the west and also flows into the Nile to the north. Above the Niger River in West Africa, Thevet follows Mercator by placing a river, the *Gher-Nubie*, that flows eastward into the Nile River.

Numerous, well-detailed placenames cover the map. *Cefala* (Sofala) is to the south of the *Cuama* River. The island of *Zanzibar* off the East African coast is relatively well-placed. *Cape de Corrientes* is on the east coast by the Tropic of Capricorn.

The esthetics of Thevet's map is hindered by using a woodcut to portray considerable cartographic detail and decorative embellishments. This results in a map that, while displaying an enormous amount of cartographic information, is somewhat difficult to read. The engraver for the map is not known.

Publication Information
The map, along with maps of the other three continents, appeared in two issues of *La cosmographie universelle* in 1575, one published by Pierre l'Huilier and one also by Guillaume Chaudiere. In the 1575 edition of Thevet's *La cosmographie universelle* published by Pierre l'Huilier, the map of Africa is in Volume I. The verso of the map is blank. Burden (1996) notes a difference at the bottom of the cartouche on the Americas map between the 1575 Chaudiere and l'Huilier editions of the book. However, in the Chaudiere edition (*New York PL*, Arents Collection # 21), the map of Africa and the cartouche are unchanged from the map that appears in the l'Huilier edition.

Chaudiere issued the third, and final, printing of this map in 1581. This is a rare map. Burden (1996) only cites a total of six examples of the 1581 Americas map; the author has only seen two examples of the 1581 Africa map: one at the Bibliothèque Nationale de France in Paris, and one with the German antiquarian book and map dealers, Antiquariat Gebr. Haas.

References
Norwich 1997: map # 12; Burden 1996: map # 46 (Americas map); Karrow 1993: 529-46; Tooley 1966: 26.

18   1577 Abraham Ortelius – Filips Galle (1), Antwerp

Africae | ta bula nou | a.
- No scale bar.
- [Antwerp : Christoffel Plantin, 1577].
- From: *Spieghel der Werelt*.
- 1 map : copperplate engraving ; 8 x 11 cm.
- No latitude and longitude coordinates.
- North at the top; no compass rose.
- Title, at bottom left, set within a three dimensional box-like cartouche with a handle at the top.

Examples: 1577 Dutch language edition: *Leiden UB*, 1192 H 24; *New York PL*, *KB 1577; 1593 Italian language edition: *New York PL*, *KB 1593. 1585 Latin language edition: *Washington LC*, G8200.1585.F (with lines of text about the map with 'AFRICA, 4' with ten lines of text below and 'B' and 'Nigrita').

Description

This map is the first in a series of numerous miniature maps of Africa that were prepared for a reduced-size version of Abraham Ortelius' extremely popular *Theatrum Orbis Terrarum*.

It follows Ortelius' folio-size map of Africa, but there are changes. Numerous placenames and particularly geographic features had to be omitted from this map, owing largely to its small size in comparison with Ortelius' 1570 folio-size map.

One major omission on this map was the removal of the smaller of the two Ptolemaic source lakes (the eastern lake of Zaflan) for the Nile River. The primary source for this change appears to be a very strict interpretation of João de Barros. De Barros in his first *Década*, published in 1552 in Lisbon and as an Italian abridgement in Ramusio's second edition of his *Delle Navigationi* in 1554, wrote the most detailed description of the hydrographical system of Central Africa with his account of a vast central lake (not lakes) as the source for the Nile, the Zaire (Congo) and the Zambere, which split to become the Cuama (*Zuama fl.* on this map) and Espirito Santo Rivers (see Relaño 2002: 207-8, for a further discussion of this point).

Also, Zanzibar is removed from the southwest coast where the name was placed in the Ortelius folio map. This map does make *Zambro* and *S[Z]imbaoe* more apparent, signifying an increased awareness of the Zimbabwe gold regions in Southern Africa.

Publication Information

As a result of Abraham Ortelius' enormously popular *Theatrum Orbis Terrarum*, a more affordable and convenient, pocket-size version was produced by the Christoffel Plantin publishing house in Antwerp. The purpose of this atlas was to make the *Theatrum* more available to the general public, particularly the expanding, business-oriented middle class.

The first reduced-size atlas of 1577 was titled *Spieghel der Werelt* (Mirror of the World). Later editions of this atlas had the names, *Le Miroir du Monde, Theatrum Orbis Terrarum Enchiridion,* and finally the newer *Epitome du Théâtre du Monde* in 1588. The name, *Epitome*, has come to be commonly used in more recent times for all of the Ortelius pocket-size atlases.

Filips (Phillipe) Galle (1537-1612), an engraver, printseller, and editor, is credited with engraving this map of Africa and the other plates in the atlas, though he was not an accomplished cartographer. Galle, and to some degree Pieter Heyns, is also credited with the idea of producing this atlas in the first place. This was a major undertaking as a total of 70 maps were involved. It is therefore assumed that Galle began this project some time before 1577. Pieter Heyns, a poet and schoolteacher in Antwerp, wrote the text in rhyme.

This book proved to be as popular as the folio size *Theatrum Orbis Terrarum* and it went through eleven editions. For the first time, an affordable atlas, with a description of the world with accompanying maps, was made easily available to the European middle class.

The map itself is unchanged in the various editions. With the exception of the Latin edition of 1585, which was on a page of text, all of the maps are oblong in shape with the maps on the recto of the right hand page facing corresponding text on the verso of the left hand page.

The following chart can be used to differentiate the Ortelius-Galle (1) Africa map in the various editions of the book in which the map appeared using the language of the text, signature, and page number. This information is from Van der Krogt (2003):

Date	Van der Krogt no.	Atlas Title	Language	Signature	Pg no.
1577	331:01	*Spieghel*	Dutch	C1r	5
1579	331:11	*Miroir*	French	A6r	5
1583-96	331:02/03	*Spieghel*	Dutch	C1r	5
1583	331:12	*Miroir*	French	B2r	5
1585	331:21	*Enchiridion*	Latin	B1r	7 (on a page of text)
1588-90	332:01/02	*Epitome*	Latin	A4r	4
1589	332:11	*Epitome*	Latin	B4r	4
1593	332:21	*Theatro*	Italian	B3r	3

References

Van der Krogt 2003: map 8600:331; Relaño 2002: 207-8; Norwich 1993: 15: map # 10a; Koeman 1967-71, III: Ort 47-55.

19   1579 Claudio Duchetti, Rome

Map 19.2.

19.1 First State 1579

[Map of Africa].
[within strapwork cartouche at lower left]:
Il disegno della Geografia moderna de tutta la parte' dell' Africa j confini della | quale stanno in questo modo. da ponente il mar oceano Computate l'isole. di capo Ver= | de et le canarie da Tramontana il stretto de Gibelterra & il mare meditterraneo, | da Siroco vna linea che principia a feramida insino al Sues & dal Sues per il mare | Rosso da leuante il mare oceano includendo l'isola di S.to lorenzo insino al capo di | Bona Speranza dall 'ostro il mare oceano, Graduata in longhezza & in larghezza. | Claudio ducheto exc. Lanno. 1579. | Henricus honius Harlemensis sculpsit.
- No scale bar.
- [Rome] : Claudio Duchetti, 1579.
- From: separately published / Italian composite atlas.
- 1 map : copperplate engraving ; 2 map sheets ; 44 x 60 cm in total.
- Latitude coordinates along left and right sides of the map inside the grid. Longitude coordinates along Equatorial Line, and also outside the grid at both the top and bottom of the map.

- North at the top; compass rose within Northern Indian Ocean.
- The title, in the form of descriptive text, is set within strapwork cartouche at bottom left.
- No ships, sea monsters, or animals on the map.

Example: *Birmingham CL,* Ref. AE 094/1580 (map no. 56 within an Italian composite atlas; part of the Cadbury Early and Fine Printing Collection).

19.2 Second State after 1585

With the addition of the new publisher and printer, *Petri de Nobilibus Formis* (Pietro de Nobili) within the frame at the bottom of the cartouche. His name is repeated at the bottom right side of the map: *Petri de Nobilibus Formis.* In all other aspects, the map is similar to the first state.

Example: *Johannesburg PL,* S Map 912(6)'1579'.

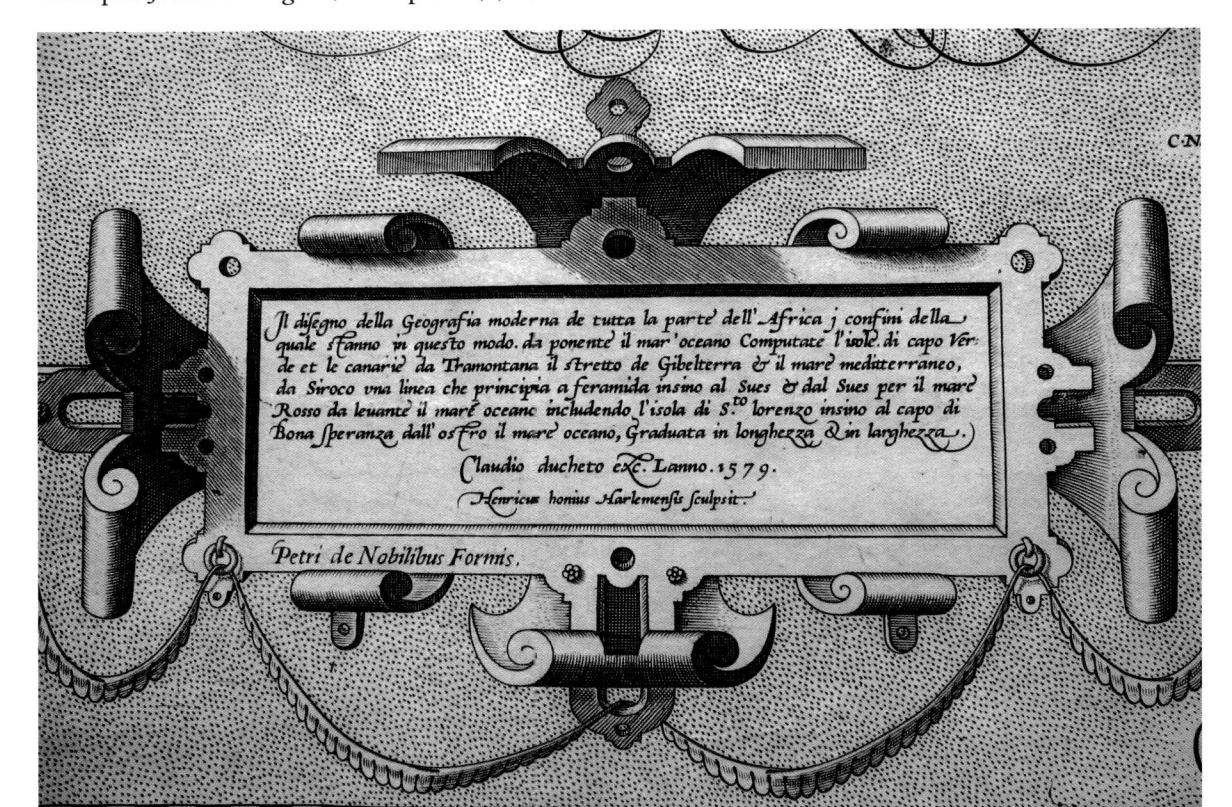

Map 19.2 Title cartouche.

Description

This is an elegantly engraved map in the Forlani-Lafreri tradition by Claudio Duchetti, who was active in Rome from 1554 to 1585. Duchetti was born in France under the name Claude Duchet. He was the nephew of and part successor to Antonio Lafreri's publishing firm upon the latter's death in 1577. Upon Duchetti's own death, the copperplates for his maps were sold to a variety of publishers.

The geography of the map appears to be a close copy of the Giovanni Francesco Camocio - Paulo Forlani map of 1563. Evidently, Duchetti acquired, or at least had access to, this Camocio-Forlani 1563 map of Africa published in Venice, which served as his model. Duchetti updated the Camocio - Forlani map with a new title cartouche. He also removed all ships and sea monsters, and replaced the image of Noah's Ark with a decorative compass rose in the Indian Ocean. The palace of Prester John is shown in East Africa just above the Equator.

The inscription within the cartouche is in Italian and describes Africa. Following this is the imprint of the mapmaker and publisher, Claudio Duchetti (*Claudio ducheto exc. L'anno. 1579.*), and the engraver, Henricus Honius (*Henricus honius Harlemensis sculpsit.*).

Publication Information

This rare map is known in two states. Only one example of each state was located by the author, though other examples may exist in various Italian composite atlases and other irregular atlases.

Norwich wrote that there is another 'state' of this map (Norwich 1997: map # 13) engraved by Henricus Hondius. The reference to the engraver on the map, 'Henricus honius Harlemensis sculpsit', does not refer to Hondius. The reference is to Henricus Honius, an engraver from Haarlem in The Netherlands who worked in Italy for Duchetti (Tooley 1999-2004, E-J: 366).

Upon Duchetti's death in 1585, much of his map stock was dispersed. Pietro de Nobili is known to have acquired and re-issued some of these plates (Tooley 1999-2004, K-P: 327-328). It appears that De Nobili acquired this Duchetti plate and added his own name (in two places) before re-issuing it, sometime after Duchetti's death.

The author has only been able to locate one example of the second state of this map at the Johannesburg Public Library in Johannesburg, South Africa. This example is as a separate map. The map may also exist in Italian composite atlases, though an example has not been found in the various composite atlases that have been examined. The example of this map depicted in the Norwich book is not part of the Norwich collection, now at Stanford University in California. Tooley (1966: 31) states that three examples of the second state were known to him in the Beans-Gotha collection, the forementioned Johannesburg example, and another in private hands.

References

Tooley 1999-2004, K-P: 327-328, and E-J: 366; Norwich 1997: map # 13; Karrow 1993: 245 (30/98.8); Tooley 1966: 31 (map # 127 for first state and map # 128 for second state with image as plate # VIII); Tooley 1939: 3: 12-21 (described as map # 75 for first state and map # 76 for second state).

Special acknowledgement to Professor Elri Liebenberg, Emeritus Professor of Geography at the University of South Africa, who located the second state of this map for the author.

20 1582 Giovanni Lorenzo D'Anania, Venice

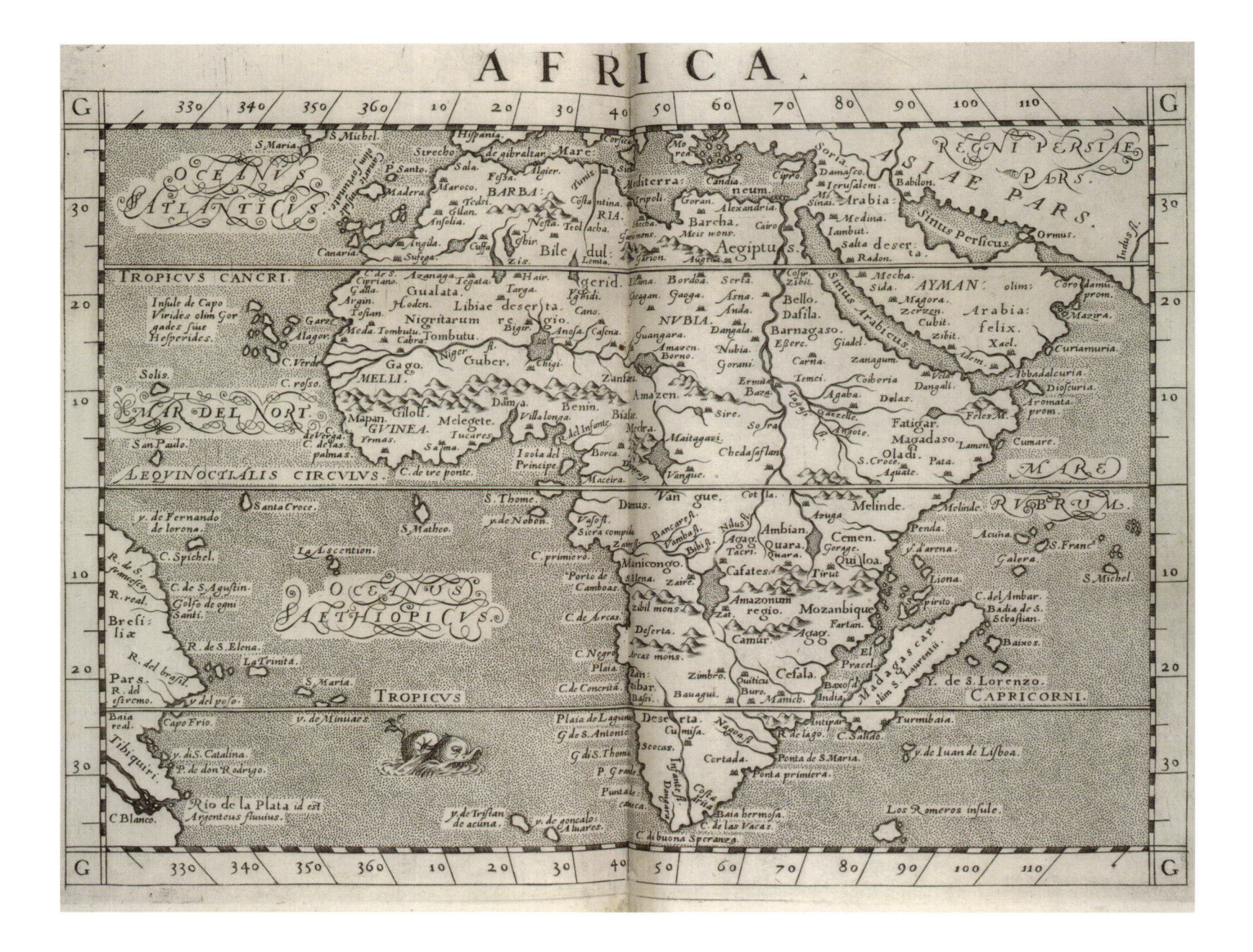

20.1 First State 1582

AFRICA.
- No scale bar.
- [Venice : Andrea Muschio, 1582].
- From: *L'Universale Fabrica del Monde overo Cosmograpfia*.
- 1 map : copperplate engraving ; 17.5 x 24.5 cm.
- Latitude and longitude coordinates on all four sides of the map.
- North at the top; no compass rose.
- Title above map image.
- There is one sea monster in the Atlantic Ocean.
- *Aegiptus.* has the letter 's' to the east of the Nile River. This map has *Dama* in West Africa above the Guinea Coast (not *Dauma* as on the Botero map found in the 1596 edition of D'Anania's *L'Universale Fabrica Del Monde)*. 'G' in the four corners of the map.

Examples: *Cambridge Harv MC,* MAP-LC G120 .A53*; *London BL,* 566.d.29; *Washington LC,* G120 .A53.

20.2 Second State 1598
Addition of an engraved line completely encircling Africa. This map is distinguished from the 1640 Botero-Giunti map, which also has an engraved line encircling Africa, by the inclusion of a period '.' in the shape of a triangle after the title of 'AFRICA' on the D'Anania map.

Example: *Author's collection* (separate map).

Description
Giovanni D'Anania (c.1525-c.1582) was the noted author of *La Universale Fabrica del Monde...* , first published in Naples in 1573. This map is from the third edition of D'Anania's *La Universale...* of 1582, published in Venice.

The geography and placenames on the map are closely based on Abraham Ortelius' 1570 map of Africa. For example *Zanzibar* is still on the southwest coast of Africa as on the Ortelius map. The hydrographical systems follow Ortelius precisely with three lakes feeding the Niger River. The sources for the other major rivers on the continent are also the same as on Ortelius' map.

Publication Information
The history of this map is long and complicated. The Africa map is similar in appearance to Giovanni Botero's map of 1595 (Map # 32) and Giovanni Magini's map of 1596 (Map # 34), which further complicates its history. What makes it extremely difficult to precisely differentiate this map and the very similar Botero map of 1595 is that both maps often seem to have been used interchangeably in the numerous editions of Botero's *Le Relationi*.

This map appeared for the first time in the 1582 edition of D'Anania's *L'Universale Fabrica del Monde overo Cosmograpfia* published in Venice. The first edition of D'Anania's *L'Universale Fabrica del Monde...* was published in 1573 in Naples and it contains maps of the continents with a different map of Africa (Map # 15). There was a second edition of D'Anania's *L'Universale Fabrica del Monde...* published in Venice in 1576 (*London BL,* 1295.l.9; *New York PL,* *KB 1576) which does not contain any maps. There was also an edition of D'Anania's *L'Universale Fabrica del Monde overo Cosmograpfia* published in 1596 that contained an example of Botero's 1595 map.

This D'Anania map was re-issued in its second state in Giuseppe Rosaccio's expanded edition of Girolamo Ruscelli's *Geografia di Claudio Tolomeo* in 1598 and again in 1599. The author has also seen this map in its second state in an edition of Botero's *Le Relationi* of 1618 published in Venice by Vecchi (*London BL,* G.6749).

First published in 1561, the earlier editions of Ruscelli's *Geografia di Claudio Tolomeo* did not contain a map of the continent of Africa and only contained separate maps of Northern and Southern Africa. The publisher for both editions of Rosaccio's expanded edition of Girolamo Ruscelli's *Geografia di Claudio Tolomeo* was the Heirs of Melchior Sessa. The only difference between the D'Anania second state map found in these 1598 and 1599 Rosaccio editions from D'Anania's 1582 map is the addition of a line completely encircling Africa. Otherwise, the plate is the same.

There are a number of similar appearing maps, using different copperplates, derived from this D'Anania map. These maps and their distinguishing characteristics are as follows:

Map # 20	1582	D'Anania	17.5 x 24.5 cm	*Aegiptu* to left of Nile and *s.* to right. *Dama* in West Africa above the Guinea Coast.
Map # 32	1595	Botero	17.5 x 24.5 cm	*Aegipt* to left of Nile and *us.* to right. *Dauma* in West Africa above the Guinea Coast.
Map # 34	1596	Magini	12.5 x 17 cm	*Aegiptu* to left of Nile and *s.* to right. *Dauma* in West Africa above the Guinea Coast.
Map # 35	1597	Keschedt	12.5 x 17 cm	*Aegiptus* all to left of Nile. *Dauma* in West Africa above the Guinea Coast.
Map # 43	1600	Africanus-Pory	17.5 x 24 cm	*Aegipt* to left of Nile and *us* to right. *Dauma* in West Africa above the Guinea Coast. No *G* in all four corners of outer gridline.
Map # 51	1605	Rosaccio	19 x 26 cm	*Aegiptu* to left of Nile and <u>no</u> *s* to right. No *Dama* nor *Dauma* in West Africa above the Guinea Coast.
Map # 69	1640	Botero-Giunti	17.5 x 25 cm	*Aegiptu* to left of Nile and *s* to right. *Dama* in West Africa above the Guinea Coast. A line appears around Africa in the ocean. <u>No</u> period in the shape of a triangle after the title of *AFRICA*.

References
Norwich 1997: map # 22; Hellwig 1994: 105-120.

21 1588 Sebastian Münster – Sebastian Petri, Basel

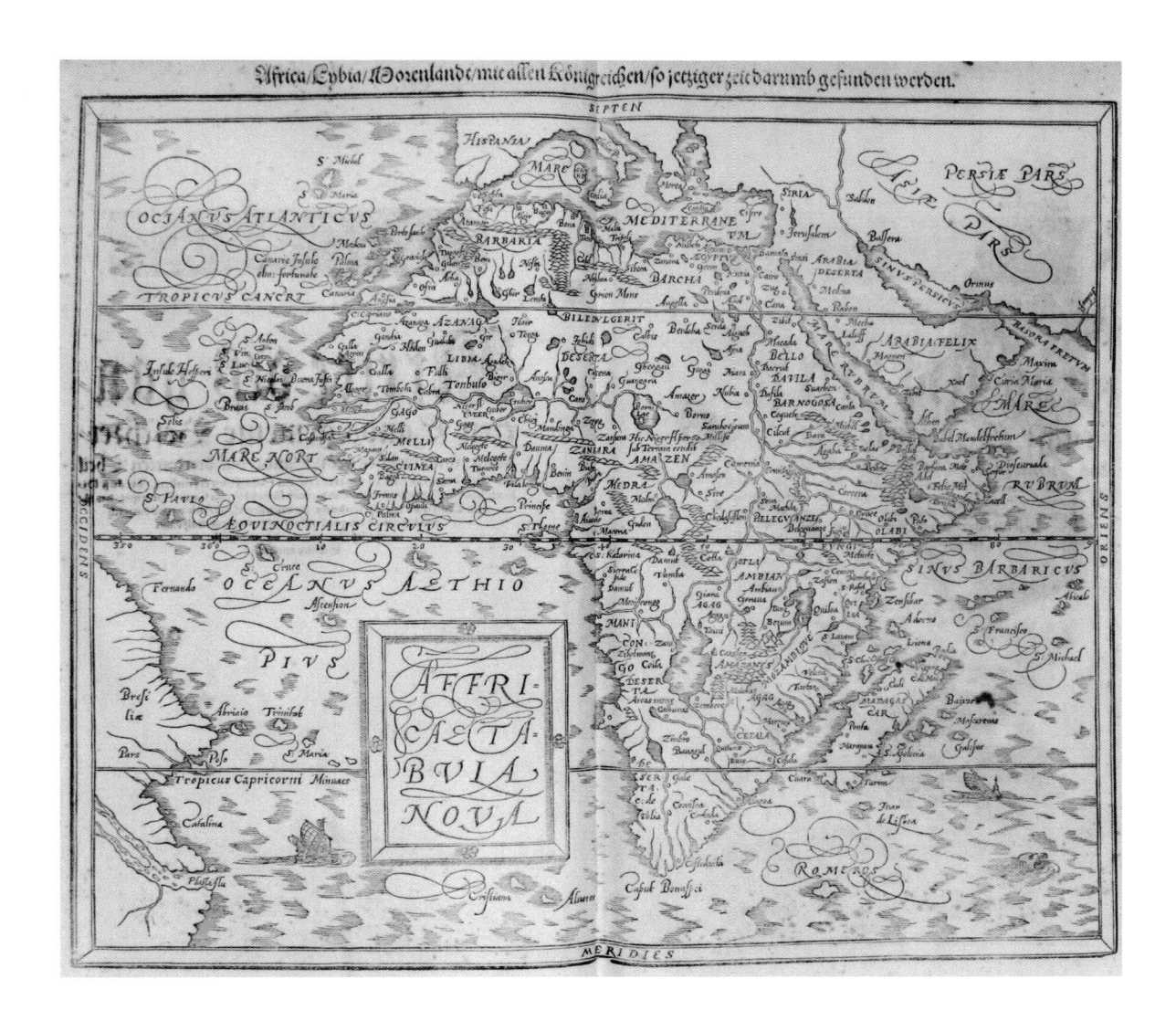

AFFRI= | CAE TA= | BVLA | NOVA.
- No scale bar.
- [Basel : Sebastian Petri, 1588].
- From: *Cosmographia*.
- 1 map : woodcut ; 31 x 36 cm (to outer black line).
- No latitude coordinates. Longitude coordinates (350°-90°) along Equatorial Line.
- North at the top; no compass rose.
- Title is set within a box at lower left side of the map.
- There is one ship in the Atlantic Ocean, and one ship and one sea monster in the Indian Ocean. No flora and fauna are found on the map.
- Above the map on one line of text is the following: *Africa/Lybia/Morenlandt/mit allen Koenigreichen /so jetziger zeit darumb gefunden werden*.

Description

Henrich Petri was the stepson of Sebastian Münster and the publisher of Münster's *Geographia* and *Cosmographia*. After Petri's death in 1579, his son and heir to the publishing business, Sebastian Petri (1546-1627) or Sebastian Henricpetri as he is sometimes identified, had this new woodcut map of Africa prepared along with other maps.

This map was used for the editions of Sebastian Münster's *Cosmographia* starting in 1588. It was cut to correct and update geographical information that had become outdated in the 48 years since the first issue of Münster's 1540 map of Africa. The map was also prepared to attempt to compete with Abraham Ortelius' enormously popular *Theatrum Orbis Terrarum* and his more up-to-date maps.

This map is often referred to as Sebastian Münster's second map of Africa, though it bears no resemblance to Münster's 1540 map of Africa. Besides attempting to reproduce the look of a copperplate engraved map using a woodcut, Petri closely followed Abraham Ortelius' 1570 map of Africa.

The style of this woodcut map is quite different in appearance from Münster's 1540 map. The unidentified preparer of this map attempted to duplicate with a woodcut the much finer detail available from a copperplate with woodcut flourishes and finer lines. The wording has been cut into the woodblock, unlike the 1540 Münster map of Africa which used insertable metal type.

Publication Information

This map appeared in editions of Sebastian Münster's *Cosmographia* from 1588 to 1628. After 1628, likely because of Sebastian Petri's death the previous year, the book and the map ceased to appear. The text on the verso of the map in all editions was only in German. The text in the earlier editions of Münster's *Cosmographia* of 1544 to 1578, with Map # 3, were in Latin, German, French, and Italian.

The map itself is known in only one state. The differences among the various editions are distinguished by an examination of the verso of the map. These differences are as follows:

1588 Verso: text begins 'AFRICA / Afrie des gan / tzen Landts gemeine Beschrei /…' At bottom, 'XXV.' Woodcut headpiece with two cherubs above text in a separate box.
Examples: *London BL*, 568.h.2; *Washington LC*, G113 .M75 1588.

1592 Verso: Same text and woodcut headpiece as 1588. Same square letter 'A' in Africe with the fleuron as on 1588 edition measuring 2.7 x 2.7 cm (Burden noted differences in text and in fleuron size with Americas map).
Examples: *London BL*, 1003.t.10.

1598 Verso: Same text as 1588. New woodcut headpiece: a decorated arch with the monogram V.G. and portraits to left and right. 'XXV' at bottom <u>with no period</u>.
'A' in Africe has precisely square shaped decorative fleuron with numerous series of swirls added. In size, the fleuron is 2.5 x 2.5 cm
Examples: *Washington LC*, G113 .M75 1598; *Basel UB*, E.U.I. 61a; *London BL*, 569.h.1.

1614 Same text as 1588. Same woodcut headpiece as 1598. 'XXV' at bottom <u>with no period</u>. In this edition, the large 'A' in Africe only has several swirls originating from the 'A'. In size, the fleuron is 3 x 3 cm.
Examples: *Washington LC*, G113 .M75 1614; *New York PL*, *KB+ 1614; Zisska & Kistner Kunstauktionshaus, May 2001 catalog, item no. 2971. Also See Sotheby's 1994 catalog.

1628 Verso: Same text as 1588. At bottom, 'XXV.'. Woodcut headpiece same as 1598 edition, but not 1588 and 1592. In this edition, the large 'A' in Africe only has several swirls originating from   the 'A'.
Examples: *London BL*, 10003.t.1; *New York PL*, *KB+ 1628; *Washington LC*; *Basel UB*, E.U.I. III.3.

Rodney Shirley has told the author that there are also editions of 1615 (with an example at Det Kongelige Bibliothek, Copenhagen as confirmed by Hendrik Dupont, curator), and of 1618. The Africa map in the edition of 1615 at Copenhagen is the same as in the edition of 1614.

References
Tooley 1999-2004, E-J: 311; Norwich 1997: map # 14.

22 1588 Livio Sanuto, Venice

AFRICAE TABVLA XII.
- Scale bar in bottom left: Milliaria 500 = 4.0 cm.
- [Venice : Damiano Zenaro, 1588].
- From: *Geograpfia di M. Livio Sanuto...* .
- 1 map : copperplate engraving ; 39.5 x 51.5 cm.
- Latitude and longitude coordinates on all four sides of the map.
- North at the top; no compass rose.
- Title on one line above map.
- There are three ships and two sea monsters in the Atlantic Ocean, and two ships in the Indian Ocean.

Examples: *Boston Afriterra; Cambridge Harv MC*, MAP-LC DT7 .S2 f*; *Washington LC*, DT7 .S2 fol.

Description

The atlas that contained this map of the continent is notable for being the first atlas devoted solely to Africa. This map only appeared in the one edition of Sanuto's atlas that was published in 1588. The atlas was not reprinted and no other uses of this map are known.

This map, like all of Sanuto's maps in the atlas, is elegantly engraved with attention to fine detail. The engraving of the map of Africa is attributed to Livio Sanuto's brother, Giulio.

Sanuto based his map's general outline and the coordinates for most placenames, especially along the coastlines, on Gastaldi's 1564 wall map, as well as Portuguese sea charts. Biasutti (1920) believed that Sanuto did not know the Gastaldi wall map but only the smaller versions produced in Venice by Forlani in 1562 and Giovanni Francesco Camocio - Paulo Forlani in 1563. This seems unlikely as the Gastaldi wall map of Africa, like Sanuto's map, was produced in Venice and was well known by 1588.

For the interior, Sanuto, like his contemporaries, had access to a vast amount of information, some fanciful and some based on fact, taken from the numerous books of the period (Cadamosto for West Africa, Leo Africanus for North and West Africa, Francisco Alvares for Ethiopia, Duarte Barbosa and João de Barros for East and Southern Africa, and Ramusio's Volume I of 1550 for the earlier accounts.

Unlike some of his contemporaries, Sanuto follows the text sources more conservatively and does not locate placenames somewhat indiscriminately over the map. The result is a map that has fewer placenames in the interior. As Relaño (2002: 213) notes, Sanuto attempts to re-locate Abyssinian placenames north of the Equator away from their southern orientation on other maps of this period. For example, the region of *Amara* is further north of the Equator on Sanuto's map.

Publication Information

Livio Sanuto (c1520-1576) was a Venetian cosmographer, mathematician, inventor of astronomical instruments, and maker of terrestrial globes. Sanuto along with his brother, Giulio (active from 1540-1588), planned to produce a comprehensive atlas with maps of all regions of the world that was intended to be more accurate than any previously published. Sanuto started his work by preparing descriptive text and a series of maps of Africa. He died, however, at the age of 56 in 1576 before the text and maps for any further areas of the world could be undertaken.

It is not precisely known when Sanuto wrote his *Geograpfia*. Skelton (1965c) theorizes that the text was prepared between 1561 and 1575. By 1561, Sanuto had established his reputation as a geographer. By that time, he had been admitted for membership into the 'Accademia della Fama' in Venice.

Sanuto's atlas *Geograpfia di M. Livio Sanuto...*, showing all parts of Africa (12 maps in total with a map of the continent and regional maps), was published posthumously in 1588 by Damiano Zenaro. It is curious that there was a delay of 12 years between the time of Sanuto's death and the publication of the atlas. Likely, as the other parts of the world were not undertaken and as Africa was of secondary interest to the new discoveries in the Americas and in the East, there was not a tremendous pressure to produce the atlas. For those seeking geographic information on Africa, the description provided by Leo Africanus was by this time readily available in French and Latin, with a soon-to-be-published English edition.

The text for the *Geograpfia* provides an exceptional description of Africa and thoroughly summarizes sixteenth century knowledge of Africa. Based on Sanuto's careful and meticulous attention to detail, coupled with the finely engraved maps, there is little doubt that had Sanuto lived and been able to complete his atlas, it likely would have ranked as one of the masterpieces of Renaissance geographical work (Skelton 1965c: viii).

References

Relaño 2002: 213; Norwich 1997: map # 15; Loeb-Larocque 1989; Skelton 1965c; Almàgia 1946; Biasutti 1920: 327-346 and 387-436.

23   1589 Jodocus Hondius – Jean Leclerc, Paris

AFRICÆ | TABVLA.
[to the right of the title cartouche]:
I. Hondius inuentor.
[below this]:
I. le Clerc excudit. | 1602.
- No scale bar.
- [Paris] : Jean Leclerc, [1589].
- From: separately published.
- 1 map : copperplate engraving ; 37 x 49.5 cm.
- Latitude coordinates on left and right sides of the map. Longitude coordinates at bottom of the map.
- North at the top; no compass rose.
- Title is in a simple strapwork cartouche at the bottom right corner of the map.
- There is a large figure of Neptune on a seashell drawn by two horses in the South Atlantic Ocean. To the left of this figure is a sailing ship. A sea monster is in the Indian Ocean to the east of Madagascar.

Examples: *London BL*, Maps C.7.c.24 (in *Theatre Geographique du Royaume de France...M.DC.XXI*

1621. This copy is listed by Pastoureau [1984] under Leclerc c.1621 with this map as one of the maps added to the atlas); *Private French Collection* (as a separate map).

Description

This is a beautifully engraved, rare map. It is distinguished by an elaborate vignette of Neptune astride a shell being pulled by two sea horses. The fine engraving style, with its distinctive depiction of the seas, is typical of works by Jodocus Hondius.

The map is not notable for its cartographic information as it generally follows the earlier Ortelius map of 1570 in terms of its geography and placenames. The shape of Africa and Madagascar is also similar to the Ortelius map. The hydrographical systems are virtually identical. Placenames and their location on the two maps, including the placement of *Zanzibar* in southwest Africa, are also the same.

The engraver for this map is Jodocus Hondius (the imprint 'I. Hondius inuentor.' is to the right of the title cartouche). The publisher's imprint of Jean Leclerc is located just below the Hondius imprint ('I. le Clerc excudit 1602.').

Publication Information

Little is known about this map or when and where it was published. Even the map's date is uncertain. The Americas map in the set of four continent maps has a date above the title of 1589 with the publisher Leclerc's name and the date of 1602 below and to the right of the title cartouche imprint ('I. le Clerc excudit 1602.'). The Africa map does not have a 1589 date above the title, however it does have a date of 1602 in the same area below and to the right of the title cartouche. Burden (1996: map # 145) provides a lengthy analysis of this Americas map yet is not able to completely confirm an absolute date for when the map was prepared. Tooley (1969: 53) dates the map of Africa as 1602, but it is possible that this date may be too recent, or that the date of 1602 only reflects its publication date in Paris (also see Schilder 2006).

If the map was actually prepared in 1602, the geographical content for the Africa map is clearly out of date. The map follows the earlier Ortelius map of 1570, and not Jodocus Hondius' own more up-to-date wall map of Africa of 1598. As Burden states, it is unlikely that Hondius would chose to produce a map in 1602 that was so out of date, when he had more recent information available, reflected in his 1598 wall map. Thus, it seems that this map was engraved earlier than the often accepted date of 1602, probably in 1589 to conform to the date on the Americas map in the set. In 1589, Hondius was still in a self-imposed exile in London escaping religious unrest in his own country. While this map was likely engraved while Hondius was a refugee in London, it seems to have come into the possession of Jean Leclerc, who was a publisher in Paris, sometime prior to 1602. Leclerc (1560-1621) was an engraver, bookseller, and publisher in Tours and then later Paris from 1594.

This copperplate for the map of Africa that Leclerc acquired from Hondius does not appear to have been prepared for use in any specific atlas, but it is sometimes found in French atlases of the early 1600s such as the *Theatre Geographique du Royaume de France* (the Africa map is not in the Library of Congress example dated 1633). Leclerc may also have been associated with Maurice Bourguereau in the production of the *Theatre Francois* in 1594, but he is mostly known for the re-publication of this atlas in 1620 as the *Theatre Geographique du Royaume de France* with various reissues. This atlas sometimes contains this map of Africa and the companion maps of the other three continents. Only a very few examples of this map are known.

References

Schilder 2006: chap. 7.5: facsimile 7; Burden 1996: map # 145 (Americas map); Schilder 1992: 44-46; Loeb-Larocque 1989; Pastoureau 1984: 295; Tooley 1969, Koeman 1967-71, II: 136.

24   1589 Heinrich Bünting, Magdeburg

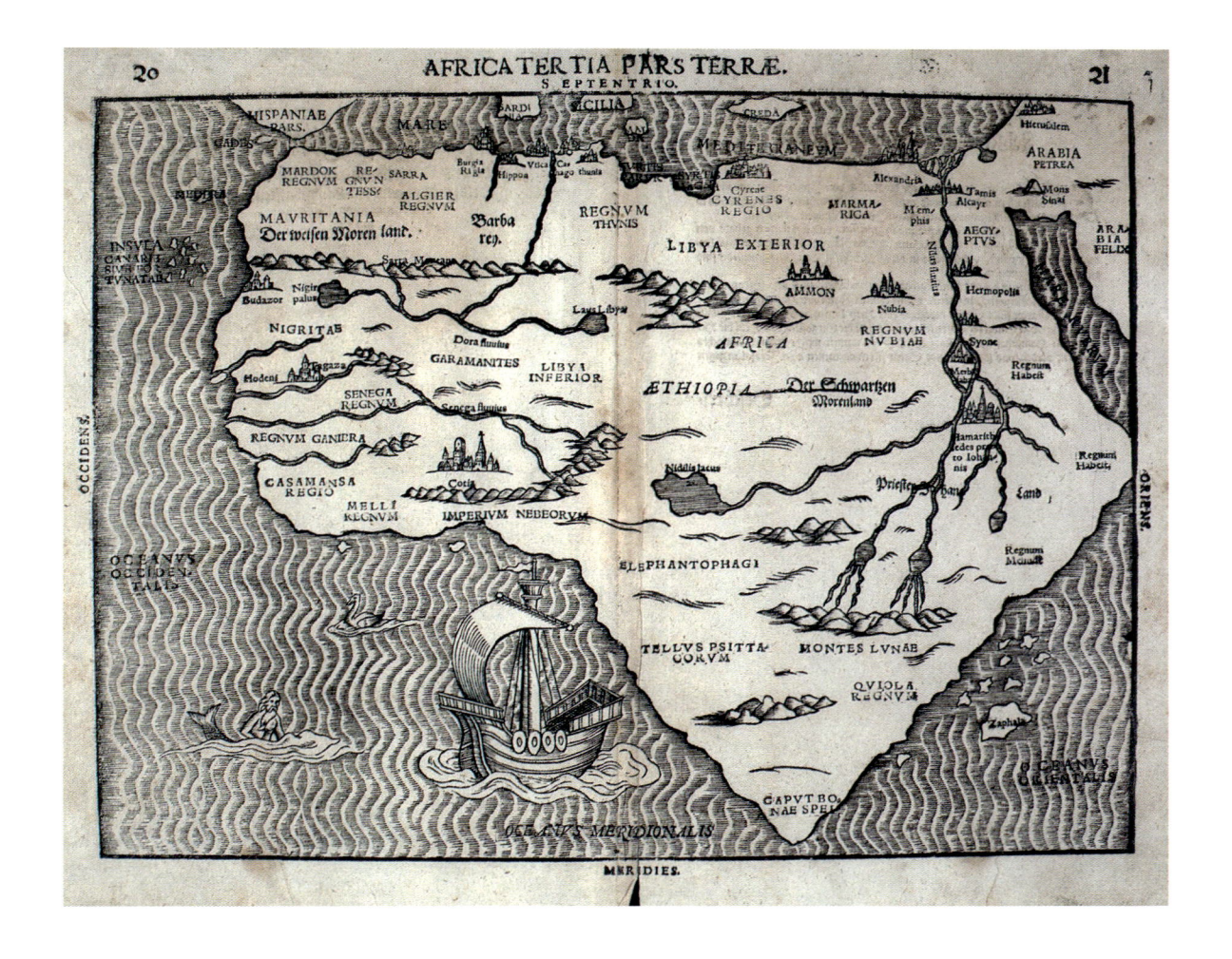

24.1 Variant 1 1589

AFRICA TERTIA PARS TERRÆ.
- No scale bar.
- [Magdeburg : Paul Donat in [verle]gung Ambrosij Kirchner, 1589].
- From: *Itinerarum Sacrae Scripturae*.
- 1 map : woodcut ; 26 x 34 cm.
- No latitude and longitude coordinates.
- North at the top; no compass rose.
- Title on a line above the map image.
- Within the Atlantic Ocean, there are a large ship, a merman, and an aquatic bird that appears to be a swan(!). The engraving for the sea is vertical in appearance rather than horizontal. Placenames on the map are in Latin and in German (some in Gothic type).
- The city of *Alcayr* (Cairo) is on the east side of the Nile River in Egypt with the city of *Memphis* to the west side of the Nile.

Examples: 1589 Magdeburg edition: *Budapest NSL*, ANT 9272. 1592 Magdeburg edition: Tooley 1969: map image no.141.

Map 24.2 Placement of cities in Egypt.

24.2 Variant 2 1597

The city of *Alcayr* (Cairo) is on the east side of the Nile River in Egypt with the city of *Memphis* to the west side of the Nile. Some of the placenames are shifted upward in such a way as to be printed almost over the geographical features (for example the words MONTES LVNAE are almost on top of the Mountains of the Moon in Central Africa rather than just below the mountains).

Example: 1597 Magdeburg edition: *Königstein, Reiss & Sohn Buch- und Kunstantiquariat Auktionen* (auction catalog of April 2005).

24.3 Variant 3 1600

In this edition, the city of *Alcayr* (Cairo) is on the west side of the Nile River in Egypt with the city of *Memphis* to the east side of the Nile (in previous editions the cities are reversed).

Examples: 1600 Magdeburg edition: *Washington LC*, BS630.B79 1601; *Author's collection* (as a separate map).

Map 24.3 Placement of cities in Egypt.

Description

This map is one of the later woodcut maps of Africa. As such, the appearance of the map, especially when compared to other maps from this period, is unusual.

Heinrich Bünting appears to generally follow Sebastian Münster's 1540 map of Africa for his basic depiction of Africa, especially with its west to east slope of Africa. Bünting presents a traditional view of the two Ptolemaic lakes as the source of the Nile River with these lakes being fed by the *Montes Lunae* (Mountains of the Moon). The source lake in the Abyssinian highlands is also shown. The river system in Southern Africa is not depicted. The shape of the continent is unusual in that it tapers to a point at the Cape of Good Hope.

Bünting identifies the Moslems in northwest Africa as *Der weissen Morenland* (the White Moors) and the Moslems in Central Africa, just above *Priester Johans Land* (the Land of Prester John), as *Der Schwartzen Morenland* (the Black Moors). Likely relying on Münster's 1540 map, Bünting places the island of *Zaphala*, among other islands, off the east coast of Africa, in a probable reference to the region from which King Solomon supposedly imported gold and silver to Jerusalem. As on the Münster map, Madagascar is absent from Bünting's map.

Most interestingly, Bünting introduces a third, even larger lake, *Nidilis lacus*, in Central Africa as a new source for the Nile. Though this may be a reference to Gerard Mercator's *Sachaf lacus* on his world map, which fed the Nile, the Congo, and the southern rivers, Bünting places the lake much further north. Like Bünting's other, more allegorical maps in his book, this map was unique. It does not fit with any previous or later model in the mapping of Africa.

As with other makers of woodcut maps, Bünting employed insertable metal type which accounts for the numerous different styles of lettering found on the map. Many examples of the maps from Bünting's *Itinerarum* are in poor condition, owing to the inferior quality of the paper that was often used. Often, examples of the book are found wanting leaves and entire sections. Frequent changes in lettering styles and placement of names seen in different editions make it hard to give a precise number of variants or to date specific maps.

Publication Information

Heinrich Bünting (1545-1606) was a Protestant Professor of Theology at Hanover. He is most known for his *Itinerarum Sacrae Scripturae*, a religious commentary on the world. The book describes the travels of the religious figures of the Old and New Testaments of the Bible as well as its geography. It was a popular book and was the most complete summary of the geography of the Bible of this period.

The book was first printed in Helmstadt in 1581 by Jacob Lucius Siebenbürger and again in 1582. Numerous other editions were issued in Magdeburg, Wittenburg, Braunschweig, Erfurt, and Leipzig. The book also appeared outside of Germany with text in Danish, Swedish, Dutch, Czech, and English. In total, Van der Heijden (1998: 53) states that over 60 editions of Bünting's *Itinerarum Sacrae Scripturae* were published in various languages between 1581 and 1757.

The early editions of the book, besides containing various allegorical maps of the world, Europe, and Asia, as well as maps of the Holy Lands, contained a map showing the old world of Africa, Asia, and Europe. These early editions did not contain a separate map showing only Africa. The separate Africa map was not in the two examples of the 1585 Magdeburg edition at the British Library (*London BL*, Maps C.26.e.5.(1) & (2), and 3105.a.14).

The first edition of Bünting's *Itinerarum Sacrae Scripturae* with a map of just the African continent (and not just a map of Europe, Africa and Asia together) that the author was able to locate was the 1589 edition, published by 'Paul Donat in [verle]gung Ambrosij Kirchner' in Magdeburg. The same map is also in a 1592 edition, published by 'Paul Donat für Ambrosij Kirchner' in Magdeburg (Tooley 1966: map no.141). In 1597, another edition of the *Itinerarum*, with variant 2 of the map, was published by 'P. Donat für A. Kirchner' in Magdeburg. The next edition in which this map is known to appear was the 1600 edition of Bünting's *Itinerarum*, published in Magdeburg for Ambrosij Kirchner (Durch Andream Duncker, in Vorlegung Ambrosij Kirchner). This edition contains variant 3 of the map.

For other editions of the *Itinerarum Sacrae Scripturae*, different woodcut maps of Africa were used. These are as follows:

1592: This edition was published in Prague by Dan. Adam z. Weleslawnja (1546-1599) in the Czech language. This is an exceedingly rare edition. This map is most easily identified by the inclusion of Czech words inserted under Latin words on the map. See Map # 26.

1605: This edition was published in 1605 by Jan Jansz., an Arnhem publisher and the father of Johannes Janssonius. There are thirteen Dutch editions of the *Itinerarum*. The first was published in Rees by Wylics van Santen in 1594. This was followed by editions in Utrecht in 1596, and in Amersfoort in 1601. See Map # 50.

References
Tooley 1999-2004, A-D: 209; Van der Heijden 1998: 49-71; Norwich 1997: map # 17; Tooley 1966: map # 141.

25   c. 1590 Giovanni Battista Mazza, Venice

AFRICA | EX MAGNÆ | ORBIS TER= | RÆ DESCRI= | PTIONE.
[at the bottom right corner of the map]:
Iouan Batista Mazza fece
- No scale bar.
- [Venice : unknown publisher, c.1590].
- From: separately published / Italian composite atlas.
- 1 map : copperplate engraving ; 36 x 47 cm.
- Latitude coordinates on left and right sides of the map. Longitude coordinates at bottom of the map.
- North at the top; one compass rose (in the shape of a directional compass with a needle) in the Indian Ocean.
- Title within a strapwork cartouche at lower left of the map.
- There are two ships and one sea monster in the Atlantic Ocean, and one ship and one fish in the Indian Ocean.
- Dedicatory text box at lower right to Francesco Moresini.

Examples: *Bern SUB*, Ryh 7601:3; *Chicago NL*: Novaco 4F 395; *Private US Collection*.

Description
This is a separately published, beautifully engraved map of Africa. It is attributed to the engraver, Giovanni Mazza whose imprint is on the map as 'Iouan Batista Mazza fece'. There are no other identifying names on the map, other than a dedicatory cartouche for Francesco Moresini. Little is known of Mazza. He may be most known for his superb engraving of Giuseppe Rosaccio's large, ten-sheet world map in 1597. All of his maps are of great rarity.

Geographically, this map is a close copy of the Ortelius 1570 map of Africa (Map # 12) which in turn followed the Gastaldi eight-sheet wall map of Africa of 1564. See Map # 12 for a discussion of the geography of this map.

Publication Information
The map was published separately, but may have appeared in composite Italian atlases of the late sixteenth century. There is no text on the verso of the map in the examples seen by the author.

The Africa map is known in one state. There is a possibility that there is a second state of the Africa map, but an example has not yet been located. This is based on the existence of a second state of Mazza's Americas map, with the removal of the imprint of Mazza and with evidence of some additional hachuring to the sea (Burden 1996: map # 73).

References
Christie's Auction Catalog, June 2000; Burden 1996: map # 73 (Americas map); Kraus 1970; Almàgia 1944-45: 118.

26   1592 Heinrich Bünting – Daniel Adam z Weleslavina, Prague

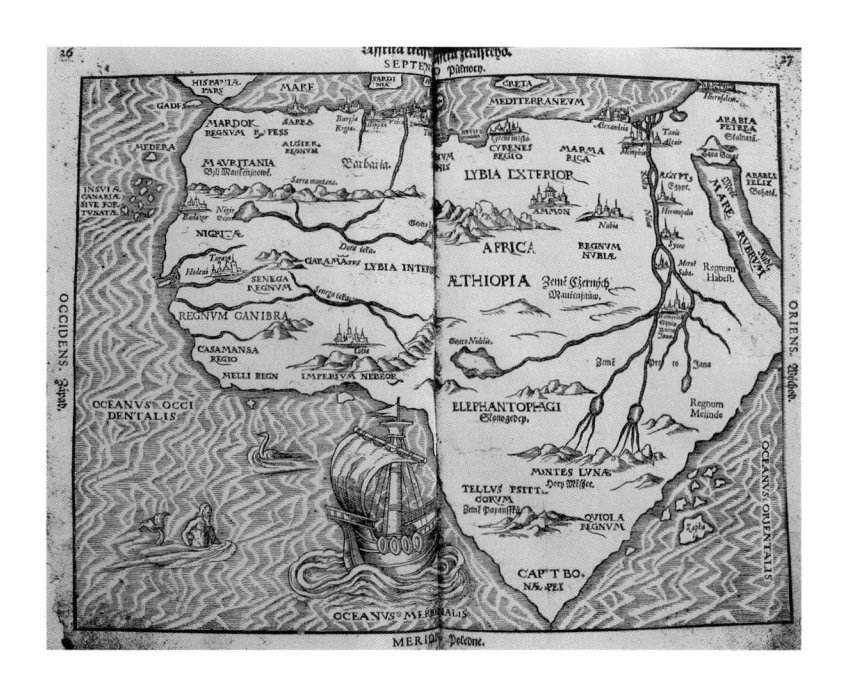

Affrica....
- No scale bar.
- [Prague : Daniel Adam z Weleslavina, 1592].
- From: *Itinerarum Sacrae Scripturae.*
- 1 map : woodcut ; 26 x 34 cm.
- No latitude and longitude coordinates.
- North at the top; no compass rose.
- Title on a line above the map image.
- Within the Atlantic Ocean, there are a large ship, a merman, and an aquatic bird that appears to be a swan(!). The engraving for the sea is vertical in appearance rather than horizontal. Placenames on the map are in Latin and in Czech (some in Gothic type).

Example: *London BL*, RB.23.b.4757.

Description

This rare map is one of the later woodcut maps of Africa. As such, the appearance of the map, especially when compared to other maps from this period, is unusual.

As with other makers of woodcut maps, the publishers of the Bünting map employed insertable metal type which accounts for the numerous different styles of lettering found on the map. Many examples of the maps from Bünting's *Itinerarum* are in poor condition, owing to the inferior quality of the paper that was often used. Examples of the book are sometimes found wanting leaves and entire sections. Frequent changes in lettering styles and placement of names seen in different editions make it hard to date specific separate maps.

Heinrich Bünting (1545-1606) was a Protestant Professor of Theology at Hanover. He is most known for his *Itinerarum Sacrae Scripturae*, a religious commentary on the world. The book describes the travels of the religious figures of the Old and New Testaments of the Bible as well as its geography. It was a popular book and was the most complete summary of the geography of the Bible of this period. For a discussion of the geography of this map, please refer to Heinrich Bünting's 1589 map of Africa, Map # 24.

Publication Information

This edition of Bünting's *Itinerarum Sacrae Scripturae was* published in Prague by Daniel Adam z. Weleslawnja (1546-1599) in the Czech language. It is an exceedingly rare edition. This map is most easily identified by the inclusion of Czech words inserted under Latin words on the map.

References
Tooley 1999-2004, A-D: 209; Van der Heijden 1998: 49-71; Norwich 1997: map # 17; Tooley 1966: map # 141.

27   1593 Cornelis de Jode, Antwerp

AFRICAE VERA FOR= | MA, ET SITVS.
[at upper right, there is a text box with 17 lines of text describing Africa. Just below this text box there is a separate circle]: Formis | haredum Ge= | rardi de Iode.
- No scale bar.
- [Antwerp] : Cornelis de Jode, [1593].
- From: *Speculum Orbis Terrae*.
- 1 map : copperplate engraving ; 32.5 x 45 cm (to outer neatline).
- Latitude coordinates in the form of a gridline along the Prime Meridian along the left side of map. Longitude coordinates along Equatorial Line.
- North at the top; no compass rose.
- Title is in a box above the map with decorative strapwork on either side of the title.
- There are two ships and four sea monsters in the Atlantic Ocean and two ships and three sea monsters in the Indian Ocean. A figure waving a small flag (?) is near the Canary Islands. Various other figures in and around Africa depict inhabitants.

Examples: *Various locations; Author's collection* (as a separate map).

Description

This is the De Jode family's second map of Africa as issued in 1593 by Cornelis de Jode (1568-1600). His father, Gerard, evidently was intending to issue a revised and expanded edition of his 1578 *Speculum Orbis Terrarum*, but his death in 1591 prevented this from happening. Upon Gerard's death, Cornelis finished the atlas, adding ten new copperplates including this map of Africa. Although this map is well-executed, Cornelis was noted more as a publisher rather than as a cartographer like his father.

When this map was engraved for Cornelis' *Speculum Orbis Terrae* of 1593, a reference to Gastaldi as the source for the map, which was on Gerard's previous map, was omitted. Instead, Cornelis seems to utilize a variation of the Mercator model, specifically the Gerard Mercator world map of 1569, in his depiction of much of the geography, including the river systems, of Africa. Cornelis shows a lake, *Sachaf lac.*, as the source for the *Zabere* (Zembere), *Cuama* and *R. d S. Spirito* Rivers in Southern Africa, rather than the western of the twin Nile source lakes as in the Gastaldi-Ortelius model. This presentation is clearly modeled from Gerard Mercator's world map, though De Jode does not connect this lake to the Nile, as Mercator did. To the north of the Niger River, De Jode follows Mercator with a river flowing into the Nile. On Mercator's map this river is first called the Giras flu and later the Nubia flu as it flows through the region of Nubia. On this De Jode map only the name *Giras flu* is present.

Other aspects of the map are taken directly from Ortelius (for example the placement of *Zanzibar* on the southwest coast). The coastline in this map is almost identical to the c.1573 Gerard de Jode map with its less pointed shape to the southern tip of Africa.

Publication Information

Only one atlas was published by the De Jode family: *Speculum Orbis Terrarum*, issued by Gerard de Jode in 1578, and then re-issued by his son Cornelis in 1593 as *Speculum Orbis Terrae*. The atlas was not a financial success as very few copies were sold in comparison to Ortelius' *Theatrum*. However, it appears to have received favorable comments from contemporaries such as Petrus Montanus (Koeman 1967-71, II: 206). After Cornelis' premature death in 1600, Jan Baptist Vrients obtained the plates of the atlas. Vrient's widow later sold the plates to Moretus, although there was no further use of the plates. Consequently, all maps by the De Jode family are quite rare.

References

Van der Krogt 2003: map 8600:32B; Norwich 1997: map # 19; Koeman 1967-71, II: Jod 2, fol.4.D.

28   1593 Abraham Ortelius – Filips Galle (2), Antwerp

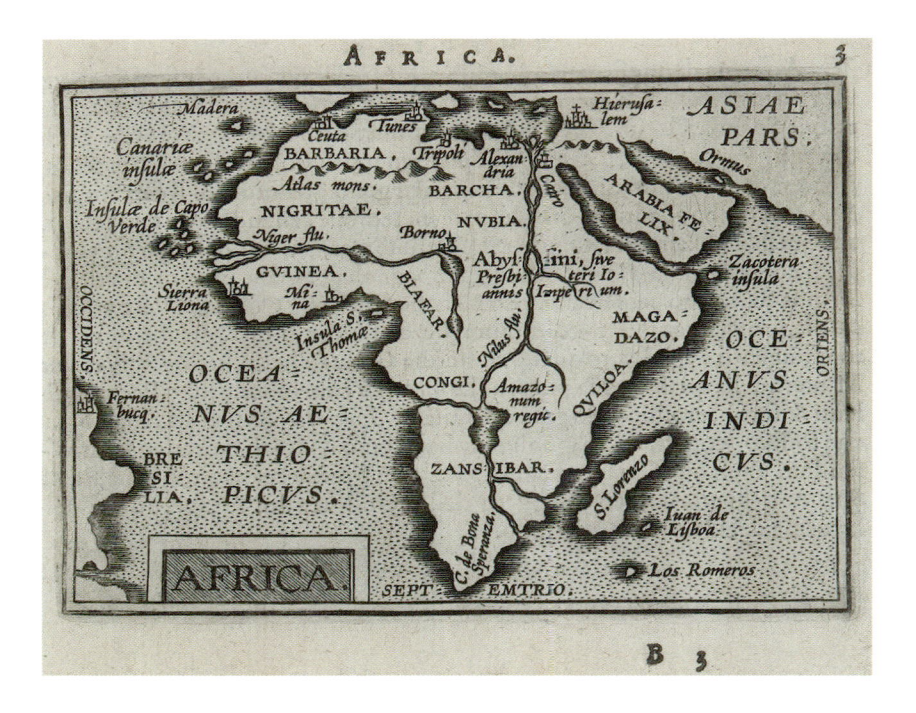

AFRICA.
- No scale bar.
- [Antwerp : Christoffel Plantin, 1593].
- From: *Theatro d'Abrahamo Ortelio.*
- 1 map : copperplate engraving ; 8 x 11 cm.
- No latitude and longitude coordinates.
- North at the top of the map, no compass rose.
- Title is set in a simple box at the bottom left of the map.
- No ornamentation on map.

Examples: 1593 Italian edition: *Amsterdam UB*, 1802 G 20; *New York PL*, *KB 1593 (Ortelius, A. Theatro); 1595 Latin edition: *London BL*, Maps C.2.b.4; 1601 Latin edition: *New York PL*, *KB 1601; 1602 English edition: *London BL*, Maps C.2.b.11.

Description

This map is a considerably simplified version of the 1577 Ortelius-Galle (1) map of Africa. From a cartographic perspective, the new map has much less geographic detail, although it is elegantly engraved.

The map only has two towns identified south of the Sahara Desert: a settlement in *Sierra Liona* (Sierra Leone) and the Portuguese fort and gold mining area in Guinea called *Mina*. The river systems of Africa reflect Ortelius, but in a simplistic fashion. However, there are some significant changes. *Zansibar* (Zanzibar) returns to its earlier prominence in the 1570 Ortelius map, in this case denoting all of Southern Africa. The second Nile River source lake of Zaflan in the east is completely removed from this map in a similar fashion to Ortelius-Galle (1). The island of Madagascar, with its earlier name, is identified on this map only as *S. Lorenzo*.

Publication Information

Beginning in 1588, Filips (Philippe) Galle began the issuance of a new pocket-size atlas under the title of *Epitome du Théâtre du Monde*. For the 1588 edition, Galle used the 1577 Abraham Ortelius–Filips Galle (1) map of Africa, though he did start the introduction of a number of new maps for use in the

Epitome. Evidently, Galle was not satisfied with the earlier maps from his *Spieghel der Wereld* of 1577. Later editions of this newer atlas carried various titles: *Epitome Theatri Orteliani* in 1595, *Abrégé du Théâtre d'Ortelius* in 1602, *Breve Compendio dal Theatro Orteliano* in 1602, and finally *An Epitome of Ortelius his Theatre* in 1602. In more recent times, all of these pocket-size Ortelius atlases came to be commonly called the *Epitome*.

This map of Africa did not appear in the newer *Epitome* until the edition of 1593, which was titled *Theatro d'Abrahamo Ortelio*. For this newer series of Galle's *Epitome*, a new copperplate was cut by Galle to replace the earlier one of 1577. It should be noted that the older Ortelius-Galle (1) map of Africa is found in some editions of the 1593 *Theatro d'Abrahamo Ortelio*.

This edition was also published by Christoffel Plantin until 1601 when the publisher became Jan Baptist Vrients. Vrients used this plate up to the 1602 edition. In about 1604, he purchased the superior, competitive plates of the Ortelius-Van Keerbergen *Epitome* series.

The following chart can be used to differentiate the Ortelius-Galle (2) Africa map in the various editions of the book in which the map appeared using the language of the text, signature, and page number. This information is from Van der Krogt (2003):

Date	Van der Krogt no	Title	Lang.	Signature	Page
1593	332:21	*Theatro d'Abrahamo Ortelio*	Italian	B3r	3
1595	332:12	*Epitome Theatri Orteliani*	Latin	B4r	4
1598	332:03	*Epitome Theatri Orteliani*	French	A4r	4
1601	332:13	*Epitome Theatri Orteliani*	Latin	B4r	4
1602	332:22	*Breve Compendio dal Theatro Orteliano*	Italian	B3r	3
1602	332:04	*Abrégé du Théâtre d'Ortelius*	French	A4r	4
1602	332:31	*An Epitome of Ortelius his Theatre*	English	A4r	4

The map is only known in one state.

References
Van der Krogt 2003: map 8600:332; Koeman 1967-71, III: Ort 55-62.

29   1594 Giuseppe Rosaccio, Ferrara

[Map of Africa].
- No scale bar.
- [Ferrara : Vittorio Baldini, 1594].
- From: *Teatro del Cielo, e Della Terra, Nel quale si discorre breuemente del centro, e doue sia ... del Mondo, e sue parti ... Delle sfere, e come girino ... di Gioseppe Rosaccio*.
- 1 map : woodcut ; 7.7 x 7.6 cm.
- No latitude and longitude coordinates.
- North at the top; no compass rose.
- There is no ornamentation on the map.

Example: *New York PL,* *KB 1594 (in Rosaccio, G. *Teatro del Cielo*).

Description
This is an uncommon, small map of Africa on one page of text. The map was prepared for Giuseppe Rosaccio's *Teatro del Cielo*. Rosaccio (c.1530-1620) was an Italian physician, an editor (the 1598 and 1599 editions of Ptolemy's *Geographia*.) and an author of geography and travel books. These books included the popular *Teatro del Cielo* and *Il Mondo* which went through a number of subsequent editions into the late seventeenth century.

This map generally follows Ortelius. Although small and roughly shaped, the map identifies the main features of Africa including *Guinea, Bona S.* (Cape of Good Hope), and *Etiopia* (Ethiopia). It shows two lakes as the source for the Nile River, with the western one also feeding the Congo River. There are several lakes feeding the large West African river. Also, unlike most maps of Africa, this map shows a large *Terra Avstrale* at the bottom, denoting what was assumed to be an excessively large, southern continent that was supposed to be a counter-balance to the landmasses to the north.

The cutter of the woodblock is not known, but is assumed to be Rosaccio as a small world map in the book at the bottom of page 53 has the signature of 'Gioseppe Rosatio .F.' and is similar in style to this Africa map.

Publication Information
Only one edition of Giuseppe Rosaccio's book, *Teatro del Cielo...*, is known by the author to contain this map. No other uses of this woodcut are known. The map is bound within a small pocket-book measuring 14 x 9.5 cm. The example of this book, *Teatro del Cielo*, at the New York Public Library is bound with Rosaccio's *Le Sei eta del Mondo*, published in Bologna by Gio. Rossi in 1594.

Above the map, on the left side, in letter press is '26' followed by 'Teatro Del Cielo'. Below the map in letterpress is 'Dell' Africa, sue Prouincie' followed by 15 lines of text beginning with 'Africa...'. The last line of text on the verso of the map is 'ne, & Maluasia'.

References
Sabin 1868: entry # 73198.

30 1594 Giuseppe Rosaccio, Florence

[Map of Africa].
- No scale bar.
- [Florence : Francesco Tosi, 1594].
- From: *Teatro del Cielo*.
- 1 map : woodcut ; 2 map sheets from two woodblocks ; 12.5 x 16 cm (in total with each sheet of map as 8 cm wide, not including line of text above map) (Though small in size, the map of Africa in this edition was printed and bound in such a way as to be in two halves with each half on a different leaf. Note: the Americas map as per Burden (1996) is on one leaf).
- No latitude and longitude coordinates.
- North at the top; no compass rose.
- There is a large sea monster in the Atlantic Ocean and one small ship in the Indian Ocean.
- Above the map of Africa on the left sheet is the letterpress number '24' and the wording 'Dell'Africa,'; on the right sheet at the top is the letterpress wording 'e sue Prouincie' and number '25'.

Example: Florence 1594 edition: *London BL*, C.27.b.17 (on the verso, the last full line of text begins 'ne Napoli di Romanis...' and last line is 'uafia' with a signature of 'B 4' below).

Description
This map is unique in appearance and represents a later use of a woodcut map. Particularly in Italy, the use of copperplate engraved maps was well established by this time.

This map was prepared for Giuseppe Rosaccio's *Teatro Del Cielo* and *Il Mondo e sue parti* (or as it is also titled, *Le sei eta del Mondo*). Rosaccio (c.1530-1620) was an Italian physician, an editor (the 1598 & 1599 editions of Ptolemy's *Geographia*), and an author of geography and travel books. These included the popular *Teatro del Cielo* and *Il Mondo* which went through a number of subsequent editions into the late seventeenth century.

This map shows all of Africa, the Arabian Peninsula, and part of Brazil. Several islands are shown in the South Atlantic, such as an early appearance of *Tristan de Cugna* (Tristan da Cunha), reflecting their importance to early shipping around the Cape of Good Hope.

The geography for this map generally follows Ortelius in its presentation of the hydrography of Africa, though there are differences. Only one river is shown flowing to the south of the western Nile source lake. Further to its south, the unnamed Spirito Santo River originates in a mountain and not in the western Nile source lake as was often shown in this period.

The placenames and geography also follow Ortelius, though owing to the map's small size and roughness of the woodcut, numerous details are omitted. The cutter of the woodblock is not known, but is assumed to be Rosaccio as a small world map in the book has the signature of 'Gioseppe Rosatio' and is similar in style to the Africa map.

Publication Information

This map appeared in the numerous editions of Giuseppe Rosaccio's *Teatro del Cielo* and *Il Mondo e sue parti* (*Le sei eta del Mondo*), published from 1594 until well into the late 1600s. The woodcut is the same in all editions of the *Teatro del Cielo* and *Il Mondo* ... up to 1688 (when a new woodblock was cut) that were seen by the author.

Besides the 1594 edition described above, the following editions of Rosaccio's *Teatro del Cielo* and *Il Mondo e sue parti* (*Le sei eta del Mondo*) were examined by the author. The information with each edition may be used as a guide to determine from which edition an individual map appeared:

1595: Above the map of Africa on the left page is the letterpress number '140' and the wording 'FIGVRA DELLA:'. On the right page is the letterpress word 'AFFRICA:' and number '141'. On the verso, the last full line of text is 'sono li monti Argentato & il Curoniza.' with 'K 2' below. The 1595 edition was published in Florence by Francesco Tosi.

Examples: *London BL*, Maps C.26.a.25; *New York PL*, *KB 1595 (in Rosaccio, G. *Mondo e sve parti cioe Evropa, Affrica, Asia, et America*).

1598 'FIGVRA DELL'AFFRICA.' above left leaf with page '32' and 'E SUE PROVINCIE.' above right leaf with page '33'. 'B' on right leaf at bottom right. On verso, the last line of text is 'Arno Micine, Monone, Corone; & Maluafia.' and below this 'DEL-'.

Example: *London BL*, Maps.32.aa.49(1.).

1599: An edition of the *Teatro* by Tosi in 1599 (not seen by the author, but referred to by Burden (1996: map # 84).

1602: Map title is 'FIGURE DELL' AFRICA ESVE PROVINCIE'. To the upper left above the map is '32' and on upper right is '33'. A 'B' bottom right of the map below the neatline. On the verso: the last two lines of text are 'done, Corone, & Mal- / uasia'.
From: *Teatro Del Cielo* published in Venice.

Example: *New York PL*, *KB 1602 (in Rosaccio, G. *Teatro del cielo*).

1604: No title above the map. No pagination on the map. Western half of the map printed on recto and eastern half printed on verso of the same leaf.
From: *Mondo Elementare et Celeste*, published by Ciotti in Trevigi.

Example: *Amersfoort, Leen Helmink* (examined by the author).

1686: No map title. No page numbers above the map. On the verso: Image of sun with a face and sunrays. '85' at the top right. 'F 3 Della' at bottom right corner. Some wear marks, especially around neatline as on 1595 and 1602, which indicates the same woodblock as for 1595 and 1602 maps.
From: *Teatro del Cielo* published by Gio. Molino in Trevigi.

Example: *New York PL*, *KB 1686 (in Rosaccio, G. *Teatro del cielo e della terra*).

In 1688, a new, larger woodblock (13 x 17 cm) was cut by Giuseppe Moretti, from one woodblock and not two as for the previous map. This map appeared in the *Teatro del mondo e sue parti*, published by Antonio Pisarri in Bologna in 1688. It also appeared in the 1693 edition of the *Teatro del Cielo e della terra*, published in Treviso. In this example, the map was bound into the book on the left side. 'Fig. 9' at top right above the map. The verso is blank (See Map # 144).

References
King 2003: 72-73; Burden 1996: map # 84 (Americas map); Sabin 1868: entry # 73194-98.

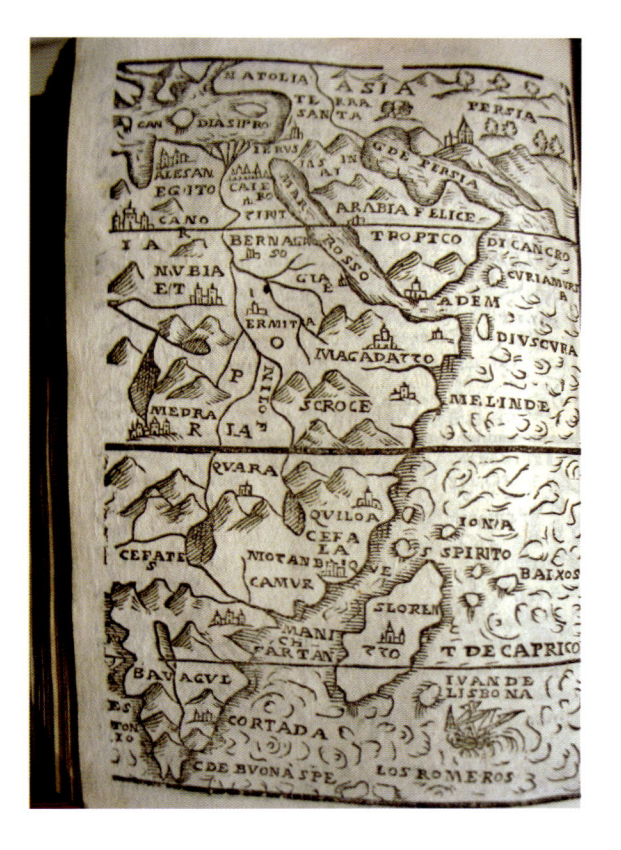

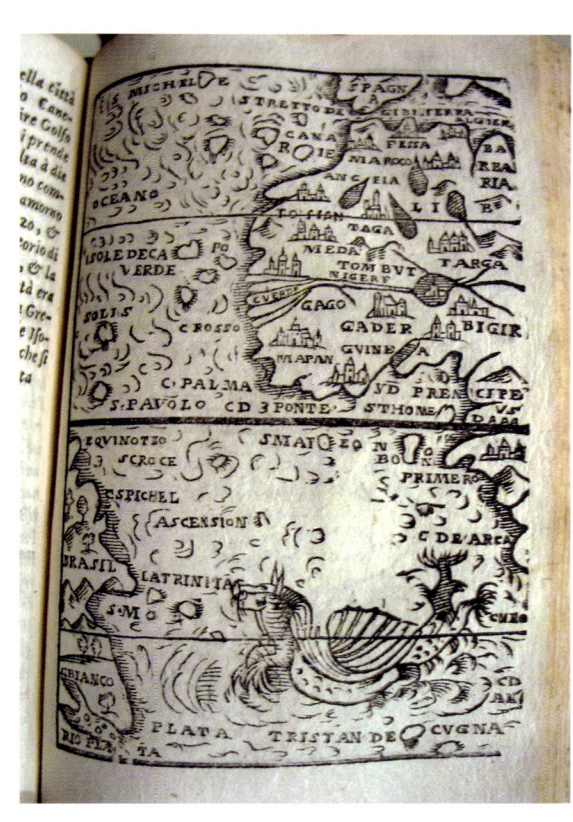

Map 30 from Mondo Elementare et Celeste *(1604)*.

31 1595 Gerard Mercator II, Duisburg

MAP 31

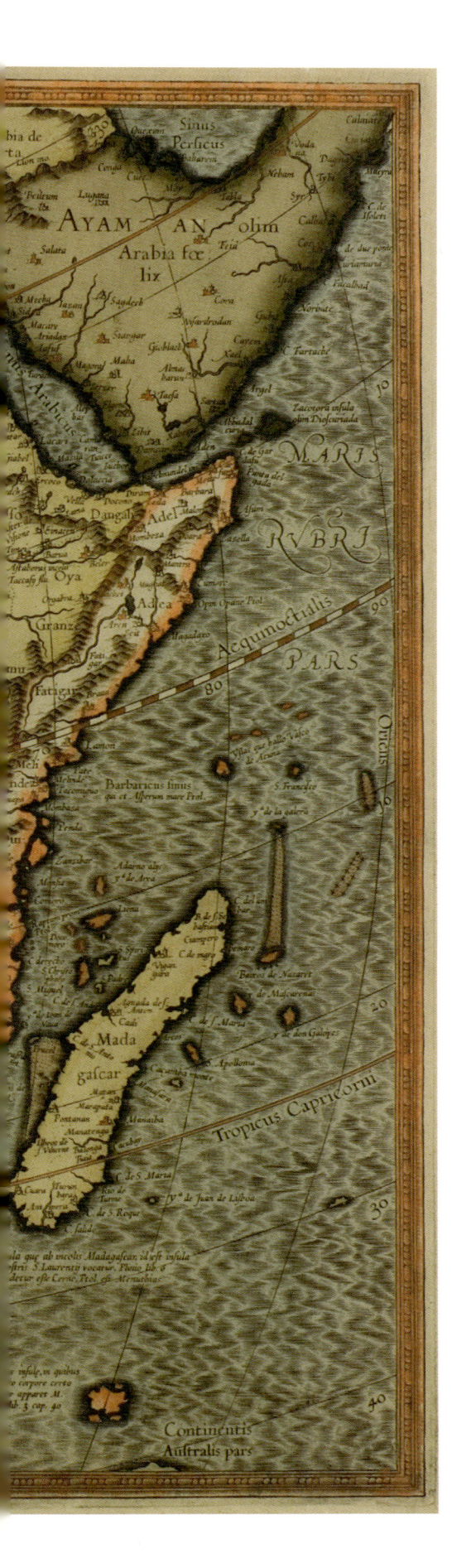

AFRICA | Ex magna orbis ter- | rę descriptione Gerardi | Mercatoris desumpta, | Studio & industria | G.M. | Iunioris.
[at bottom left corner of the map]:
Cum Priuilegio.
- Scale bar below title cartouche on two lines: Miliaria Germanica communia quorum 15 gradum unum | latitudinis constituunt
  150 = 4.8 cm.
- [Duisburg : Rumold Mercator, 1595].
- From: *Atlas sive Cosmographicae Meditationes de Fabrica Mundi et Fabricati Figura...* .
- 1 map : copperplate engraving ; 38 x 47 cm.
- Latitude coordinates across map at left and right sides. Longitude coordinates along the Equatorial Line.
- North at the top; no compass rose.
- Title set within an elaborate cartouche at bottom left of the map.
- The only figurative representation on the map is an image of the legendary priest-king Prester John in Abyssinia.

Examples: *Various locations*.

Description

This is a classic and important map of Africa, especially as it is directly based on Gerard Mercator's rare wall map of the world of 1569.

This map of Africa was prepared by Gerard Mercator the younger ('G.M. Iunioris' on title). Gerard Mercator II (c. 1565-1627) was the grandson of Gerard Mercator, the famous sixteenth century cosmographer. Gerard Mercator II seems to honor his grandfather by preparing this map.

Gerard Mercator (1512-1594) was noted for his rigorous and methodical approach to the preparation of his maps. He and his family members were responsible for the research and development of the maps, and much of their engraving, printing, and publication. This is in sharp contrast to Ortelius who, through his business acumen, created a prolific mapmaking empire by freely borrowing the maps of other cartographers and using various map engravers and colorists to issue his atlases.

To prepare the Africa section of the world map of 1569, Gerard Mercator relied on the Martin Waldseemüller modern maps of North and South Africa of 1513, the Gastaldi wall map of 1564, and the other maps produced by Gastaldi (the 1548 maps of North and South Africa, and the Ramusio-Gastaldi map of 1554) for the general shape and coastal features. The map showed a fair degree of accuracy especially along the coastline that was unsurpassed by most sixteenth century cartographers, though he portrays Madagascar as slightly too elongated to the northwest.

Mercator, like many others, used the writings of Ramusio, Leo Africanus, and others for additional details. He also relied on classical sources for his depiction of the interior of Africa. These included the work of Herodotus to show the river systems of West Africa with the top river (*Giras* and then *Nubia flu*) flowing into the Nile river and the lower river (which is labeled as the *Nil* on the 1595 folio map) flowing due westward into the Atlantic.

Mercator continued the evolving Ptolemaic tradition with his placement of the *Lun´ Montes* (Mountains of the Moon) as the headwaters of the two Nile source lakes in the far southern area of Africa. Ignoring the Ramusio-Gastaldi 1554 map of Africa, Mercator portrays the source of the Cuama River and the Spirito Santo River as being a common river, the *Zambere flu*, originating in Sachaf Lacus (1569) or *Sac.| haf lac* (1595) and not the western Ptolemaic lake of *Zembere*. This third lake, [sic] Sacaff Lacus, first appeared on Waldseemüller's 1507 world map, though without a connection to the southern flowing rivers. Mercator places this third lake, *Sachaf Lacus*, to the west of the *Lun´ Montes* (Mountains of the Moon). Interestingly, this third lake also becomes a source for the Nile River as well as the Zaire (Congo) River in the west.

| 163

Partly owing to the importance of the Mercator name, *Sachaf Lacus* was to appear in maps of Africa well into the late seventeenth century.

Importantly, Mercator also introduces a river above the Cuama, the *Zuama flu*, though too far north. In the late seventeenth century maps of Africa, particularly those by the French, this Zuama River was to become a more accurately placed Zambezi River (Ortelius also placed a *Zuama fl.* on his map, but this river appears to be the Cuama River placed too far to the south.).

Mercator seems to have had access to Portuguese texts as well. He shows European advances into the interior, for example, Portuguese exploration up the Cuama River into the interior of South-central Africa in the region of Monomotapa, or *Benamataxa*, as he names it on this map. *Ca. Portogal* (the Portuguese Fort) is placed on the map within the junction of the Spirito Santo and Cuama rivers.

Mercator tried to adhere to a more scientific approach and did not include the numerous sea creatures or beasts of other cartographers. The only figure within Africa is the seated, legendary priest-king Prester John in Abyssinia. In the 1595 folio Mercator map the finely engraved cartouche is embellished with fruit and two satyrs. The map has the uniquely-engraved, moiré-pattern sea of the Mercator maps and superb italic script, which Gerard Mercator was the first to design for maps.

Publication Information
This map appeared for the first time in the *Atlantis Pars Altera* (the Other Book of the Atlas) of the *Atlas* of 1595, published less than one year after the death of Gerard Mercator in December 1594. The atlas was published by Mercator's son, Rumold.

Mercator had originally intended his atlas to be an all encompassing work describing the world. The first part was published in 1585 and 1589 with maps of Europe and, with Mercator's death in 1594, the second part including maps of the continents was issued by his heirs in 1595 with sections lacking, notably Spain. The heirs published an unchanged edition in Duisburg in 1602 with the same map of Africa, but ended any further attempt at a more comprehensive atlas. This was likely due in part from the competition provided by the immensely successful Ortelius atlas, but probably also due to the deaths of Rumold in 1599 and Michael in 1600. In 1604, Mercator's heirs sold the copperplates to Jodocus Hondius and Cornelis Claesz. In 1606, Jodocus Hondius, along with Cornelis Claesz., introduced a completely revised edition of Mercator's atlas using this map of Africa, but also including a new map of Africa prepared by Hondius, new maps of the other continents, and 36 new, regional maps to fill gaps in Mercator's 1595 edition.

This map is known to exist in only one state from its first publication in 1595 to its final appearance in a French edition of the Mercator-Hondius atlas (*Gerardi Mercatoris Atlas*) of 1633.

The following chart can be used to differentiate the Mercator Africa map in the various editions of the book in which the map appeared using, from the verso, the language of the text, signature, and page number. This information is from Van der Krogt (1997):

Date	Van der Krogt no.	Language:	Signature:	Page Number:
1595-1602	1:1:011/12	Latin	3 C	{no pg no.}
1606	1:101	Latin	4 G	37-38
1607-08	1:102	Latin	4 G	35-36
1609	1:111	French	4 G	35-36
1611-12	1:103	Latin	4 G	35-36
1613-19	1:104	Latin	4 K	35-36
1613-16	1:112	French	4 H	41-42
1619	1:113	French	4 G	35-36
1623	1:105	Latin	4 K	35-36
1628	1:114	French	4 Q	63-64
1630	1:107	Latin	4 K	35-36
1633	1:311	French	4 Q	63-64

References
Van der Krogt 1997: map 8600:1A; Norwich 1997: map # 21; Karrow 1996; Klemp 1968.

32   1595 Giovanni Botero, Venice

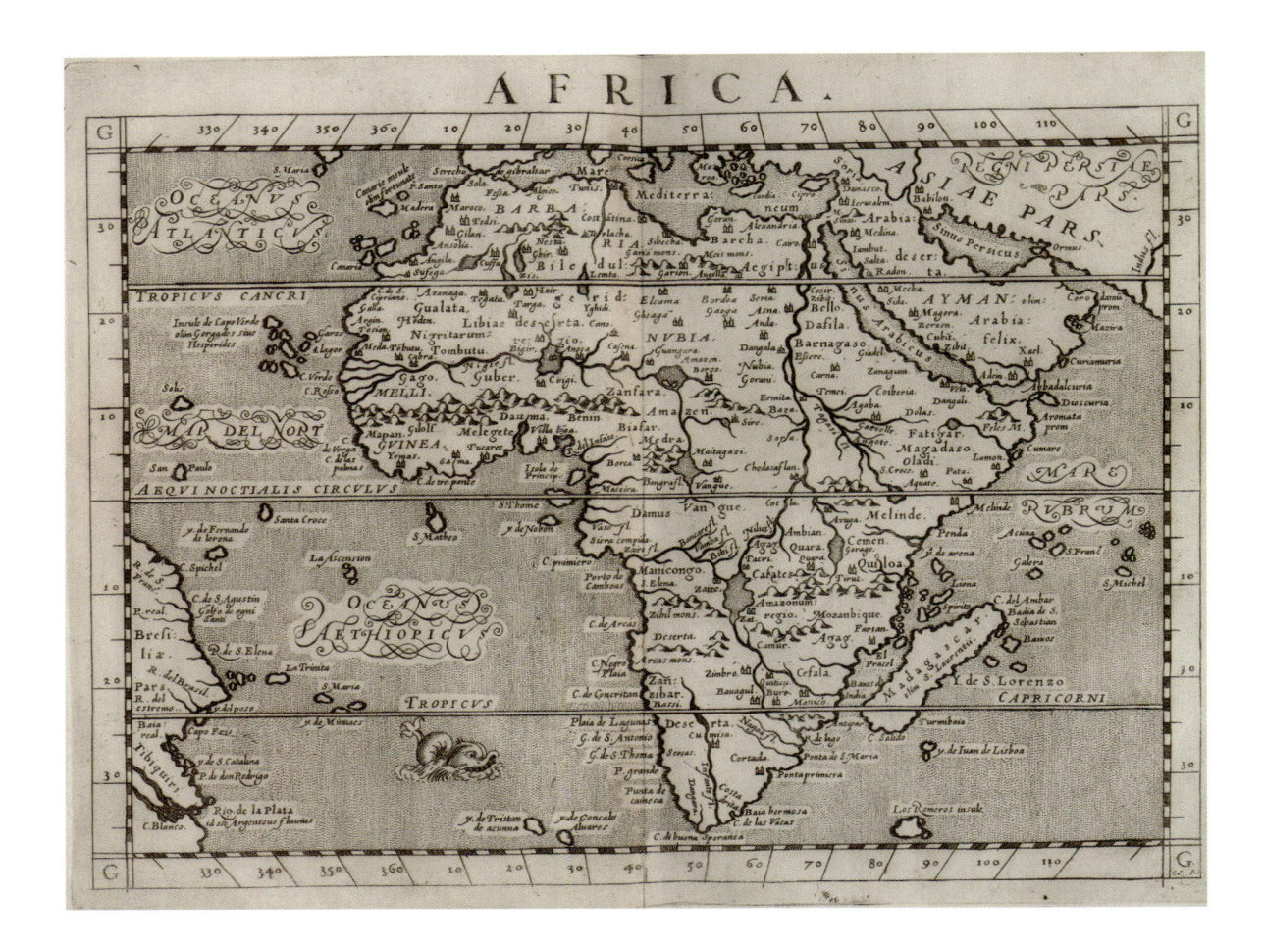

32.1 First State 1595

AFRICA.
- No scale bar.
- [Venice : Giorgio Greco, 1595].
- From: *Le Relationi universali di Giovanni Botero Benese: divise in tre parti.*
- 1 map : copperplate engraving ; 17.5 x 24.5 cm.
- Latitude and longitude coordinates on all four sides of the map.
- North at the top; no compass rose.
- Title above the map.
- One sea monster in South Atlantic Ocean ; no other ornamentation on the map.
- *Aegiptus.* has both 'u' and 's' to the east of the Nile River; *Dauma* in West Africa near the Guinea coast (not *Dama* as in 1582 D'Anania map). 'G' in the four corners of the map. At the bottom right corner below 'G' is *Con Pri.*

Examples: 1595 Botero: *Newport News MM*, G 120.B73 C; *London BL*, 1295.f.6; D'Anania 1596: *New York PL*, *KB 1596 Anania; *Washington LC*.

32.2 Second State 1599
Addition of 'Poro' in lower left below 'G' in the map grid.

Examples: This state appeared in an edition of Botero's *Le Relationi* of 1599: *Chicago NL*; *London BL*, 793.f.6.

32.3 Third State 1600
Addition of binder's instructions of 'Fac – 153' above map grid in top left corner.

Examples: This state of the map appeared in an edition of Botero's *Le Relationi* of 1600 by Giorgio Angelieri in Venice: *London BL; New York PL*, *KB 1600 75-879.
This same state of the map also appeared in an edition of Botero of 1602 published by N. Polo in Venice: *New York PL*, *KB 1602 Botero.

Description

This map is from Giovanni Botero's *Le Relationi*. A popular work that was reprinted over many years and in a number of languages, the *Le Relationi* was a geographical and political discourse on the world. The book contained maps of the four continents. Giovanni Botero (1540-1617) was known as a priest, geographer, and Secretary to the Duke of Savoy in Italy.

The map of Africa is closely modeled after the D'Anania map of 1582, which in turn was modeled on the Abraham Ortelius map of 1570. Many of the placenames on the map are directly from Ortelius. The hydrography also follows Ortelius. For example, the unnamed Zambere River on this map has its source in the western of the two Nile source lakes in Central Africa and then splits into the unnamed Cuama and Spirito Santo rivers.

This map was likely engraved by Girolami Poro of Padua (1520-1604), whose imprint appears at the bottom left corner of this map in its second state. As a result, it is sometimes referred to as the Poro map of Africa.

Publication Information

Botero's *Le Relationi* was first published in Rome in 1591, but this first edition did not contain the map of Africa. It appeared for the first time in the first Venice edition of Giovanni Botero's *Le Relationi Universali*, published by Giorgio Greco in 1595. This map also appeared in Giovanni Lorenzo D'Anania's *L'Universale* of 1596 and in Giovanni Botero's *Le Relationi* of 1599, 1600, and 1602.

The history of the map is long and complicated. What makes this map especially difficult to precisely catalog is that, on occasion, either the 1582 D'Anania or the 1595 Botero map of Africa was inserted into various editions of Botero's *Le Relationi*. The author has seen both a later state of this Botero 1595 map as well as the D'Anania 1582 map used in Botero's 1595 *Le Relationi*. Only through a continual examination of various examples of the 1595 Botero edition was it possible to determine this map's first appearance and its subsequent states. The Africa map is similar in appearance to both D'Anania's map of 1582 and Magini's map of 1596, which further complicates its history.

There were numerous editions of Botero's *Le Relationi* that did not contain maps. A map of Africa did not appear in editions of Botero's *Le Relationi* of 1595, published by Ventura in Bergamo, nor in the 1595 Rome edition, nor in the Brescia edition. Also, the map did not appear in the English edition of Botero of 1630.

To distinguish this map from similiar appearing maps, using different copperplates, by D'Anania (1582), Magini (1596), Keschedt (1597), Africanus-Pory (1600), Rossaccio (1605), and Botero-Giunti (1640), please refer to the table under the 1582 D'Anania map (Map # 20).

References
Burden 1996: map # 86 (Americas map).

33   1596 Cornelis de Jode, Antwerp

AFRICÆ NOVA DELINEATIO.
- Scale bar at bottom to left of Southern Africa: Scala Miliarium Italicorum 800 = 13.5 cm.
- [Antwerp : Cornelis de Jode], 1596.
- From: separately published.
- 1 wall map : copperplate engraving ; 11 map sheets ; 102 x 122 cm (the map is printed from eight large copperplates for the main map and three small and narrow ones for the title strip at the top).
- Latitude coordinates across map at center and at left and right sides. Longitude coordinates along Equatorial Line.
- North at the top; Large compass rose in South Atlantic Ocean.
- Title is on a separate strip above the map with a scene on either side.
- There is a large text box at the bottom left with 'new' and 'ancient' placenames. The date of 1596 is

within a very decorative medallion above the textbox. A large text box is at the top right of the map describing Africa with a reference to Gastaldi's 1564 wall map of Africa as the source for this map. Various smaller, decorative text boxes are also on the map.
- There are a number of other decorative elements on the map. To the left of the title is an allegorical representation of Africa and to the right is a view of a procession of Prester John (*Presbyter Iohannes magnus Aethiopiae Imperator*). There are twelve sailing ships and smaller coastal vessels in the Indian Ocean and 13 sailing ships in the Atlantic. There are five ships in the Mediterranean. The coat of arms of Portugal is off the southwest coast of Africa. Animals and people are placed throughout Africa.

Examples: The sole complete example of the 1596 map is at the Staatsarchiv, Nürnberg (Nürnberger Karten und Pläne no. 1260). There is a fragment (bottom left sheet) of the 1596 map in the Plantin-Moretus Museum, Antwerp.

Description
This wall map of Africa by Cornelis de Jode is actually the second state of a map originally prepared by Cornelis' father, Gerard. By 1569, Gerard de Jode published in Antwerp a re-engraving of Giacomo Gastaldi's 1564 wall map of Africa, so as to capitalize on the fame of Gastaldi's map. It is possible that Gerard acquired, or at least closely examined, an example of the Gastaldi wall map at the Frankfurt Book Fair, where De Jode exhibited and was known to often acquire Italian maps and prints during the active, continent-wide trade that occurred at this fair.

The geography for this map is a faithful copy of Giacomo Gastaldi's map. De Jode did add a number of decorative, attractively-engraved elements. Of note, are two lines indicating the route followed by a delegation of Japanese nobility from Japan to Rome (*Continuatio itinerationis Legatorum Iaponensium versus Romam*) to the east of Madagascar and around the Cape of Good Hope in 1582-85. See the Gastaldi-Licinio 1564 map (Map # 9) for a discussion of the geography.

The map was engraved by the renowned copperplate engravers Joannes van Doetecum the elder (?-d.1600) and his brother Lucas (active from 1558-c.1579).

Publication Information
The first state of Gerard de Jode's 1569 wall map is not known to exist. This second state, dated 1596, of the 1569 Gerard de Jode wall map appears to have survived in only one complete example (Karrow 1993: 244. Schilder 1996: 8). This example has some slight modifications to the 1569 plates of Cornelis father, Gerard. These are changes to the wording in the large bottom left text box translated as follows: *with the same industry as he edited the continents, Cornelis de Jode not only <u>extended this map</u> but also decorated it beautifully and published it at his own expense*. Schilder also observed that the title for Cornelis' wall map appears to have been affixed by him and was not present in the first state of this map by Gerard de Jode.

As noted by Burden (1996: 61), while the sole De Jode example of the Africa map bears the date of 1596, Cornelis could have prepared the map in 1595. Cornelis de Jode makes reference to this wall map in a 1595 publication of his that was intended to be an accompanying pamphlet to the set of wall maps of the four continents in 1595 entitled *Introductio Geographia in Tabulas Europae, Asiae, Africae et Americae*; no examples of the pamphlet survive. The copperplates for the wall map eventually passed to the Officina Plantiniana as these plates are recorded in their possession in an inventory of existing stock upon Jan Baptist Vrients death in 1612. No further use of the copperplates is known.

References
Schilder 1996; Burden 1996: 61 (map # 47); Karrow 1993: 24: 30/98.6; Schilder 1981: 2-8.

34 1596 Giovanni Antonio Magini, Venice

AFRICA.
- No scale bar.
- [Venice : Heirs of Simonis Galignani de Karera, 1596].
- From: *Geographiae Universae tum Veteris tum Novae...illustratus est à Io. Antonio Magino Patavino... .*
- 1 map : copperplate engraving ; 12.5 x 17 cm.
- Latitude and longitude coordinates on all four sides of the map.
- North at the top; no compass rose.
- Title is above map.
- This map is generally devoid of ornamentation except for one small sea monster in the South Atlantic Ocean. No flora or fauna in Africa. 'G' in the four corners of the map.

Examples: *Cambridge Harv MC*, MAP-LC G1005 1596* (In the 1596 edition on the verso of the map, *CRETAE* is at the top with *184* on top left; line of text at bottom ends *ne urbi Candiae ad Septentrionem iacens*. Bottom right: *AFRICAE.*); *London BL*, Maps C.1.b.7; *New York PL*, *KB 1596 (in Ptolemaeus, C. *Geographiæ Universæ*); *Author's collection* (as a seprate map).

Description
This is Giovanni Magini's modern map of Africa from his edition of Claudius Ptolemy's *Geographia*. It is basically a slightly-reduced version of the D'Anania 1582 map of Africa, which in turn was modeled after Ortelius' 1570 map. The shape of Africa on the map is quite accurate for this period, and considerable geographical details along the African coasts are shown.

Giovanni Magini (1555-1617), originally from Padua, was a professor of astronomy at the University of Bologna. Magini was the author of the 1596 edition of Ptolemy's *Geographia* as well as the first printed atlas of Italy, among his numerous other publications. The finely-detailed copperplate for this map was engraved by Girolamo Porro of Padua (1550-1604).

Publication Information

The Africa map appeared in Magini's editions of Ptolemy's *Geographia* in 1596 and 1598, published in Venice, and in 1621, published in Padua. The Padua edition was the final edition of this book. Besides translating and editing Ptolemy's text, Magini also wrote additional text describing the modern maps which followed the section by Ptolemy. This section was titled *Novae Geographicae Tabulae*.

Besides the 1596 edition described above, the following editions of Magini's *Geographia* were examined by the author. The information with each edition may be used as a guide to determine from which edition an individual map appeared:

1598 Map set within a page of text. Text above map *XXIII. | DESCRITTIONE DELL' AFRICÆ | QVELLA DELL' ISOLE,| CHE A LEI S'ASPETTANA..* Map title *AFRICA* above map. Text below map starting with large 'L' set within a woodcut engraving with a woman, bird, and vines. Last, word on page *Africani*. Verso: At top, *DELL' ISOLA DI CRETA* with *130* at top right corner. At bottom, *DE-*. Last line of text is *ne dirimpeto alla citta di Candia*.

This edition was translated from Latin into Italian by Leonardo Cernoti and printed by the Galignani
brothers (*Appresso Gio Battista & Giorgio Galignani, Fratelli*).

Examples: *London BL,* 216.d.5; *New York PL,* *KB+ 1598 (in Ptolemaeus, C. *Geografia*).

1621 This is the final edition of Magini's *Geographia*. In this edition, the map is set within a page of text. Letterpress text above map begins *XXIII. | DESCRITTIONE DELL' AFRICA| E QVELLA DELL' ISOLE,|.* Map title *AFRICA* above map. Verso: At top, *DELL' ISOLA DI CRETA* with *130*. At bottom, *DE-*. Last line of text is *Settentrione dirimpeto alla citta a di candia*.

This is also an Italian text edition, published in Padua by Paolo & Francesco Galignani.

Examples: *London BL*, Maps C.1.c.8; *New York PL*, *KB 1621 (in Ptolemaeus, C. *Geografia*).

This Magini plate of Africa had an extremely long life. The plate was later used to produce the map of Africa for editions of Abraham Ortelius' *Theatro del Mondo*, published in Venice by Stefano Curti in 1679, 1683, and 1684. These editions only contained the four maps of the continents, all with blank versos (Van der Krogt 2003). Finally, the Magini map's last known appearance was in Raphael Savonarola's (or as he is commonly known, Lasor a Varea) *Universus Terrarum Orbis Scriptorum*, published in Padua in 1713. In this work, the map appears set within a page of text.

To distinquish this map from similiar appearing maps, using different copperplates, by D'Anania (1582), Botero (1595), Keschedt (1597), Africanus-Pory (1600), Rossaccio (1605), and Botero-Giunti (1640), please refer to the table under the 1582 D'Anania map (Map # 20).

References
Van der Krogt 2003: atlas 33A:21-23; Norwich 1997: map # 24; Mickwitz et al. 1979-1995 Vol. 2:190-193: entry 223; Tooley 1966.

35   1597 Petrus Keschedt, Cologne

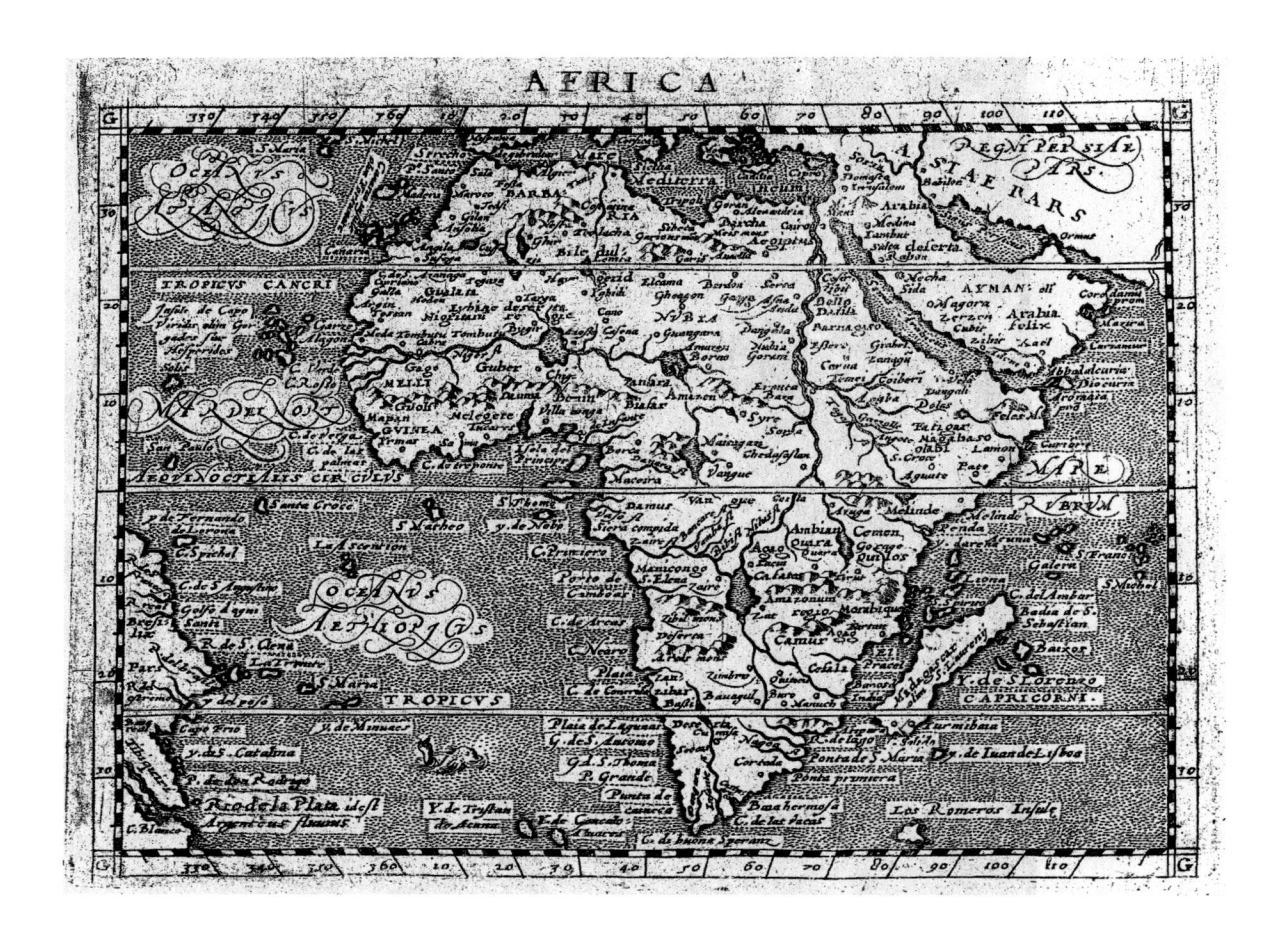

AFRICA.
- No scale bar.
- [Cologne : Petrus Keschedt, 1597].
- From: *Geographiae Universae tum Veteris tum Novae...Auctore eodem O. Ant. Magino Patavino... Excudebat Petrus Keschedt.*
- 1 map : copperplate engraving ; 12.5 x 17 cm.
- Latitude and longitude coordinates surround the map.
- North at the top; no compass rose.
- Title is above map.
- The map is generally devoid of ornamentation. There is one small sea monster in the South Atlantic Ocean. No flora or fauna in Africa. 'G' in the four corners of the map.

Examples: *Amsterdam UB,* 1802 E 3; *Chicago NL; London BL,* Maps C.1.b.8; *New York PL,* *KB 1597 (In the 1597 edition, the verso of the map has *184* in the upper right corner. Capital *C* on a formal arabesque design. The last line of text on the page begins *gione*. At the bottom right is *AFRICAE.*).

Description
This map was produced by Petrus Keschedt for his edition of Ptolemy's *Geographia*. It is a close copy of Giovanni Antonio Magini's 1596 map of Africa for Magini's own edition of Ptolemy. The minor differences between the two maps are noted in the table under Map # 20. Like the Magini map, the primary model for this map was the D'Anania 1582 map of Africa which in turn was modeled on the Ortelius 1570 map.

An extensive part of Keschedt's book is devoted to the modern, non-Ptolemaic geography, 'Novae Geographicae'. The text is in Latin. The engraver for this map is not known.

Publication Information

Although Petrus Keschedt acknowledges Magini as the author in the title of his edition of Ptolemy's *Geographiae Universae* that he published in Cologne in 1597, it appears that this edition was issued without Magini's consent or knowledge. Regarding this unauthorized use, Ortelius informed Magini on 4 November, 1597 about this Keschedt publication, after which the Keschedt book was not re-published until 1608 (Meurer 1988: 158). It is possible that Magini attempted to stop the re-publication of the Keschedt issue of the book, but that is not known.

In any event, a second edition was published by Keschedt in 1608 in Cologne. Sometime after 1608, the copperplate of the Africa map, along with the other copperplates in the book, was acquired by the publishing firm of Jan Jansz., the father of Johannes Janssonius. Jansz. published another edition of Magini's version of Ptolemy's *Geographia* in 1617 in Arnhem using this map of Africa.

Besides the 1597 edition described above, the following editions of Keschedt's *Geographia* were examined by the author. The information with each edition may be used as a guide to determine from which edition an individual map appeared:

1608: On verso of the map is *155* in the upper right corner. Last line of text begins with *rij;*. At the bottom right is *AFRICA*.

Examples: *Cambridge Harv MC,* MP 2.1608. *Chicago NL; London BL,* Maps C.1.b.10.

1617: On verso of the map is *176* in the upper right corner. Last line of text begins with *diae*. At the bottom right is *AFRICA*.

Examples: *Cambridge Harv RB; Chicago NL.*

The author also observed the use of this map in an edition of Matthias Quad's *Enchiridion Cosmographicum: das ist, ein Handbüchlein, der ganzen Welt...,* published in Cologne by Wilhelm Lützenkirchen in 1599 (*Wien ON*, ÖNB/Kar: 364044-B.Kar., on page 200-201 in the book).

To distinquish this map from similiar appearing maps, using different copperplates, by D'Anania (1582), Botero (1595), Magini (1596), Africanus-Pory (1600), Rossaccio (1605), and Botero-Giunti (1640), please refer to the table under the 1582 D'Anania map (Map # 20).

References

Van der Krogt 2003: map 8600:381; Meurer 1988: 155-161, Mickwitz et al. 1979-1995: Vol. 2: 193-198: entries # 224 and 225.

36   1597 Fausto Rughesi, Rome

AFRICA.
[within the title cartouche and just below the title are eleven lines of text]:
Trium terrae partium, uiribus minima, Mari Ru: | bro Mediterraneo, et Oceano terminatur. Qua mari noi: | tro, et Oceano Atlantico alluitur, omnibus fere, quibus Eu | ropa bonis affluit. Vlteriora tum palmae, et cameli tum ele: | phantes, et ignota nobis animalia tenent, pleriq.eius amnes, | maxime qui Aethiopiam alluunt, statis temporibus ex: | undant, agrosq, tum rigant tum fecundant ad | incredibilem frugum ubertatem. Africae | incolae ultra Girim fluuium | ignota naturae ui, uario | nigrore colorantur
[in the lower right hand corner of the map]:
Romae | CIƆIƆ XCVII | Cum Privilegio.
- No scale bar.
- Rome : [Fausto Rughesi], 1597.
- From: separately published.
- 1 map : copperplate engraving ; 2 map sheets ; 53 x 68.5 cm.
- Latitude and longitude coordinates on all four sides of the map. Longitude coordinates across the

Equatorial Line.
- North at the top; no compass rose.
- Title is set at the top of a large, decorative cartouche in the lower left of the map.
- This is a very decorative map with four sea monsters, four ships, and a group of two mermen and one mermaid in the Indian Ocean, and five sea monsters, one ship, and one shipwreck in the Atlantic Ocean. There are no flora or fauna within Africa. A dedicatory cartouche to Vincenzo Gonzaga, the Duke of Mantua, is at the upper left of the map. An unidentified coat of arms is at the upper right of the map.

Examples: The two known examples are at: *Austin HR*; *Roma BVat*.

Description

This exceedingly beautiful, finely-engraved map of Africa is by Fausto Rughesi, an architect from Rome. While Rughesi is not well-known as a cartographer, this map exhibits an understanding of cartographic concepts. How this understanding was developed is not known and in fact little else is known about Rughesi.

This map of Africa is generally modeled on Ortelius' map of Africa of 1570. The shape of Africa on the Rughesi is most similar to the Ortelius with its more pointed southern end, and most of the hydrography follows Ortelius. On the Rughesi, the larger, western Nile source lake of *Zaire et Zenbre* in Central Africa is considerably further to the south than the eastern lake of *Zaflan*. The lake in the Abyssinian highlands depicted as the source for the Blue Nile is on the Equator. The western Nile source lake is the source for the *Zaire* (Congo River) and the unnamed Zembere River, which then divides into the *Cuama* and *Spirito Santo* Rivers further south. The Niger River system is the same as on the Ortelius map.

However, Rughesi seems to partially follow the Mercator 1595 map of Africa. He places a lake (the *Loana*?) in the Congo region with several small tributaries flowing into the Atlantic Ocean. Though, unlike Sachaf Lacus on the Mercator map, this lake does not empty into the Congo River to the northwest, the *Zembere* River to the south, or the Nile River to the north. Also, the island of Madagascar is shorter and wider than as shown on the Mercator map, and it is named *Lorenzo*, not Madagascar as on both Ortelius's and Mercator's maps.

Publication Information

Fausto Rughesi, originally from Montepuliciano in the Tuscan part of Italy, was well-known as an architect in Rome. In 1597, he produced a map of the world and a map of each of the four continents. He is not known to have produced any other maps. All of his maps are exceedingly rare.

There is the possibility of an earlier proof or published state that is unrecorded. The University of Texas set of maps of the world and the four continents all bear imprint erasures. Also, as noted by Burden (1996: map # 108), there are some nomenclature alterations on the Americas map.

36.1	Possible First State ?		Date on map of ?.
36.2	Second State	1597	Date on map of: CIƆ I ƆXCVII. Note: The date as represented on the map means that CIƆ equals M, and IƆ equals D; thus the map's date as expressed more conventionally is MDXCVII or 1597.

The only known examples of the Africa map by Rughesi are at the Harry Ransom Humanities Research Center of the University of Texas at Austin, and at the Bibliotheca Apostolica Vaticana. Almagia (1944-55: 69-71) has a reproduction of the Vatican example. The Texas set was originally part of the Hauslab-Liechtenstein Collection. The Texas example has a watermark of a kneeling figure with a halo, holding a cross (similar to Briquet 1997: entry 7628).

References
Briquet 1997; Burden 1996: map # 108 (Americas map); Kraus 1956: 94; Almàgia 1944-55.

37   1597 Cornelis Claesz.- Barent Langenes, Middelburg

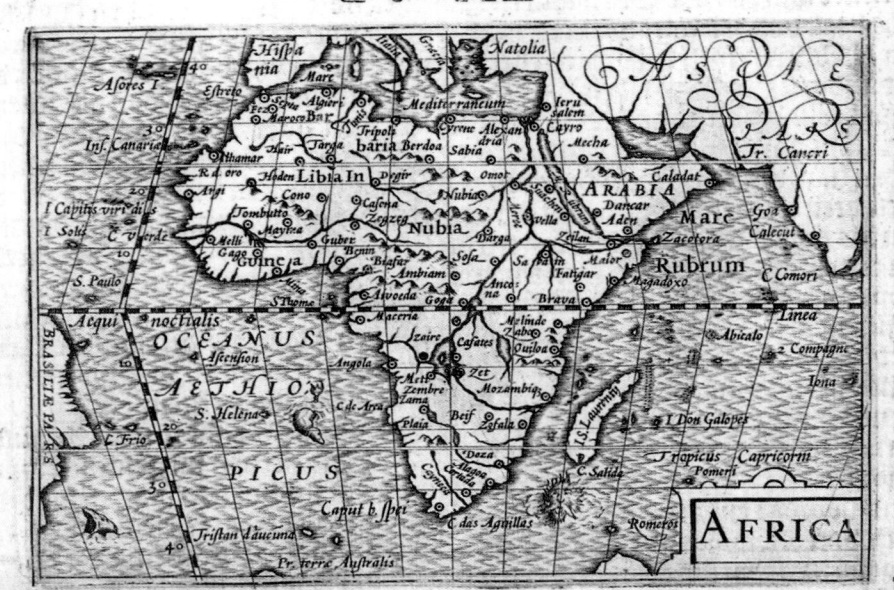

37.1 First State 1597

AFRICA.
- No scale bar.
- [Middelburg : Barent Langenes, 1597].
- From: *Verhael vande Reyse by de Hollandtsche Schepen gedaen naer Oost-Indien.*
- 1 map : copperplate engraving 8.5 x 12.5 cm.
- Latitude coordinates along the Prime Meridian. No longitude coordinates.
- North at the top; no compass rose.
- Title cartouche as a simple box at lower right.
- Generally devoid of ornamentation except for a face of an animal or sea monster in the Southern Indian Ocean and two sea monsters in the Southern Atlantic Ocean.

Examples:
1597   *Verhael vande Reyse by de Hollandtsche Schepen gedaen naer Oost-Indien: Rotterdam MM*, ARCH 3.A.30 (The map is on a page of Dutch text with letterpress above the map of 'Het punt van Africa.').
1598   *Iournal du voyage de l'Inde Orientale: New York PL,**KB 1598 (Langenes. Iovrnal Du Voyage...).
1598   *Iournal du Voyage de l'Inde Orientale,* journal published in Middelburg for Adrien Perier of Paris: *New York PL*: *KB 1598 (Langenes *Iovrnel du Voyage*).

1598 *Caert-Thresoor*, Middelburg, published by Barent Langenes: *London BL*, Maps 39.a.2.
1599 *Caert-Thresoor*, Amsterdam, Cornelis Claesz.: *London BL*, Maps C.39.a.3.
c.1600 *Thrésor de Chartes*, Den Haag, Albert Hendricks for Cornelis Claesz.: *Amsterdam UB*, 1802 G 2.
1602 *Thrésor de Chartes*, Leiden, Christoffel Guyot for Cornelis Claesz.: *London BL*, Maps.C.39.a.4.
1609 *Handboeck of Cort Begrijp der Caerten*, Amsterdam, Cornelis Claesz.: *Amsterdam UB*, 1802 G 3.
c.1609 or c.1610 *Thrésor de Chartes*, Frankfurt, Matthaeus Becker for Hendrick Laurensz.: *Rotterdam MM*, WAE 10; *Washington LC*, G.1015.L292 1610 Vault.

37.2 Second State 1649
Addition of engraved 'g . 1' just below the title of AFRICA in bottom right corner of the map.

Example: *London BL*, Maps C.39.a.10 (1649 *Tabulae* by Visscher).

Description
This is an elegant, well-engraved map of Africa, particularly given its relatively small size. By the late 1590's, a large amount of cartographic information on the newly discovered lands in Africa, Asia, and America was starting to flow into The Netherlands. This map reflects this expansion in cartographic knowledge, including information supplied by Jan Huygen van Linschoten in his *Itinerario* of 1596.

This map is an amalgam of the Gastaldi-Ortelius and the Mercator models. Like Mercator, this map shows a river system above the Niger in West Africa flowing into the Nile. Like Ortelius, this map shows the Congo River originating in the western Ptolemaic lake of Central Africa. However unlike Ortelius' southward flowing Zembere river which splits into the Cuama and Spirito Santo Rivers, on this map the unnamed river flowing southward does not divide. The information provided by these basic models seems to have been supplemented with information on the more recent discoveries in the Congo and on the Southeast coast of Africa, notably from the Van Linschoten maps of Southwest and Southeast Africa.

A unique feature is a single lake at the Equator into which both the two Ptolemaic lakes of Central Africa flow. This third lake seems to have been based on the third lake on the Al-Idrisi map of the world, which shows a similarly positioned third lake. Al-Idrisi's *The Book of Roger* first appeared in Rome, in Arabic as *Kitab nuzhat al-mushtaq...*, in 1592, and it is possible that Claesz. had access to this text.

The engraving of the map is likely an early work by Jodocus Hondius, who was employed by Cornelis Claesz. and who, along with Pieter van den Keere, engraved the plates for the 1598 *Caert-Thresoor*.

Publication Information
This map is often attributed to Barent Langenes though recent research by Günter Schilder (2003: 457-488) and Peter van der Krogt (2003: 373-375) suggests that the map may have actually been prepared in collaboration with Cornelis Claesz. (1560-1609) and should be equally attributed to him. Unlike Claesz., little is known about Langenes. He is known as a printer and publisher in the town of Middelburg, the capital of the province of Zeeland.

Langenes was the editor of the text as well as the publisher of the brief book in which this map of Africa first appeared in 1597. This book, *Verhael vande Reyse by de Hollandtsche Schepen gedaen naer Oost-Indien, haer avontuer ende succes, met de beschryvinghe der Landen daer zy gheweest zijn, der Steden ende Inwoonderen met Caerten ende Figueren verlicht seer ghenoechlijck om lesen - Middelborgh : Ghedruckt voor Barent Langenes...*, describes the accounts of the famous Dutch expedition around the Cape of Good Hope to the East under the command of Cornelis de Houtman in 1595-1597. The book was compiled and published by Barent Langenes from journals kept by an anonymous sailor aboard the lead ship, *Hollandia*. De Houtman's was the first Dutch expedition to round the Cape of Good Hope and establish trading contacts on Java. The book was published as early as November 1597.

There was an expanded and corrected version of the *Verhael vande Reyse by de Hollandtsche Schepen*, a Dutch language edition of the journal published in Middelburg in March 1598 entitled *Iournael vande*

Reyse der Hollantsche Schepen ghedaen in Oost Indien. However, a Claesz.- Langenes map of only Southern Africa and not one of the entire African continent was used in the 1598 *Iournael vande Reyse der Hollandtsche Schepen*, based on an examination of the Middelburg edition in the New York Public Library.

A French language edition was also published in Middelburg for Adrien Perier, though likely later in 1598 with the title of *Iournal du Voyage de l'Inde Orientale*. This edition contained the Cornelis Claesz.- Barent Langenes map of Africa on a page of text. A Latin edition of the book was also published in 1598 (Schilder 2003: 258).

Besides appearing in the 1597 *Verhael vande Reyse...*, this map also appeared in the *Caert-Thresoor*, a reduced-format atlas of the world with 169 maps. The *Caert-Thresoor* was first published by Barent Langenes in Middelburg in 1598. Cornelis Claesz. was a co-seller in Amsterdam of the first edition of the *Caert-Thresoor* in 1598 and Claesz. and his successors were the publishers of the subsequent editions of the *Caert-Thresoor* from 1599. It should be noted that there is some conjecture that there was an edition of the *Caert-Thresoor* published by Claesz. in Amsterdam in 1597, but no example has yet been located. See Van der Krogt (2003) for a further discussion of this point.

As noted by Van der Krogt (2003: 373), what makes this atlas particularly interesting is that, unlike many from this period and later, the text of the *Caert-Thresoor* was adapted to fit the maps and not the other way around.

Besides this map's appearance in the *Verhael* and the *Caert-Thresoor* , it also appeared in the *Tabulae Geographicae Contractae*, initially published in Amsterdam by Cornelis Clasez. along with Jan Jansz. in 1600, 1602/3, 1606 (published by Cornelis Claesz. in Amsterdam), 1612 (published by Hendrick Laurensz in Frankfurt), and 1650 (published by Johannes Janssonius in Amsterdam, using map sheets left from the 1612 edition). For the *Tabulae Geographicae Contractae* the earlier *Caert-Thresoor* text was replaced and reordered with text written by Petrus Bertius, using the maps as illustrations for the text. This map made its final appearance, in a second state, in Claes. Jansz. Visscher's *Tabulae Geographicae Contractae*, published in Amsterdam in 1649.

The following chart may be used as a guideline to determine from which atlas edition a separate example of this map appeared. This information is based on Van der Krogt (2003: 879) with his information on the language of the text, signature and pagination:

Date	Van der Krogt no.	Title	Language	Signature	Pg no.
1598	8600:341::01	*Caert-Thresoor*	Dutch	B8r	31
1599	8600:341:02	*Caert-Thresoor*	Dutch	B8r	31 (2M6r 91)
1600	8600:341:11	*Thrésoor de Chartes*	French	Clr	33
1600	8600:341:51	*Tabulae*	Latin	C6r	43
1602-03	8600:341:52	*Tabulae*	Latin	2G4v	472
1602-12	8600:341:12/13	*Thrésoor de Chartes*	French	Clr	33
1606	8600:341:53	*Tabulae*	Latin	2G4v	472
1609	8600:341:03	*Handboek*	Dutch	Clr	33; (2S3r 643)
1612-50	8600:341:61/62	*Tabeln*	German	2N6r	571
1649	8600:341:54	*Tabulae* (by Visscher)	Latin	g.1	

References
Schilder 2003: Vol. VII: 457-488; Van der Krogt 2003: map 8600:341; Werner 1998; Warnsinck 1939; Burger 1929: 289-304 and 321-344; Tiele 1867.

38   1597 Johannes Matalius Metellus, Cologne

AFRICA.
- No scale bar.
- [Cologne : Lambert Andreas, 1597].
- From: *Amphitheatridion hoc est parvum amphitheatrum, cui pauca mundi theatra compararae vix ullum anteponere Amphitheatrum possis.*
- 1 map : copperplate engraving ; 19 x 24 cm.
- No latitude and longitude coordinates.
- North at the top; no compass rose.
- Title within round cartouche at bottom left.
- Devoid of ornamentation.

Examples: *München SB*, Res/4 Geo.u. 6 k (1597 edition of Botero); *Stuttgart WL*; *Wien ÖNB*; *Author's collection* (as a separate map).

Description

This scarce map shows all of Africa with part of Arabia. It is relatively austere and does not contain any ornamentation except a plain, circular title cartouche in the lower left of the map.

Aspects of the basic geography are derived from Gerard Mercator's world map of 1569, particularly in the general shape of Africa and the island of Madagascar. Also following the Mercator map of Africa, Metellus does show European influence in the interior by including the Portuguese (*Portogal*) settlement up the unnamed Cuama River. *Buro mina Zimbaos*, to show the mines at Zimbabwe, is placed in the same area as *Zimbaos* on Mercator's map between the Cuama and Spirito Santo rivers.

However, there are a number of unusual depictions. The unnamed Spirito Santo River originates in the mountains in Southern Africa and not in a lake. The unnamed source lake for the Cuama River is probably the same as Mercator's Sachaf Lac, since it also feeds the Congo River to the west and the Nile to the north, as on the Mercator. What is unique on this map is that it does not depict the western of the Ptolemaic source lakes for the Nile, although the four small tributaries which flow into this lake on the Mercator map are clearly depicted in the same region and flow together in the same pattern toward the Nile. It seems that the author of this map inadvertently left out this lake while copying Mercator's model.

Publication Information

This map first appeared in the Latin edition of 1597 of *Amphitheatridion hoc est parvum amphitheatrum....* This work was translated and abridged from Part Two of Giovanni Botero's *Le Relatione Universale*. It was published in Cologne by Lambert Andreas, who sometimes has the attribution for this map.

The map was prepared by Johannes Matalius Metellus, or Jean Matal as he was known in his native France, who was a well-esteemed cartographer in his day. Metellus (c1520-1597) spent much of his working life in Louvain, but moved to Cologne after about 1560. Metellus' atlas, the *Speculum Orbis Terrae* published in 1602, appeared in parts from 1594 onwards. The complete edition of the *Speculum*, issued in 1602, was the most comprehensive atlas of its time with 261 maps. It appears that Matthias Quad, who was the editor for works by Metellus, Acosta, and Botero, also used Metellus' map of Africa in editions of Botero's *Mundus Imperiorum...* . It should be noted that some editions of Botero's book, published in Cologne, such as those of 1596, 1599, and 1602, only contained regional maps of Africa.

Upon Metellus' death in 1597, the work on the unfinished maps was completed by a colleague; Meurer (1998: 162-196) identifies him as Conrad Loew, a pseudonym for Matthias Quad.

The map of Africa appears to have been issued in only one state.

Besides appearing in the 1597 edition of Botero's *Amphitheatridion hoc est parvum amphitheatrum....*, this map appeared in the following books (with information on date, author, book title, publisher, and location):

1598: Botero, *Mundus Imperiorum...* published by B. Buchholtz in Cologne.

Example: *New York PL,* *KB 1598 Botero.

1600: Metellus, *Africa ad Artis Geographicae Regulas. published by* C. Sutorius in Ursel.

Examples: *Grenoble BM,* B 2503-2505; *München SB,* Mapp.121; *Cambridge Harv RB* (in this example, the Africa map was inserted into a Metellus book section in *America sive Novvs orbis*).

1602: Metellus, *Speculum Orbis Terrae* published by C. Sutorius in Ursel.

1603: Botero, *Mundus Imperiorum...* published by C. Sutorius in Ursel.

Example: *London BL,* 582.c.13.

References

Burden 1996: maps # 115-122 (Americas); Meurer 1988: 66-83 (describing Botero) and 162-196 (describing Metellus); Koeman 1967-71, III: 2; Church 1951; Sabin 1868.

39   1598 Abraham Ortelius – Pietro Maria Marchetti, Brescia

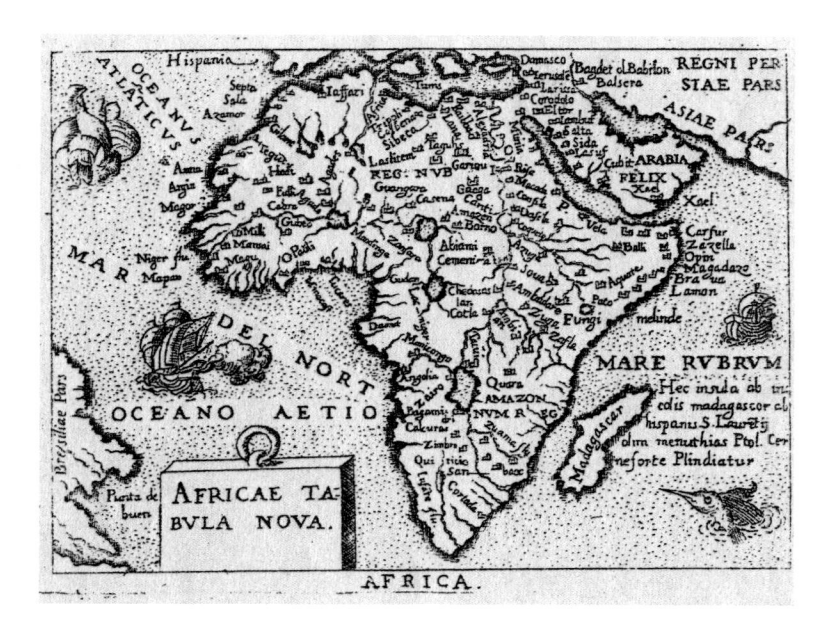

AFRICAE TA: | BVLA NOVA.
- No scale bar.
- [Brescia : La Compagnia Bresciana, 1598].
- From: *Il Theatro del Mondo di Abraamo Ortelio.*
- 1 map : copperplate engraving ; 8 x 10.8 cm (to outer black line).
- No latitude and longitude coordinates.
- North at the top; no compass rose.
- Title, at bottom left, set within a three-dimensional, box-like cartouche with a handle at the top. There is a secondary title, *AFRICA.*, engraved onto the map at bottom just below the neatline.
- There are two ships in the Atlantic Ocean, and one ship and one sea monster in the Indian Ocean.

Example: *London BL*, Maps. C.2.b.6.

Description

The great success of the 1577 pocket-size version of Ortelius' *Theatrum Orbis Terrarum* generated a number of imitators. This map of Africa is from the first issue of the pocket-size Ortelius atlas in Italy. The atlas was translated into Italian and edited by Pietro Maria Marchetti in Brescia, quite probably without Ortelius' approval.

This map of Africa is directly copied from the Ortelius-Galle (1) map of 1577 (Map # 18). Marchetti possibly chose to not use the more recent maps from the 1588 Ortelius pocket-size atlas (Ortelius-Galle (2), Map # 28), as Galle was still in the gradual process of preparing new plates, and the Ortelius-Galle (2) maps had much less cartographic detail. The particular edition of Ortelius-Galle (1) that Marchetti used as the model is not known. While the geography is identical to the Ortelius-Galle (1) map, the engraving is slightly different. For example, the oceans are done in a stippled (dotted) effect in the Ortelius-Marchetti map rather than with waves as on the Ortelius-Galle (1) map, and there is a somewhat less-polished style overall in the quality of the engraving.

Publication Information

The first issue of Marchetti's *Il Theatro del Mondo* atlas was in Brescia in 1598. The page with the map of Africa has 13 lines of text in letterpress below the map. At the top of the page in letterpress is *DEL MONDE* and to the right a page number *9*.

Shirley (2004: 668) mentions an edition of *Il Theatro del Mondo* of 1608, not in Koeman (1967-71, III: 82) that was also published in Brescia, but the author has not been able to locate this edition to determine if it contains this map of Africa. A subsequent edition of *Il Theatro del Mondo* was published in Venice by Giovanni Turrini in 1655, and a final edition was published in Venice by Scipion Banca in 1667, both with this map of Africa.

This map also appeared in a 1599 edition of Giovanni Botero's *Le Relationi* published in Brescia by Bresciana Compagnia. The Sotheby's catalog of February 6, 1990 (lot no. 292) lists a 1618 edition of Botero's *Le Relationi*, published in Venice by Alessandro Vecchi, with this same map of Africa, according to King (2003: 79).

The map appears unchanged in all editions of Marchetti's *Il Theatro del Mondo* and Botero's *Le Relationi*. Besides appearing in the 1598 edition of Marchetti's *Il Theatro del Mondo* described above, this map appeared in the following books (with information on date, author, book title, publisher, and location):

1599: Giovanni Botero's *Le Relationi*, published by Compagnia Bresciana in Brescia.
This map of Africa is set within a text page. In this example, the last line of text on the recto is *situa*. On the verso of the map, there is *286* at top left and the last line of text at the bottom ends *e di poco*.

Examples: *London BL*, 793.f.6; *Chicago NL*.

1655: *Il Teatro del Mondo* published by Giovanni Maria Turrini in Venice.

Example: *Amsterdam UB*, 1802 G 9.

1667: *Il Teatro del Mondo* published by Scipion Banca in Venice.
The Africa map in the last edition of 1667 is within an 8° book with no text on the page.

Example: *London BL*, Maps C.2.b.14.

A new map of Africa appeared in 1697 attributed to Domenico Lovisa (Map # 164), also based on the Ortelius-Galle (1) map. Though using a new copperplate, this Lovisa map is similar in appearance to the Marchetti map. The following differences on the two maps are noted below:

Difference 1:
Ortelius-Marchetti 1598 map: *OCEANVS ATLATICVS*, with insertion of an 'n' in *ATLATICVS* above the second letter 'A'.
Ortelius-Lovisa 1697 map: *OCEANVS ATLANTICVS*. Generally, the Lovisa is more roughly engraved with placenames often squeezed onto map; for example, the wording for *Bagdet al Babilon* intrudes into *REGNI PE*.

Difference 2:
Ortelius-Marchetti 1598 map: The Marchetti map has two sailing ships in the Atlantic.
Ortelius-Lovisa 1697 map: The Lovisa map only has one sailing ship in the Atlantic Ocean.

Difference 3:
Ortelius-Marchetti 1598 map: The Marchetti map has a sea monster in the lower right with horizontal hachuring.
Ortelius-Lovisa 1697 map: The Lovisa map has a sea monster in the lower right with generally vertical hachuring on the body.

References
Shirley 2004: 668-669; Van der Krogt 2003: atlas 33A:01-03; King 2003: 79 and 199; Burden 1996: map #126 (Americas map); Mickwitz et al. 1979-1995: Vol. 2: 129-131; Koeman 1967-71, III: Ort 69-71.

40 1598 Zacharias Heyns, Amsterdam

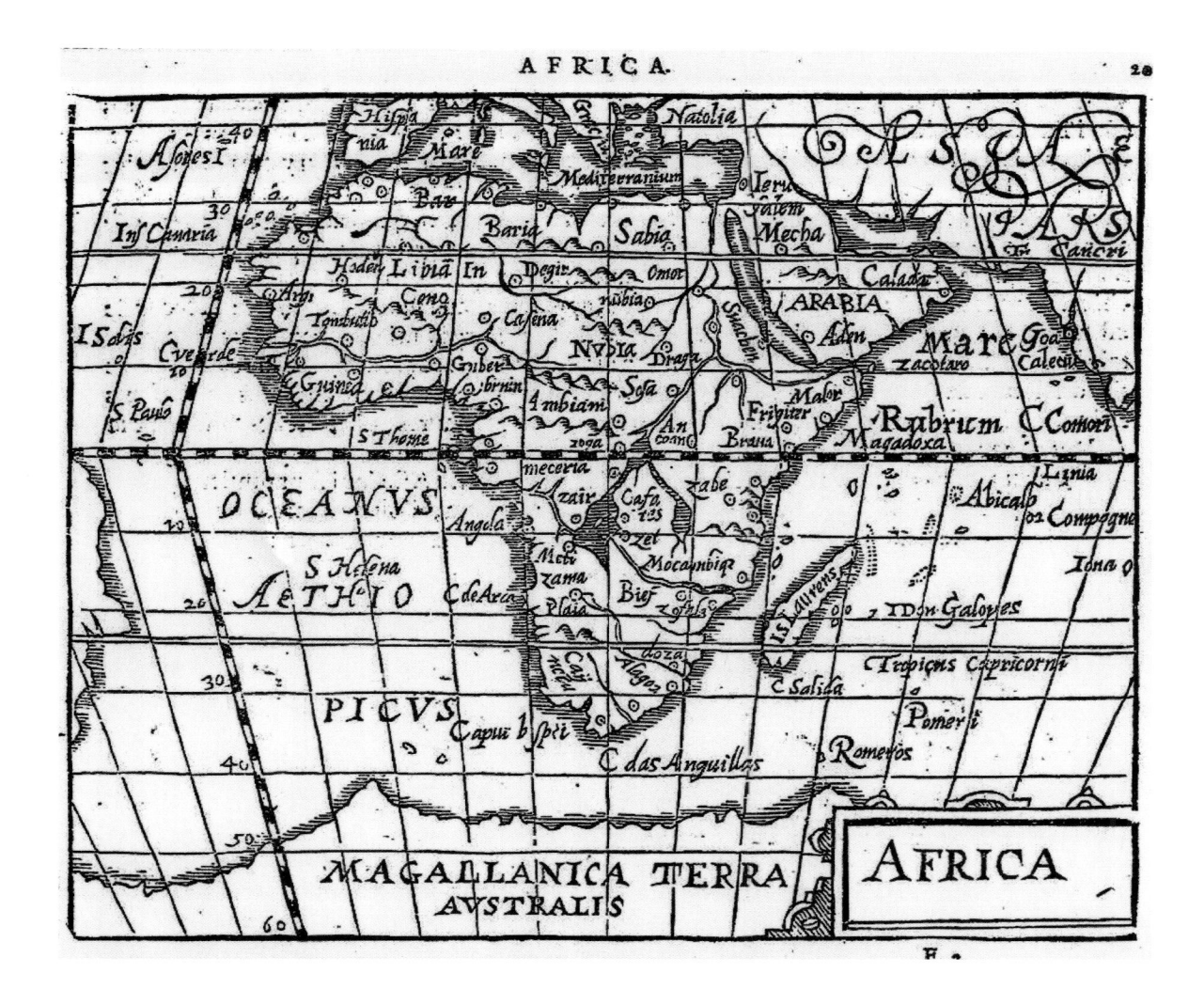

AFRICA.
- No scale bar.
- [Amsterdam : Zacharias Heyns, 1598].
- From: *Le Miroir du Monde ou, Epitome du Theatre d' Abraham Ortelius.*
- 1 map : woodcut ; 14 x 17.5 cm.
- Latitude coordinates along Prime Meridian gridline at left of map. No longitude coordinates but a gridline across Equatorial Line.
- North at the top; no compass rose.
- Title within a strapwork box at lower right.
- No decorative elements except for a box-shaped title cartouche at lower right of the map.
- The map is also distinguished by the following: AFRICA. above map in letterpress and to its right, the page number 20. At the bottom right of the page, F2.

Examples: *Amsterdam SMA*, A-III-29; *Amsterdam UB*, 1804 E 9; *Antwerpen PM*; *London BL*, Maps C.2.b.5; *Paris BN* (on the verso of the map, the last line of text begins *les precieux cloux de gyrosles...*)

Description
This map is by Zacharias Heyns (1570-c1640). He was the son of Pieter Heyns, a poet and school-teacher in Antwerp, who wrote the text in rhyme for the pocket-size version of Ortelius' atlas, *Spieghel der Werelt*, published in 1577, which contained the Ortelius-Galle (1) map of Africa (Map # 18). In 1598, after his father's death, Zacharias Heyns tried to duplicate his father's highly successful book by publishing his own atlas *Le Miroir du Monde...* with this map of Africa.

In appearance, the Heyns map is quite different from the earlier Ortelius-Galle (1) map, having the longitude lines in a trapezoidal pattern. Heyns does reintroduce the eastern Nile source lake, inexplicably omitted in the Ortelius-Galle (1) map. He shows the western Nile source lake as the source for only one unnamed river (Cuama?) flowing to the south instead of two rivers. Madagascar is now *IS Laurens*. Heyns shortens the Niger River in West Africa, and the three lakes from Ortelius comprising the Niger River are not shown here. Unlike most other maps of Africa during this period, Heyns extends his map far to the south (60° S) and shows the vast, supposed southern continent, *Magallanica Terra Avstralis*. Many placenames do not correspond to Ortelius-Galle (1) with the introduction of *Tombutto* (Timbuktu) in West Africa and *Alagoa* (Algoa Bay) in Southern Africa as examples.

Like the Claesz.-Langenes map of Africa (Map # 37), Heyns shows a single, third lake above the Equator into which both the two Ptolemaic lakes of Central Africa flow. This third lake seems to have been based on the third Nile lake on the Al-Idrisi map of the world, which shows a similarly positioned third lake. Al-Idrisi's *The Book of Roger* first appeared in Rome, in Arabic as *Kitāb nuzhat al-mushtāq...*, in 1592, and it is possible that Heyns had access to this text.

Though this map of Africa is generally based on the Ortelius–Galle (1) map of Africa, with aspects of the Mercator model, such as an eastward flowing river above the Niger in West Africa, different sources were clearly used. Koeman (1967-71, I: 32) believed that maps in Heyn's atlas were based on unknown German source maps of c.1580. Meurer (2001: 354) believes these source maps, while German, may be of an even earlier period before 1570.

Publication Information

Owing to the intense competition from the Cornelis Claesz.- Barent Langenes atlas, *Caert-Thresoor*, published in the same year, and to the various Ortelius pocket-atlases in circulation, the Heyns atlas did not enjoy commercial success. There were no further editions of this atlas beyond the 1598 issue.

All of the maps in this book are woodcuts. This was one of the last atlases to use woodcuts since copperplates, by this time, had become dominant. The text describing the map is in rhyme on the page opposite the map.

Although Heyns did not publish another edition of his atlas, Jan Janszoon (the father of Johnanes Janssonius) re-issued this atlas in 1615 in Arnhem. Jansonnius obtained these woodcuts sometime prior to 1615. The only example known to the author is at the Bibliothèque Nationale in Paris (*Paris BN*, G 3110), as per Koeman and the Bibliothèque Nationale on-line catalog. This sole example has been missing since 1991.

References

Shirley 2004; Van der Krogt 2003: map 8600:334; Meurer 2001; Koeman 1967-71, II: Z Hey 1.

41 1598 Jodocus Hondius, Amsterdam

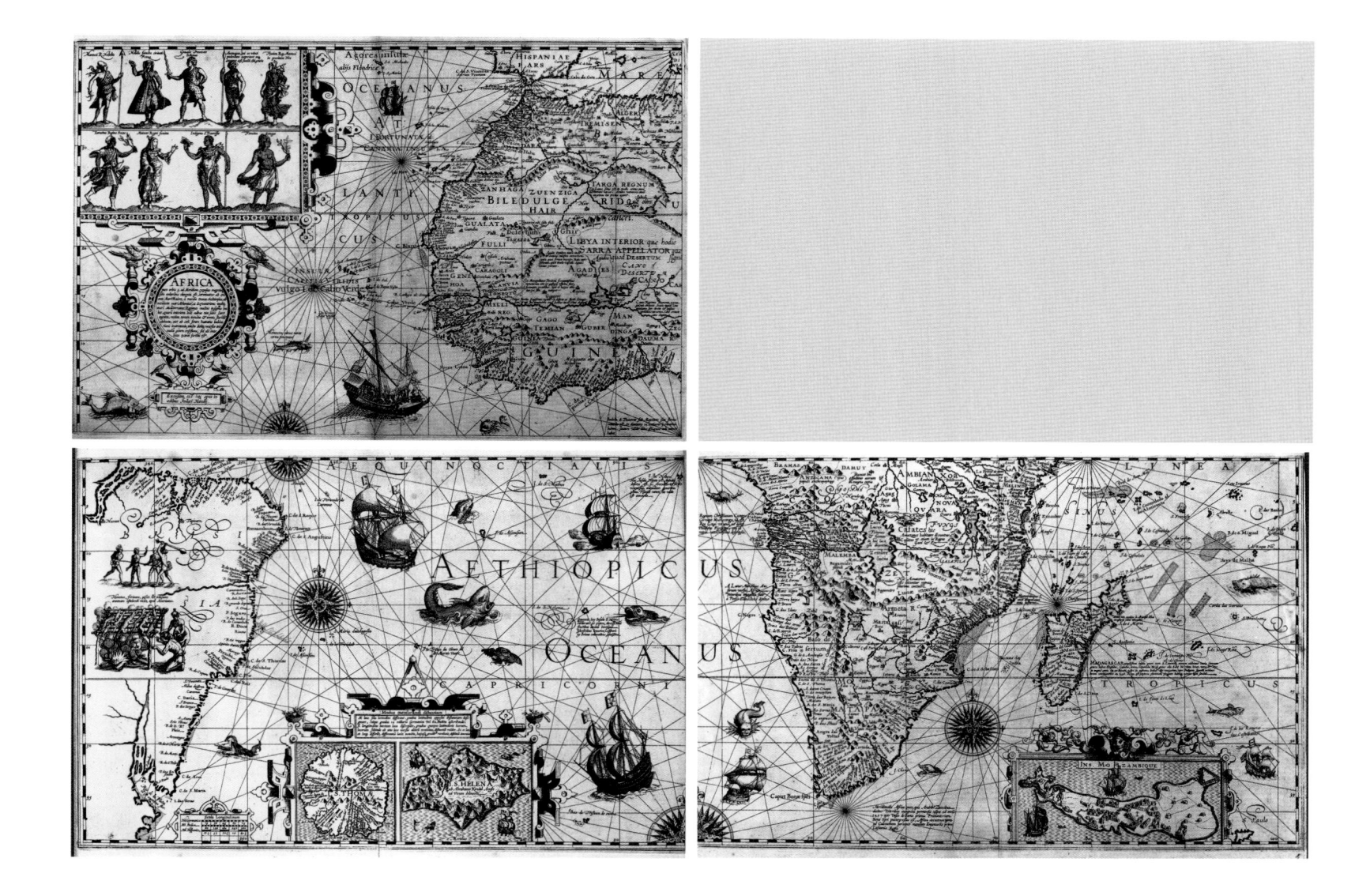

AFRICA | pars orbis 3. ad Meridiem exposita magnisque | solis ardoribus obnoxia est, terminatur ab ori: | ente Mari Rubro, a meridie Oceano Aethiopico, ab | occidente mari Atlantico, a Septentrione undique | mari Mediterraneo: Regiones multas insignes ha: | bet, quarum interiora loca adhuc non satis sunt | cogni- ta, multa tamen inculta & arenis sterilib | obducta, aut ob cæli situm humanæ habita | tioni incommoda, multo denique, venefico ani: | malium genere infestata, sed ubi inco: | liture eximie fertilis est.
[in a separate cartouche attached to and just below the above title cartouche]:
Excusum est hoc opus in | ædibus Judoci Hondij.

- Scale bar on bottom left sheet:
  Scala Longitudinum | Mil Germanica 75 = 4 cm | Mil Italica 300 = 4 cm | Mil Hispania 87 $^1/_2$ = 4 cm.
- [Amsterdam] : Jodocus Hondius, [1598].
- From: separately published.
- 1 wall map : copperplate engraving ; 4 map sheets ; 70 x 106 cm (35 x 53 cm for each of existing 3 sheets from the original 4 sheets).
- Latitude and longitude coordinates on all four sides of the map. Longitude coordinates along Equatorial Line.
- North is at the top; a total of four compass roses on three sheets of the map.
- Title in elaborate round cartouche on upper left sheet off West Africa.
- The map is extensively decorated. Above the title cartouche are nine different African figures, primarily from North Africa, with one from sub-Saharan Africa: Femina Senege. There are three inset maps of Sao Thomé, St. Helena, and the Island of Mozambique, all important re-supply points for expeditions around

Africa to Asia. On sheet 1, there are two ships and three sea monsters. On sheet 3, there are three ships and five sea monsters. On sheet 4, there are two ships and five sea monsters. Various animals are in Africa.

Example: *Rotterdam MM*, K-64 (1,3,4) (The only known example; missing sheet 2).

Description
This is Jodocus Hondius' (1563-1612) wall map of Africa and represents the beginning of Dutch designed and produced wall maps of the world and the four continents. Previous efforts in The Netherlands were limited to reproductions of the Giacomo Gastaldi 1564 wall map of Africa by the De Jodes, and other similar wall maps. The Hondius wall maps were also the oldest wall maps based on Mercator's projections published in The Netherlands.

Shortly after Jodocus Hondius' return to Amsterdam in 1593 from a self-imposed exile in London to escape religious unrest, he began work on a set of wall maps. An early reference to the Hondius wall maps of the continents is in a resolution of the Amsterdam Admiralty of November 25, 1598 in which Hondius was to receive 200 guilders for the supply of maps for navigation (Schilder 1996: 53). There were other references to these maps by Edward Wright in his *Certaine Errors in Navigation*, London, 1599, in which he accuses Hondius of using his ideas without permission in the production of these wall maps. It was thought that these maps were lost, but in 1956, Hondius' wall map of Europe (three of the original four sheets, missing sheet three) and this wall map of Africa (three of the original four sheets) were mentioned as recently being acquired in the Annual Report of that year of the Maritiem Museum 'Prins Hendrik, Rotterdam. Burden (1996: map # 112) described the two northern sheets of Hondius four-sheet wall map of America (1598) at the New York Historical Society Library as the only known example of the Americas wall map. Excitingly, Schilder (1996: 60-61) located a copy of the Hondius' wall map of the world (1595/96) in eight sheets in Dresden in 1993.

Jodocus Hondius drew on a variety of sources for this map of Africa, including elements of both the earlier Gastaldi-Ortelius and Mercator models. The coastline of Africa seems to reflect aspects of Ortelius, especially in its triangular shape of Southern Africa, and in its depiction of Madagascar. This has been modified largely by the reports of more recent European explorations along the coast and into the interior by the Portuguese and others.

Van Linschoten's *Itinerario* of 1596 with its two maps of the southern part of Africa, based on Portuguese source material, was also used. The hydrographical systems that Hondius depicts, particularly in Southern Africa seem to use Mercator with his Sachaf Lacus as a source for the unnamed Zembere which splits into the also unnamed Cuama and Spirito Santo Rivers. However, Hondius does not show this lake feeding the Congo River and the Nile River, as Mercator did. This presentation is similar to the Van Linschoten 1596 map of southwest Africa.

The detailed geography of the Congo region is largely from the work of the Italian writer Filippo Pigafetta in his *Relatione del Reame di Congo* of 1591, based on the Portuguese explorer, Duarte Lopes. The latest reference for a source on the map is a notation to the east of Madagascar which refers to the first Dutch expedition to the East under Cornelis de Houtman in 1595-1597. Sheet 2, depicting North-central and Northeast Africa, is lost so sources used for this area are not known.
All in all, this wall map of Africa by Hondius is a seminal map and represents a bridge into the seventeenth century and further refinement in the mapping of the African continent, particularly as represented by Willem Blaeu's 1608 wall map of Africa.

Publication Information
This is a four-sheet wall map of Africa on Mercator's projection. In the only known example, the top right sheet showing North-central and Northeast Africa is missing.

There has been some speculation that Hondius published a set of wall maps of the continents in 1602, as claimed by Anthony Linton in his *Newes of the Complement of the Art of Navigation* (London, 1609). Schilder (1996: 61) states that this can not be proven at this time due to a lack of evidence.

References
Schilder 2006: chap. 12.2 and facsimile 17[1-3]; Schilder 1996: Vol. V: 49- 75, with map image: 55-57; Burden 1996: map #112 (Americas map).

42   1598 Hernando de Solis, Valladolid

AFRICA.
- No scale bar.
- [Valladolid : Heirs of Diego Fernandez de Cordova, 1603].
- From: *Relaciones Universales del mundo...* .
- 1 map : copperplate engraving ; 36 x 49 cm.
- Latitude coordinates on gridlines at left and right sides of the map. Longitude coordinates across the bottom of the map.
- North at the top; no compass rose.
- The title, in cursive script, is set within a strapwork cartouche at bottom left.
- There are three sea monsters in the Atlantic Ocean, and two ships and one sea monster in the Indian Ocean.
- No flora or fauna in Africa.

Examples: *Chicago NL*, Ayer 7 B7; *Bedburg-Hau, Antiquariat Gebr. Haas* (two examples of book); *Author's collection* (as separate map).

Description

Hernando de Solis prepared this map for an edition of Giovanni Botero's *Le Relationi Universali* that he edited and translated into Spanish with the title of *Relaciones Universales del mundo*. It is a close copy of the popular Ortelius map of Africa of 1570. De Solis even includes the three Ortelius sea monsters in the Atlantic. He does replace Ortelius' large sea battle scene at the bottom right with a sea monster and two ships. Also, on this map, many of the placenames are in Spanish.

The engraver and mapmaker for this map is Hernando de Solis, as his imprint is on the similarly engraved Americas map, but not on this map of Africa.

Publication Information

This is a rare map in a very rare book. It is only known to have appeared in the Spanish edition of Giovanni Botero's *Le Relationi Universali*, published in 1603 in Valladolid. The *Le Relationi Universali* was a geographical and political discourse on the world and first appeared in Italy in 1590, though not with maps until 1595. The book is in two parts.

This map is assigned a date of 1598, based on the date of 1598 on the Americas map from this book. It is possible that Diego Fernandez de Cordova intended to publish this book in or shortly after 1598, but that his death interrupted this publication. The book was published by the Heirs of Diego Fernandez de Cordova in 1603.

The example examined at the Newberry Library has a date of 24 November, 1600 on the recto of the second leaf, although the book appeared in 1603. According to Sabin (1868: entry no. 6809), the second part of the book that he examined had a date of 1599, though the author has not located this example. The author has seen two examples of this book at the German antiquarian book and map dealers, Antiquariat Gebr. Haas. One example has 'Ano 1600' on the recto of the second leaf and the other example has 'Ano 1602' on the second leaf. In all of these examples, the map is unchanged.

The map is only known in one state.

References

Norwich 1997: map # 25; Burden 1996: map #129 (Americas map); Sabin 1868: entry # 6809.

43   1600 Leo Africanus - John Pory, London

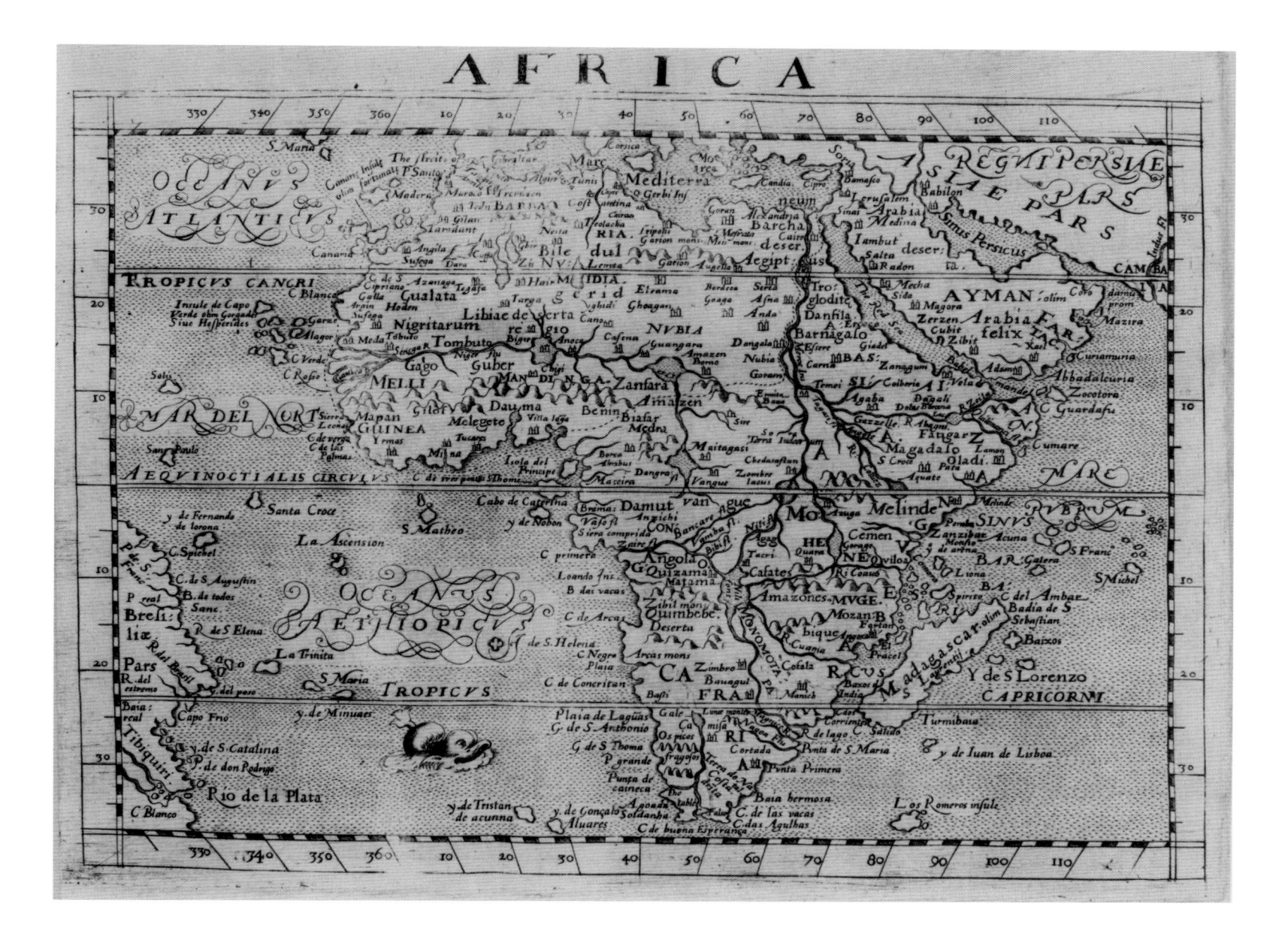

AFRICA.
- No scale bar.
- [London : George Bishop, 1600].
- From: *A Geographical Historie of Africa, written in Arabicke and Italian... by John Leo...Translated and collected by J. Pory.*
- 1 map : copperplate engraving ; 17.5 x 24 cm.
- Latitude and longitude coordinates on all four sides of the map.
- North at the top; no compass rose.
- Title above map.
- There is one sea monster in the South Atlantic Ocean; no other ornamentation on the map.

Examples: *London BL*, G.4258; *Washington LC*, DT7 L5.

Description

This map is from the first English translation of Leo Africanus' description of Africa by John Pory (1572-1636).

Leo Africanus (c.1483-1552), or al Hassan Ibn Muhammad al Wazzan as was his proper name, was a noted Arab traveler and geographer, who visited much of Northern and Eastern Africa and wrote on much of the rest of Africa based on his interactions with other Africans. Africanus' work proved to be an important and influential book on the history, geography, language, customs, and natural history of Africa. Although Africanus' original 1526 manuscript does not exist, Giovanni Ramusio evidently had access to Africanus' text and published a description by Africanus in his *Descrittione dell 'Africa* of 1550 with subsequent editions. Africanus' information on Africa was well-received and continued to appear in various issues well into the seventeenth century, often without a map of Africa.

Pory's edition of the Africanus text contains a new map of Africa, modeled after the 1582 D'Anania map of Africa, which in turn was modeled after Ortelius' 1570 map of Africa. However, there are some unusual, confusing elements in this map related to the lakes and river systems. Most notably, there are two rivers, and not one, exiting the western Nile source lake, here called *Nils Fons*. Instead of the Zambere flowing southward from the western Nile source lake and splitting into the Cuama and the Spirito Santo rivers as on the Ortelius and D'Anania maps, the Pory map has both the *R. Cuania* (Cuama) and the unnamed Spirito Santo exiting separately from *Nils Fons*. Also, positioned on the Equator is a third source lake for the Nile River. This third lake, here named *Zembre Lacus*, was somewhat similar to the third lake in the 1597 Cornelis Claesz.-Barent Langenes and the 1598 Heyns maps of Africa. There are other features unique to this map such as a new river, *Ri Coauo*, flowing due east from *Nils Fons* to the Indian Ocean. Clearly, Pory interprets the sources for his map, João de Barros and other Portuguese sources, in a unique way. Finally, the lettering on the map represents an amalgam of languages and engraving sizes and styles.

Publication Information

This edition was translated into English from the Latin in 1600 by John Pory, who was supported by Richard Hakluyt in this work. The particular edition that was used for this English translation is not known; it may well have been one of the editions without a map such as those in the Johannes Florian Latin editions from 1556.

The book was published in London by George Bishop. The beginning part of the book has a section covering the southern regions of Africa not covered by the Africanus-Temporal edition of 1556.

The following are the different maps that were used to illustrate the Leo Africanus text with information on the date, editor for the text, and the map used:

Date	Editor	Map used
1554	Giovanni Ramusio	Gastaldi-Ramusio map of 1554
1556	Jean Temporal	Africanus-Temporal map of 1556
1600	John Pory	Pory map of 1600
1665	Arnout Leers	Cloppenburch map of 1630

To distinquish this map from similiar appearing maps, using different copperplates, by D'Anania (1582), Botero (1595), Magini (1596), Keschedt (1597), Rossaccio (1605), and Botero-Giunti (1640), please refer to the table under Map # 20, 1582 D'Anania.

References
Tooley 1999-2004, K-P: 453; Moreland and Bannister 1989: 151-2.

44   1600 Arnoldo di Arnoldi, Siena

AFRICA.
[12 lines of text below title within the title cartouche]:
Tengono alcuni che l'Africa ... descendente | ... | populata.
- No scale bar.
- [Siena : Matteo Florimi, 1600].
- From: separately published.
- 1 map : copperplate engraving ; 37.5 x 48.5 cm.
- Latitude and longitude coordinates on all four sides of the map.
- North at the top; no compass rose.
- Title cartouche is at the bottom left of the map. At the top of the title cartouche there is an unidentified coat of arms.
- In the Indian Ocean, there are two ships and one large sea monster. In the Atlantic Ocean, there are one ship and one large sea monster. No flora and fauna are shown on the map. Various text legends are found on the map and in the oceans.

Examples: *London BL*, Maps c.7.e.3.(4) (map # 4 within volume one of a composite atlas); *Providence JCB*; *Private US Collection.*

Description

This rare map is one of a set of the four continents. Arnoldo di Arnoldi (? - 1602) was a Flemish cartographer who worked in his later years with Giovanni Magini in Bologna from 1595-1600 and with Matteo Florimi in Siena from 1600-1602. Arnoldi also produced a world map in ten sheets and one in two sheets.

There is no reference to Arnoldi on this map of Africa. The name of Arnoldi is on the similar map of Europe and it is thus assumed that Arnoldi also prepared this map.

The map generally follows Ortelius' 1570 map of Africa. The hydrographical systems and the outline for Africa are the same on both maps. Placenames faithfully follow Ortelius. For example, *Zanzibar* is placed on the west coast of Africa along with descriptive text, as on the Ortelius map.

There is a lengthy text passage describing Madagascar just below the island in the area where Ortelius placed his sea battle. Arnoldi does place additional text passages on this map such as a reference by the Cape of Good Hope to Da Gama's voyage around the Cape in 1497.

Above the elaborate title cartouche at the lower left, there is a coat of arms that is not identified on the map. From the map of Europe, this coat of arms is identified as belonging to Arnoldi's patron, the Tuscan writer, Scipione Bargaglia. Also on the Europe map is the imprint of 'Arnoldi di Arnoldi DD' along with the printer's name: Matteo Florimi. Along with this information, there is a reference on the Europe map to the geographical information being taken from Ortelius' *Theatrum*.

Publication Information

The map was only issued as a separate map and is not found in any standard atlas. It is only known in one state.

References

Norwich 1997: map # 26; Burden 1996: map #138 (Americas map); Van der Heyden 1993, map # 4; Sotheby's 1985; Almàgia 1934.

45   c. 1600 Luis Teixeira - Joannes van Doetecum, Amsterdam

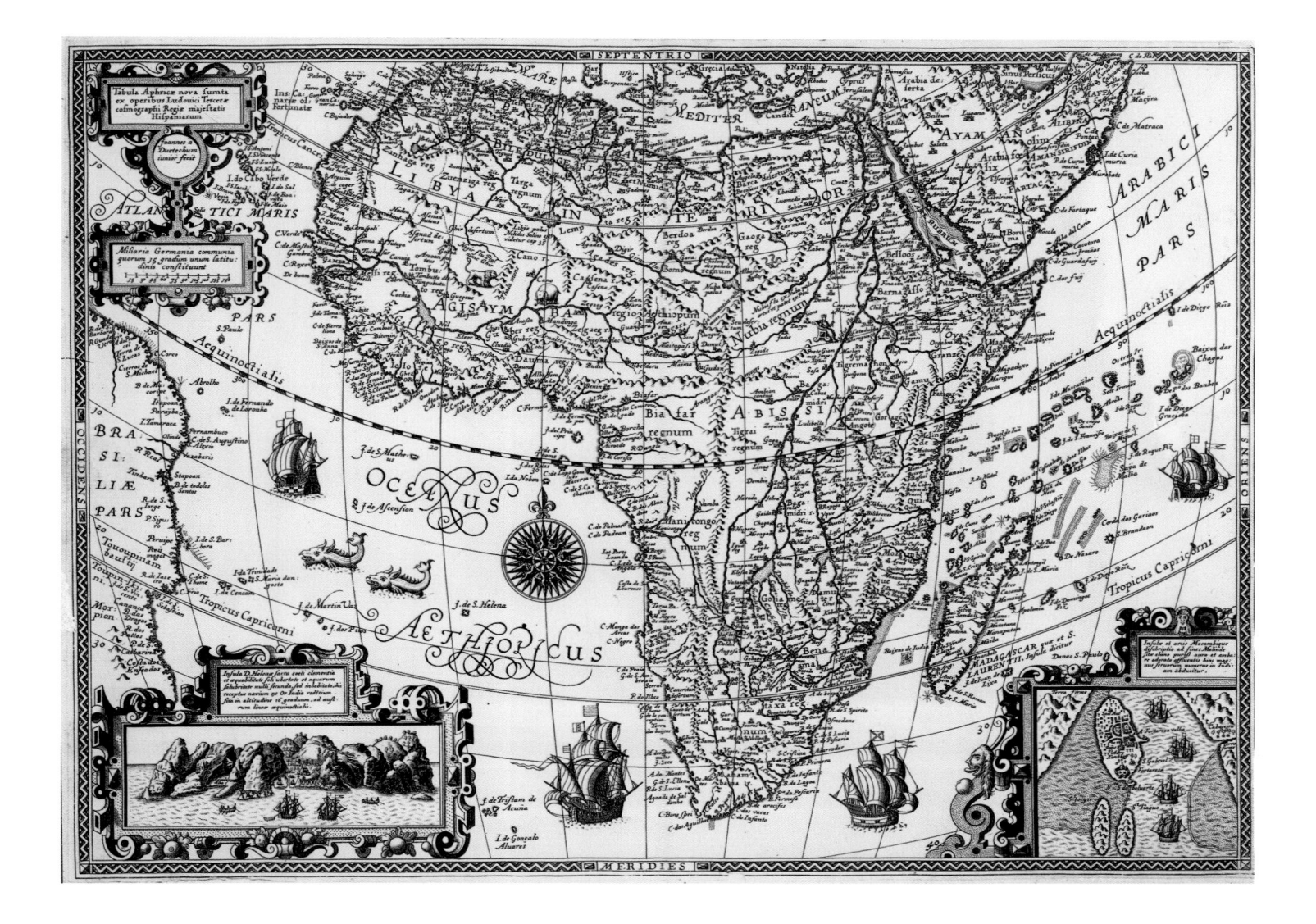

45.1 First State c.1600

Tabula Aphricæ nova sumta | ex operibus Ludouici Tercerz | cosmographi Regiæ majestatis | Hispaniarum.
[in oval just below title]:
Joannes à | Duetechum | iunior fecit.
- Scale bar in a box just below oval: Miliaria Germania communia | quorum 15 gradum unum latiitu= | dinis constituunt 150 = 4.7 cm.
- [Amsterdam : Cornelis Claesz., c.1600].
- From: separately published.
- 1 map : copperplate engraving ; 38.5 x 53.5 cm.
- Latitude coordinates on left and right sides of the map. Longitude coordinates as gridline along the Equatorial Line.
- North at the top; compass rose in the South Atlantic.
- Rectangular strapwork title cartouche at top left with the imprint of the engraver, Joannes van Doetecum Jr. (Joannes à | Duetechum | iunior fecit) just below the title cartouche in a strapwork oval, and the scale bar below that in a rectangular strapwork box.
- Inset view of St. Helena at bottom left with text box above the view describing St. Helena; Inset map of Mozambique Island at bottom right with text box above this map describing Mozambique Island.

- There are two ships (with the larger of the two off the west coast of the Cape of Good Hope flying a flag with three crosses from the Amsterdam coat of arms) and two sea monsters in the Atlantic Ocean. There are two ships (with Portuguese flags) in the Indian Ocean. A lion and an elephant are placed in West Africa. The image of Prester John on a throne, with a notation below it, is shown in Northeast Africa.

Examples: *Amsterdam UB*, O.K. 126; *London BL*, Maps 63510 [19].

45.2 Second State before 1620
Addition of the following: 't'Amsterdam gedruckt bij Dauit de Meijne inde weeelt cart'. This state was published by David de Meijne.

Examples: According to Schilder (2003: 296-298), the map was shown in this second state at the Exhibition Cartographique Ethnographique et Maritime (Antwerp, 1902) as item 647, lent by the antiquarian bookshop of W.P. van Stockum & Zoon from The Hague. Its present location is not known.

Description
This is a beautiful map, prepared and engraved by Johannes van Doetecum II, who was active in Amsterdam from 1592 to 1630, and who was the younger son of Johannes van Doetecum, the elder.

As the map title indicates, this map relied on maps initially prepared by the Spanish mapmaker, Luis Teixeira. Teixeira was the leading mapmaker in Portugal in the later part of the sixteenth and beginning of the seventeenth century. Teixeira's cartographic output was well known in Amsterdam during this period. Ortelius, Jodocus Hondius, Claesz. and others used information presented by Teixeira as source material for their own maps.

Gerard Mercator II's map of Africa of 1595 served as a model for the interior of Africa, supplemented by the wall map of the world that his grandfather published in 1569. The shape of Africa and Madagascar tend to follow the Mercator map of Africa. The hydrographical systems also precisely follow Mercator. This is evident with *Sachaf Lake* in Southern Africa as a source for the Congo, *Cuamal S Spirito* and Nile Rivers, and with its *Giras/Nubia* River north of the Niger River in West Africa.

Most of the placenames for the interior are from the Mercator map as well, including the Portuguese settlement, *Ca Portogal*, up the Cuama River in South-central Africa. The image of Prester John is directly taken from the Mercator 1595 map of Africa and is placed in Northeast Africa exactly as it is on the Mercator map. The depiction and nomenclature for the coast of Africa does rely to some degree on the works of Teixeira.

Cortesão and Teixeira da Mota in *Portugaliae Monumenta Cartographica* (1960) date this map as c.1600. This date is based on the use of the two inset engravings of St. Helena and Mozambique, which were clearly taken from Van Linschoten's *Itinerario* of 1596. The Latin titles of these two insets appear to be completely copied from the *Itinerario*, including the errors in the Latin text. While the map uses the information from the 1596 *Itinerario*, it does not use the more detailed information available in a separate Teixeira - Van Doetecum map of West Africa, which is dated 1602; thus the date on this map of c.1600.

Publication Information
This separately issued map was published by Cornelis Claesz. and is known in two states. The map did not appear in a standard atlas.

References
Schilder 2003: 296-298 and facsimile 31; Cortesão and Teixeira da Mota 1960: Vol. III: 41-84 for Teixeira and 66-67: plate 362C for this map.

46   1600 Matthias Quad, Cologne

A= | PHRI= | CA.
[below and to the left of the title cartouche]:
Johan Bussemecher ex- | cudit in Vbioru Coloniæ.
- No scale bar.
- Cologne : Johann Bussemacher, [1600].
- From: *Geographisch Handtbuch*.
- 1 map : copperplate engraving ; 21 x 29.5 cm (to outer black line beyond text box on left).
- Latitude and longitude coordinates as gridlines on all four sides of the map. Longitude coordinates along Equatorial Line.
- North at the top; no compass rose.
- Title is set within a strapwork cartouche at the bottom left of the map.
- Text box, set sideways on the left side of the map, describing Africa in nine lines of Latin text.

Examples: *London BL*, Maps C.39.c.1 (in *Geographisch Handtbuch*, 1600); *New York PL*, *KB+ 1611 (in Botero, *Allgemeine Historische Weltbeschreibung*, 1611).

Description

Matthias Quad (1557-1613) was a German cartographer active at the end of the sixteenth century in Cologne. After training and employment in Holland, Quad moved to Cologne in 1587 to work as an engraver and geographer. Besides working under his own name, Quad also used several pseudonyms, Cyprian Eichovius and Konrad Loew, possibly as a result of difficulties due to his Protestant faith (Meurer 1988: 243-4). Quad was also the editor of works by Metellus, Acosta, and Botero.

This map of Africa is from his *Geographisch Handtbuch* of 1600, a general atlas of the world published by Johann Bussemacher. Bussemacher was also the engraver of this map. Below and to the left of the title cartouche is the imprint of the publisher, Bussemacher (Johan Bussemecher ex- | cudit in Vbioru Coloniæ.). On the verso is a description of Africa in German.

The map generally follows the Mercator world map of 1569 and the later Mercator Africa map of 1595. Like the Mercator map, an unnamed southern lake (*Sachaf lac.* in the Mercator map) is the source for the major rivers flowing southward into the Indian Ocean, the *Zaire Rio* (Congo) flowing into the South Atlantic, and even as a source for the Nile River. The Quad map has a few changes to the Mercator map such as a shortening and widening of Madagascar.

Publication Information

This map first appeared in Quad's *Geographisch Handtbuch* of 1600. It also appeared in Quad's *Fasciculus Geographus* of 1608, published in Cologne by Johann Bussemacher, and in an edition of Giovanni Botero's *Allgemeine Historische Weltbeschreibung* of 1611, published in Munich by Nicolaus Henricum. The map is only known in one state.

Quad had previously issued a travel book, *Europae totius terrarum orbis...* in 1596, which was reported to contain this map. However, a map of Africa is not present in this book.

References

Norwich 1997: map # 20; Meurer 1988: 197-235; Mickwitz et al. 1979-1995: Vol. 2: entry # 239 & 240; Bonacker 1969; Sabin 1868: entry # 6808 and 66894.

47   1601 Abraham Ortelius – Johannes van Keerbergen, Antwerp

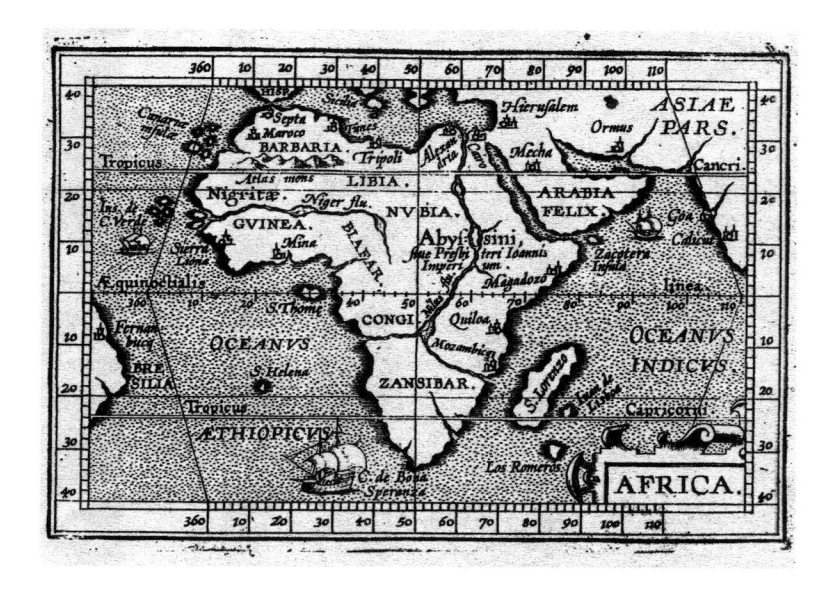

AFRICA.
- No scale bar.
- [Antwerp : Johannes van Keerbergen, 1601].
- From: *Epitome Theatri Orbis Terrarum Abrahami Ortelij.*
- 1 map : copperplate engraving ; 8.5 x 12.5 cm.
- Latitude and longitude coordinates on all four sides of the map as gridlines. Longitude coordinates across Equatorial Line.
- North at the top; no compass rose.
- Title is set within simple strapwork cartouche at the lower right.
- There are two ships in the Atlantic Ocean and one ship in the Indian Ocean.

Examples: 1601 edition: *London BL*, Maps C.2.b.7; 1603 edition: *London BL*, Maps C.2.b.10; 1604 edition: *New York PL*, *KB 1604 (in Ortelius, A. *Auszzug auss des Abrahami Ortely Theatro orbis Teutsch*).

Description

In 1601, Johannes van Keerbergen produced a new version of the successful, smaller-sized Ortelius atlas. The text was provided by Michel Coignet and a new series of copperplates were cut by Ambrosius and Ferdinand Arsenius. This map is sometimes referred to as the Coignet map, although Coignet does not appear to have been involved in the map's preparation.

The Van Keerbergen map of Africa has borrowed heavily from the Abraham Ortelius – Philippe Galle (2) map of Africa, though there are noticeable changes. These include the introduction of latitude and longitude gridlines on the sides of the map and the delineation of the Tropic of Cancer, the Equator, and the Tropic of Capricorn. This map also adds three ships.

The map generally follows the geography of Africa depicted in the 1570 Ortelius map. For example, as on the 1570 Ortelius map, this map places *Zansibar* (Zanzibar) in Southern Africa. It appears that Ortelius is using Zanzibar in a traditional way to describe a people and area of East and Southern Africa. However, like the Ortelius-Galle (1) and Ortelius-Galle (2) maps, this map does not show the eastern Nile source lake (Zaflan).

The primary source for these three maps appears to be a very strict interpretation of João de Barros. João de Barros in his first *Década*, published in 1552 in Lisbon and as an Italian abridgement in Ramusio's second edition of his *Delle Navigationi* in 1554, wrote the most detailed description of the hydrographical system of Central Africa with his account of a vast central lake (not lakes) as the source for the Nile, the Zaire (Congo) and the Zambere, which split to become the Cuama and Espirito Santo Rivers (see Relaño 2002: 207-8, for a further discussion of this point).

Publication Information

The map first appeared in Van Keerbergen's *Epitome Theatri Orbis Terrarum Abrahami Ortelij* of 1601 in Antwerp. All of the subsequent editions were published in Antwerp, except the London and the Frankfurt editions. The London English language edition, *Epitome of the Theater of the Worlde...*, was printed for James Shawe in 1603. This atlas and the Abraham Ortelius–Filips Galle atlas, *An Epitome of Ortelius his Theatre...* , of 1602 were the two earliest world atlases to be published in England and the earliest world atlases with English text (Skelton 1968b: notes).

The Frankfurt edition was jointly published by Van Keerbergen and Levinus Hulisus in Frankfurt in 1604 and is titled *Ausszug auss des Abrahami Ortely Theatro Orbis Teutsch...* . For this edition, Hulsius translated the text into German. To aid in identifying this map in this Frankfurt edition: At the top right of the map is page '3'. Bottom right is 'Der'. On verso, last line of text: 'bracht', and at the lower right 'BARBA-'.

Sometime after 1604, Jan Baptist Vrients acquired these plates and continued publishing this book until 1612.

The following chart can be used to differentiate the Ortelius-Van Keerbergen Africa map in the various editions of the book in which the map appeared using the language of the text, signature, and page number. This information is from Van der Krogt (2003):

Date	Van der Krogt no.	Title	Lang.	Signature	Page
1601	333:01	*Epitome*	Latin	A4r	4
1602	333:11	*L'Epitome*	French	A4r	4
1602 & 1612	333:21/22	*Breve Compendio*	Italian	B3r	3
1603	333:31	*Epitome*	English	A4r	4
1604	333:41	*Ausszug*	German	C8r	3
1609 & 1612	333:02/03	*Epitome*	Latin	B7r	4

References

Van der Krogt 2003: map 8600:333; Relaño 2002: 207-8; Koeman 1967-71, III: Ort 63-68A; Skelton 1968b: notes.

48 1602 Levinus Hulsius, Nürnberg

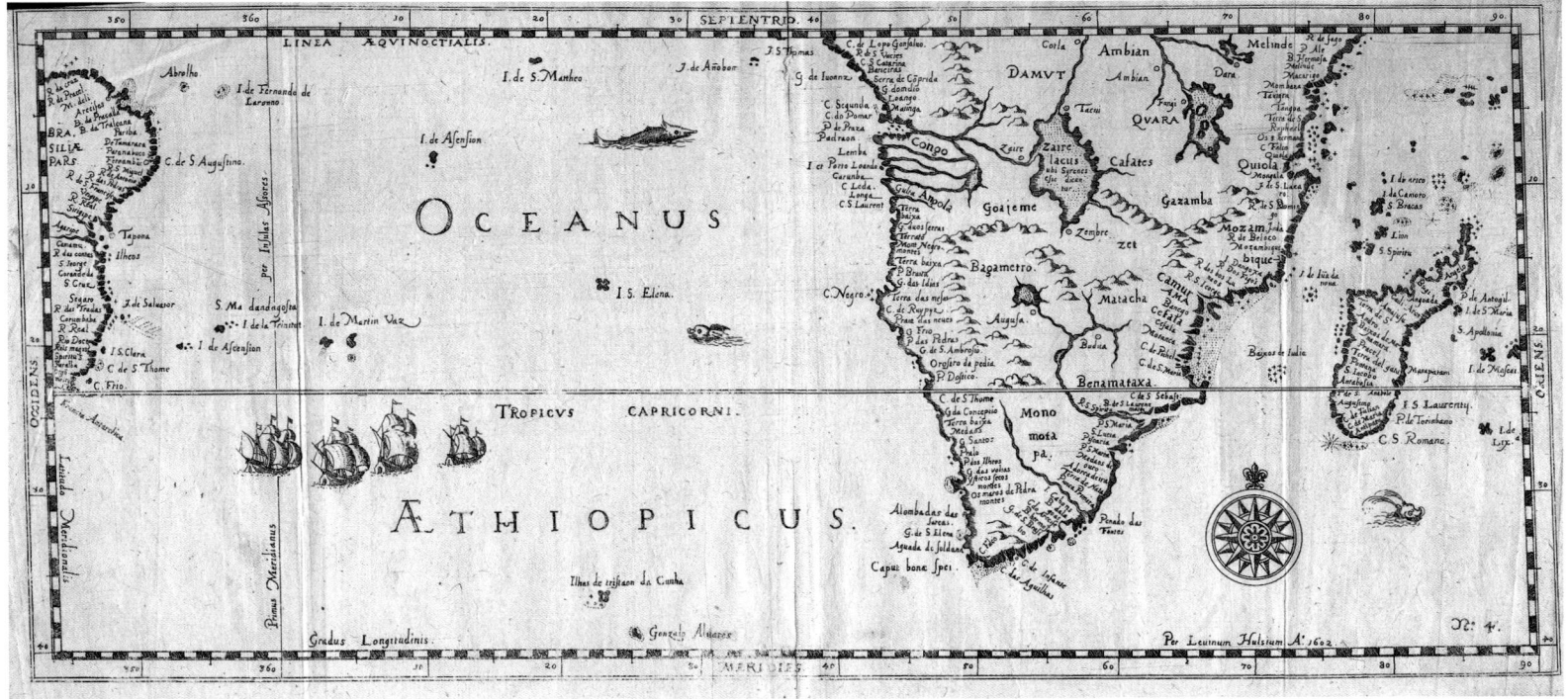

Map 48.2.

48.1 First State 1602

[Map of Africa].
[on top sheet at the bottom left, and on the bottom sheet at the bottom right]:
Per Leuinum Hulsium A° 1602
- Scale bar on top sheet at the bottom left: Millaria Germanica 150 = 3.7 cm.
- [Nürnberg] : Levinus Hulsius, 1602.
- From: *Sechste Theil and Siebende Schiffart* [from the] *Sammlung von Sechs und Zwanzig Schiffahrten in Verscheidene Fremde Länder durch Levinus Hulsium...* .
- 1 map : copperplate engraving ; 2 sheets ; 21.5 x 36 cm (top sheet), and 16.5 x 36.5 cm (bottom sheet) to neatline.
- Latitude and longitude coordinates on all four sides of each map sheet. A gridline surrounds both map sheets. The line of the Prime Meridian is on the map and is identified on the bottom sheet as *Primus Meridianus per Insulas Assores*.
- North at the top; two compass roses: one on top sheet to the left of West Africa, and one on bottom sheet to the right of Southern Africa.
- Both maps contain an imprint for Hulsius: 'Per Leuinum Hulsium A° 1602.' The northern sheet has 'No. 1' at bottom left corner and the southern sheet has 'No. 4' at bottom right corner, signifying their place in the order of the maps. Between these two maps are maps of South America and of India Orientalis.
- On the top sheet there are four smaller ships by the Azores Islands, and three larger ships sailing towards The Netherlands. On the bottom sheet, there are four ships sailing around the Cape of Good Hope, and three sea monsters. All of the ships possibly represent the ships of the first Dutch expedition to the East under the command of Van Houtman.

Example: *Washington LC*, G159.H8 (Part V of the book dated 1603 on title page).

48.2 Second State 1606

Changes to the top sheet as follows: at the bottom of the top sheet, to the left and below the gridline by *Meridies*, the coastline is extended and addition of *C. de lopo Gonsalves*. Same date of 1602 on the map as in the first state.

Examples: *Washington LC*, G159.H8 (Part VII of the book dated 1606 on title page) *and* G419.K87 (Part VI dated 1626 on title page); *Author's collection* (as separate map).

Description

This is a rare two-sheet map of Africa by Levinus Hulsius (c.1546-1605). Hulsius was a publisher, geographer, and instrument maker from Frankfurt who published a series of accounts of voyages. The books were published in Frankfurt and in Nuremberg.

The geography for the map has elements from a number of previous cartographers. West Africa seems to rely on the Mercator map of Africa of 1595 with a river flowing eastward toward the Nile and, below it, a simplified version of the Niger River flowing due west into the Atlantic. In Central Africa, Hulsius follows a conventional presentation based on the 1570 Ortelius map with a dominant western lake, *Zaire lacus*, feeding the Zaire and Nile Rivers, and with an unnamed second Nile source lake. Unusually, Hulsius shows a third lake to the east of the twin Nile source lakes, in this case, just south of the Equator flowing into the river which leaves the second Nile source lake. In the Ortelius model, this third lake is a source for the Blue Nile and flows through Abyssinia. In another variation on the Mercator and Ortelius models, in Southern Africa, a small unnamed lake is the source for only one river, the *S. Spirito*. There is no indication of the Cuama River, which is unusual given this river's importance, particularly in relation to the East African port of *Cefala* (Sofala).

The placenames for the coastal areas of this map are provided in detail and likely represent fairly up-to-date knowledge. This recent information came from the travel accounts of Duarte Lopes in the Congo, as written by Pigafetta in 1591, and by various Portuguese and Dutch narratives.

Publication Information
Eight years after Théodore de Bry began the publication of his *Grands Voyages*, Hulsius began publica-

tion of a similar collection of voyages in 1598, hoping to emulate the success of De Bry's publication. According to Church (1951: 601), Hulsius is credited with exercising better judgment in his selections and translations than De Bry, though Tiele (1867) says that the plates are imitations of those from the De Bry edition.

Hulsius' work comprises twenty-six parts. With its various editions and issues of the parts, the books were published, initially by Hulsius and then by his widow and family, at intervals from 1598 to 1663, the date of the fifth edition of Part V. As the books were published in German and were of a convenient size, they became quite popular.

Part VI of Hulsius' work contains this two-sheet map of Africa: one sheet north of the Equator as well as a second sheet south of the Equator. This work covers the first four circumnavigations of the world: Magellan in 1519-1522, Drake in 1577-1580, Cavendish in 1586-1588, and Olivier van Noort in 1598-1601. Part VI was initially published in 1603 in Nuremberg. A second edition of Part VI was published in 1618 and a third edition in 1626.

Part VII of the book also contains the sheet of the map showing Africa north of the Equator. It covers the discoveries and establishment of the Dutch on the coast of Guinea from November 1600 to 1602 by Pieter de Marees. This part was initially published in 1603 in Frankfurt. A second edition was published in 1606, with a new state of the map of North Africa, and a third and final edition was published in 1624. There were no subsequent editions of Part VII, although there were later editions of other parts. The narrative appeared in De Bry's *Petit Voyages* a few weeks before the appearance of the first edition of Part VII by Hulsius in 1603.

References
Church 1951: 601-745; Lenox 1877; Asher 1839; Tiele 1867.

49   1602/09 - 1617 Cornelis Claesz. – Johannes Janssonius, Amsterdam

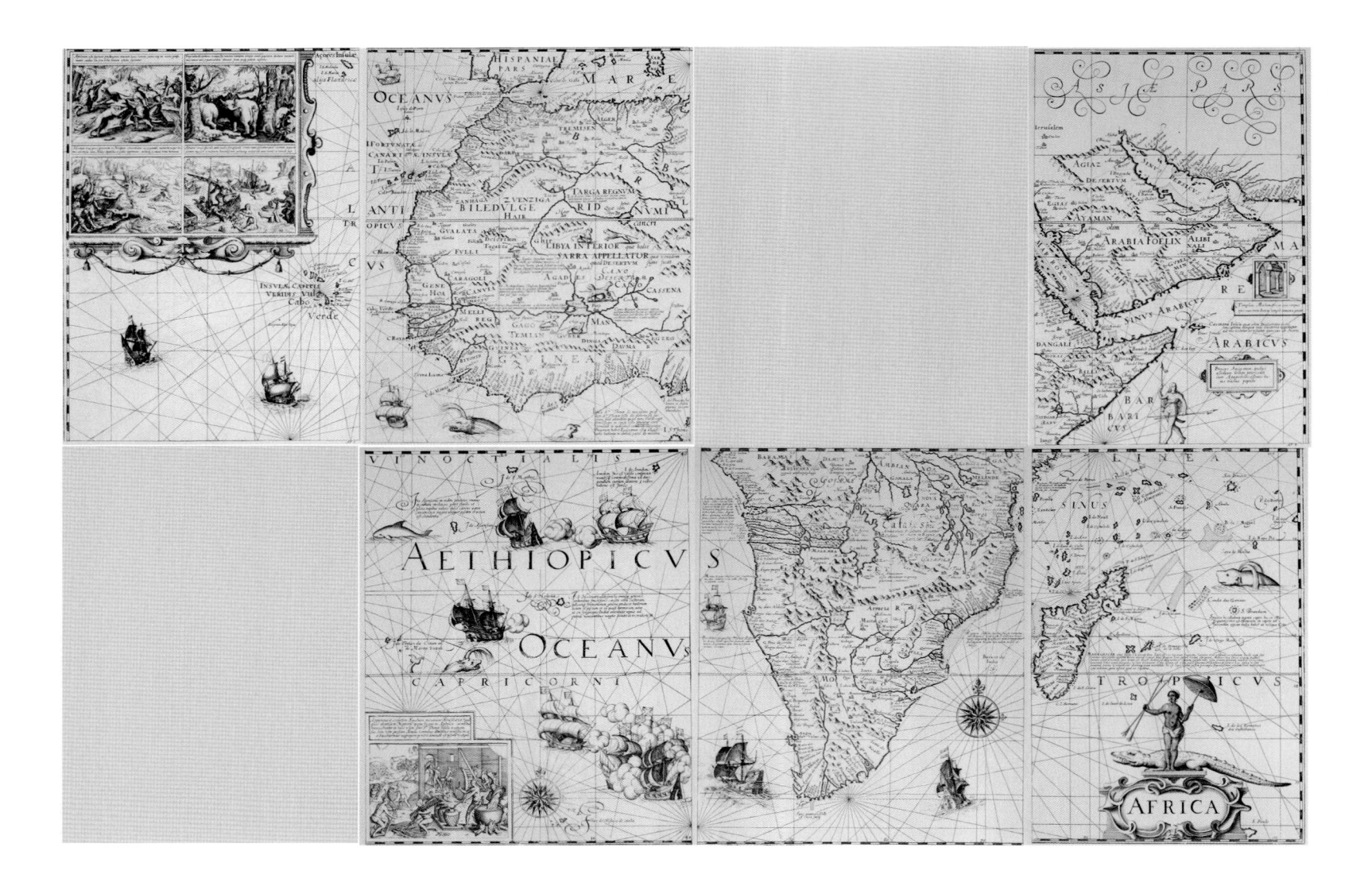

[Describes second state of this map; first state not extant]:

AFRICA.
- No scale bar.
- [Amsterdam : Johannes Janssonius, 1617].
- From: separately published.
- 1 wall map : copperplate engraving ; 8 sheets ; 91 x 135 cm (in total).
- Latitude coordinates along right side of map (sheets 4 & 8). Longitude coordinates along top of map (sheets 2 & 4) and along Equatorial Line at top of sheets (sheets 6-8).
- North at the top; one compass rose on sheet 6, and one compass rose on sheet 7.
- Title cartouche is on the bottom of sheet 8 (bottom right sheet), with an allegorical representation above it of Africa as a woman with a parasol and a crocodile.
- Two ships and one inset of four scenes on sheet 1; one ship and two sea monsters on sheet 2; one figure of an African and a scene of a mosque on sheet 4; four ships, a sea battle, two sea monsters, and one inset of a scene on sheet 5; three ships on sheet 7; and one sea monster on sheet 8.
- Animals within Africa.

Example: *Paris BN*, Ge DD 5081 {1-6} (wanting sheets 3 & 5).

Description

This beautiful and rare wall map of Africa by Johannes Janssonius (1588-1664) is known in only one example at the Bibliothèque Nationale de France in Paris. It is actually the second state of a now lost wall map by Cornelis Claesz. of 1602-1609. The example at the Bibliothèque Nationale de France is incomplete, missing the top sheet depicting Egypt and much of East-central Africa (the third map sheet) and the bottom left sheet of the South Atlantic and the east coast of South America (the fifth map sheet).

The top left sheet (first sheet) of the map is dominated by four decorative scenes: the two top scenes show hunting for lions and elephants respectively; a hunt for crocodiles is in the bottom left scene; and the bottom right scene shows Africans diving for coral. The fourth sheet shows the interior of a *Temple of Mohammed* and a woman carrying a weapon near the Horn of Africa. The sixth map sheet shows a mill for crushing sugar cane. The eighth sheet contains a title cartouche with the word *AFRICA*. Above the title is an allegorical scene of Africa with an African woman carrying a parasol and two spears with a crocodile behind her.

The map is very similar to the Hondius wall map of the continent of 1598; some of the legends are identical. The interior of Africa is traditionally divided into large regions, which are in turn divided into smaller sub-regions. In the northeast, these are *Barb*[aria], *Biledulgerid* and *Libya Interior*.

The *Niger fluuius* (Niger) and *R. Canaga al senaga* (Senegal) form one large river system as together they flow into the Atlantic, continuing a long erroneous depiction. In West Africa, little current information is used, especially from Teixeira and from Pieter de Marees' map of the region of 1602 as published by Cornelis Claesz. The Central African region of the Congo is largely based on the travels of Duarte Lopez, as translated and published as the *Relatione del Reame di Congo* by Filippo Pigafetta in 1591, for which a two-sheet map of Southern and Northeastern Africa was prepared. Claesz. had published a Dutch translation of this book in 1596.

The two sources of the Nile River originate in *Zaire Lacus/Zembre lacus* in the west and *Zaflan lacus* in the east. *Zaire Lacus/Zembre lacus* is also the source of the *R. Zaire-R. Congo* flowing west to the Atlantic. In Southern Africa, the *Lunae montes* (Mountains of the Moon) clearly separate the Southern African river systems from those north of the *Lunae montes*. Both the *Cuama flu* and the *R. de Spirito Santo* begin in a common, unnamed river which originates in *Sachaf Lacus*. The map shows an unusually placed river flowing due west from *Sachaf Lacus* into the Atlantic Ocean by *Cape Frio*. A small text legend describes this river's placement from the English geographer *Molentis* (Molyneux).

The southern extreme of Africa still retains its earlier Portuguese placenames and does not show the newer Dutch placenames (for example *S. Bras* rather than the newer Mossel Bay) from Cornelis de Houtman's voyage of 1595-97. Three of the Portuguese ports on the East African coast are shown and described on the map: *Coffala* (Sofala), *Mozambique Island*, *Melinde* (Malindi) from south to north.

Publication Information

The catalog of works owned by Cornelis Claesz., the *Const ende Caert-Register*, printed in 1609 contains listings for four wall maps of the continents including Africa made up of eight sheets each (Schilder 1990). These maps, except the map of Europe which is dated 1604, may have been printed as early as 1602 (Schilder, 1990: 61). Claesz.' wall map of Africa as mentioned in the *Register* is not known to have survived.

After the death of Cornelis Claesz., the copperplates of the four continents came into the possession of Johannes Janssonius, who reissued them in 1617. The map of Africa contains no reference to the author, place of publication, or date on the remaining sheets of the map. The assumption that this is the second state of the Claesz. map of Africa is based on a comparative analysis of the style of the Claesz. Asia wall map in its first and second states, and particularly the title cartouches, which are similar (Schilder 2003: 346-357).

References
Schilder 2003 Vol. VII: 357-360 and facsimile 38[1-6]; Burden 1996: map # 148 (Americas map); Schilder 1990 Vol. III: 32 and 56-63; Kraus 1970: 10-11.

50 1605 Heinrich Bünting - Jan Jansz., Arnhem

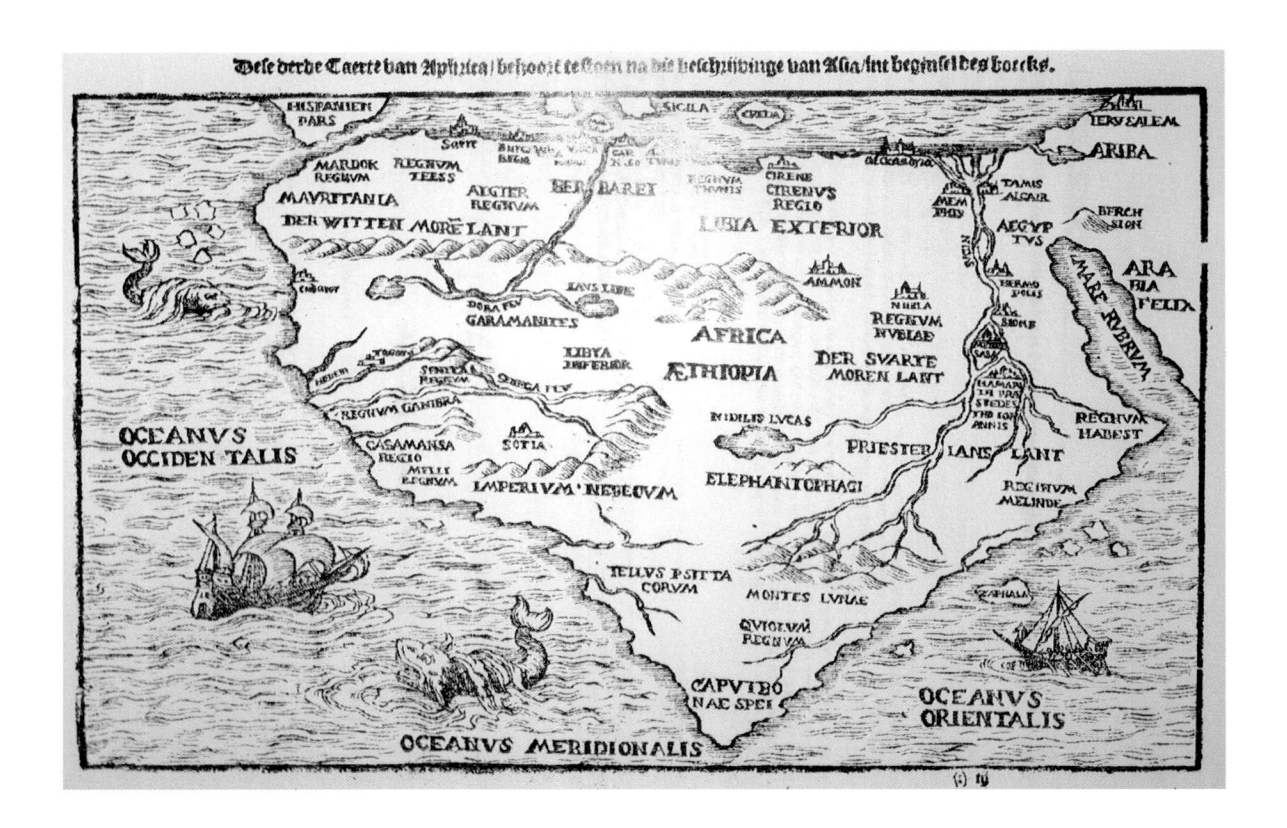

50.1 Variant 1 1605

[Map of Africa].
[above map on one line, in letterpress]:
Dese derde Caerte van Aphrica, behoort te staen na die beschrijvinge van Asia, int beginsel des boecks.
- No scale bar.
- [Arnhem : Jan Jansz., 1605].
- From: *Itinerarum Sacrae Scripturae*.
- 1 map : woodcut ; 15.5 x 25 cm.
- No latitude and longitude coordinates.
- North at the top; no compass rose.
- Bookbinder's instructions on a line above the map image.
- Within the Atlantic Ocean, there is one large ship and two sea monsters. Within the Indian Ocean, there is one ship. The engraving for the sea is horizontal, unlike the vertical appearance in the German and Czech editions. Placenames on the map are in Latin and in Dutch.
- The city of *Alcair* (Cairo) is on the east side of the Nile River in Egypt with the city of *Memphis* to the west side of the Nile.

Example: *Private European collection*.

50.2 Variant 2 1649

This map also appeared in 1649 in a small (4°) book, *De kleyne wonderlijcke werelt bestaende in dese keyserrijcken...*, written by Jacob Joosten and published by Dirk Uittenbroek in Amsterdam. This example has new bookbinder's instructions above the map of *Dese Kaerte moet staen achter Fol 50. De Stadt Ierusalem. 51 Den Tempel Salomon. 52.* There were later editions of this book published in Amsterdam in 1651, 1662, 1670, and 1717, and in Utrecht in 1684.

Example: *London BL*, 10105.aaa.5.

MAP 50

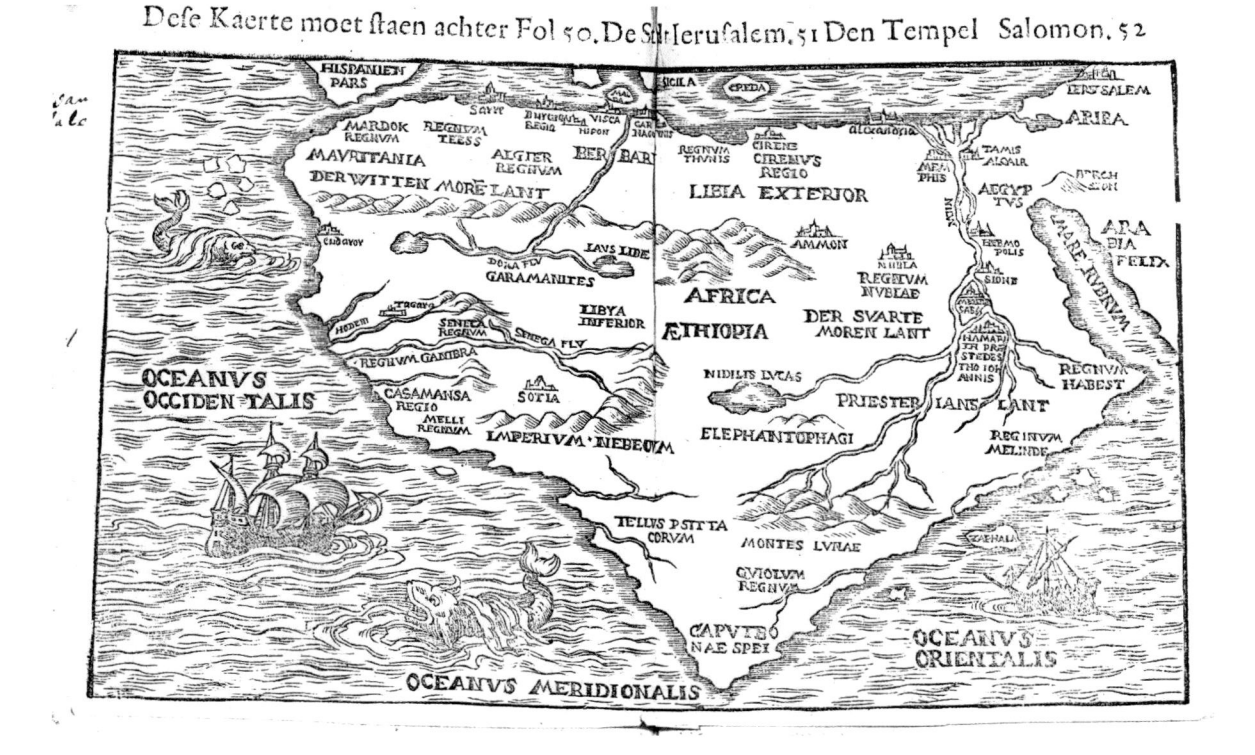

Variant 2.

Description

This is a late woodcut map of Africa from a Dutch edition of Bünting's *Itinerarum*. As such, the appearance of the map, especially when compared to other maps from this period, is unusual.

This map is considerably different from the German and Czech editions of Bünting's *Itinerarum*. For example, there is a different style of ship in the South Atlantic Ocean and two large sea monsters in the Atlantic Ocean. Another ship is added in the Indian Ocean. The map is considerably smaller than the German and Czech maps at 15.5 x 25 cm. Also, the title on the earlier editions of AFRICA TERTIA PARS TERRAE is replaced with *Dese derde Caerte van Aphrica behoort te staen na die beschrijvinge van Asia int beginsel des boecks*. These are binder's instructions that are translated as follows: This third Map of Africa needs to be placed after the description of Asia at the beginning of the book.

Heinrich Bünting (1545-1606) was a Protestant Professor of Theology at Hanover. He is most known for his *Itinerarum Sacrae Scripturae*, a religious commentary on the world. The book describes the travels of the religious figures of the Old and New Testaments of the Bible as well as its geography. It was a popular book and was the most complete summary of the geography of the Bible of this period. For a discussion of the geography of this map, please refer to Heinrich Bünting's 1589 map of Africa, Map # 24.

Publication Information

There are thirteen Dutch editions of the *Itinerarum*. The first was published in Rees by Wylics van Santen in 1594. This was followed by editions in Utrecht in 1596 and Amersfoort in 1601. This 1605 edition is the first located by the author containing a map of Africa. It was published by Jan Jansz., an Arnhem publisher and the father of Johannes Janssonius.

References
Tooley 1999-2004, A-D: 209; Van der Heijden 1998: 49-71; Norwich 1997: map # 17; Tooley 1966: map # 141.

51   1605 Giuseppe Rosaccio, Venice

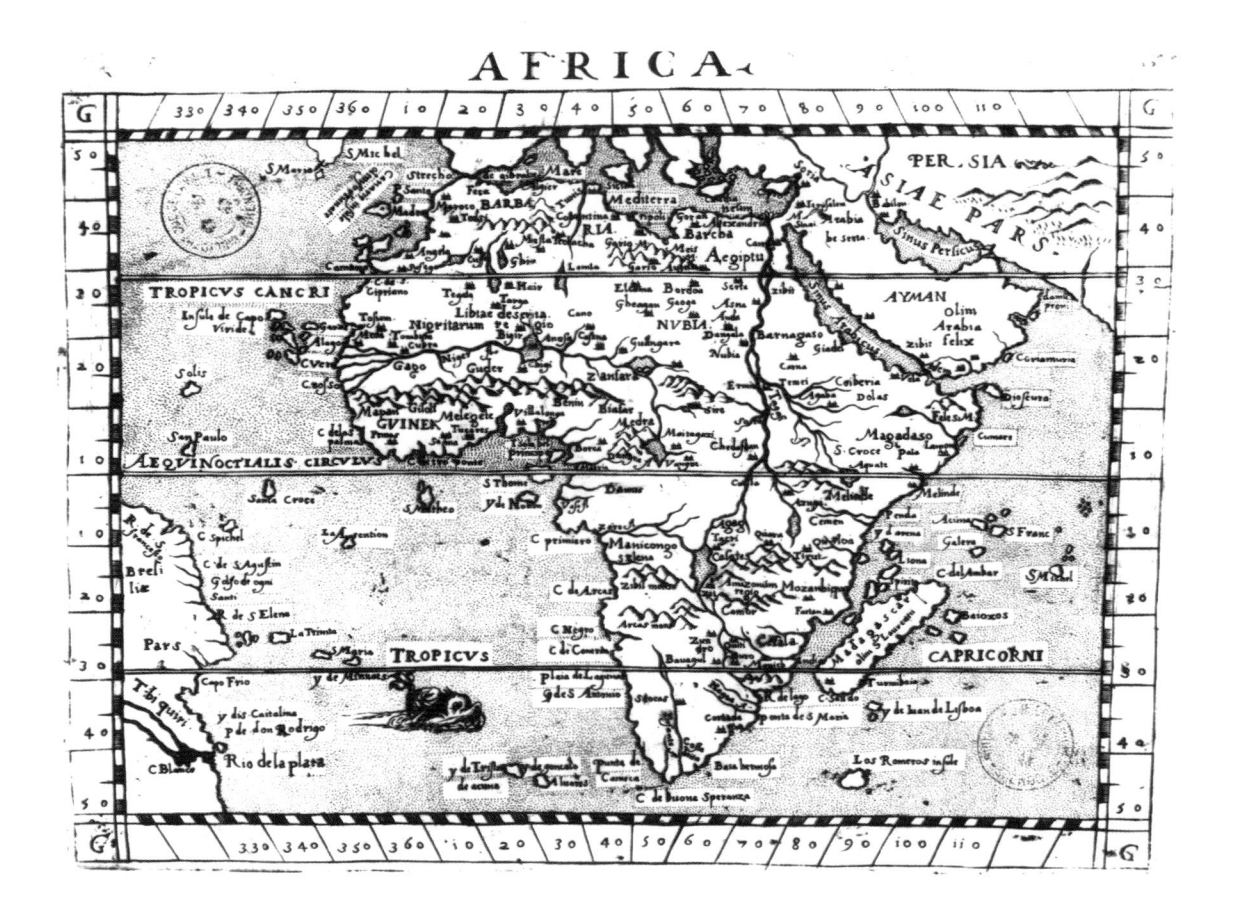

AFRICA.
- No scale bar.
- [Venice : Agostino Angelieri, 1605].
- From: *Le Relationi Universali.*
- 1 map : copperplate engraving ; 19 x 26 cm.
- Latitude and longitude coordinates on all four sides of the map.
- North at the top; no compass rose.
- Title is engraved above the map.
- Devoid of all decorative elements except for one sea monster in the South Atlantic.
- There are no titles for the Atlantic or Indian Oceans.
- *Aegiptu* with no 's' to the west of the Nile River. 'G' is on all four corners of the map.

Example: *Grenoble BM*, E 28421 (1605 edition).

Description

This map is from the extremely popular and widely published *Le Relationi Universali,* a geographic and political treatise on the world written by Giovanni Botero. The first edition of 1590, did not contain maps. A map of Africa first appeared in the 1595 Venice edition of Botero, which contained the Botero 1595 map (Map # 32).

These previous Italian editions of Botero's work, *Le Relationi Universali*, contained either the Giovanni Lorenzo d'Anania map of 1582 or the Giovanni Botero map of 1595. For the 1605 Venice edition of Botero's *Le Relationi Universali,* published by Agostino Angelieri, a new set of copperplates was engraved.

This is an enlarged, somewhat crudely engraved map of Africa when compared to the earlier d'Anania and Botero maps. Geographically, it closely follows the earlier Botero map of 1595, with the omission of some placenames and mis-spellings of many others (for example *Zundro* rather than *Zimbro*).

The author has attributed this map to Giuseppe Rosaccio. The Asia map in the 1605 *Le Relationi* in which this Africa map appears is signed by Gioseppe Rosatio (Rosaccio) as the engraver, and as the engraving for the Africa and Asia maps are similar in style, this map of Africa is also attributed to Rosaccio.

Publication Information

The map is known in one state. Sometime before 1618 cracks developed on the copperplate that produced this map. One enters the map on the right side by 30°N latitude with two other cracks to the plate at the bottom by 85° and 120° longitude. These cracks in the copperplate became increasingly more noticeable in the later editions.

Besides appearing in the 1605 Venice edition of *Le Relationi* described above, this map also appeared in other editions Botero's *Le Relationi Universali* in the 1600s:

1618 edition of Botero's *Le Relationi Universali*.
This map is known to have appeared in an edition of Botero of 1618, published by Alessandra Vecchi in Venice. Example: *Chicago NL*; *London BL*, Maps 63510 [19].

1659 edition of Botero's *Le Relationi Universali*.
This map was also in an edition of Botero of 1659, published in Venice by Bertani. Examples: *London BL*; *New York PL*, *KB 1659 Botero.

This map may also have appeared in an edition of 1671, published again by Bertani, though this edition has not been examined.

To distinquish this map from similiar appearing maps, using different copperplates, by D'Anania (1582), Botero (1595), Magini (1596), Keschedt (1597), Africanus-Pory (1600), and Botero-Giunti (1640), please refer to the table under Map # 20, 1582 D'Anania.

References
Burden 1996: map #149 (Americas map); Shirley 1993: 233-4.

52   1606 Jodocus Hondius, Amsterdam

NOVA | AFRICÆ | TABULA | AUCTORE | Jodoco Hondio.
[and at bottom right of map]:
Excusum in aedibus Auctoris | Amsterodami.
- No scale bar.
- Amsterdam : Jodocus Hondius [and Cornelis Claesz., 1606].
- From: *Gerardi Mercatoris Atlas.*
- 1 map : copperplate engraving ; 37.5 x 50 cm.
- Latitude and longitude coordinates on all four sides of the map. Longitude coordinates across Equatorial Line.
- North at the top; no compass rose.
- Decorative title cartouche at the bottom left of the map.
- There are three large sailing ships along with one smaller, single mast boat and two sea monsters in the Atlantic. In the Indian Ocean there is one sailing ship and one sea monster.
- Within Africa, there is one each of a camel, an elephant and a monkey. There are various legends in Latin within Africa, including one by *Amara* on the Equator in Northeast Africa, which relates a legend about Prester John keeping his sons captive on the mountain top of *Amara*.

Examples: *Various locations.*

Description

This is an attractive map with numerous sailing ships and sea monsters in the oceans in a finely engraved style, typical of Jodocus Hondius (1563-1612). It appears in various editions of the *Gerardi Mercatoris Atlas*, published by Jodocus Hondius and later his sons from 1606 to 1630. As such, this Hondius map of Africa is an important transitional map, bridging the earlier work of Ortelius, Mercator, and others, with the more modern maps of Willem Blaeu in 1608, Claesz. in 1617, and others.

The geography for this map is taken from a variety of sources. Some of the interior of Africa is based on Mercator's 1595 map of Africa, but numerous additions and changes have been made. Hondius had access to the books of exploration of Africanus, Pigafetta (based on the travel account of Duarte Lopez), Ramusio, De Barros, and others. These were supplemented by the more recent Dutch sources such as Van Linschoten's *Itinerario* with the two maps of the Southern portion of Africa based on Portuguese sources.

Hondius follows Pigafetta in his detailed depictions of the Congo Region. He provides greater detail in West Africa and in Southern and East Africa, likely reflecting new information from the Dutch and the Portuguese reports from these areas. The Southern African lake, *Sachaf Lacus*, is not depicted as a source for the *Zaire Rio / R. Congo* River or the Nile, but is only the source for the unnamed Cuama and Spirito Santo Rivers.

In Northern Africa, Hondius retains the Mercator northern river, *Nubia Fluvius*, flowing eastward into the Nile. In West Africa, the *Niger fluvius* (Niger River) flows in a more straight line into the Atlantic, though Hondius uses Ortelius' presentation of *Niger Lacus* in Central Africa as the source for the Niger. In other instances, Hondius follows Ortelius as well. For example, Hondius places Prester John's *Amara* on the Equator as in the 1570 Ortelius map, and not below the Equator as on the 1595 Mercator map.

The East coast of Africa and the island of Madagascar are considerably updated. Hondius slightly corrects the longer taper of Southern Africa with a westward bulge at the Cape of Good Hope and a more defined bay to the north.

Publication Information

When Jodocus Hondius decided to re-issue Gerard Mercator's atlas in 1606, he had this map engraved for inclusion in his atlas, *Atlas sive Cosmographicae Meditationes...*, or, as it is more commonly called, the Mercator-Hondius Atlas.

Mercator had originally intended his atlas to be an all-encompassing work. The first part was published in 1585 and 1589, and after Mercator's death in 1594, the second part was issued by his heirs in 1595 with sections lacking, notably Spain (See Mercator II, 1595, Map # 31 for the Africa map prepared by his grandson, Gerard). The heirs re-issued an unchanged edition in 1602, but ended any further attempt at a more comprehensive atlas, likely in part from the competition provided by the immensely successful Ortelius atlas, but probably also due to the deaths of Rumold in 1599 and Michael in 1600. In 1604, Mercator's heirs sold the copperplates to Jodocus Hondius and Cornelis Claesz. While it appears that this map, along with the atlas, was primarily printed by Hondius ('Excusum in aedibus Iudoci Hondij' on the title page), the atlases were jointly sold by Hondius and Cornelis Claesz.

This map of Africa was issued along with Gerard Mercator II's 1595 map, *Africa ex magna orbis terre...* in the various editions of the Mercator-Hondius Atlas up to 1630.

After Jodocus' death in 1612, the atlas continued under his name but published by his widow, Colette van den Keere Hondius, his son, Jodocus Jr., and his second son, Henricus, along with their brother-in-law, Johannes Janssonius. With the death of Colette and then Jodocus Jr. in 1629, many of the copperplates prepared by Jodocus Jr. were sold by his estate and purchased by Willem Blaeu, who issued an *Appendix* to the Mercator-Hondius Atlas using Jodocus Jr.'s plates as well as new plates. With the publication of the last edition of the Mercator-Hondius Atlas in 1630, and in response to the increasing competition from the new Blaeu *Appendix*, Henricus Hondius in collaboration with Janssonius published a new *Appendix* to the Mercator-Hondius Atlas in 1631. For this *Appendix*, Henricus ceased using this older copperplate of his father's and prepared a new plate (See Jodocus Hondius' 1619 map, Map # 58, for further information on this map).

This map is known in only one state.

The following is a list of the various editions in which this map appeared. The list may be used as a guideline to determine from which atlas edition a separate example of this map appeared. This information is based on Van der Krogt (1997: 711) with his information on date, language, signature and pagination:

Date	Van der Krogt no.	Language	Signature	Pg no.
1606	8600:1B:101	Latin	122	
1607-08	8600:1B:102	Latin	6V	311
1609	8600:1B:111	French	6X	313
1611-12	8600:1B:103	Latin	6Y	315
1613-19	8600:1B:104	Latin	7D	320
1613-16	8600:1B:112	French	7D	329
1619	8600:1B:113	French	7H	333
1623	8600:1B:105	Latin	7L	332
1628	8600:1B:114	French	7S	621-624
1630	8600:1B:107	Latin	7S	346

References
Van der Krogt 1997: 8600:1B: Vol. I: 711; Norwich 1997: map # 27; Koeman 1967-71, II: 282.

53 1607 Jodocus Hondius, Amsterdam

AFRICÆ | DESCRIPTIO.
- Scale bar at bottom right: Miliaria Germanica 250 = 2.9 cm.
- [Amsterdam : Jodocus Hondius, Cornelis Claesz. and Jan Jansz., 1607].
- From: *Atlas Minor Gerardi Mercatoris a Hondi...*
- 1 map : copperplate engraving ; 15 x 20 cm.
- Latitude and longitude coordinates on all four sides of the map.
- North at the top; no compass rose.
- Title set in strapwork cartouche at bottom left.
- There are two ships engaged in a sea battle in the South Atlantic. There is one sea monster in the Indian Ocean. Except for the title cartouche at the bottom left of the map, no other ornamentation is on the map.

Examples: 1607 edition: *Washington LC*, 422; 1613 edition: *New York PL*, *KB 1613; 1625 *Purchas*. *London BL*, 679.h.11-14; 1625 edition of Purchas: *London BL*, 213.d.2; 1626 edition of Purchas: *London BL*, W 2111/5; *Author's collection* (as a separate map).

Description

The year after first publishing a folio edition of Mercator's Atlas in 1606, Jodocus Hondius prepared a set of small-scale plates for a pocket-sized atlas. This effort was done in collaboration with Cornelis Claesz. and Jan Jansz., the father of Joannes Janssonius, who was to become the son-in-law of Jodocus Hondius. The decision to produce an *Atlas Minor* was likely in response to the increasing popularity and financial success of the pocket-sized atlases based on Ortelius and then Langenes-Claesz., and others that had been appearing since 1577. In actuality, the Hondius Atlas Minor was larger than the Ortelius-Galle *Spieghel, Epitome,* and other smaller pocket-sized atlases, though smaller in size than the Hondius folio atlas of 1606.

The geography for this map was taken directly from Hondius' folio map of Africa of 1606. By necessity, some placenames and other features were omitted due to the reduced size of this map in comparison with the larger 1606 map.

Although there is no indication, Koeman (1967-71, II: 509) states that the copperplates were probably engraved by Jodocus Hondius himself.

Publication Information

A number of editions were published containing this map of Africa, as outlined below and based on Van der Krogt (2003). While it appears that this map, along with the atlas, was printed by Jodocus Hondius ('Excusum in aedibus Iudoci Hondij' on the title page), the atlases were sold by Hondius, Cornelis Claesz., and Jan Jansz. ('veneunt etiam').

The final Dutch edition of 1621 was published by Jan Jansz. of Arnhem. Sometime after 1621, the copperplates were acquired by unknown London booksellers. The Africa map was reprinted in the books *Purchas His Pilgrimage,* published in 1624-26 by William Stansby for Henry Featherstone, and in *Historia Mundi,* published in 1635, 1637, and 1639 by Thomas Cotes for Michael Sparke and Samuel Cartwright.

The map is known in only one state.

The following is a list of the various editions in which this map appeared, generally based on Van der Krogt (2003). The list can be used to determine the date and edition for a separate map by examining the language, signature and page number on the verso of the map:

Date	van der Krogt no.	Language	Signature	Page	Signature Notes
1607	351:01	Latin	B2r		
1609	351:11	French	B2r	11	
1609	351:21	German	B2r		
1610	351:02	Latin	B2r	11	
1613-1614	351:12/13	French	B2r	11	'L'AFRIQVE. 11' (above map)
1620-21	351:03	Latin	B2r	11	

English publications with this map, with signature and pagination:

Date	van der Krogt no.	Title	Signature	Page	Signature Notes
1625		*Purchas*		748 (at top left).	'Hondivs his Map of Africa.' in letterpress above map
1626		*Purchas*		620 (at top left)	'Hondivs his Map of Africa.' mis-printed onto the map image.

1635	351:03	*Historia Mundi* Clr	13	'AFRICKE.' sideways to left of map.
1637		*Historia Mundi* Clr	13	'AFRICKE.' sideways on to left of map.
1639		*Historia Mundi* Clr	13	'AFRICKE.' sideways to left of map.

The map of Africa appears to have been re-issued in other publications in England during this period. Burden (1996: 186) mentions that the maps of the continents appear in an untitled atlas by Robert Walton.

Author's example from 1626 Purchas edition.

References
Van der Krogt 2003: map 8600:351; Norwich 1997: map # 27a; Burden 1996: 186; Plak 1989: 55-77; Mickwitz et al. 1979-1995: Vol. 2: 48-53; Skelton 1968; Koeman 1967-71, II: 509.

54   1608 Willem Janszoon (Blaeu), Amsterdam

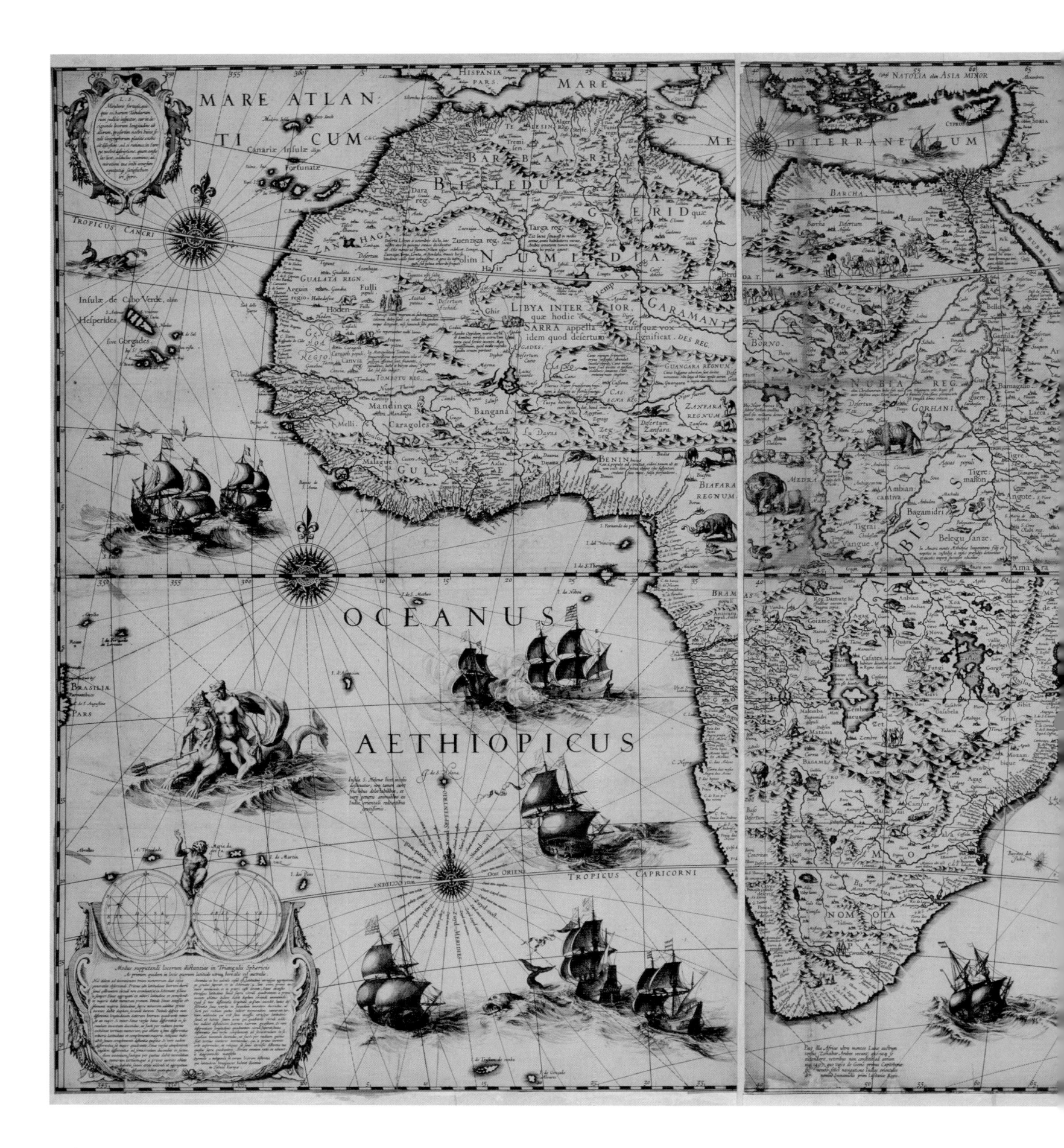

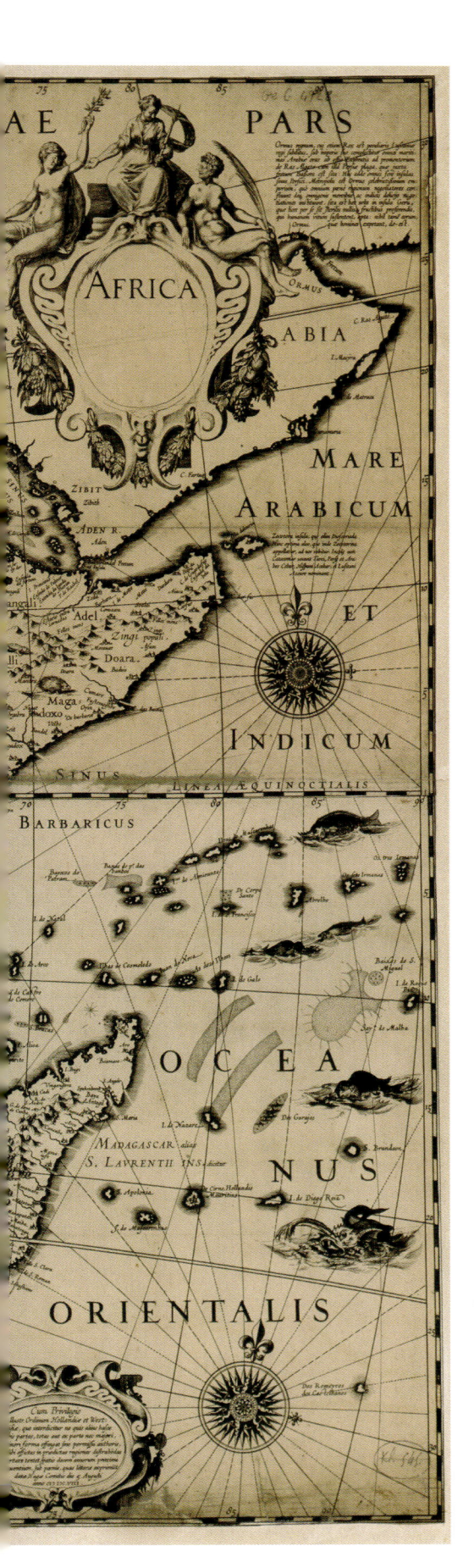

(Describing the second and later states, as the first state at the Bibliothèque Nationale in Paris only contains the map sheets without the surrounding decorative borders and text:)

AFRICA.
[within a cartouche with ten lines of text on the bottom right map sheet]:
Cum Privilegio | Illustr. Ordinum Hollandiæ et West= | frisiæ, ... 5. Augusti | anno CIƆ IƆ C. VIII.
- No scale bar.
- Amsterdam : 1608, Willem Jansz. [Blaeu].
- From: separately published.
- 1 wall map : copperplate engraving and etching ; 4 map sheets ; 84 x 111 cm (4 map sheets only) ; 128 x 176 cm (including 4 map sheets, decorative borders, letterpress text, and title strip).
- Latitude and longitude coordinates on all four edges of the map. Longitude coordinates across Equatorial Line.
- North at the top; there are five compass roses on the map: two in the Atlantic Ocean, two in the Indian Ocean and one in the Mediterranean Sea; one windrose in the South Atlantic.
- Title at upper right of the map is set within a decorative oval cartouche with three figures at the top. Title text strip across the top of the map is in second and subsequent states as described below.
- As described in detail below: In the lower left corner, there are two circular diagrams at the top of a cartouche with text; in the upper left corner, there is a cartouche with text.
- In Indian Ocean, one ship and five sea monsters. In Atlantic, eight ships, three sea monsters, vignette of Neptune and mermaid. In Mediterranean, one smaller ship and one sea monster.
- Numerous animals and figures of Africans are found on the map of Africa.
- As described in detail below: eight groups of costumed figures on each side of the map. twelve town views across the bottom of the map. Various other decorative elements on the map.

54.1 First State. 1608
Without the imprint of the engraver, Josua van den Ende at the bottom left map sheet: I[a]. van den Ende sculp.

Examples: Only known in one example with all four map sheets, but without surrounding decorative borders and text: *Paris BN*, Ge C 4928 (from ex Klaproth Collection, no. 545). The two lower map sheets only are at *Chicago NL*, Novacco Collection.

54.2 Second State. 1612
With the addition of the engraver, Josua van den Ende to the right of the cartouche at the bottom left map sheet: I[a]. vanden Ende sculp.

NOVA AFRICAE GEOGRAPHICA ET HYDROGRAPHICA DESCRIPTIO, auct: G: Ianss. / I[a]. vanden Ende sculp.
- AMSTELODAMI, | Ex officina Guilielmi Ianssonij, sub signo Solarij deaurati, | Anno à Christo nato M.DC.XII.
- 128 x 176 cm, including decorative borders and letterpress text, which are described above. Four map sheets, with a title strip.

The map consists of four main map sheets with decorative borders along the sides and the lower edge, including a Latin description. The title with the name of Willem Jansz. [Blaeu] in Latin runs along the entire upper edge. The privilege in the lower right corner is still dated 1608. The geography is unchanged from the 1608 example. The only addition is the inclusion of the name of the engraver (Josua van den Ende) to the right of the cartouche in the lower left corner: I. vanden Ende sculp.
Examples: *Burgdorf RV; Dresden SH,* Schr. II, Mappe 32 b, No. 7.

54.3 Third State. 1624
With the imprint for Henricus Hondius.

GEOGRAPHICA ET HYDROGRAPHICA NOVA AFRICÆ DESCRIPTIO, auct: G: Ianss. / AMSTELO-DAMI, | Ex officina HENRICI HONDII [1624]. / Iª. vanden Ende sculp.
128 x 176 cm, including decorative borders and text.

The map is the same in composition to the second state: the original title with the name of Blaeu is at the top; Josua van den Ende name remains as the engraver; the privilege remains with its original date of 1608. The geography is unchanged. In the example of this state at the Herzogin Anna Amalia Bibliothek in Weimar the surrounding descriptive text has been reset, but is similar in content to the other third state. The title of the surrounding text begins with: 'NOVA AFRI- | CAE DESCRIPTIO' and ends with 'Ex officina HENRICI HONDII.'.

Examples: *[Breslau (Wroclaw), Stadtbibliothek,* Rolle 81 (destroyed during World War II)]; *Weimar HAAB*.

54.4 Fourth State. before c. 1652
With the imprint now Claes Jansz. Visscher (*CIVisscher excudit* to the right of the cartouche containing the privilege).

[deducted from an incomplete title strip]: NOVA AFRICÆ GEOGRAPHICA ET HYDROGRAPHICA DESCRIPTIO, auct: G: Blaeu
- [Amsterdam]: Visscher excudit, [s.a.] / Iª. vanden Ende sculp.
- 84 x 111 cm for map only.

The title strip along the upper edge has been replaced. The text concerning the privilege is unchanged, but the year of 1608 in the last line was removed. To the right of the cartouche containing the privilege is: Visscher excudit.

Examples: *Paris BN*, Ge DD 5080 (with decorative borders but lacking title and text); *Regensburg SB* (the title is incomplete and the text, which framed the map, has been lost).

54.4.A Fourth State. Variant A 1644
With the imprint for Pieter Verbist and the date of 1644.

The variant mentioned by Schilder as having been described by Denucé (1937) is now held in the Royal Museum for Central Africa in Tervuren, Belgium:
NOVA AFRICAE GEOGRAPHICA ET HYDROGRAPHICA DESCRIPTIO, auct: G: Blaeu / Antuerpiae apud Petrum Verbist sub signo Americae | in platea qua vulgo Lombardorum moenia dicitur | an 1644 / Iª. vanden Ende sculp.

In this Tervuren example, the text within the dedication cartouche has been replaced by a pasted-over printed text with a dedication to the Spanish King Philip IV. The four main map sheets and the decorative borders are the same as described. The surrounding texts in Latin (left side), Dutch (at bottom), and French (right side) are very close to the texts of the two maps in State 5 but have been reset.

Example: *Tervuren RMCA*, 50.62 (with the imprint of Pieter Verbist. This is the example referred to by Denucé (1937), who was aware of an example of this state that had been sold by Pieter Verbist in Antwerp,1644.).

54.5 Fifth State. possibly as early as 1656
With the imprint of Nicolaas Visscher at the end of text sections within the surrounding text panels.

NOVA AFRICÆ GEOGRAPHICA ET HYDROGRAPHICA DESCRIPTIO, auct: G: Blaeu / Iª. vanden Ende sculp.
- Amsterdam: Visscher excudit, [s.a.].
- 128 x 176 cm including decorative borders and text. 84 x 11 cm for map only.

This state is by Nicolaas Visscher, son of Claes Jansz. Visscher. Claes Visscher died in 1652 and his son reprinted the map, inserting his imprint below each of the text sections. The map may have been printed as early as c.1656.

The decorative borders with the eight groups of figures and the twelve views of African towns at the bottom edge are now surrounded by an entirely new descriptive text in Latin (left side), Dutch (at bottom), and French (right side). Each text includes Nicolaas Visscher's address at the end.

Examples: *Amsterdam SMA*; [*München, Bayerisches National Museum* (destroyed during World War II)]; [*Stuttgart WL* (destroyed during World War II)]; *Washington LC* (cataloged as part of Leo Belgicus exhibit; no call number. This example contains the imprint of Claes Jansz. Visscher on the map sheet, but with the addition of Nicolaas Visscher on the surrounding text panels).

Map 54.5 Visscher's imprint.

Description

The appearance of this wall map was a major landmark in the production of maps in The Netherlands and for the Blaeu publishing firm. The map was of such importance that for the next 100 years numerous mapmakers throughout Europe diligently copied this map, both in design and in content.

Willem Janszoon or Willem Jansz. Blaeu (1571-1638), as he later called himself, and as he is more commonly known, was one of the most noted Dutch cartographers and map publishers of the seventeenth century. R.V. Tooley stated that Blaeu's maps are 'esteemed by collectors for their decorative quality, historical importance, and as the highest expression of Dutch cartographic art during the period of its supremacy' (Tooley 1969: 28).

After preparation as a pupil of the great Danish astronomer, Tycho Brahe, Blaeu initially concentrated on globemaking and separately published maps, including this wall map of Africa of 1608. From 1608, he dominated the market for sea atlases and in 1630 produced his terrestrial atlas. These atlases were continued by his sons, culminating in the great *Atlas Maior* of 1662 in 11 volumes with subsequent editions.

Josua van den Ende (c.1583/84 - ?) was responsible for engraving the four copperplates that comprise the map image. Of the four continent maps, only this map of Africa, in its second and subsequent states, bears his name as the engraver. Van den Ende was also later responsible for the engraving of some of Blaeu's folio-size maps, including that of Africa of 1617. The decorative borders and other decorative elements for this map were etched by Hessel Gerritsz (1580/81-1632). He also was responsible for the overall design of the map. The shields of the map bear the names of the engravers: Josua | vanden | Ende |et Hessel | Gerritsz. | sculpserunt.

The only known example of the 1608 map lacks the decorative borders and the title strip above the map sheets. Based on the 1612 map, it would normally have been framed on both sides by eight groups of costumed figures of various Africans, and along the lower edge by twelve views of African towns as in the 1612 second state of the map. These decorative elements were printed from two copperplates. On the left side, these figures are (1) SENAGENSES; (2) MERCATOR, DOMINVS, SERVVS in Guinea; (3) Guinearum MVLIERES et ANCILLA; (4) CABO LOPES GONSALVI Accolae; (5) MAGNATES IN CONGO; (6) MILES CONGENSIS cum FEMINA; (7) PROMONTORII BONÆ SPÆI HABITATORES; (8) REX, cum Subditis in MADAGASCAR. On the right side, these figures are (1) MAROCCHI; (2) FEZANI; (3) Femina ex MALTA, TRIPOLI, et ALGIER; (4) VIRGO, VIDVA, MVLIER AFRA. AETHIOPS; (5) ÆGYPTII; (6) PEREGRINI euntes ad MECCAM; (7) ABISSINI; (8) CAFRES IN MOZAMBIQVE.

Below the map image are twelve town views of what were deemed to be the most important towns of Africa: (1) *TANGER*; (2) *ÇEVTA*; (3) *ALGER*; (4) *TVNES* (5) *ALEXANDRIA*; (6) *ALCAIR*; (7) *QVILOA*; (8) *MOZAMBIQVE*; (9) *CEFALA*; (10) *S. GEORGIVS della MINA*; (11) *CANARIA*; (12) *TZAFFIN*.

For the figures, Hessel Gerritsz. did not have access to actual models and depended on information from the available published works, most notably Pigafetta's work on the Congo of 1591 and De Bry's subsequent *Petits Voyages* of 1598 for the Congo region, as well as numerous other books. Sources for the town views go back much earlier to Braun and Hogenberg's *Civitates* and their primary sources.

The title for the 1612 example is on a separate strip above the entire top edge of the map. The map contains four cartouches. A second title with only the name, AFRICA, is in a cartouche in the upper right corner. In the lower right corner is a cartouche with a reference to the privilege granted to Willem Jansz. [Blaeu] by the States of Holland and West Friesland on August 5, 1608. The lower left corner includes two diagrams with a cartouche containing detailed instructions on how to measure distance. The upper left corner contains a cartouche in which Blaeu defends his method of calculating longitude.

The geography for the Africa map is similar to Blaeu's 1605 double hemisphere wall map of the world, which is not surprising given the close proximity in dates. By this time, there was a thriving exchange of ideas and cartographic information between Northern Europe and Italy, and to some degree Portugal. This was likely facilitated in part by the regular Frankfurt Book Fair which sat astride most north-south travel routes, but also by direct commercial contacts. Blaeu probably had access to Gastaldi's 1564 detailed wall map of Africa and those by the Northern European imitators such as De Jode. He also had access to the books of exploration of Africanus, Pigafetta (based on the travel account of Duarte Lopez), Ramusio, De Barros, and others. These were supplemented by the more recent Dutch sources such as Van Linschoten's *Itinerario* of 1596. The Van Linschoten maps in turn were based on Portuguese sources.

The geographical divisions within Africa are similar to those of earlier sixteenth century maps of Ortelius, Gastaldi, etc. In West Africa (*Barbaria, Biledulgerid, Libya Interior*, etc.), *Niger fluvius* (the Niger River) combines with *R. Senega* (the Senegal River) to make a vast river system and flows directly westward into the Atlantic Ocean. The source for the Niger continues to be a lake, *Niger lacus*, in Central Africa. North of the Niger River is *Tombotu* (Timbuktu), a major trading center linking North Africa with the minerals and other trade goods of sub-Saharan West Africa. The depiction of the Gold Coast in Guinea is based on a drawing from a trip to *El Mina* which was offered to Luis Teixeira in Lisbon (Schilder 1996: 155, Cortesão 1960: III, 67-90). This Teixeira map found its way to Amsterdam where it was used for a 1602 map drawn by Pieter de Marees for Cornelis Claesz. and was the source for this as well as other maps of the region.

In Central Africa, the geography for the Congo Region is mainly from a map that was inserted into Ortelius' *Theatrum* from 1595 (*Fessae et Marocchi...* , with the Congo map inset), which was in turn derived from travel accounts of Duarte Lopez, as published by Pigafetta in 1591.

Along the Southern African coast, the recent voyage to the east by De Houtman is reflected by references to Dutch names: *Vleijs baij, Vis baij,* and *Mossel baij*. The interior of Southern Africa uses placenames from Portuguese explorations up the river systems from their bases at *Mozambique* Island and *Coffala* (Sofala), most notable being *Cast[ellum] Portogal* (Castle Portugal) and various African kingdoms known to Europeans: e.g. *Butua* and *Monomotapa*.

The sources for the Nile River are faithful to Ptolemy with its two lakes of *Zaire lacus* & *Zembre lacus* in the west and *Zaflan lacus* in the east. The *Lunae Montes* (Mountains of the Moon) serve as a divider between the Nile source lakes and the major southern river systems of *R de Spirito Santo* and *Cuama Fluvi*. Both of these two latter rivers have a common source in a lake southwest of the Mountains of the Moon, *Sachaf lacus*, and an initially common river in the *Zambere flu* flowing from it. This follows Mercator, who, in his 1569 world map, placed a large Sachaf lacus as the source for the southern rivers.

In Abyssinia, Blaeu based his work on Ortelius' map of the region (the Prester John map, *Presbiteri Iohannis, Sive Abissinorum...*) and numerous texts of Portuguese exploration along the coast and into the interior. As with most writers and geographers of that day, Prester John's kingdom was greatly exaggerated in size.

The Prime Meridian is placed at a peak, called *Pico,* on the island of *Tenerife* in the Canary Islands off the West African coast.

Publication Information

As noted earlier, the Blaeu wall map of Africa was an important new model that was to be imitated to varying degrees by numerous mapmakers through the rest of the seventeenth century. From the first appearance of the Africa map in 1608 up to the early 1700s, the Blaeu map influenced how mapmakers and the public viewed Africa. Though by the time of the Italian and French imitators, the geography was generally out-of-date, the Blaeu name and the decorative elements of the map continued to be important. The wall maps of Africa by these Italian and French imitators, Godefridus de Scaicki (Map # 59, c.1620-35), Stefano Scolari (Map # 76, 1646), Pietro Todeschi (Map # 117, 1673), Giovanni Giacomo de Rossi (Map # 105, 1666), and Alexis Hubert Jaillot (Map # 112, 1669), are described separately.

References

The primary source for information on the Blaeu wall map of 1608 and its later states is Günter Schilder. Of particular importance is Schilder's Vol. V of *Monumenta Cartographica Neerlandici*, which covers the Blaeu wall maps in detail. This volume contains a full size reproduction of the third state of 1624 in facsimile 3^{1-14}. Additional information on the Verbist example provided by Wulf Bodenstein. Zisska 2001: no. 3518; Schilder 1993: Vol. IV, 1996: Vol. V, 2000: Vol. VI; Keuning 1973; Kraus 1970: 10-11; Ristow 1967: 3-17; Wieder 1925-33: 70.

55 1614 Pieter van den Keere, Amsterdam

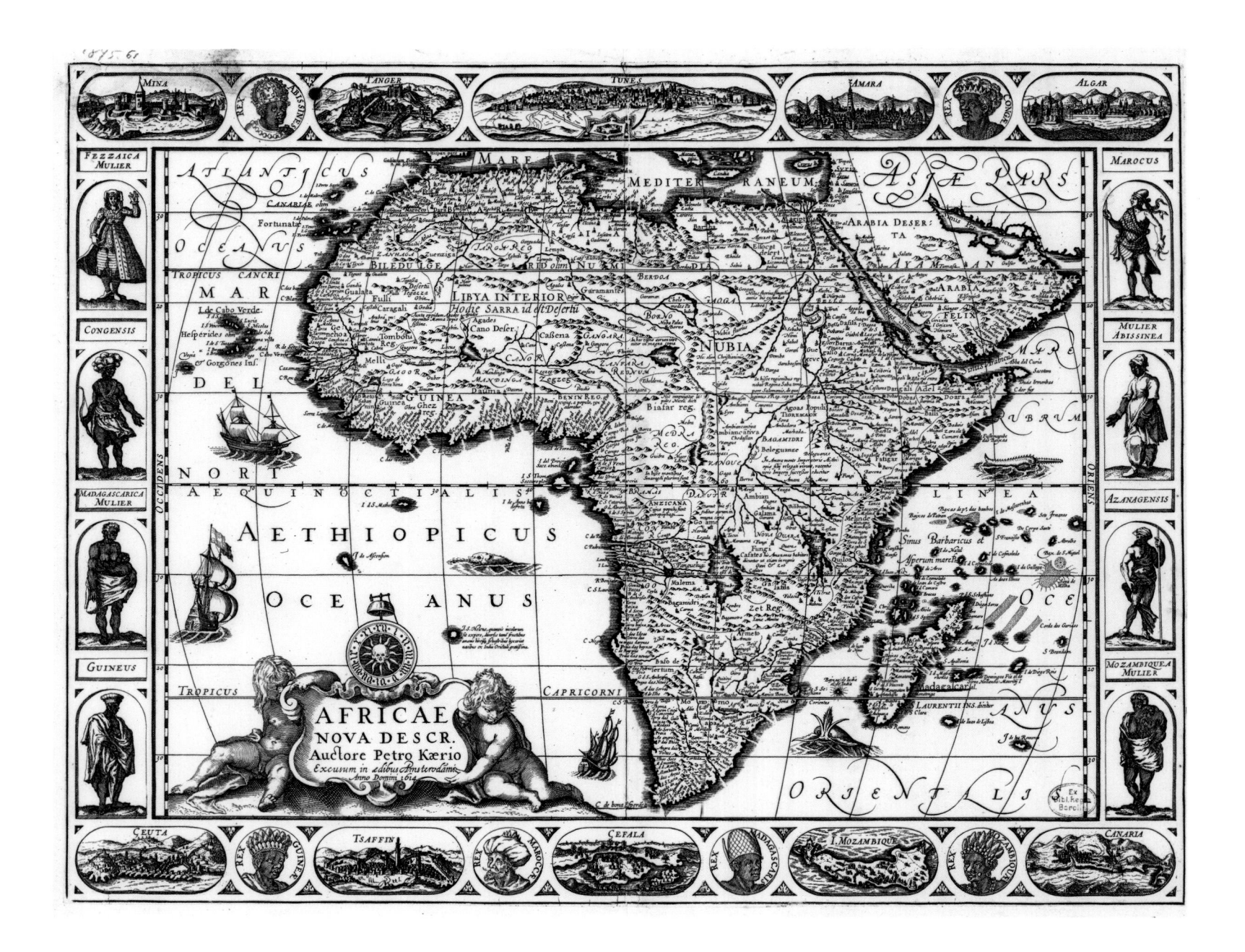

55.1 First State 1614

AFRICAE | NOVA DESCR. | Auctore Petro Kærio | Excusum in edibus Amsterodami | Anno Domini 1614.
- No scale bar.
- Amsterdam : Pieter van den Keere, 1614.
- From: separately published.
- 1 map : copperplate engraving ; 43.5 x 56 cm (35 x 48 cm for map only without borders).
- Latitude coordinates along left and right sides. Longitude coordinates across Equatorial Line.
- North at the top; no compass rose.
- An elaborate title cartouche is at the bottom left of the map with a shield held by two large putti. Above the cartouche is a clock with the face of a 'death's head'.
- In the Indian Ocean, there are two sea monsters. In the Atlantic Ocean, there are three sailing ships and one sea monster. No animals within Africa. Various notations within Africa denoting information such as *Amara* and the legend of Prester John.

- Decorative borders on all four sides of the map: four figures on left and right sides; five town views and two faces at top; and five town views and four faces at the bottom, as further described below.

Examples: *Berlin SB*, C 75; *Washington LC*, G8200 1614.K;

55.2 Second State 1631
Entirely new title cartouche with no reference to Van den Keere. The title is now *AFRICAE | NOVA DESCR. | PER | Nicolaum Iÿ. Visscher. | Anno 1631.*

Example: *Private Foundation, formerly collection Stopp.*

Sometime before 1631, Claes Jansz. Visscher acquired the copperplate for this map and removed the Van den Keere imprint. The title cartouche in the second state is now an allegorical depiction of the continent: an African woman sitting on a crocodile. There is no change to the geographical content on the map.

55.3 Third State 1636
The date on the title cartouche is changed to 1636.

Example: *Paris BN*, Ge D 11371.

55.4 Fourth State 1648
The date on the title cartouche is changed to 1648.

Example: *Private Foundation, formerly collection Stopp.*

This fourth state was only recently discovered in 2005 at a fair in Paris and is part of a set of maps of the four continents by Visscher. The Americas and Europe maps in the set are dated 1650 on the title cartouche. The map of Africa is only known in this one example.

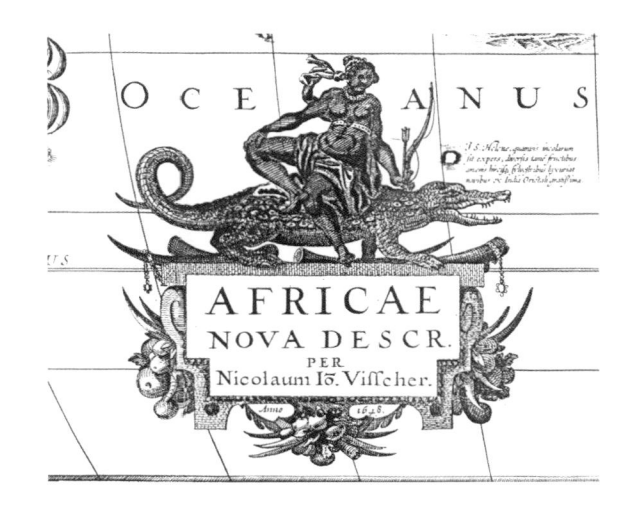

Map 55.4 Title cartouche.

55.5 Fifth State 1652
The date on the title cartouche is changed to 1652.

Examples: *Amsterdam UB*, 33-17-46; *Boston Afriterra*; *Private Foundation, formerly collection Stopp.*

Description
This is the earliest, folio-sized map of the continent of Africa framed with decorative borders. Pieter van den Keere (1571- c.1646), or as he is sometimes known by his Latinized name Petrus Kaerius, was a noted publisher, cartographer, and engraver of numerous folio and pocket-size maps. Van den Keere had a number of map publishing connections through marriage. His sister married Jodocus Hondius and Van den Keere was the brother-in-law of Petrus Bertius. He was also the uncle of Abraham Goos.

This is a superbly designed and engraved map. The engraver for this map is not identified, but the Americas map contains the imprint of Abraham Goos as the engraver, while Van den Keere is identified as the engraver for the Europe map. The title cartouche is at the bottom left with a shield held by two seated putti. Schilder (2000: 113) states that this decorative shield was taken from a Willem Blaeu chart of 1608 as engraved by Hessel Gerritsz. Above the shield, Van den Keere adds a clock face with a skull in the center and a bell above the clock. This is Van den Keere's rather interesting way to depict the fleeting nature of human existence.

Across the top and bottom borders of the map are views of African towns. At the top from left to right are the following town views: *MINA* (The Mine at St. George in Guinea), *TANGER* (Tangiers), *TUNES*, *AMARA* (the home of Prester John), and *ALGAR*. At the bottom from left to right are the following town views: *ÇEUTA*, *TSAFFIN*, *CEFALA*, *I. MOZAMBIQUE*, and *CANARIA* (in the Canary Islands). These town views are interspersed by faces of various kings of Africa: at the top, REX ABISSINES and REX CONGÆ, and at the bottom, REX GUINEÆ, REX MAROCCÆ, REX MADAGASCARIS, and REX MOZAMBIQUÆ.

The left and right borders show various costumed African figures. On the left side from the top, are FEZZAICA MULIER, CONGENSIS, MADAGASCARICA MULIER, GUINEUS. On the right side from the top are: MAROCUS, MULIER ABISSINEA, AZANAGENSIS, and MOZAMBIQUEA MULIER.

The town views at the bottom and top borders are mostly taken from Volume I and II (1572 and 1575) of Georg Braun and Frans Hogenberg's *Civitates Orbis Terrarum*, a collection of views of famous towns of the world. The information for these views was mainly from unknown Italian and Portuguese sources.

Van den Keere's map is a faithful copy of the geography of the Jodocus Hondius map of 1606, except for the mis-copying of a few placenames (for example *Sansibar* rather than Zanzibar). As on the Hondius map, Van den Keere places the mouth of the Cuama River too far to the south resulting in the placement of the trade port of *Sefala* (Sofala) as north of the river. *Cast. Portugal* (at 20° south and about 60° degrees east) shows the extent of Portuguese settlements up the Zambezi River from the Mozambique coast.

Publication Information
The map was issued separately and is not found in any standard atlases. There were thought to be four states of this map until the recent discovery of a new state dated 1648.

This Van den Keere map proved popular, especially in its later states by Visscher, as evidenced by similar maps using new copperplates by Jacques Honervogt in 1640, Robert Walton in 1658, and John Overton in 1668.

References
Tooley 1999-2004, K-P: 10-11; Schilder 2000: Vol. VI:112-115 and facsimile 8; Briels 1974; Keuning 1960: 66-73. The author is grateful to Paul and Stephan Haas, Antiquariat Gebr. Haas, for bringing the newly discovered fourth state of this map to his attention.

56   1616 Petrus Bertius – Jodocus Hondius Jr., Amsterdam

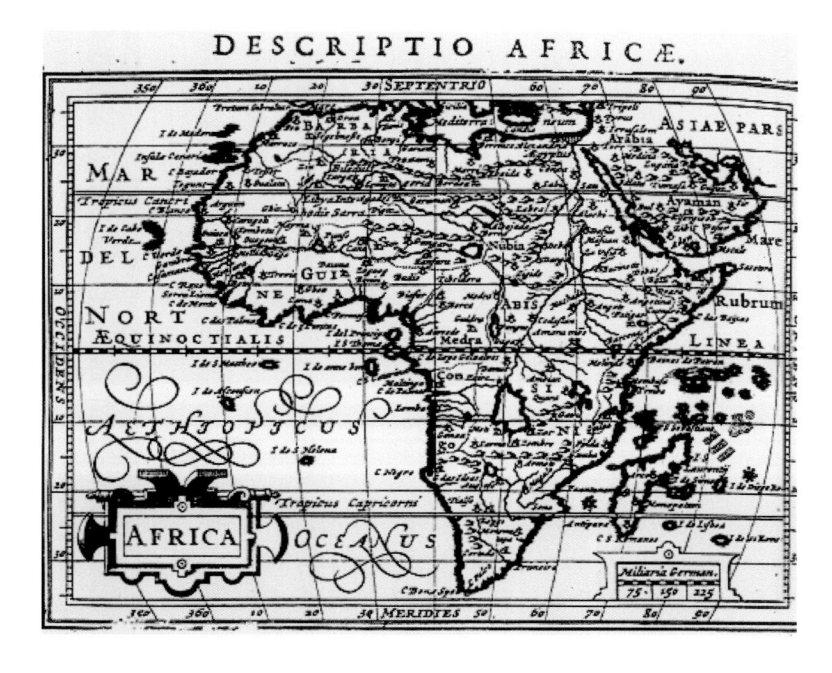

AFRICA.
- Scale bar at bottom right within box: Miliaria Germanica 225 = 1.8 cm.
- [Amsterdam : Jodocus Hondius Jr., 1616].
- From: *P. Bertij Tabularum Geographicarum Contractarum ... Libri.*
- 1 map : copperplate engraving ; 9.5 x 13.5 cm.
- Latitude and longitude coordinates along all four sides of the map.
- North at the top; no compass rose.
- Title set within a strapwork cartouche at bottom left of map.
- The map is devoid of all ornamentation except for the title cartouche and the scale bar.

Examples: For 1616 edition: *London BL*, Maps C.39.a.8; For 1621 *Wereld Spiegel*: Bedburg-Hau, *Antiquariat Gebr. Haas;* For 1637 edition: *London, Burden Collection; Chicago, Baskes Collection; Providence JCB*, Z B632 1637.

Description

The previous version of Petrus Bertius' atlas minor, *Tabularum Geographicarum Contractarum... Libri*, had appeared in Amsterdam in 1600, 1602, 1606, and 1609, and in Frankfurt in 1612. All of these editions included the Cornelis Claesz.- Barent Langenes copperplates, including the Claesz.- Langenes map of Africa of 1597 (Map # 37).

After the death of Cornelis Claesz. in 1609, Jodocus Hondius Jr. (1593-1629) became the new publisher of Bertius' *Tabularum*. For some reason, Hondius either was not able to acquire the original plates, or he simply decided to issue new maps using a new set of copperplates. He prepared an expanded number of new maps and made them slightly larger in size than the earlier Claesz.-Langenes maps. With the new plates and with revisions to Bertius' text, the work was definitely improved and enjoyed a growing demand from the public, which up to that time had the choice between the Mercator-Hondius *Atlas Minor* of 1607 and Bertius' earlier *Tabularum*.

Petrus Bertius (1565-1629) was born in Flanders but as a refugee settled first in Amsterdam and then in Leiden, where he was a professor of mathematics and a librarian at the University of Leiden. Afterwards, he moved to Paris, where in 1618 he became cosmographer to King Louis XIII of France. Bertius died in Paris in 1629. Through marriage, Bertius was related to both Pieter van den Keere (his sister married Van den Keere) and Jodocus Hondius (Van den Keere's sister married Hondius).

The geography for this map is based on the Jodocus Hondius atlas minor map of Africa of 1607, which in turn was taken from the Hondius 1606 folio-size map of Africa.

Publication Information

The editions in which this map appears are the 1616/18 *Tabularum Geographicarum Contractarum… Libri* with Latin text and the 1618 *La Geographie Racourcie*, also published in Amsterdam, though with French text. Sometime after the 1618 edition and possibly in 1629, with the sale of Jodocus Hondius Jr.'s folio-size copperplates by his widow upon Hondius' death, these plates came into the possession of Willem Blaeu. Blaeu's son, Joan, published an extremely rare atlas, *Atlas Minor sive Tabulae Geograhicae*, in 1637, using these Hondius plates. This is the only *Atlas Minor* produced by the Blaeus. Very few examples of this Blaeu *Atlas Minor* are known to exist, all in private hands with the exception of one at The John Carter Brown Library (*Providence JCB*, Z B632 1637).

This map also appears in *Wereld Spiegel*, published by Jan Evertsz. Cloppenburch in Amsterdam in 1621. Van der Krogt (2003: 442) theorizes that Cloppenburch probably borrowed the copperplates from Hondius before these plates were sold to Blaeu.

The map is known in only one state. There were changes to the letterpress title above the map as noted below.

The following may be used as a guideline to determine from which atlas edition a particular separate example of this map appeared. This information is based on Van der Krogt (2003 Vol. IIIB: 879) with his information on the date, atlas title, signature, pagination, and other characteristics:

Date	Van der Krogt no.	Title	Page	Signature	and Notes
1616 & 1618	342:01	*Tabulae*	616	2Q4v	'DESCRIPTIO AFRICAE.' in letterpress above the map.
1618	342:11	*Geographie*	616	2Q4v	with change in letterpress title to 'DESCRIPTION D'AFRIQVE' above the map.
1637	342:21	*Atlas Minor*	192	none	with map title of AFRICA within title cartouche. No text on the verso of the map.

References

Van der Krogt 2003: map 8600:342; Koeman 1967-71, I: 60-62 and II: 258-260; Mickwitz et al. 1979: Vol. 1: 30-34; Tiele 1867: 101.

57   1617 Willem Janszoon (Blaeu), Amsterdam

Map 57.3.

57.1 First State 1617

AFRICÆ | nova descriptio. | Auct. Guil: Janssonio.
[at bottom left corner of the map]:
Cum privilegio | ad decennium.
- No scale bar.
- [Amsterdam] : Willem Jansz. (Blaeu), [1617].
- 1 map : copperplate engraving ; 41 x 55.5 cm (including decorative borders) ; 36 x 46 cm (without decorative borders).
- From: separately published ; with later states in standard Blaeu atlases from 1630.
- Latitude coordinates along left and right sides. Longitude coordinates along Equatorial Line.
- North at the top; one compass rose in Atlantic Ocean.
- Title cartouche at top right.
- There are two Dutch ships and one sea monster in the Indian Ocean, and seven Dutch ships, four sea monsters, one seahorse and four flying fish in the Atlantic Ocean. Various animals are within Africa.

- As described further below: nine town views above the map and five pairs of Africans on each side border.

Examples: *Amsterdam UB*, 1800 A 7; *London BL*, Maps C.3.c.9 {36*} *and* Maps 63510 [27] (without the town views); *Paris BN*, Ge DD 1284 [5]; *Washington LC*, G8200 1617.J (within composite Tavernier atlas dated c.1635); *Private Foundation, formerly collection Stopp*.

Map 57.1 Cartouche.

57.2 Second State 1621
For this second state, Blaeu's name in the title cartouche is changed to 'Guiljelmo Blaeuw.'

This map in its second state was initially issued as a separate map from 1621. It first appeared in an atlas in this second state in Blaeu's *Atlantis Appendix* of 1630 with no text on the verso. From Blaeu's 1631 *Atlantis Appendix* onward, the map had text on the verso in the various language editions of Blaeu's atlases (Latin, Dutch, French, German, or Spanish).

Examples: *Leiden UB*, COLLBN Port 179 N 8; *München SB*, Mapp. XX, 2 *and* Mapp. 105 (within a composite atlas, an atlas factice); *Rotterdam MM*, WAE 49.

57.2. A Second State Variant A 1633
This is a variant of the second state with no decorative borders, but in all other aspects the same as in the second state. For this variant, the decorative borders were masked during the printing process to enable the map to fit into a book smaller in size than Blaeu's folio-size atlas. An example in this format appeared in Emanuel van Meteren's *Warhafftige Beschreibung aller denckwurdigsten Geschichten*, published by Blaeu in 1633.

Example: *Amsterdam UB*, O.K. 203a (printed on vellum).

57.3 Third State c. 1647
In the third state, there is evidence of frequent re-engraving of the copperplate showing new, additional hachuring, most notably below the sea monster near the left border and the latitudinal line 10°S.

Examples: This is the most common state of this map and is located in numerous libraries and with private collectors.

MAP 57

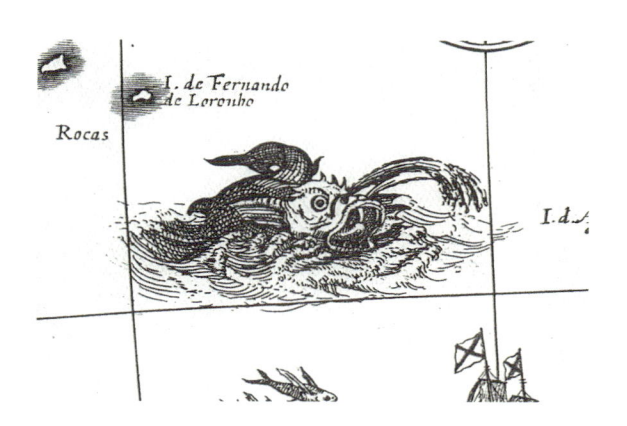

Map 57 Hachuring around the sea monster in Map 57.1 and 57.3.

Description

This is a cornerstone, folio-size map of Africa and is one of the better known, more decorative maps of Africa of the seventeenth century. It is part of a set of four continent maps that Blaeu issued separately; only the map of Europe is dated. This map is best known in its second and third states in editions starting in Blaeu's *Atlantis Appendix* of 1630 and ending with his son, Joan's, *Atlas Maior* in German of c.1670.

Willem Janszoon, or Willem Jansz. Blaeu (1571-1638) as he was to be later more commonly known, was one of the most noted Dutch cartographers and map publishers of the seventeenth century. After preparation as a pupil of the great Danish astronomer, Tycho Brahe, Blaeu initially concentrated on globemaking and separately published maps, including wall maps of the continents. From 1608, he dominated the market for wall maps (see his wall map of Africa of 1608) and sea atlases, and in 1630 produced his terrestrial atlas. These atlases were continued by his sons Joan (c.1599-1673) and Cornelis (1610-1644), culminating in the great *Atlas Maior* of 1662 in 11 volumes by Joan Blaeu with subsequent editions. It appears that this map was not re-issued after the February 23, 1672 fire at the offices of the Blaeu publishing firm that effectively ended the successful Blaeu business. Joan died after the fire in 1673, and, though Joan II continued the business for a time, many of the plates were auctioned shortly thereafter. This map was not reprinted after 1672 as the Africa copperplate is not known to have survived the fire. The Spanish edition of the *Atlas Maior* was in the process of being printed, including the section on Africa with this map of Africa, when the fire occurred.

Across the top border of the map are views of African towns. These views are from left to right: *TANGER* (Tangiers), *CEUTA, ALGER* (Algiers), *TUNIS, ALEXANDRIA, ALCAIR* (Cairo), *MOZAMBIQUE* (the island), *S. GEORGIUS della MINA* (The Mine at St. George in Guinea), and *CANARIA* (in the Canary Islands).

The left and right borders shows various costumed African figures. On the left side from the top, these are Marocchi (Moroccans), Senagenses (Senegalese), Mercatores in Guinea (traders in Guinea), Cab:lopo Gonsalvi Accolæ and Miles Congensis (Congolese). Along the right side from the top are Aegyptij (Egyptians), Abissini (Ethiopians), Cafres in Mozambique (Mozambicans), Rex in Madagascar (King in Madagascar) and Cap:bonæ Spæi habitatores (inhabitants of the Cape of Good Hope). These decorative borders are based on the border scenes found on later states of Blaeu's 1608 wall map of Africa.

The geography on this map is essentially a reduction of Blaeu's famous and influential wall map of 1608. Owing to the size difference, it does not have as much detail as the wall map. For example, on the wall map, there is an extra island at the top of the western Ptolemaic Lake of *Zaire lacus/Zembre lacus*. It is somewhat surprising that it took Blaeu almost ten years before issuing this folio-size map of Africa along with the other continents. As well, it took another thirteen years before this map appeared in Blaeu's first atlas, the *Appendix* of 1630.

One notable difference on the 1617 map, when compared with Blaeu's 1608 wall map, is that this map breaks the long-standing tradition of the single southern river, the *Zambere*, originating in *Sachaf Lacus*, and then dividing into the Cuama and Spirito Santo Rivers. This map shows the *Cuama* River as originating in the highlands to the south of the *Lunæ Montes* (Mountains of the Moon) and not connected to the *Zambere* River.

Publication Information

The following is a list of the various editions of the Blaeu atlas in which this map appeared, based on Van der Krogt (2000). The list can be used to determine the date and edition for a particular separate map by examining the language, signature and page number on the map. The Africa map never appeared with Spanish text as the Africa volume of the Spanish edition of the *Atlas Maior* was not printed before the 1672 fire.

Date	Van der Krogt no.	Language	Signature and page no. on verso
1630	2:011	Latin	Verso blank (Appendix)
1631	2:021	Latin	F 10
1631	2:022	Latin	F
1634	2:131	German	4 D
1635	2:132	German	4 D
1635	2:101	Latin	4 F
1635-38	2:111-12	French	4 F
1635	2:121	Dutch	75 A
1638	2:113	French	71 A
1640-43	2:201	Latin	A 1
1640-43	2:211	French	A
1642-43	2:221	Dutch	A 1
1641-42	2:231	German	A 1
1643-50	2:212-3	French	A 1
1644-55	2:202-3	Latin	A 1
1647-48	2:222	Dutch	A 1
1647-49	2:232	German	A 1
1649-50	2:223	Dutch	A 1
1658-61	2:224	Dutch	A 1
1662	2:601A	Latin	A 1-2
1662	2:601B	Latin	A 1-2 (with different placement in the atlas than 2:601A)
1663	2:611	French	A 1-2
1664	2:621	Dutch	A 1
c.1670	2:631	German	A 1

References

Schilder 2000: Vol. VI: 116-119 and facsimile 9; Van der Krogt 2000: map 8600:2; Norwich 1997: map # 32 (third state of the map); Schilder 1993: Vol. IV: 304-308 and facsimile 21; Douwma 1979: 2 and 104; Keuning 1973; Wieder 1925-33: Vol. III: 70.

58   1619 Jodocus Hondius Jr., Amsterdam

58.1 First State 1619

AFRICÆ | nova Tabula. | Auct. Jud: Hondio.
- No scale bar.
- [Amsterdam : Jodocus Hondius Jr. 1619].
- 1 map : copperplate engraving ; 47.5 x 60.5 cm (including decorative borders), 37.5 x 49.5 cm (map without decorative borders).
- From: separately published / *Atlantis Maioris Appendix* and later Hondius-Janssonius atlases (in third, fourth, and fifth states).
- Latitude coordinates along left and right sides of map. Longitude coordinates along Equatorial Line.
- North at the top; no compass rose.
- Title cartouche at upper right of the map.
- There are six Dutch ships, three sea monsters, six flying fish and Neptune with a mermaid in the Atlantic Ocean. In the Indian Ocean, there is one Dutch ship and one sea monster. Various animals are within Africa.

- As described further below: six town views above and below the map, and five pairs of Africans on each side border.

Examples: *München SB*, Mapp. XX 1b; *Paris BN*, Ge D 11364 (bound into a composite atlas) *and* Ge DD 4796 [98] map 3; *Private Foundation, formerly collection Stopp.*

58.1.A First State, Variant A c.1619
In this variant, there are no decorative borders, but in all other aspects it is the same as the first state.

For this variant, the decorative borders were masked during the printing process. Van der Krogt (1997: 711) lists this map as not being in a regular atlas. This author's example is without text, but it does have a binders stub indicating that it must have appeared in a composite atlas.

Examples: *Amsterdam UB*, O.K. 151; *Author's collection* (as a separate map).

58.2 Second State 1631
Within the title cartouche, a date has been added and the mapmaker now as 'Auct: Hen. Hondio | 1631.'

The name of the mapmaker was changed to Henricus Hondius and the date of 1631 was added for this state. The decorative borders remain the same on all four sides.

Examples: No copy of this state is known to exist, but as Schilder postulates (2000:120-122) the map did exist as there are surviving, similar maps of Asia and Europe that were part of a set of maps of the four continents.

58.3 Third State 1631
Decorative borders are removed, but in all other aspects it is the same as the second state.

In this state, the decorative borders were masked to enable the map to fit into the folio-size atlas, *Atlantis Maioris Appendix*, prepared by Henricus Hondius for publication in 1631. This third state of the map, and states four and five, are the most common states of this map that can be found.

Examples: *Various libraries and private collections.*

The following is a list of the various editions of the Hondius atlases in which this third state of the map appeared, based on Van der Krogt (1997: 711). The list can be used to determine the date and edition for a particular separate map by examining the language, signature and page number on the map:

Date	van der Krogt no.	Language	Signature	Page No.
1631	1:203	Latin	no text on verso	
1633	1:311	French	7S	621-624
1633	1:312	French	7S (from *L'Appendice*)	621-624
1633	1:321/22	German	C	9-12
1634	1:331	Dutch	8E	351-352
1636	1:341	English	9K	425-426
1636	1:323	German	C	9-12
1638	1:401	Latin	4G	
1639-44	1:411/12	French	4t	
1646-49	1:415	French	4t	

58.4 Fourth State 1641
The date in the title cartouche has been changed to 1641. There is an addition of the Janssonius imprint at the bottom right of the map: 'Amstelodami, apud Ioannem Ianssonium'.

Examples: *Various libraries and private collections.*

The following is a list of the various editions of the Hondius-Janssonius atlases in which this fourth state of the map appeared, based on Van der Krogt (1997: 711). The list can be used to determine the date and edition for a particular separate map by examining the language and signature on the map:

Date	van der Krogt no.	Language	Signature	Page No.
1644-58	1:424/27	German	2A	
1645-58	1:433/36	Dutch	a	
1646-57	1:403/05	Latin	2A	
1675	1:406	Latin		

58.5 Fifth State c.1644
The date in the title cartouche has been removed.

Examples: *Various libraries and private collections.*

The following is a list of the various editions of the Hondius-Janssonius atlases in which this fifth state of the map appeared, based on Van der Krogt (1997: 711). The list can be used to determine the date and edition for a particular separate map by examining the language and signature on the verso of the map:

Date	van der Krogt no.	Language	Signature	Page No.
1644-58	1:424/27	German	2A	
1645-58	1:433/36	Dutch	a	
1646-57	1:403/05	Latin	2A	
1652-58	1:416/17	French	A	
1653-66	1:441	Spanish	o	
1666	1:407	Latin		
c.1680	1:408	Latin *(Atlas Novus)*		

Description

This highly decorative map of Africa is part of a set of four continent maps prepared by Jodocus Hondius Jr. (1593-1629) in 1619. The map is dated as 1619 based on the map of Europe which is the only map with a date in the set.

This map is best known in its third, fourth, and fifth states without the decorative borders when it appeared in atlases starting with the folio atlas of Jodocus' brother, Henricus Hondius (c.1596/7-1651), in 1631 and ending with the atlas of the heirs' of his brother-in-law, Johannes Janssonius, in c.1680.

The geography of Africa is significantly different than in his father's 1606 map of Africa. The geography for this map is copied from Willem Blaeu's folio-size map of 1617 with the *Cuama* River originating in mountains and not in the *Sachaf Lacus*. Overall, however, it was based on a reduction of the Blaeu wall map of 1608.

While Blaeu's 1617 folio map only has decorative borders on the top and the left and right sides, Hondius added a decorative border on the bottom side of the map. This allowed for twelve slightly larger town views than the nine town views on the Blaeu folio map.

The town views across the top are of major North African towns: *Alcair* (Cairo), *Alexandria*, *Alger* (Algiers), *Tunis*, *Tanger* (Tangiers), and *Ceuta*. The town views across the bottom are of major towns in sub-Saharan Africa: *S.Georgius della Mina* (The Mine at St. George in Guinea), *Mozambique* (the island), *Canaria* (in the Canary islands), *Quiloa*, *Tzaffin*, and *Cefala* (Sofala). The views at the bottom and top borders are mostly taken from Volume I and II (1572 and 1575) of Georg Braun and Frans Hogenberg's *Civitates Orbis Terrarum*, a collection of views of famous towns of the world.

The left and right borders shows various costumed African figures. On the left side from top are mostly people from regions on the west side of Africa: MAROCCHI (Moroccans), SENAGENSES (Senegalese), MERCATORES IN GUINEA (traders in Guinea), CAB:LOPO GONSALVI ACCOLÆ, and, MILES CONGENSIS (Congolese). Along the right side from the top are mostly people from the east and south side of Africa: ÆGYPTII (Egyptians), ABISSINI (Ethiopians), CAFRES IN MOSAMBIQUE (Mozambicans), REX IN MADAGASCAR and CAP:BONÆ SPEI HABITATORES (inhabitants of the Cape of Good Hope). These decorative borders are based on borders found on later states of Blaeu's 1608 wall map of Africa as were those on Blaeu's 1617 folio map, but here they are engraved in reverse from the Blaeu 1617 figures. The Blaeu figures in turn were taken from mainly Italian and Portuguese sources, some of which are now not known.

Publication Information

The map appears in a number of different states as described above, and was closely copied by later mapmakers. To aid in the identification of a number of similar maps, using new copperplates, the following is a list of the derivatives of the Jodocus Hondius Jr. map of 1619. All lack decorative borders as did the third and later states of the Hondius 1619 map:

Map #	Mapmaker & Date	Ships in Atlantic:	Other Features:
Map # 58	Hondius, 1619	Six ships	Cartouche touching top border.
Map # 61	Bertius, 1624	Three ships	Dated 1624, 1640, or 1646. Dated 1670 with no decorative elements in the oceans.
Map # 63	Bertius-Tavernier, 1627	Five ships	Bertius' name added to title.
Map # 66	Boisseau, c.1636	Six ships	Boisseau's imprint added to cartouche at bottom left. In second state, no decorative elements in the oceans.
Map # 89	Berey, 1658	Five ships	Map surrounded by text.

References

Schilder 2000: 120-122 and facsimile 10; Van der Krogt 1997: map 8600:1D.1-4; Norwich 1997: map # 34 (fourth state of the map); Stopp 1974: 6; Koeman 1967-71, II: 139-144,

59   c. 1620-35 Godefridus de Scaicki, Rome

NOVA | TOTIVS | AFRICAE | TABVLA | Auctore Guillel. Iansonio | Gottifredus de Scaicki | excudebat.
[within a cartouche at the bottom right]:
Extant ROMAE | Expensis Gottifredi Scaicki Vltraiecten | sis in via | … | ad Signia | Aquilae Imperalis | Sup.^{um} pmissu.
- No scale bar.
- Rome : Godefridus de Scaicki, [c.1620-35].
- From: separately published wall map.
- 1 wall map : copperplate engraving ; 4 map sheets ; 83.9 x 108.5 cm in total.
- Latitude and longitude coordinates on all four borders of the map. Longitude coordinates across the Equatorial Line.
- North at the top; there are five compass roses on the map: two in the Atlantic Ocean, two in the Indian Ocean and one in the Mediterranean Sea ; one windrose in the South Atlantic.
- Title at upper right of the map is set within a decorative oval cartouche with three figures at the top.
- In the lower left corner, there are two circular diagrams at the top of a cartouche with text; in the upper left corner, there is a cartouche with text.

- In Indian Ocean, one ship and five sea monsters ; In Atlantic, eight ships, three sea monsters, vignette of Neptune and mermaid. In Mediterranean, one smaller ship and one sea monster.
- Numerous animals and figures of Africans are found on the map of Africa.
- Eight groups of costumed figures on each side of the map. Twelve town views across the bottom of the map. Various other decorative elements on the map as on the Blaeu 1608 wall map.

Examples: Only two examples of the de Scaicki Africa map are known to the author. One is possibly in a private collection in France, and one is in the Mac Lean Collection in the United States.

Description

This map of Africa is noteworthy as being one of the earliest Italian wall maps to be inspired by Northern European cartography. Up to this point and with some exceptions, wall maps of Africa were often copied in The Netherlands using Italian models, the most famous being the Gastaldi wall map of Africa. The De Scaicki wall map is the first of what would be numerous imitations in Italy and France of the Willem Blaeu wall map of Africa. De Scaicki clearly acknowledges Blaeu as the author of this map within the title cartouche: 'Auctore Guillel. Iansonio'.

De Scaicki based his map of Africa on the second state of Blaeu's 1608 wall map of Africa, printed in Amsterdam in 1612. The one addition to the second state of the Blaeu map is the inclusion of the name of the engraver (Josua van den Ende) to the right of the cartouche in the lower left corner: 'I. vanden Ende sculp'. De Scaicki's copyist includes Van den Ende's name on his map, but this has been erased in the example seen by the author, though there remains evidence of Van den Ende's name.

As with the second state of the Blaeu map, the geography for this De Scaicki map is unchanged from the 1608 Blaeu wall map. Other aspects of the maps are also the same. There is a title cartouche in the upper right corner; the publisher's imprint is in a cartouche in the lower right corner. The street name on the imprint on the map has been deliberately erased. Other identical map elements are: secondary text cartouches, placement of ships and sea monsters, compass roses, a wind rose, animals within Africa, and text passages surrounding and within Africa.

This map is part of a series of wall maps of the four continents known to have been published by Godefridus de Scaicki. Little is known about De Scaicki or his work. He was originally from Utrecht and was probably named Geert (Gotfried) van Schayck before moving to Rome. De Scaicki was active in Rome between 1620 and 1635. As a result, this wall map is dated c.1620-35.

Few other examples of work by De Scaicki are known. Two notable ones are the large 1606 Blaeu panorama of Amsterdam, published by De Scaicki in 1620, and a bird's eye view of ancient Rome (A. Baynton-Williams, from notes).

Publication Information

The map is exceedingly rare. Two sets of De Scaicki's wall maps are known to have survived. One set was sold in Paris in the 1960s and is apparently in a private collection in France. The other set is in a private collection in the US.

The map has no date, but may very well have been published in the period c.1620-1635, when De Scaicki was active, and based on the evidence on the 1666 De Rossi wall map of Africa. Schilder believes that De Rossi may have acquired the copperplates for his wall map of Africa from De Scaicki.

The map is only known in one state as described above.

References

Schilder 1996: 77, 202 and 208.
The author is grateful to Ashley Bayton-Williams for sharing his information on this map.

60 1623 Jodocus Hondius Jr., Amsterdam

60.1 First State 1623

AFRICÆ | nova Tabula | Auct. Jud. Hondio.
[set within decorative cartouche at lower left]:
Sumptibus et | typis æneis | Judoci Hondij | Amstelodami. 1623.

- No scale bar.
- Amsterdam : Jodocus Hondius Jr., 1623.
- From: separately published / *Newer Atlas Oder Weltbeschreibung* and later Hondius-Janssonius atlases (in fourth state).
- 1 map : copperplate engraving ; 46 x 56 cm; 36 x 47 cm (map without decorative borders).
- Latitude coordinates along left and right sides of map. Longitude coordinates along Equatorial Line.
- North at the top; no compass rose.
- Title cartouche at top right.
- Highly decorative cartouche at the bottom left that appears to be very similar to the cartouche in Van den Keere's 1614 map of Africa. Above the cartouche is a clock with the face of a death's head.
- There are two Dutch ships, one sea monster, and six flying fish in the Atlantic Ocean and one Dutch ship and one sea monster in the Indian Ocean.
- As described further below: six town views above and six below the map, and five pairs of Africans on each side border. Various animals are within Africa.

Examples: *Cambridge Harv MC*, 2375.1623 (as separate map); *Fulda LB*, Geogr. B 54 [147]; *Leiden UB*, 004-08-019; *London RGS*: 264.H.16 (plate 3); *Stanford SC* (part of the Norwich Collection).

CARTOBIBLIOGRAPHY

Map 60.2.

60.2 Second State 1623
Decorative bottom left cartouche now with imprint of 'Sumptibus et | typis æneis | Joannis Ianssonij | Amstelodami. 1623.'. Same four decorative borders.

Examples: *Cambridge Harv MC*, 2375.1623; *Paris BN*, Ge D 11364 *and* Ge DD 4796 [137-139] (within Mercator Hondius atlas dated 1628); *Private Foundation, formerly collection Stopp*; *Gent, Antiquariaat Sanderus*.

60.3 Third State 1632
Decorative bottom left cartouche now has a new date of 1632 (with some evidence of erasure of the earlier date). Same four decorative borders.

Example: *Private Foundation, formerly collection Stopp*.

60.4 Fourth State 1632
The bottom decorative border is removed. The map is still dated 1632.

Examples: *Chicago NL*, Novacco Collection 4 F 396; *Paris BN*, Ge D 11365; *Author's collection*.

The following is a list of the various editions of the Hondius-Janssonius atlases in which this fourth state of the map appeared, based on Van der Krogt. The list can be used to determine the date and edition for a particular separate map by examining the language and signature on the verso of the map:

Date	van der Krogt no.	Language	Signature
1638-42	1:421/2	German	2L
1638-44	1:431	Dutch	a
1645-58	1:433/36	Dutch	a
1652-58	1:416/17	French	A

60.5 Fifth State c.1652
This fifth state lacks a date and a bottom border.

This fifth state has no text on the verso and likely appeared irregularly in the Mercator-Hondius-Janssonius atlases after 1652-58.

Examples: *München SB*, 2° Mapp. 190 (within a composite atlas, an atlas factice); *Private Foundation, formerly collection Stopp.*

Description
Jodocus Hondius Jr. issued this map as part of a new set of maps of the four continents in 1623. These maps closely followed his set of maps of the continents of 1619. It is unclear why these new maps were issued as the 1619 maps were attractive, finely engraved maps, and the information in these 1623 maps was not significantly updated from the information in his 1619 maps. Like the 1619 map of Africa, the geography for this map is copied from Willem Blaeu's folio-size map of 1617 with the *Cuama* River originating in the mountains and not in *Sachaf Lacus*. Overall, however, it was based on a reduction of Blaeu's wall map of 1608.
Based on Schilder (2000: 123), Pieter van den Keere either engraved this map of Africa or was involved in the selection of the engraver. This would likely explain why this map has a similar decorative cartouche to Van den Keere's 1614 map of Africa.

The town views at the top and bottom borders are also almost identical to those on Hondius' 1619 map, except for smaller lettering and different placement of the town names in some cases.
The town views across the top are of major North African towns: *Alcair* (Cairo), *Alexandria, Alger* (Algiers), *Tunis, Tanger* (Tangiers), and *Çeuta*. The town views across the bottom are of major towns in sub-Saharan Africa: *S. Georgius Mina* (The Mine at St. George in Guinea), *Mozambique* (the island), *Canaria* (in the Canary islands), *Quiloa, Tzaffin* and *Cefala* (Sofala). The views at the bottom and top borders are mostly taken from Volume I and II (1572 and 1575) of Georg Braun and Frans Hogenberg's *Civitates Orbis Terrarum*, a collection of views of famous towns of the world.

The left and right borders show various costumed African figures. On the left side from top are mostly people from regions on the west side of Africa: MAROCCHI (Moroccans), SENAGENSES (Senegalese), MERCATORES in GUINEA (traders in Guinea), CAB:LOPO GONSALVI ACCOLæ, and, MILES CONGENSES (Congolese). Along the right side from the top are mostly people from the east and south side of Africa: ÆGYPTII (Egyptians), ABISSINI (Ethiopians), CAFRES IN MOZAMBIQUE (Mozambicans), REX IN MADAGASCAR and C. BONÆ SPEI HABITATor (inhabitants of the Cape of Good Hope). The depictions of the costumed figures on the left and right borders are mirror images of Hondius' 1619 map suggesting that these figures were copied directly from the earlier map. In turn, these decorative borders are based entirely on borders found on Blaeu's 1608 wall map and 1617 folio map of Africa.

Publication Information
The map was first issued as a separate map and was not part of any atlas until its appearance in its fourth state in Volume II of *Newer Atlas Oder Weltbeschreibung* of 1638 by Johannes Janssonius. This map is known in five states as described above.

References
Schilder 2000: 123-125 and facsimile 11; Van der Krogt 1997: map 8600:1C.1 and 2; Norwich 1997: map # 29; Keuning 1948: 69.

CARTOBIBLIOGRAPHY

61   1624 Petrus Bertius, Paris

Map 61.3.

61.1 First State 1624

Carte de | L'AFRIQVE | Corrigeé et augmenteé | dessus toutes les aultres | cy deuant faictes P.Bertius | L'anneé 1624.
- No scale bar.
- [Paris] : Petrus Bertius, 1624.
- From: separately published.
- 1 map : copperplate engraving ; 37.5 x 50 cm.
- Latitude coordinates along left and right sides of map. Longitude coordinates along Equatorial Line.
- North at the top; no compass rose.
- Title cartouche at the top right.
- There are two sea monsters and three sailing ships in the Atlantic Ocean and one sea monster in the Indian Ocean. Various animals are within Africa.

Example: Assumed to exist but location is not known (the title above is based on Tooley's description of the Americas' map (Tooley 1973: 299).

Map 61.3. Cartouche.

61.2 Second State 1640

The map is now dated 1640 as 'L'anneé | A° 1640' (with evidence of erasure behind the date and behind 'L'anneé'). The second state has Bertius' name removed.

The second state is attributed to Michel van Lochom, as two of the four continent maps (not Africa nor America) bear Van Lochom's name. Van Lochom (1601-1647), born in Antwerp, was known as an engraver and printer in Paris. As Bertius died in 1629 and this state was not published until 1640, it is unclear if Van Lochom purchased the plates directly from Bertius or through some intermediary for his issue of the maps in 1640. The second state of the Africa map appears in some examples of Pierre D'Avity's *Description Generale*, published by Sonnius and Bechet in 1643.

Examples: *Private French Collection.*

61.3 Third State 1646

The map is now dated 1646 (with evidence of erasure behind the date and behind 'L'anneé').

The third state appeared in two editions of Pierre Mariette's *Theatre geographique de France* in 1650 and 1653. The map also appeared in D'Avity's *Description Generaie.*

Examples: *London BL*, 568.H.10 (in D'Avity's *Description Generale*, dated 1643); *Bedburg-Hau, Antiquariat Gebr. Haas; Author's collection* (as separate map).

Map 61.4.

61.4 Fourth State 1670

The map is now dated 1670. All evidence of the decorative elements in the oceans has been erased. Also, there is an addition of the imprint of 'Par | F. delapointe | 1670' below the title in the title cartouche.

The fourth state is also very rare. The author located the fourth state of this map in an example of

D'Avity's *Description Generale de l'Afrique, Seconde partie dv Monde* at the Bibliothèque Nationale de France (*Paris BN*, Ge DD 1520). Little is known of the mapmaker, François de la Pointe. The book was published by Denys Bechet and Lovis Billaire in Paris in 1660.

Examples: *Paris BN*, Ge DD 1520.

Description

This Petrus Bertius map is closely based on the Jodocus Hondius Jr. maps of Africa of 1619 and 1623, published in Amsterdam as separate maps, but without the decorative borders. Like the two Hondius maps, the geography for the 1624 Bertius map is copied from Willem Blaeu's folio-size map of 1617, which was based on a reduction of Blaeu's wall map of 1608.

Petrus Bertius (1565-1629) was born in Flanders, but as a refugee from political and religious unrest settled first in Amsterdam and then in Leiden, where he was professor of mathematics and librarian at the University of Leiden. After this, he moved to Paris, where in 1618 he became cosmographer to King Louis XIII of France. He died there in 1629. Through marriage, Bertius was related to both Pieter van den Keere (his sister married Van den Keere) and Jodocus Hondius (Van den Keere's sister married Hondius). It is possible that through these connections, Bertius had ready access to the works of the Hondius family so as to obtain the models for his own maps.

Publication Information

This map is presumed to exist in four states. The first state of 1624 has not been located and is not reported in the literature. The existence of this map of Africa in its first state of 1624 is based on the supposed existence of the Bertius' map of America, Carte de l L'AMERIQVE... , and dated 1624, as described by Tooley (1973: 299) and reported by Burden (1996: map # 209).

The second state of the Bertius Africa map, dated 1640, clearly shows reworking of the title cartouche, especially of the date, indicating evidence of an earlier date for this map. This is also reported by Burden on the Americas map.

To aid in the identification of this Bertius map and a number of similar maps using different copperplates by Jodocus Hondius Jr. (Map # 58), Bertius-Tavernier (Map # 63), Boisseau (Map # 66), and Berey (Map # 89), please refer to the chart under Map # 58, the Jodocus Hondius map of 1619. All of these similar maps are derivatives of the Jodocus Hondius Jr. map of 1619. All lack decorative borders as did the third and later states of the Hondius 1619 map.

References

Burden 1996: map # 209 (for Americas map); Pastoureau 1984: 345; For de la Pointe, see: Loeb-Larocque 1989: 26; Tooley 1980: 299.

62 1626 John Speed, London

62.1 First State 1626

AFRICÆ, described, | the manners of their Ha | bits, and buildinge: newly | done into English by I. S. | and published at the cha: | rges of G. Humble Ano 1626.
[at the lower left]:
Abraham Goos | Sculpsit.
- No scale bar.
- [London] : George Humble, [1627].
- From: *A Prospect of the Most Famous Parts of the World.*
- 1 map : copperplate engraving ; 39.5 x 51.5 cm.
- Latitude coordinates along left and right sides of map. Longitude coordinates along Equatorial Line.
- North at the top; no compass rose.
- Title cartouche at the top right.
- There are two sea monsters and four sailing ships along with seven flying fish above the Equator in

the Atlantic Ocean and one sea monster and one sailing ship in the Indian Ocean. Various animals are within the map of Africa.
- As further described below: each side border contains five costumed figures of Africans, and the top panel contains eight town views. English text on map (and on the verso).

Examples: *London BL*, Maps C.7.e.13.(1); *Author's collection* (as separate map); *Various locations.*

62.2 Second State 1662
Erasure of reference to Humble and also the date of 1626 within the title cartouche. Addition at bottom of 'Are to be Solde by Roger Rea the elder and younger at ye Golden Crosse in Cornhill against the Exchange.'.

Examples: *London BL*, 118.e.8.(1); *Various locations.*

62.3 Third State 1676
Erasure of imprint for Rea at the bottom of the map and the replacement with 'Are to be sold by Tho Baßet in Fleet Street, and Ric: Chiswell in St Pauls Churchyard. ' at the bottom of the title cartouche.

There is no date on the map, but the map is dated 1676 since it appears in *A Prospect of the Most Famous Parts of the World*, published by Basset and Chiswell in 1676.

Examples: *London BL*, Maps C.7.e.5.(2); *Tervuren RMCA*; *Various locations.*

Description
This is the first English map of the continent of Africa. Its author is John Speed (c.1552-1629) a noted English historian and mapmaker who worked from about 1592 at St. Paul's Churchyard in London. He is mostly known as the author of a history of Great Britain with accompanying maps and *A Prospect of the Most Famous Parts of the World* with maps of the world, the four continents, and regions of the world.

This map of Africa is closely based on the 1619 and 1623 maps of Africa by Jodocus Hondius Jr. published separately in Amsterdam. Many of the Hondius text legends which are in Latin on his map appear in English on this map. Like the two Hondius maps, the geography for Speed's map is copied from Willem Blaeu's folio-size map of 1617 which was based on a reduction of Blaeu's wall map of 1608. The geography for this map of Africa remained unchanged in all editions of Speed's *Prospect* from 1626 to 1676.

While the Blaeu and Hondius folio maps had pairs of figures within five boxes on the left and right side borders of the map, Speed only shows single figures within five boxes on the left and right sides. These are from the top left: MAROCCHIAN, SENAGENSIAN, Marc: in GVINEA, Cab: Iopo Gonsalvi Accolae and soul: CONGENSIS. From the top right, they are: ÆGYPTIAN, ABISSINIAN, MOZAMBIQVEAN, K: MADAGASCAR and habi:of cape of good hope. Speed copied one figure from each of the Blaeu 1617 pairs, but in reverse.

Across the top from the left are eight African town views of: *TANGER, CEVTA, ALGER, TVNIS, ALEXANDRIA*, ALCAIR, *MOZAMBIQVE* and *CANARIA*. These views were copied from the Blaeu 1617 town views with the omission on the Speed map of S. GEORGIVS della MINA.

This map of Africa was engraved by Abraham Goos ('Abraham Goos | Sculpsit'). Goos (c.1590-1643), from Amsterdam, was a very active engraver, mapmaker, and publisher.

Publication Information
The copperplate that produced this map changed owners a number of times over the course of the seventeenth century. It was prepared in 1626 as noted on the publisher's imprint on the map title and was first published in the first edition of Speed's *Prospect* in 1627 by George Humble. Although prepared for Speed's *Prospect*, this book and the map of Africa are also often found bound together with later editions of Speed's *Theatre of the Empire of Great Britaine*, first published in 1611.

There was a later edition of the *Prospect* published by George Humble in 1631-2. After Humble's death in 1640, his son William published another edition in 1646. For this edition, his father's name, G. Humble, continued to appear on the map as publisher.

In March 1659, the copperplate for the Africa map, along with the other copperplates of the *Prospect*, was sold to William Garrett, who seems to have sold them directly to the publisher Roger Rea. The Roger Reas, father and son, issued an edition of Speed's *Empire of Great Britaine* in 1662; this atlas was often bound with the *Prospect* (still dated 1646) containing the map of Africa. Burden (1996: map #268) states that the *Prospect* was issued with the Rea imprint in 1665. This edition is scarce as most examples were lost in the Great London Fire of 1666. One further edition of the *Prospect* was published by Rea in 1668.

Sometime after 1668, all of the Speed plates were acquired by Thomas Bassett and Richard Chiswell. Bassett and Chiswell reissued this map in 1676 in its most common version, along with Speed's world, the three other continents, and regional maps. The 1676 *Prospect* can be found bound with Speed's *Theatre* containing the British Isles maps, as in the 1676 British Library example cited above. There were no further editions issued after 1676.

Sometime before 1690, the original copperplates from the *Prospect* were acquired by Christopher Browne. It is known that Browne re-issued some of the Speed maps separately. Burden notes a Speed map of Virginia and Maryland with the Browne imprint (Burden 1996: map # 217), but not one of the Americas. A map of Africa with the Browne imprint is not known.

Shirley notes (2004, I: 968) that some Speed maps are also known without text on the verso, indicating that the maps were sold separately on occasion.

The following is a list of the various editions of Speed's *Prospect* in which the map of Africa appeared. The list is primarily based on Tooley (1969: 113-114) and confirmed by the author. The list can be used to determine the date and edition for a particular separate map by examining the text changes on the verso of the Africa map in the various editions of the *Prospect*:

1627 Verso: Text begins with a capital letter 'A' set within floral box with crown and rose within the 'A'. Last line of penultimate paragraph reads 'belonging to Africa.'

1631-2 Verso: Same design of capital letter as 1627. Last line of penultimate paragraph reads 'Ilands belonging to Africa.'

1646 Verso: Text begins with a 'capital 'A' in printer's type set into a square arabesque design'.

1662 Verso: Text commences with a capital 'A' with two figures and a net.

1676 Verso: Text commences with a capital 'A' on a background of flowers and leaves springing from a pot.

References
Shirley 2004, I: 968; Norwich 1997: map # 30; Skelton 1996b; Tooley 1969: 113-114.

63   1627 Petrus Bertius – Melchior Tavernier, Paris

Map 63.2.

63.1 First State 1627

Carte de | L'AFRIQVE | Corrigeé, et augmenteé, deßus toutes | les aultres cy deuant faictes par | P. Bertius.
- No scale bar.
- [Paris : Melchior Tavernier, 1627].
- From: separately published / *Theatre Geographique du Royaume de France*.
- 1 map : copperplate engraving ; 38 x 50 cm.
- Latitude coordinates along left and right sides of map. Longitude coordinates along Equatorial Line.
- North at the top; no compass rose.
- Title cartouche at the top right.
- There are five Dutch ships, two sea monsters, six flying fish and Neptune with a mermaid in the Atlantic Ocean. In the Indian Ocean, there is one Dutch ship and one sea monster. Various animals are within Africa.

Example: *Amsterdam UB, 33-18-01; London BL, C.26.c.12; Washington LC; Private French collection.*

Map 63.1, Title cartouche. *Map 63.2, Title cartouche.*

63.2 Second State 1640
The date of 1640 is added within the title cartouche.

Example: *Author's collection* (as a separate map).

Description
This rare map is one of a set of four known continent maps that Melchior Tavernier had engraved, with or without the permission of Petrus Bertius, whose imprint as the map's maker appears on the title cartouche.

Petrus Bertius (1565-1629) was born in Flanders but as a refugee from political and religious unrest, settled first in Amsterdam and then in Leiden where he was professor of mathematics and librarian at the University of Leiden. After this, he moved to Paris where in 1618 he became cosmographer to King Louis XIII of France. He died there in 1629.

While in Paris, Bertius likely came into frequent contact with Tavernier. Melchior Tavernier (1594-1665) was one of the most important mapsellers and publishers in Paris in the first half of the seventeenth century.

As the Asia map has the imprint of Cornelis Danckerts (1603-1656) as the engraver, it is supposed that Danckerts engraved the other three continent maps, including this map of Africa. The date of 1627 for this map is taken from that of the Europe map, which contains the date of 1627.

While Tavernier cites Bertius as the source for this map, it is a clear copy of the Jodocus Hondius maps of 1619 and 1623, like the Bertius 1624 map. As on the 1619 Hondius map, the geography for the 1623 map is copied from Willem Blaeu's folio-size map of 1617 which was based on a reduction of Blaeu's wall map of 1608. It is not known whether the information on this map came to Tavernier via Bertius, who was related to Hondius, or if Tavernier simply obtained an example of the Hondius map and copied it.

Publication Information
Initially sold as a separate map, this map appeared in Tavernier's *Theatre Geographique* in 1632, and then again in 1634, 1637, 1638, and 1643. Although this atlas was primarily devoted to France, it contained maps of the continents at the beginning of the atlas. Christophe Tassin also used this map in his atlas *Les Cartes Generale de toutes provinces de France* in 1634 and in later editions in 1637 and c.1640.

To aid in the identification of this Bertius map and a number of similar maps using different copperplates by Jodocus Hondius Jr. (Map # 58), Bertius (Map # 61), Boisseau (Map # 66), and Berey (Map # 89), please refer to the chart under Map # 58, the Jodocus Hondius map of 1619. All of these similar maps are derivatives of the Jodocus Hondius Jr. map of 1619. All lack decorative borders as did the third and later states of the Hondius 1619 map.

References
Koeman 1967-71, I: 60-61 (under Bertius) and Vol. II: 88 (under Danckerts), Burden 1996: map # 218 (Americas map).

64 1628 Johannes Janssonius, Amsterdam

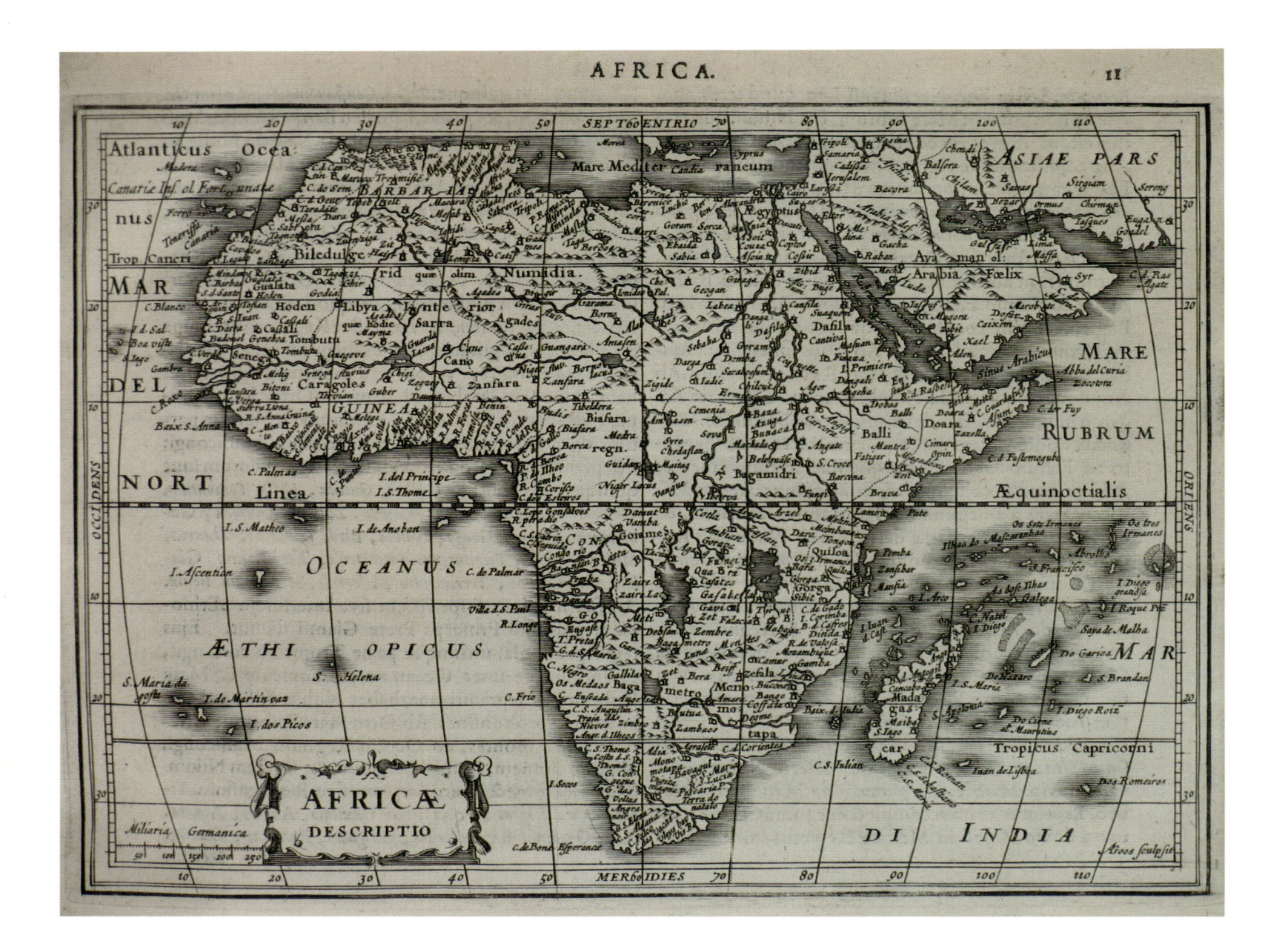

64.1 First State 1628

AFRICÆ | DESCRIPTIO.
[at bottom right on map edge]:
AGoos Sculpsit
- Scale bar to left of title cartouche at bottom left: Miliaria Germanica 250 = 2.7 cm.
- [Amsterdam : Johannes Janssonius, 1628].
- From: *Atlas Minor Gerardi Mercatoris*.
- 1 map : copperplate engraving ; 14.5 x 20 cm.
- Latitude coordinates and longitude coordinates along all four sides of the map.
- North at the top; no compass rose.
- Title cartouche at bottom left.
- No decorative elements on the map.

Examples: *Leiden UB*, COLLBN Port 179 N 10 1634 *and* 468 B 22 (1628); *London BL*, C.3.a.4; *Washington LC*, G.1007.A.9 1628 vault.

The following is a list of the various editions of Janssonius' *Atlas Minor Gerardi Mercatoris* in which the the first state of the map of Africa appeared, based on Van der Krogt (2003). The list can be used to determine the date and edition for a particular separate map by examining the language, pagination, and signature notes in the various editions of the Atlas:

Date	van der Krogt no.	Atlas Title	Language	Page	Signature	Notes
1628	352.01	*Atlas Minor*	Latin	11	B2r.	Africa above map in letterpress.
1630	352.11	*Atlas Minor*	French	11	B2r.	L'Afrique above map in letterpress.
1630	352.21	*Atlas Minor*	Dutch	11	B2.	Beschryvinghe van Africa above map.
1631	352.31	*Atlas Minor*	German text	9	Blr.	Von der Beschreibung Africæ above map.
1634	352.02	*Atlas Minor*	Latin text	11	B2r.	Africa above map in letterpress.
1648 & 1651	352.01	*Atlas Minor*	German text	361	2Zlr.	Beschreibung Africæ...

Map 64.2.

64.2 Second State c.1714

A larger cartouche replaces the first one and the title becomes 'L'AFRIQUE.' with a new scale bar within the cartouche at bottom: 'Lieues d'Allemagne' 300 = 3.2 cm. The signature of the engraver 'AGoos' is removed. A compass rose and a different surrounding grid are added.

Example: *Author's collection.*

The following is information on this map's appearance in its second state in Pieter van der Aa's *L'Atlas Soulagé de son gros & pesant fardeau* (Van der Krogt 2003: Vol. III: 879).

Date	van der Krogt no.	Atlas Title	Language	Page	Signature Notes
c.1714	352.51.1	*Atlas Soulagé… / Nouveau Petit Atlas*	French	-	No text on verso.

Description

Since the final edition in 1621 in Amsterdam of the Jodocus Hondius' *Atlas Minor* of 1607, there was no smaller-format atlas available for the Dutch market. Johannes Janssonius decided to meet this need by producing the *Atlas Minor Gerardi Mercatoris* with newly engraved maps, including this map of Africa.

The engravers for the maps in the atlas were Abraham Goos and Pieter van den Keere, two of the most notable Dutch engravers of this period. Goos engraved this map of Africa (AGoos Sculpsit).

Obvious effort went into the development of this map. The geography in the Janssonius map borrows some aspects from Jodocus Hondius' 1606 folio map of Africa and Hondius' reduced-size map of 1607, though Janssonius seems to rely more on Willem Blaeu's folio map of 1617 and his wall map of 1608. There are numerous additional placenames throughout this map, when compared to the slightly larger 1607 Hondius map. The configuration of the interior river systems closely follows Blaeu's wall map of 1608. For example, the major Southern African rivers of Spirito Santo and Cuama (here unnamed), which originate in Sachaf lacus (also unnamed) follow a pattern copied from the Blaeu wall map. Also, the outline of Africa is much more refined than as presented on Hondius' 1606 folio map and his 1607 *Atlas Minor* map and seems to follow Blaeu's view of the shape of Africa.

Publication Information

The copperplate that produced this map had an extremely long life. It appeared in a number of editions of the Janssonius series of the *Atlas Minor* up to 1651, as detailed below.

In 1661, Janssonius published an edition of Philipp Cluver's popular *Introductionis In Universam Geographicam*. He illustrated this book with some of the plates from his *Atlas Minor*, including this map of Africa.

Some time after Janssonius' death in 1664, the copperplates were acquired by Pieter van der Aa. Along with this map of Africa, they were included Van der Aa's extremely rare atlas. Part One of the atlas was called *Nouveau Petit Atlas* and Parts Two through Nine were called the *L'Atlas soulagé de son gros & pesant fardeau*. The atlas was published in Leiden in c.1714. In this edition, the Africa map has no text on the verso. The map of Africa appeared both in Part One and Part Nine of the atlas.

There were five similar maps produced around the same period of time, derived from the Janssonius 1628 map; the last four of which were produced in Paris. For comparison purposes, the differences are as follows:

Map # 64	Janssonius	1628	Longitude coordinates on top grid with the number '60' between T and E in SEPTENTRIO.
Map # 72	Boisseau	1643	60° longitude line through the 'r' in Septentrion with longitude coordinate numbers on top grid.
Map # 81	Berey	1651	60° longitude line through the 'r' in Septentrion with <u>no</u> longitude coordinate numbers on top grid.
Map # 82	Picart (1)	c.1651	60° longitude line through the 't' in Septentrion with longitude coordinate numbers on top grid.
Map # 92	Picart (2)	c.1659	60° longitude line between the 't & r' in Septentrion with longitude coordinate numbers on top grid in the first state.

References

Van der Krogt 2003: map 8600:352; Koeman 1967-71, I: 9-11 (Aa 4, no. 209, and Aa 5, part 9, no. 209) and Vol. II: 520 (Me 194, no. 5); Phillips 1909-92: entry 437.

65   1630 Jan Evertsz. Cloppenburch, Amsterdam

65.1 First State 1630

AFRICÆ | nova Tabula | Auct J. Hondio.
- No scale bar.
- [Amsterdam : Jan Evertsz. Cloppenburch, 1630].
- From: *Gerardi Mercatoris Atlas*.
- 1 map : copperplate engraving ; 18.5 x 25 cm.
- Latitude coordinates and longitude coordinates along all four sides of the map.
- North at the top; no compass rose.
- Title cartouche at the top right.
- There are two sea monsters and two sailing ships in the Atlantic Ocean.

Examples: 1630 edition: *London BL*, C.3.b.3; *Washington LC*, G 1007.A84 1630 [vault]; *Author's collection* (separate map); 1632 edition: *Washington LC*, G 1007.A8 1632 [vault].
1673 edition: *Leiden UB*, COLLBN Atlas 63; *London BL*, Maps C.3.b.5; *Chicago, Baskes Collection*.

65.2 Second State 1734
Addition of engraved page number '236' in top right corner of the map.

Examples: 1734 edition: *Basel UB*, Ew 84; *London BL*, Maps C.29.c.7 (bound in one volume); *Washington LC*, G 1015.D87 1735 [vault]; c. 1738 edition: *London BL*, Maps C.29.c.7; *Washington LC*, G 1015.D87 1738 [vault].

Description

This uncommon map is from the Amsterdam publisher Jan Cloppenburch's reduced-size edition of the *Gerardi Mercatoris Atlas*, first published in 1606. The Cloppenburch *Atlas Minor* was of a convenient size for users of its day, somewhat larger than Hondius and Janssonius' *Atlantes Minores* of 1607 and 1628 respectively, and reasonably priced when compared with folio-size atlases of this period.

This is a finely engraved, attractive map of Africa. The copperplate that produced it was engraved by Pieter van den Keere. While the title refers to Jodocus Hondius as the source for this map (Auct J. Hondio.), it is not at all similar to the elder Jodocus Hondius' 1606 and 1607 maps. In fact, this map bears close conformity to Willem Blaeu's 1617 map and Jodocus Hondius Jr.'s maps of 1619 and 1623. Much of the interior, including the hydrography and the placenames, are directly based on Blaeu's 1617 map of Africa. The outline of Africa is also more closely aligned to Blaeu. One notable exception is an excessively wide bay where the unnamed Spirito Santo River empties into the Indian Ocean.

Publication Information

The Cloppenburch version of the Mercator-Hondius *Atlas Minor* was only published in 1630, 1632, and 1636. Cloppenburch was either forced out of business or Johannes Janssonius bought his copperplates, as the plates later appeared in a 1673 edition of *Atlas* by Johannes Janssonius van Waesberge, the son-in-law of Janssonius. A second edition of this atlas appeared in 1676 in Dutch and Latin.

While Cloppenburch's publication record was short, this map continued to appear for over the next 100 years. In an example in the Library of Congress, the map, in its first state, appeared in Arnout Leers' edition of Leo Africanus in 1665, titled *Pertinente beschryvinge van Africa....* 1665 (*Washington LC*, DT7.L48).

In 1676, the map was also used for an edition of Phillip Cluver's *Introductionis in Universam Geographicam* (*London BL*, 1477.DD.27).

Finally, in 1734, the map of Africa appeared in its second state, along with other Cloppenburch maps, in the *Atlas Portatif* by Henri du Sauzet, published in Amsterdam. Burden (1996: map # 224) notes an untitled and incomplete atlas in the Universiteitsbibliotheek in Leiden, which appears to be based on the old Cloppenburch edition and which he dates as c.1738. The author has not seen this example to determine if it contains a map of Africa.

The following is a list of the various editions of books in which this Cloppenburch map of Africa appeared, based on Van der Krogt (2003). The list can be used to determine the date and edition for a particular loose map by examining the language, pagination, and signature notes on the map:

Date	van der Krogt no.	Atlas Title	Language	Page	Signature	Notes
1630 & 1636	353:01/02	*Atlas*	French	11	B2r.	'L'Afrique.' and '11' above map.
1632	353:11	*Atlas*	Latin	11	B2r.	'11' above map.
1665		*Pertinente beschryvinge*	Dutch		none	no print above map.
1673	353:21	*Atlas*	Latin		none	no print above map.
1676	353:31	*Uytbeeldinge*	Dutch		none	'Africa' 'Afrijcke' above map
1676		*Introductionis in Universam*	Latin		none	no page number.
1734	353:41	*Atlas Portatif*	French		none	
c.1738	353:42	*Atlas Portatif*	French		none	'236' engraved in top right corner of the map.

References

Van der Krogt 2003: map 8600:353; Burden 1996: map # 224 (Americas map); Koeman 1967-71, II: Me 198, no. 3 and Me 205-208.

66 c. 1636 Jean Boisseau, Paris

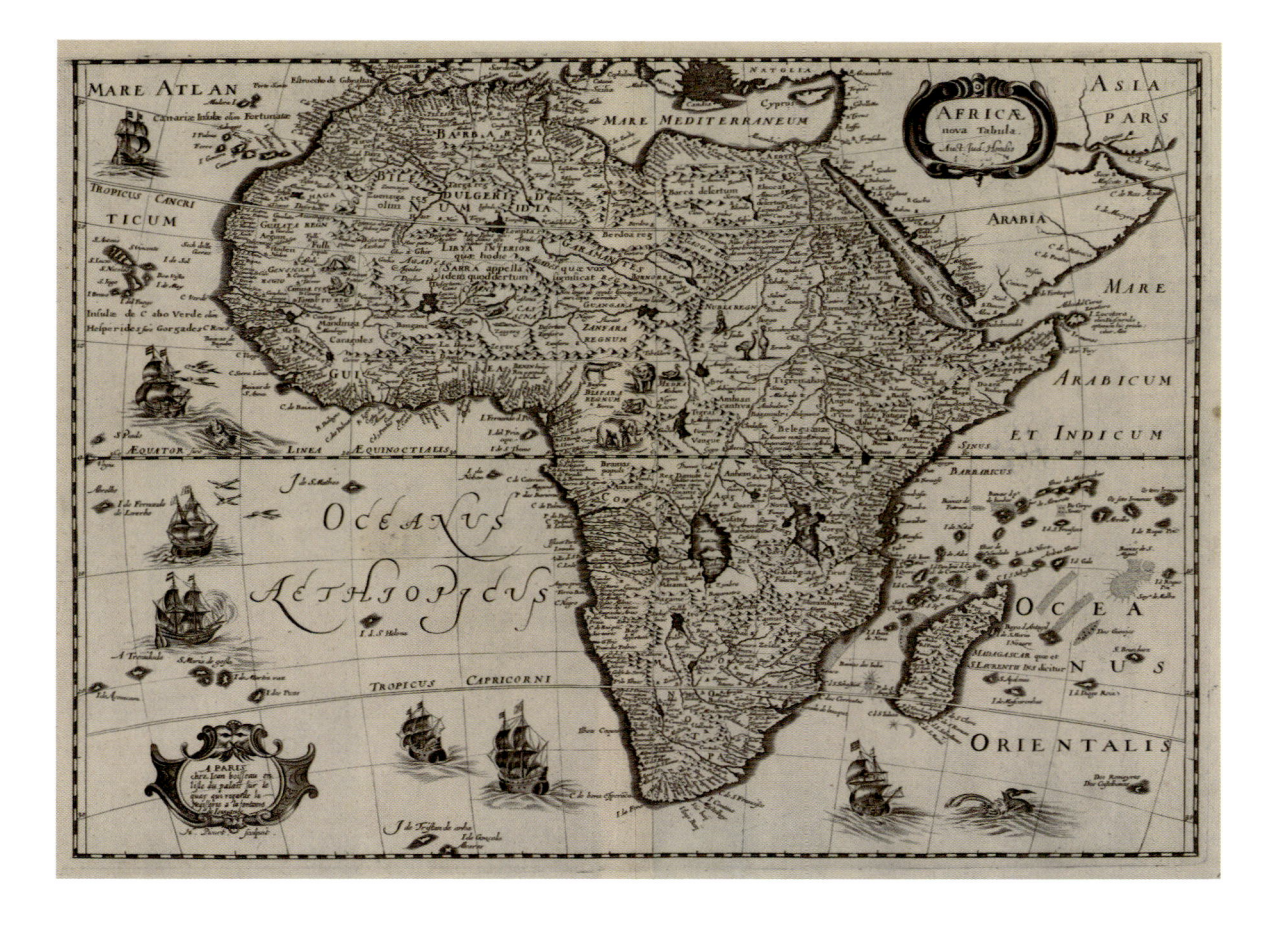

66.1 First State c.1636

AFRICÆ | nova Tabula. | Auct: Jud: Hondio.
[Within the bottom left cartouche]:
A PARIS | chez Iean boisseau en | lisle du palais sur le | quay qui regarde la - | Megisserie a la fontaine | de Iouuance.
[below the bottom left cartouche]:
H. Picart scalpsit.
- No scale bar.
- Paris : Jean Boisseau, [c.1636].
- From: separately published / *Theatre des Gaules* of 1642.
- 1 map : copperplate engraving ; 37.5 x 49 cm.
- Latitude coordinates at the sides. Longitude coordinates along Equatorial Line.
- North at the top; no compass rose.
- Title cartouche at the top right.
- Cartouche with publisher's imprint at the bottom left.
- There are six sailing ships, one sea monster and five flying fish above and below Equator in the Atlantic Ocean, and one sea monster and one sailing ship in the Indian Ocean.

Examples: *London BL*, Maps 39.e.6.[4] (in *Theatre Geographique du Royaume de France* of 1648); *Paris BN*, Ge D 11368; *Washington LC*, G8200 1640.H6 (as a separate map);

After 1648, little is known of Boisseau. It is known that part of Boisseau's business was taken over by Lovis (Louis) Boissevin (c.1610-1685). It is generally assumed that this map, and the other continent maps, were not re-issued. However, the April 2005 internet catalog of Altea Gallery in London shows a separate map of what is a later state of the Boisseau map.

Map 66.2.

66.2 Second State after 1648
Removal of the publisher's cartouche and imprint at the bottom left, and also removal of the engraver's imprint below this cartouche. Removal of all ships and sea monsters from the sea. As in the first state, the map still retains evidence of an erasure just below Auct: Jud: Hondio in the title cartouche.

Example: *London, Altea Gallery* (April 2005 internet catalog).

Description

This map is attributed to the Parisian publisher, Jean Boisseau (active from 1637-1658), who published this map separately and in his *Theatre de Gaules*. The beautiful engraving for this map is by Hughes Picart, whose imprint, 'H. Picart sculpsit', is below the bottom left cartouche for the publisher.

The map title refers to Hondius as the source for this map (Auct: Jud: Hondio). This map has a number of elements taken directly from Jodocus Hondius Jr.'s 1619 Africa map. The decorative elements (ships and sea monsters) within the seas are identical to those on the Hondius map. A notable exception is the cartouche at the bottom left that replaces two sea monsters and the figures of Neptune and a mermaid on the Hondius map. There is also one less flying fish on this map.

Much of the interior, including the hydrography, is based on Blaeu's 1617 folio-size map of Africa, which was also the basis for the Hondius 1619 and 1623 maps.

Boisseau follows Hondius in his placement of the Prime Meridian. On this map, the Prime Meridian is placed thorough the Cape Verde Islands, rather than through the Canary Islands as on the Blaeu map. As a result, the Cape of Good Hope is at a longitude of approximately 48°E, rather than about 41°E as on the Blaeu map.

The map of Africa is not dated. As the accompanying World and Europe maps in the set are dated 1636, it is likely that this map of Africa and the other two continents were prepared in c.1636.

Publication Information
Although it was separately issued, this map also appeared in Jean Boisseau's *Theatre des Gaules* of 1642. There is no text on the verso of the map.

There was a further edition of the *Theatre des Gaules* by Boisseau in 1648, under the title *Theatre Geographique du Royaume de France*, using the first state of this map.

Below 'Auct: Jud: Hondio' in the title cartouche, there is evidence of an erasure of the words Alexandre de Meufves *('A de meufues ex')*. It is not certain why this erasure occurred. It seems that Meufves was a business partner of Boisseau. Meufves (the dates of his birth and death are not known) may have died prior to the issuance of the map, or there may have been a splitting of the business partnership just before the initial publication of the map, necessitating the removal of his name from the title.

To aid in the identification of this Boisseau map and a number of similar maps using different copper-plates by Jodocus Hondius Jr. (Map # 58), Bertius (Map # 61), Bertius-Tavernier (Map # 63), and Berey (Map # 89), please refer to the chart under Map # 58, the Jodocus Hondius map of 1619. All of these similar maps are derivatives of the Jodocus Hondius Jr. map of 1619. All lack decorative borders as did the third and later states of the Hondius 1619 map.

References
Burden 1996: map # 249 (Americas map); Shirley 1993: 363; Loeb-Larocque 1989: 22-3; Pastoureau 1984: 67 and 69-70.

67 1638 Matthäus Merian, Frankfurt

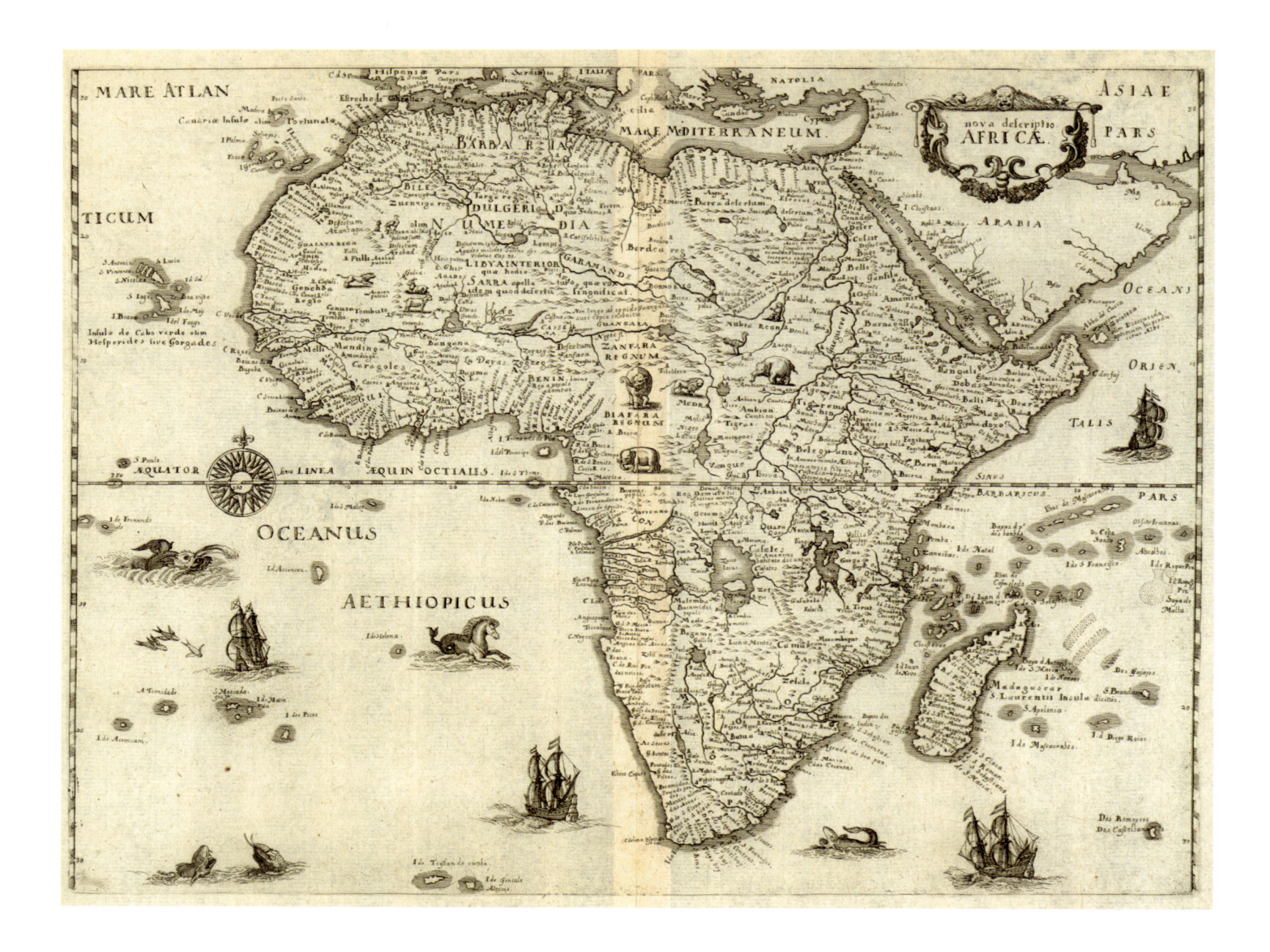

nova descriptio | AFRICÆ.
- No scale bar.
- [Frankfurt : Matthäus Merian, 1638].
- From: *Neuwe Archontologia Cosmica... durch J.L. Gottfried.*
- 1 map : copperplate engraving ; 27 x 36 cm.
- Latitude coordinates along left and right sides. Longitude coordinates along Equatorial Line.
- North at the top; one compass rose on Equatorial Line within Atlantic Ocean.
- Title cartouche at the top right.
- There are three sea monsters, two sailing ships, one seahorse, and three flying fish below the Equator in the Atlantic Ocean, and one sea monster and two sailing ships in the Indian Ocean. Various animals are within Africa.

Examples: 1638 edition: *London BL*, 568.h.8; *Washington LC*, D18.A96 1638 *and* G1028.A9 1649 (1646 edition of *Neuwe Archontologia*); *Author's collection.*

Description

Matthäus Merian (1593-1650), originally from Basel, was a publisher and engraver active in Frankfurt where he worked with Théodore de Bry, whose daughter he later married. Besides publishing his own books, Merian is also noted for completing the later parts and editions of De Bry's *Grands Voyages* and *Petits Voyages*, originally started by De Bry in 1590.

The geography for this map is based on Willem Blaeu's map of Africa of 1617. The Merian map has many of the decorative elements in the oceans and in the interior of Africa, in some cases with slightly different placements. It does not have Blaeu's decorative borders and is reduced to about 2/3 the size of the Blaeu. The verso of the map is blank.

Publication Information

This map of Africa was published by Merian for an edition of his *Neuwe Archontologia* of 1638. The *Neuwe Archontologia* was a German translation by Johann Ludwig Gottfried of Pierre d'Avity's *Les empires du Monde*. The map of Africa also appeared in later editions of *Neuwe Archontologia* in 1646 and in 1695. It appeared in other works of the period published in Frankfurt, including the *Theatrum Europaeum*.

Based on the similarly engraved Merian maps of the Americas and the World which have Merian's name as the engraver (M. Merian fecit.), it is assumed that he engraved this map of Africa as well. Unlike the Americas and World maps, the Africa map in both the editions of 1638 and 1646 does not have the Merian imprint on it. The Africa map appears to be in only one state.

References

Norwich 1997: map # 33; Burden 1996: map # 251 (Americas map); Shirley 1993: 369; Mickwitz et al. 1979-1995: Vol. 1: 167-170 (under Gottfried); Tooley 1969: 76.

68   1639 Petrus Bertius - Melchior Tavernier (attr.), Paris

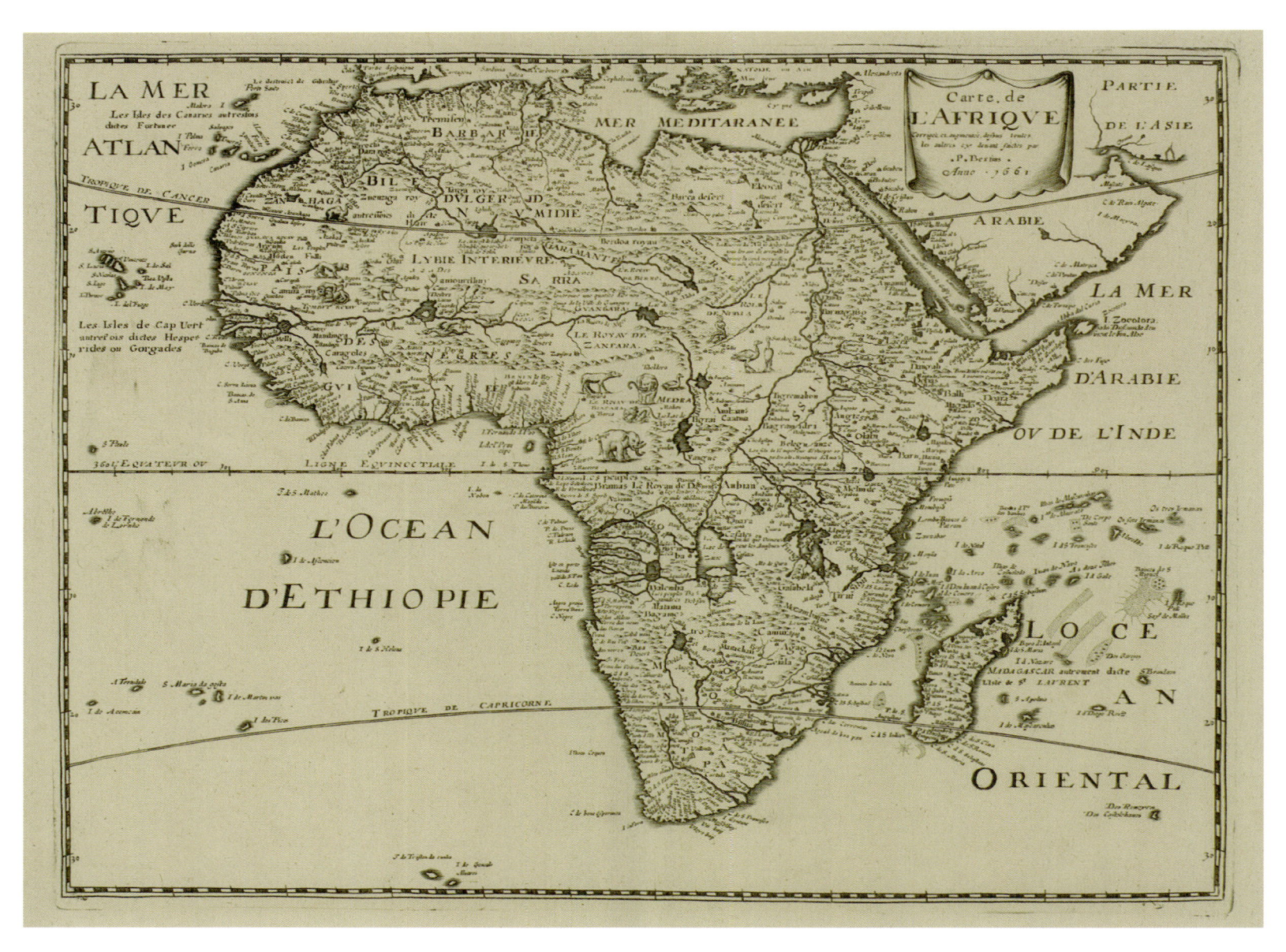

Map 68.2.

68.1 First State 1639

Carte. de | L'AFRIQVE | Corrigeé, et, augmenteé, deßus toutes | les aultres cy deuant faictes par | P.Bertius . | Anno · 1639.
- No scale bar.
- [Paris : Melchior Tavernier (attr.)], 1639.
- From: separately published.
- 1 map : copperplate engraving ; 27 x 36 cm.
- Latitude coordinates along left and right sides. Longitude coordinates along Equatorial Line.
- North at the top; no compass rose.
- Title cartouche at the top right.
- No decorative elements within oceans. Various animals within Africa.

Example: Assumed to exist; locations not known.

68.2 Second State 1661
A new date of 1661 within title cartouche.

The map is known bearing a later date of 1661 in some editions of D'Avity's *Description generale de l'Afrique*.

Example: *Author's collection* (as a separate map).

Description

This uncommon map is attributed to Melchior Tavernier. It is known as part of a set of the four continents and the world. Only the double hemisphere world map in the set has the imprint of Tavernier and the date of 1639 (Shirley 1993: 373: map no. 349).

Tavernier (1594-1665) was one of the most important mapsellers and publishers in Paris in the first half of the seventeenth century. He modeled this map of Africa after his 1627 map which was derived from Jodocus Hondius Jr.'s maps of 1619 and 1623. These in turn were based on Willem Blaeu's folio-size map of 1617. Considering the reduced size of this map and the inclusion of a considerable amount of geographical detail, the quality of engraving for this map is excellent.

Publication Information

The map in its first state of 1639 was issued separately. The author has not yet located an example of the first state of the map of Africa, but it is assumed to exist, based on the existence of the maps of the World and the Americas (without the Tavernier imprint), both dated 1639.

The Africa map is assumed to be in two states. Tooley (1973: 300) mentions a third state of this map of 1662, but it appears that this state does not exist. It is possible that there was some confusion with the second italic script '1' in 1661, which can be mis-read as an italic '2'. The '6' and '1' were engraved in a different style for the second state.

References

Correspondence with Joseph Walker to clarify that the supposed third state does not exist.
Burden 1996: map # 256 (Americas map); Shirley 1993: 373 (for the World map of 1639 as a separate map at *Paris BN*, Ge.D.12382); Tooley 1980: 300 (for Americas map); Borba de Moraes 1958: Vol. 1: 53.

69 1640 Giovanni Botero - Lucatonio Giunti, Venice

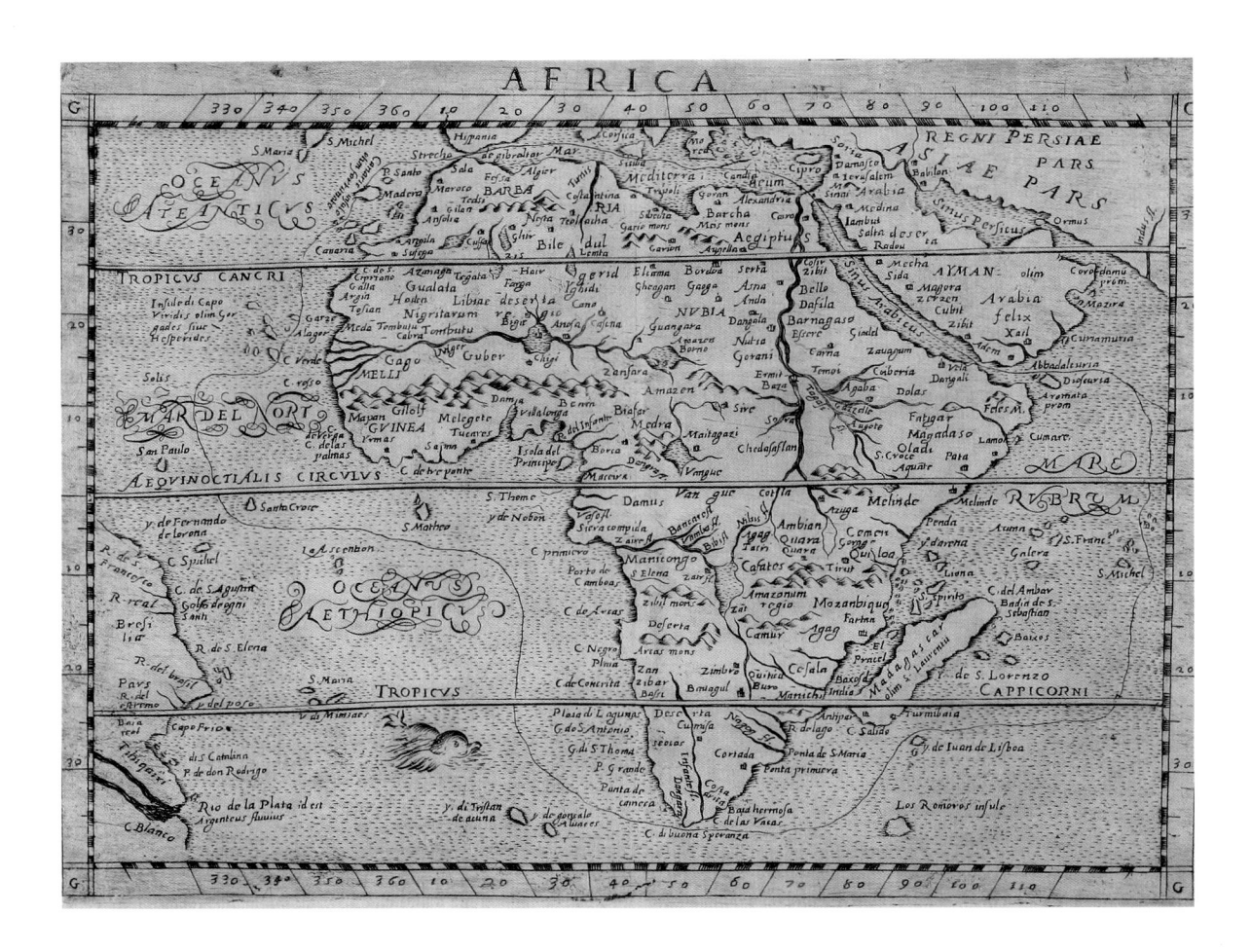

AFRICA.
- No scale bar.
- [Venice : Giunti, 1640].
- From: *Relationi Universali di Giovanni Botero*.
- 1 map : copperplate engraving ; 17.5 x 25 cm (to outer black line).
- Latitude coordinates and longitude coordinates along all four sides.
- North at the top; no compass rose.
- Title above map image.
- No decorative elements on the map except for one sea monster in South Atlantic below Tropic of Capricorn.
- An engraved line encircles all of Africa and its surrounding islands.

Examples: 1640: *New York PL*, *KB 1640 Botero*; *Washington LC, G 120.B732 1640*; 1650: *Providence JCB*.

Description
For the 1640 Venice edition of Giovanni Botero's *Le Relationi*, a new plate was cut. This map of Africa is one of four maps of the continents that appear in this book.

Geographically, the map is unchanged from the 1582 Giovanni D'Anania map, which was used as the model for the first edition of Botero's *Le Relationi* that contained maps, the 1595 Venice edition published by Grecco.

Publication Information

This map is known to appear in Giovanni Botero's *Le Relationi Universali*, published by Lucatonio Giunti and his heirs in Venice in 1640. According to Burden (1996: map # 258), there is a later edition of 1650 at the John Carter Brown Library, but the author has not seen this edition.

This map has a circle surrounding all of Africa in a similar manner to the 1582 D'Anania map in its second state that was used in Rosaccio's expanded edition of Girolamo Ruscelli's *Geographia* of 1598 and 1599. The Botero-Giunti map is distinguished from the second state of the D'Anania map of 1582 by the omission of the period '.' which appeared after the title of 'AFRICA.' in the D'Anania map. This map is known only in one state.

To distinquish this map from similiar appearing maps, using different copperplates, by D'Anania (1582), Botero (1595), Magini (1596), Keschedt (1597), Africanus-Pory (1600), and Rossacio (1605), please refer to the table under Map # 20, 1582 D'Anania.

References
Burden 1996: map # 258 (Americas map).

70 1640 Jacques Honervogt, Paris

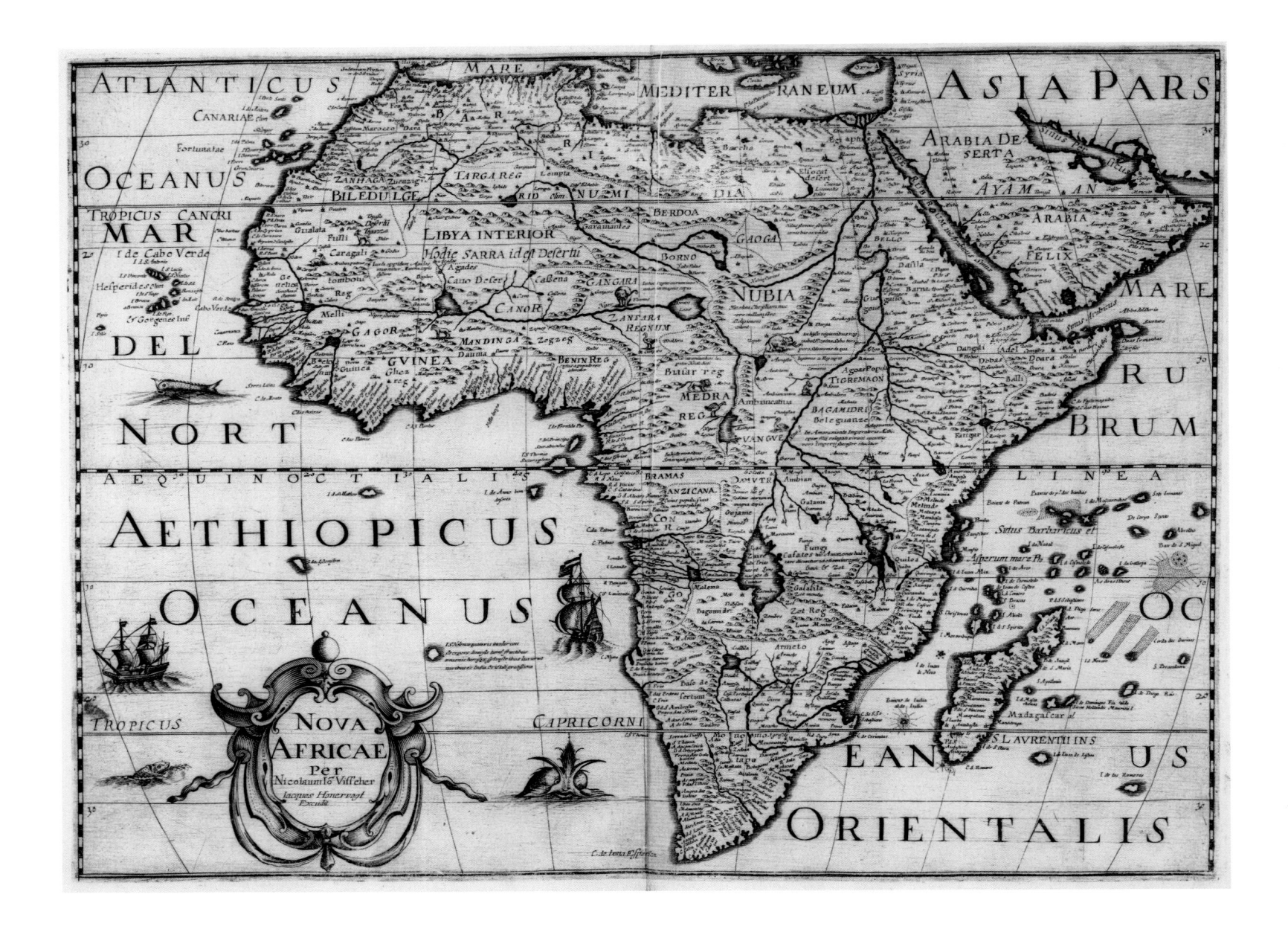

70.1 First State 1640

NOVA | AFRICAE | Per | Nicolaum Iō Vissener *[sic]* | Iacques Honervogt | Excudit.
- No scale bar.
- [Paris] : Honervogt, [1640].
- From: separately published.
- 1 map : copperplate engraving ; 36 x 46.5 cm.
- Latitude coordinates along sides of the map. Longitude coordinates along Equatorial Line.
- North at the top; no compass rose.
- Title cartouche at bottom left of map.
- There are three sea monsters and two sailing ships in the Atlantic Ocean. Various animals are within Africa.

Example: *Paris BN*, Ge DD 6150 (Pl 4).

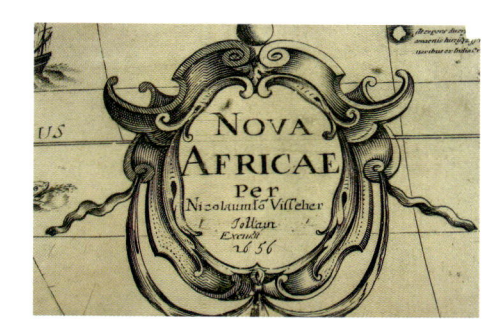

Map 70 Title cartouches of the first, second and third states.

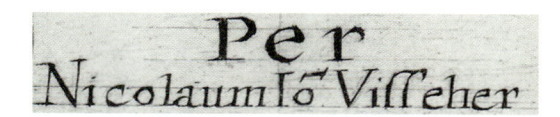

Map 70 Incorrect spelling of Visscher's name.

70.2 Second State 1656
Within the title cartouche, the Jollain imprint with a new date: 'Jollain | Excudit | 1656'. Some evidence of erasure of Honervogt name.

Examples: *London BL*, Maps C.27.g.6 (1&2) (Van Loon composite atlas); *Washington LC*, G8200 1656.v5 (with imprint of Jollain).

70.3 Third State 1666
Within the title cartouche, a new date: 'Jollain | Excudit | 1666'. Still with some evidence of the erasure of the Honervogt name.

Example: For an image of this map, see Tooley, 1972, *Map Collectors' Circle* No. 82: 9.

Description
This is an extremely rare map by Jacques Honervogt. The map is part of a set of the four continents produced by Honervogt. As the Europe map in the set is dated 1640, this map is also given that date.

Born in Cologne, Jacques Honervogt (c.1590-c.1663) established his own publishing firm in Paris around 1624. Sometime after 1654, he was in partnership with Gerard Jollain (? – 1683), whose imprint appears in the second state of this map. Jollain took over the business following Honervogt's death in about 1663. Jollain was a printseller and publisher working from 'Rue St. Jacques a 'Enfant Jesus' in Paris.

This map is modeled after the Pieter van den Keere 1614 map in its third state of 1636, as issued by Claes Jansz. Visscher. Visscher had acquired the Van den Keere map sometime before 1631. The geography for this map closely follows the 1636 Van den Keere - Visscher map. Honervogt does not include the decorative borders in his re-issue of the Visscher map.

Publication Information
This separately issued map of Africa is known in three states.

References
Tooley 1999-2004, E-J: 450-1; Burden 1996: map # 259 (Americas map); Loeb-Larocque 1993: Map # 3; Loeb-Larocque 1989: 18-19; Tooley 1972: 9; Tooley 1969: 119.

71   1641 Philipp Cluver – Conrad Buno, Brunswick (Braunschweig)

Map 71 with manuscript notes and underlinings.

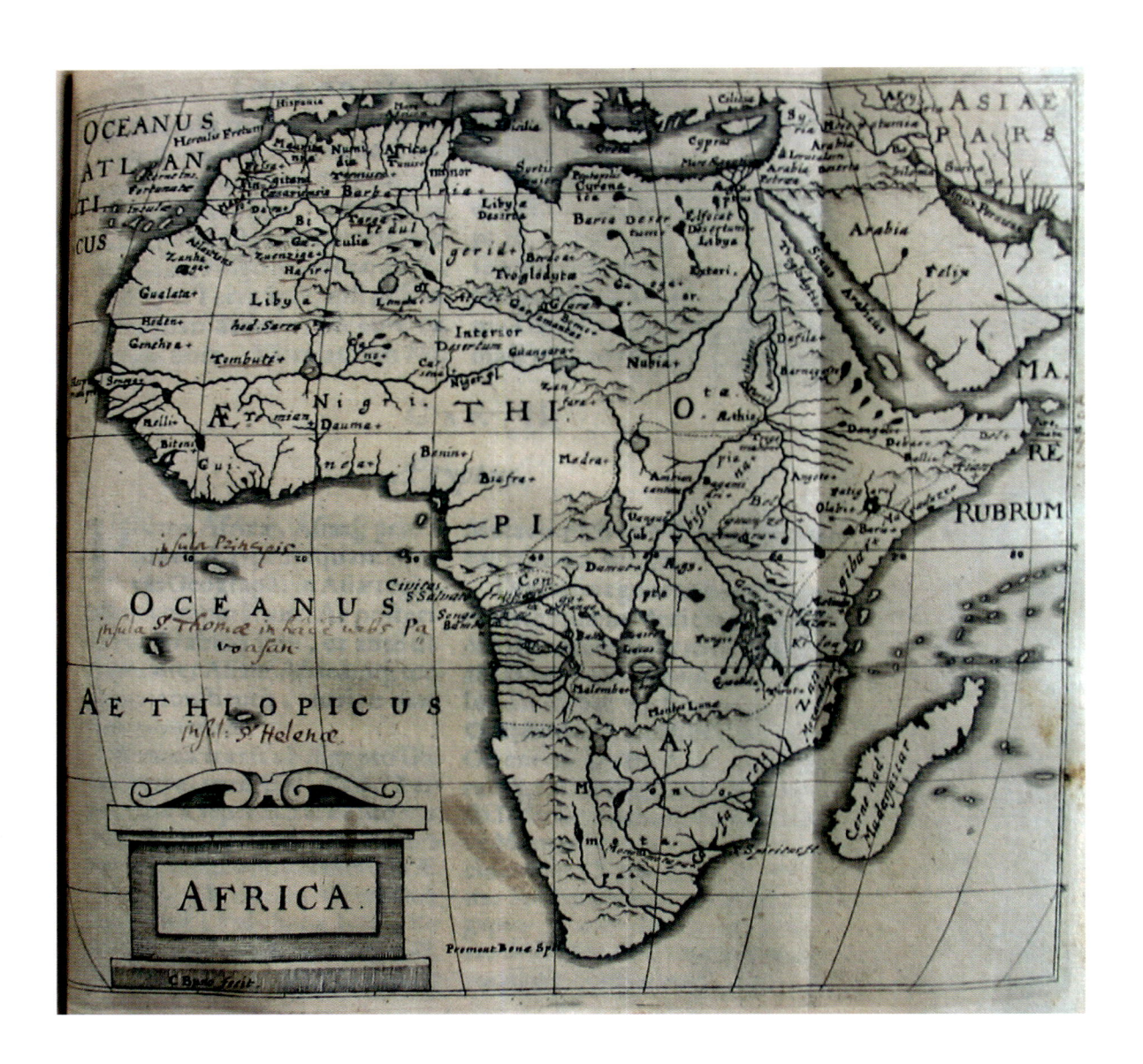

AFRICA.
[at the base of the title cartouche]:
C Buno fecit
- No scale bar.
- [Braunschweig: Gottfried Müller, 1641].
- From: *Introductionis in Universam Geographiam.*
- 1 map : copperplate engraving ; 15 x 16.5 cm.
- Latitude coordinates along sides. Longitude coordinates along Equatorial Line.
- North at the top; no compass rose.
- Title cartouche at bottom left of map.
- No other decorative elements on the map.

Examples: 1641 edition: *Author's collection*; 1652 edition: *München SB*, Res 4 Geo.u.28; *Wolfenbüttel HAB*, Ca 171.

Description

Philipp Cluver (1580-1623) was a noted geographer of the early seventeenth century. He is credited with authoring numerous works on geography, the most famous being his *Introductionis in Universam*

Geographicam, which appeared over a considerable number of years into the eighteenth century. Cluver was born in Gdansk and eventually settled in Leiden. The first edition of Cluver's *Introductio*, published in Leiden in 1624, did not contain maps. This edition of 1641 is the first known to contain a map of the continent of Africa.

The two men principally involved in the German editions were Conrad Buno (c.1613-1671), and his brother, Johann Buno (1617-1697). Johann was a theologian and pedagogue in Lüneburg. Conrad was an engraver and publisher at the court of Wolfenbüttel. The engraver for this map of Africa is Conrad Buno, whose imprint appears as 'C Buno fecit' at the base of the title cartouche.

The map of Africa appears to be modeled on Blaeu's 1608 wall map and, more immediately, the Blaeu folio map of Africa of 1617. The Cluver-Conrad Buno map shows the unnamed Cuama River as originating to the south of the *Montes Lune* (Mountain of the Moon) as on the Blaeu folio map, whereas on the Blaeu wall map, this river joins upstream with the Spirito Santo and the Zambere rivers flowing from Sachaf Lachus.

This map has a more limited number of placenames when compared to other, similar-sized maps of this period. Many of the Portuguese ports (for example Sofala) on the East African coast are not shown. Numerous rivers, such as the Cuama, are not named. The *Montes Lune* (Mountains of the Moon) stretch across Southern Africa. The Cluver-Conrad Buno map identifies much of interior Africa as being part of Ethiopia. Broadly across the middle of the map in large capital letters is *Æ.THI.O.\PI.* with the last 'A.' below the Nile source lakes in Southern Africa. This map provides the same sub-divisions of Africa commonly used during this period.

Publication Information

This 1641 edition is the first edition of Cluver's *Introductionis in Universam Geographiam* that is known to contain a map of Africa. Numerous maps were included for the first time in this German edition of Cluver's *Introductio*, published by Gottfried Müller with Balthasar Gruber, or later Andreas Duncker as the printer.

This map of Africa also appeared in a 1652 Brunswick edition of Cluver's *Introductio*. It is possible that it may have been used for further editions of Cluver's *Introductio*, though these have not been located.

There were numerous other editions of Cluver's *Introductio* that were published using different maps of Africa. Shirley (2004: 345) states that over 45 editions of the *Introductio* appeared over a 100 year period. Not all editions contained maps. Those editions that did contain a map of Africa used one of the following: Cluver-Conrad Buno from 1641; Cluver Elsevier from 1659; Cluver-Johann Buno from 1661; Cluver-Buno-Mosting with the Herman Mosting imprint from 1686; Johannes Janssonius in 1661, using his 1628 Goos map; Janssonius van Waesberge in 1676 using the 1630 Cloppenburch map; and Cluver-Wolters from 1697.

The following is a list of the different-appearing maps of Africa that were used for the various editions of Cluver's *Introductio* and the distinguishing characteristics of these maps:

Map # 71	1641, Cluver-Conrad Buno	15 x 16.5 cm	At base of title cartouche, 'C. Buno fecit.'.
Map # 93	1659, Cluver-Elsevier	12 x 12.5 cm	No name on base of title cartouche.
Map # 98	1661, Cluver-Johann Buno	21.5 x 26 cm	Title within a simple drape cartouche at top right.
Map # 64	1661, Janssonius	14.5 x 20 cm	Uses 1628 Janssonius map with imprint 'AGoos Sculpsit' at the bottom right.
Map # 65	1676, Janssonius van Waesberge	18.5 x 25 cm	Uses 1630 Cloppenburch map with 'Auct. J. Hondio' within title cartouche.
Map # 140	1686, Cluver-Buno-Mosting	21 x 25.5 cm	Addition of imprint 'Herman Mosting Sculp: Luneburg' on line at the bottom left.
Map # 166	1697, Cluver-Wolters	21.5 x 25.5 cm	At the base of the cartouche is a lion.

References
Peter Meurer 2006: notes to author; Shirley 2004: 345-350; Burden 1996: map # 335 (Americas map); Shirley 1993: 379-380; Sabin 1868: entry # 13805.

72 1643 Jean Boisseau, Paris

Nouuelle description | DAFRIQVE.
- Scale bar at the bottom to left of title cartouche: Lieues d'Allemag. 200 = 12.3 cm.
- [Paris : Jean Boisseau, 1643].
- From: separately published / *Trésor des cartes géographiques.*
- 1 map : copperplate engraving ; 14 x 19.5 cm.
- Latitude coordinates and longitude coordinates on all four sides of the map.
- North at the top; no compass rose.
- Title cartouche at bottom left of map.
- No decorative elements on the map.
- The map has a '32' engraved on lower right corner within outer grid, showing the map's placement within the atlas.

Examples: *Washington LC,* G1015.B593 1643 (1643 edition of *Trésor,* with *La Clef De La Geographie Generale,* dated 1645) *and* G1015.B593 1653 (1653 edition of *Trésor,* with *La Clef De La Geographie Generale,* dated 1654); *London, Burden* Collection (in atlas dated 1654).

Description

Following the success of the Dutch small-format editions of Ortelius' *Epitome,* the Claesz.-Langenes *Caert-Thresoor,* and the Mercator-Hondius *Atlantes Minores,* Jean Boisseau published the first French small-format atlas, the *Trésor des cartes géographiques.* Burden (1996: map # 262) states that, although there were earlier French atlases containing maps of the continents, this is the first true world atlas produced in France containing a comprehensive collection of maps. This map of Africa was included in Boisseau's atlas.

The geography of the map is modeled after the Africa map in Janssonius' *Atlantes Minores* of 1628. The configuration of the interior river systems closely follows Blaeu's wall map of 1608. For example, as on the wall map, the Cuama River joins upstream with the Spirito Santo and the Zambere rivers flowing from Sachaf Lachus. Also, the outline of Africa seems to follow Blaeu's view for the shape of Africa.

Publication Information

Jean Boisseau was very active as a publisher in the French map market in the 1630s and 1640s. In 1643 Boisseau produced this rare atlas of thirty-eight maps. This work was prepared for the childhood education of the future king of France, Louis XIV.

The atlas, *Trésor des cartes géographiques*, is sometimes found bound with another work published by Boisseau, *La clef de la geographie generale*, which has a date of 1645. The map of Africa also appeared unchanged in a second edition of *Trésor des cartes géographiques* by Louis Boissevin, dated 1653 and often bound with an edition of *La Clef* dated 1654. The maps were also found in two examples of Abbe Philippe's *La Geographie Royalle* in 1646 (Burden 1996: map # 262). The map is known in one state.

There were five similar maps produced around the same period of time, all using different copperplates, derived from the Janssonius 1628 map; the last four of which were produced in Paris. These maps are this Boisseau map of 1643, Berey of 1651, Picart (1) of c.1651, and Picart (2) of c.1659. To differentiate these maps, refer to Map # 64, the Janssonius map of 1628.

References

Burden 1996: map # 262 (Americas map); Mickwitz et al. 1979-1995: Vol. 1: 74; Phillips 1909-1992: entries # 462 & 3424.

73   1644 Nicolas Picart, Paris

73.1 First State 1644

AFRICA | nova Tabula | Auct Jud Hondio.
N. Picart Excudit | Veu et Corigé des | Meillieurs esprit | du Temps. 1644.
- No scale bar.
- [Paris] : Nicolas Picart, 1644.
- From: separately published.
- 1 map : copperplate engraving ; 40.5 x 54 cm (35 x 45 cm, map only).
- Latitude coordinates along left and right sides above Equator. Longitude coordinates along Equatorial Line.
- North at the top; no compass rose.
- Title cartouche at top right of the map.
- A shield at the bottom left surrounded by two putti has the imprint of the publisher's information and the date. Above this shield, there is a clock face, directly taken from Van den Keere's 1614 map, with a skull in the center and a bell above the clock.

- There are two sailing ships, one sea monster and six flying fish in the Atlantic Ocean. There are one sailing ship and one sea monster in the Indian Ocean.
- As described below, there are five pairs of African figures on each side and six town views at the top of the map.

Examples: *Paris BN*, Ge B 1694 (in composite atlas containing both the first and second states); *Private French Collection*.

73.2 Second State 1659

Picart's imprint in the cartouche at bottom left has been erased and 'Fol. 611' has been added in upper left corner.

Examples of the second state can be found on thin paper and show traces of being folded into a small book. The second state also appears in Pierre d'Avity's *Les estats, empires, royaumes...due monde*, Paris, 1659.

Examples: *Paris BN*, Ge B 1694 (in a composite atlas containing both the first and second states); *Private German Collection*; *Private French Collection*.

Map 73.1

Map 73.2

Description

This rare, separately issued map by Nicolas Picart is likely his first known cartographic work. It is not known to have appeared in any regular atlas. This map was produced in 1644 along with maps of the other three continents.

Not much is known about the maker of this map, Nicolas Picart. He is known as an engraver and mapmaker, working from 'au mont St. Hillaire pres le puits Certain' in Paris (Tooley 1999-2004, K-P: 428). He is also known to have published editions of *Trésor des cartes géographiques* from 1651 to 1662, with two different small maps of Africa (Picart (1) and Picart (2)). The publisher's imprint ('N. Picart Excudit | Veu et Corigé des | Meillieurs Esprit | du Temps. 1644.') is located at the bottom left of this map within an elaborate decorative cartouche surrounded by two cherubs.

This map is an exact copy of Jodocus Hondius Jr.'s 1623 map of Africa, in its fourth state of 1632, when the bottom border was cut off. There are some slight name variations between the Hondius map and this map. For example, the bottom set of figures of the right border are captioned here 'G.R. SPEI HABITATE' rather than 'C. BONAE SPEI HABITAT'. The geography is the same in both maps.

Like the rest of the elements of this map, the cartouche at the bottom left is taken from the Hondius map of 1623 which in turn was very similar to the cartouche from the Van den Keere map of Africa of 1614.

The map is distinguished by its decorative borders on the sides and at the top. The town views at the top border are identical to those on Jodocus Hondius Jr.'s 1623 map of Africa. The town views across the top are from the left: *Alcair* (Cairo), *Alexandria, Alger, Tunis, Tanger*, and *Ceuta*.

The left and right borders show various costumed African figures. Along the left side from the top are: MAROCCHI (Moroccans), SENAGENSES (Senegalese), MERCATORES in GUINEA (traders in Guinea), CAB:LOPO GONSALVI ACCOLæ, and, MILES CONGENSES (Congolese); and along the right side from the top are: AEGYPTII (Egyptians), ABISSINI (Ethiopians), CAFRES IN MOSAMBIQUE (Mozambicans), REX IN MADAGASCAR, and G.R.SPEI HABITATE (inhabitants of the Cape of Good Hope).

Publication Information
Besides publishing this map, Picart also engraved it. Picart does not appear to have been an accomplished engraver, based on the rough engraving style of this map in comparison to the much finer engraving of the Hondius map. In addition to publishing maps of the four continents modeled after Hondius, Picart engraved maps for Jean Boisseau's *Tresor des cartes geographiques des principaux estats de l'univers* (Paris, 1634). A second edition of this work was published under his own name around 1659 (Schilder 2000: Vol. VI: 426. Pastoureau 1984: 499).

The map is known in two states. In its first state, the map is not known to have occurred in any standard atlas of the period.

Tooley (1972: 5) attributes this Picart map to Hondius and dates the map as 1640. Evidently, he did not realize that this was the second state of the Picart map, with the publisher's imprint removed.

References
Schilder 2000: 424-425; Tooley 1999-2004, K-P: 428; Burden 1996: map # 265 (Americas map); Loeb-Larocque 1989: 27; Tooley 1972: 5.
[N.B.: As recorded by Burden (1996: map # 265) and Schilder (2000: 426), there are two states of the Americas map at the Bibliothèque Nationale in Paris. These are State 1 dated 1644 with the imprint of *N. Picart* above the cartouche and State 2 dated c.1665 with the imprint of *Jollain excudit* engraved above the cartouche).

74   1645 Pierre Duval, Paris

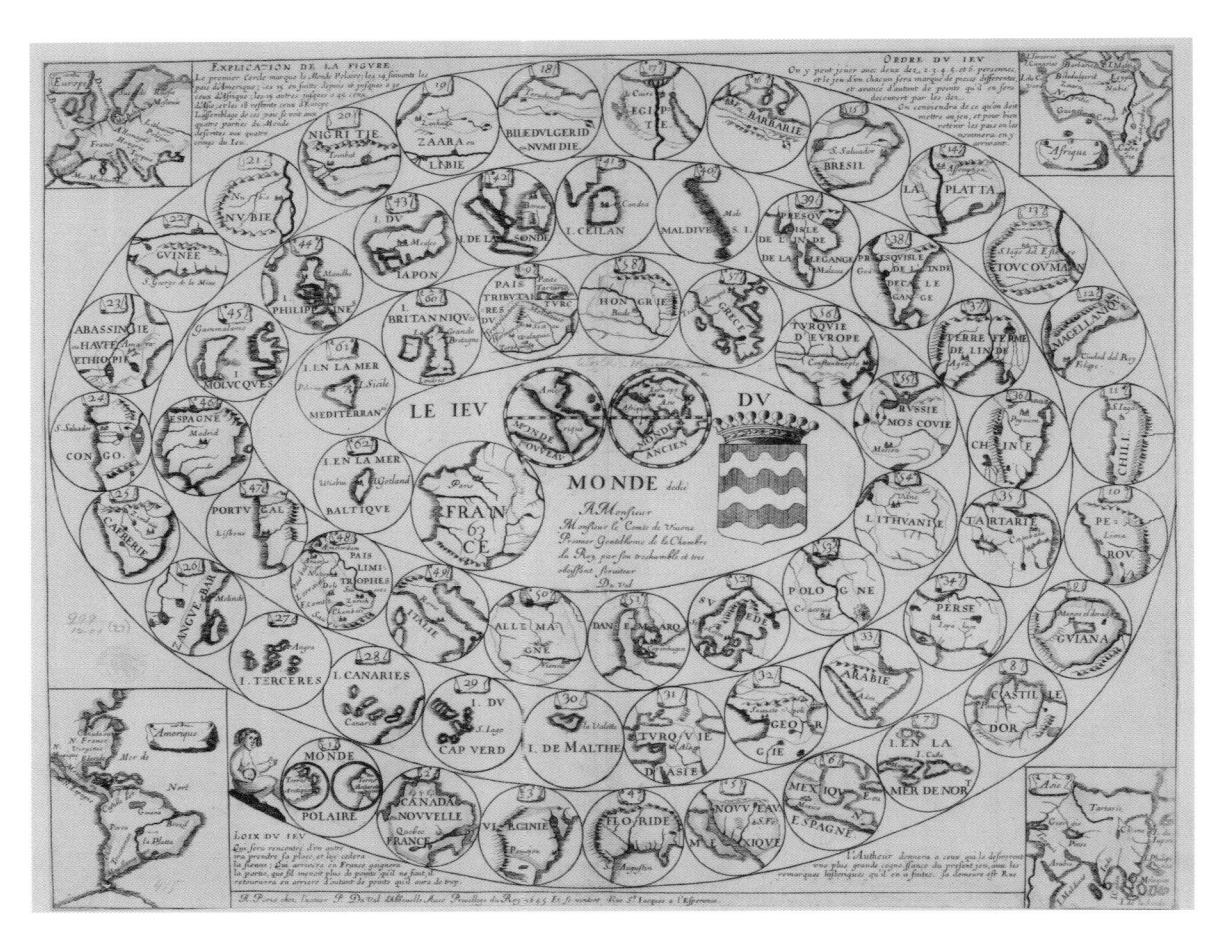

Afrique.
[at the center of the game]:
LE IEV | DV | MONDE dedié | A Monsieur | Monsieur le Comte de Viuone | Premier Gentilhome de la Chambre | du Roy, par son tres humble et tres | obeissant serviteur | Du Val.
[along the bottom of the map sheet]:
A Paris chez l'auteur P. Du Val d'Abbeville Avec Privillege du Roy 1645 Et se vendent Rue S.t Iacques a l'Esperance.
- No scale bar.
- Paris : Pierre Duval, 1645.
- From: separately published.

- 1 map : copperplate engraving ; 5.9 x 7 cm (for Africa map), 40.5 x 51.5 cm (for total sheet).
- The map of Africa is placed at the top right corner of the map board game. The four continents, clockwise from upper left corner: *Europe*, 6.8 x 6.0 cm; *Afrique*, 5.9 x 7 cm; *Asie*, 6.7 x 6.8 cm; and *Amerique* 10.5 x 8.2 cm.
- No latitude or longitude coordinates.
- North at the top; no compass rose.
- The title is at the bottom left of the map on a simple drapery cartouche.

Examples: *London BL*, *Maps 999. (27.); *Princeton UL*, with Cotsen Collection of educational games.

Description

This map of Africa is on a map board game, *Le Jeu du Monde*. At the center of the sheet for the game, there is the main title for the game and a dedication to Comte de Viuone as follows: LE IEV | DV | MONDE dedié | A Monsieur | Monsieur le Comte de Viuone | Premier Gentilhome de la Chambre | du Roy, par son tres humble et tres | obeissant serviteur | Du Val.

This map game may be one of if not the earliest work attributed to Pierre Duval (1618-1683). Duval was about 27 years old when he prepared this game. As Duval was the nephew and pupil of Nicolas Sanson, he likely had an early access to the necessary information to make this map.

Duval was born in Abbeville in France and moved to Paris along with many other academics and intellectuals at the urging of the French King. Here he eventually became Geographe Ordinaire du Roy. He published a prodigious number of maps with various states which appeared in many different publications and over a number of years. He died in Paris and was succeeded by his wife, Marie Desmarests, and his two daughters, Marie-Angélique and Michèle. One of the daughters inherited the business at their father's last business address, at Rue St. Jacques au Dauphin, though which daughter is not certain.

This circular, map board game has 63 moves, each being a small map of a different country or region of the world. Each map within the circle is 4.7 cm in diameter, beginning with number 1, Monde Polaire, and ending with France, number 63. The game is spiral in shape and is played using dice, with the winner being the first person to reach the end, represented by France at the center of the game. Besides being made for entertainment, the game was intended to provide the users with geographical knowledge on regions of the world.

The map of Africa, which is placed on the game at the upper right corner, is not part of the game, but is intended to provide decoration and geographic information along with maps of the other three continents at the other corners. The map of Africa is depicted in basic outline with some common geographic features of its time: the Nile, Congo, and Niger Rivers, and the Nile source lakes. Also identified on the map are the commonly used geographical divisions for Africa during this period.

Game circles 16 through 29 show regional maps of Africa. In order these are *Barbarie* (as circle 16), *Egipte, Biledvlgerid ou Nvmidie, Zaara ou Libie, Nigritie, Nvbie, Gvinee, Abassinie ou Havfe Ethiopie, Congo, Cafrerie, Zangvebar, I. Terceres, I. Canaries,* and *I. Dv Cap Verde* (as circle 29).

Publication Information

This separately issued map is known in one state. It was not part of any standard atlas. The game is of such extreme rarity that only two examples are known.

According to Adrian Seville (2005: 24-27), Duval invented the first educational map games. Duval is known to have produced three board map games in a similar format. After this game of the world, Duval published a similar map game of France, *Jeu de France*, in 1659. In 1662, Duval produced *Le Jeu des Princes de L'Europe* with small maps of regions or countries of Europe, again with France as the winning point. This game of Europe was published by Nicolas Berey.

References

Correspondence with Jason Hubbard, Jill Shefrin, and Adrian Seville.
Seville 2005: 24-27; Willoughby 2003; Shefrin 1999a and 1999b; Tooley 1999-2004, A–D: 406; D'Allemagne 1950: 218.

75 1646 William Humble, London

AFRICA | Petrus Kærius Cælavit.
- No scale bar.
- [London : William Humble, 1646].
- From: *A Prospect of the Most Famous Parts of the World...* .
- 1 map : copperplate engraving ; 8.5 x 12.5 cm.
- No latitude coordinates. Longitude coordinates on Equatorial Line.
- North at the top; no compass rose.
- Title cartouche at the bottom left of the map.
- No decorative elements on the map.

Examples: *Various locations.*

Description

This is a reduced-size version of John Speed's folio map of Africa. The publisher of this map was William Humble, the son of George Humble, the publisher for the 1627 folio edition of Speed's *Prospect* who had died in 1640.

The engraver for this series of maps was Pieter van den Keere or, in the Latinized form of his name, Petrus Kaerius. These maps are among the last to be engraved by Van den Keere, who at this time was about seventy-five years old. Possibly due to his age, this map of Africa is not up to the extremely high standards exhibited by Van den Keere's previous works. In this map, the placenames are often roughly engraved and, on occasion, some of the placenames are at odd angles to the rest of the names.

Though it is generally assumed that the geography for the map is based on Speed's 1626 folio-size map of Africa, there are notable variations which can not be attributed to its reduced size. This map by Humble shows both the unnamed Cuama and Spirito Santo rising in an unnamed lake, Sachaf Lake, in southern Africa. The 1626 Speed, following Blaeu's 1617 folio map, shows the Cuama rising in the southern side of the Mountains of the Moon. This Humble map also omits the small source lake directly to the west of Zaire/Zembre Lake. The Congo region lacks the details so apparent on the Speed and other Blaeu-derived maps of Africa. The use of certain placenames with the omission of other, more important, placenames (for example Mozambique and Sofala) is also puzzling.

Publication Information

Only the Americas map and the World map in the atlas bear the date of 1646, though there is evidence of an artfully done correction of the second '6' from a '2'. It is likely that the engraving of the maps for the atlas was begun as early as 1642.

The atlas that contains this map of Africa is frequently found bound with Speed's *England, Wales, Scotland and Ireland described*, often with the 1646 title page. In March 1659, the copperplate for the Africa map, along with the other copperplates, was sold to William Garrett, who it seems sold them directly to the publisher, Roger Rea. The Reas, father and son, issued an edition of Speed's *England, Wales...* in 1662; this atlas was commonly bound with the *Prospect* (still dated 1646) containing the map of Africa.

There are two distinct issues of the Great Britain atlas with a date of 1666, one just before the Great Fire of London and one just after the Fire, according to Burden (1996: map # 268). However, the *Prospect* is only known bearing the date of 1665. These editions are scarce as most examples were lost in the Fire. One further edition of the *Prospect* was published by Rea in 1668.

Sometime after 1668, all of the Speed plates were acquired by Thomas Bassett and Richard Chiswell. Bassett and Chiswell reissued this map, along with Speed's other world maps and the British maps, under a combined title of *An Epitome of Mr. John Speed's Theatre of the Empire of Great Britain and of His Prospect...* dated 1676. The *Prospect..* section of the atlas bears a date of 1675. There were no further editions issued after 1676. The map is only known in one state.

The map of Africa in the various editions of *A Prospect of the Most Famous Parts of the World...* are distinguished by changes on the verso of the map as follows:

1646: last line of text 'Parents.'
1665 ?
1668 last line of text 'of our first Parents.'
1675/6 last line of text 'rents.'

References

Norwich 1997: map # 31; Burden 1996: map # 268 (Americas map); McLaughlin 1995: entry # 8.

CARTOBIBLIOGRAPHY

76   1646 Stefano Mozzi Scolari, Venice

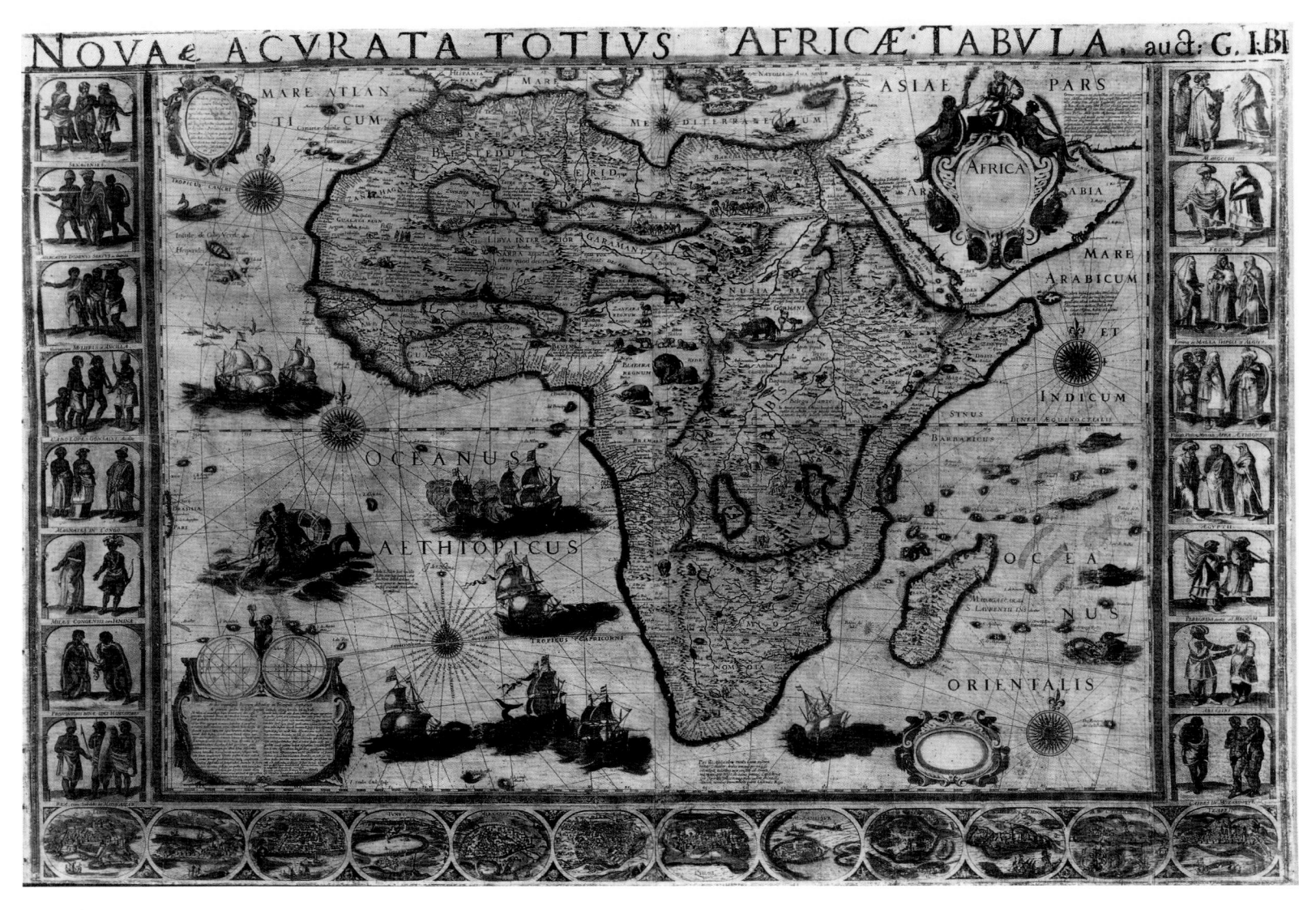

Map 76.2.C.

76.1 First State before 1646

[Describes second state of this map; first state not located]:

AFRICA.
[at bottom left corner]:
I. vanden Ende sculp.
- No scale bar.
- Venice : Stefano Scolari, 1646.
- From: separately published.
- 1 wall map : copperplate engraving ; 4 map sheets ; 2 sheets with costumed figures and town views) ; 85 x 110 cm (map only); 120 x 163 cm overall size (including woodcut title strips, decorative borders and text panels).
- Latitude and longitude coordinates on all four sides of the map.
- North at the top; there are five compass roses on the map: two in the Atlantic Ocean, two in the Indian Ocean, and one in the Mediterranean Sea ; one windrose in the South Atlantic.
- Title at upper right of the map is set within a decorative oval cartouche with three figures at the top. Second title in woodcut across the top of the map.

- In the lower left corner, there are two circular diagrams at the top of a cartouche with text; in the upper left corner, there is a cartouche with text.
- In Indian Ocean, one ship and five sea monsters ; In Atlantic, eight ships, three sea monsters, vignette of Neptune and mermaid. In Mediterranean, one smaller ship and one sea monster.
- Numerous animals and figures of Africans are found on the map of Africa.
- Eight groups of costumed figures on each side of the map. Twelve town views across the bottom of the map.
- Various other decorative elements on the map as on the Blaeu 1608 wall map.

Example: Assumed to have existed (Schilder 1996: 190). No known example.

76.2 Second State 1646
As described above.

The following are the known examples of this map, with variations in the title strips above the map, uses of decorative borders, and text panels:

76.2.A. *Perugia BU.* Title strips above map of: NOVA AFRICÆ GEOGRAPHICA ET HYDROGRAPHICA DESCRIPTIO, auct: G: Ianss.

76.2.B. *Washington LC.* Without the text border and with the map's author at top border as G.I., that is Guilielmus Ianssonius. Title strips above map of: NOVA ET ACVRATA TOTIVS AFRICÆ TABVLA [auct:] G.I.

76.2.C. *Greenwich NMM.* With no text borders and with a hastily executed title strip. Title strips above map of: NOVA & ACVRATA TOTIVS AFRICÆ · TABVLA , auct: G.I. Bl [eau].

76.2.D. *Greenwich NMM.* With text borders in Latin interspersed with illustrations in woodcut. Title strips above map of: NOVA AFRICÆ GEOGRAPHICA ET HYDROGRAPHICA DESCRIPTIO, auct Blaeu.

76.2.E. *Austin HR.* With no text borders. Title strips above map of: NOVA ET ACVRATA TOTIVS AFRICÆ TABVLA, auct: Blaeu.

Description
The Willem Blaeu wall map of Africa of 1608 in its later states had a significant impact on mapmakers of the seventeenth century. From 1608 up to the early 1700s, the Blaeu map influenced how all mapmakers and the general public viewed Africa. Even though, by the time of the Italian and French imitators, the geography of the Blaeu wall map was partially out-of-date, the Willem Blaeu name and the decorative elements of the wall map continued to attract attention and be of importance.
There were a number of Italian derivatives of the Willem Blaeu wall map of 1608. One imitator was Stefano Scolari (1598-1650), who produced this wall map of Africa in 1646 or shortly before this date. Scolari was known in Venice as a publisher of prints and maps, working at 'all'insegna delle Tre Virtu a S. Zulian'. Few examples of his maps remain.

In this wall map of Africa, the text columns are numbered on the left 1-3, on the bottom 4-13, and on the right 14-16. The text ends with 'Venetiis' and the year 1646. The letterpress text begins at the upper left with NOVA | AFRICAE | DESCRIPTIO and ends at lower right with: Venetiis MDCXLVI.

The cartouche in the lower right, which on Blaeu's original map contained the text of the privilege, is blank in this derivative. The copier also included the name of the original engraver, 'I. vanden Ende sculp'. Thus, it seems that the copier used Blaeu's second state of 1612 as his model. This second state contains the imprint of the engraver, Joshua van den Ende, whereas the first state of the map of 1608 does not contain Van den Ende's name.

This is a close, well-executed copy of the Blaeu map which carefully duplicates Blaeu's geography, placenames and decorative borders. Only by a close comparison with Blaeu's original can differences be noted. There are small differences in the long legends in the cartouches; the Scolari examples did not include the same number of words in the space provided or the same precision of regular spacing between words.

For a detailed discussion of the geography depicted on this map, please refer to the entry for Willem Blaeu's 1608 wall map of Africa (Map # 54).

Publication Information
Günter Schilder surmises that there was a first state of this map of Africa along with the other continent wall maps, but this first state is not known to still exist (Schilder 1996: 190). The wall maps of Asia and the Americas show re-engraving and traces of engravings, particularly the new information provided by Le Maire's voyage of a new sea link at the southern end of South America. The author did not notice erasures or changes in the maps of Africa that he examined.

References
Schilder 1996: 190-195.

77 1647 Cornelis Danckerts, Paris

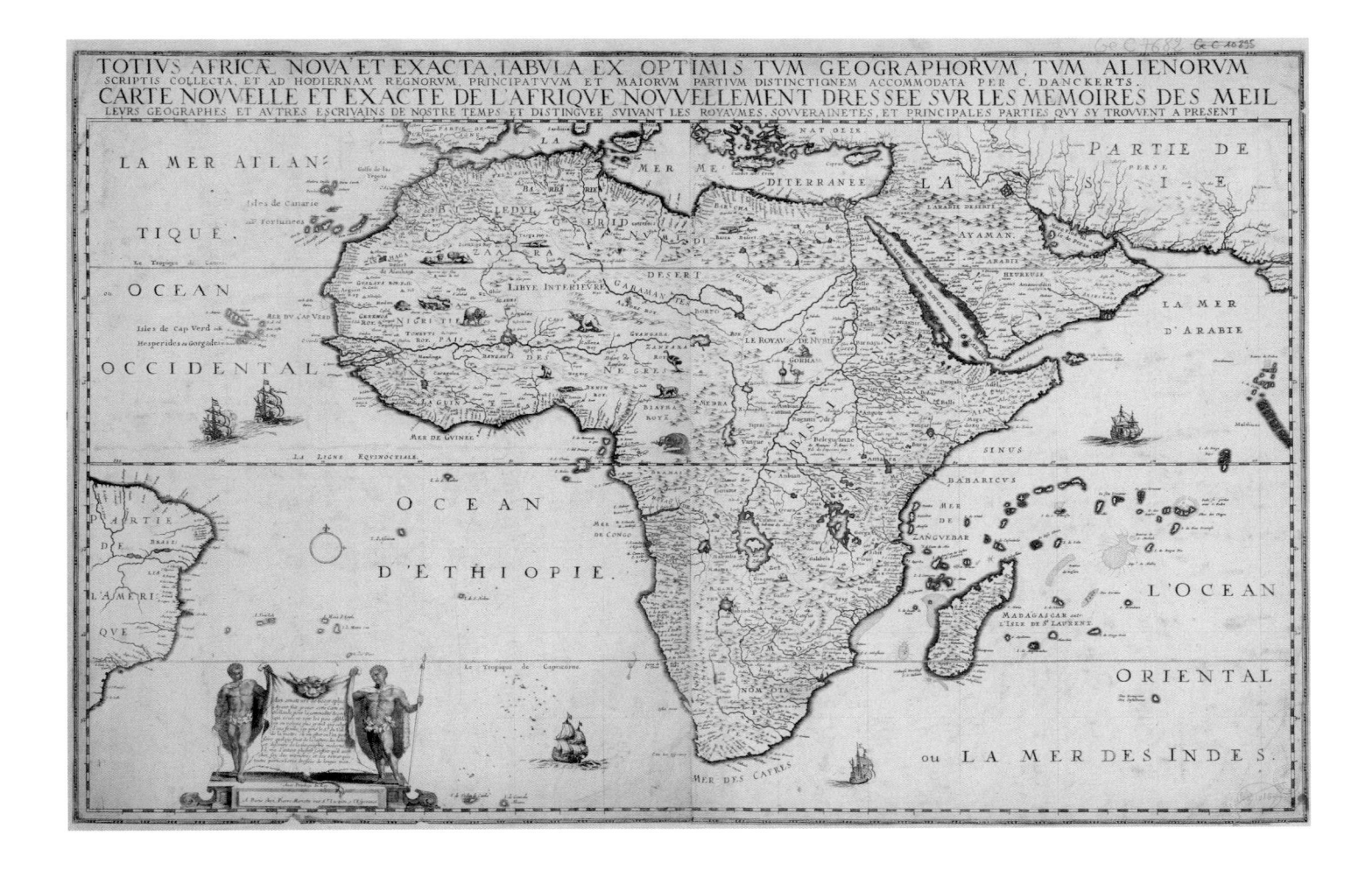

TOTIVS AFRICÆ NOVA ET EXACTA, TABVLA EX OPTIMIS TVM GEOGRAPHORVM, TVM ALIENORVM | SCRIPTIS COLLECTA, ET AD HODIERNAM REGNORVM, PRINCIPATVVM ET MAIORVM PARTIVM DISTINCTIONEM ACCOMMODATA PER C. DANCKERTS.
| CARTE NOVVELLE ET EXACTE DE L'AFRIQVE NOVVELLEMENT DRESSEE SVR LES MEMOIRES DES MEIL | LEVRS GEOGRAPHES ET AVTRES ESCRIVAINS DE NOSTRE TEMPS ET DISTINGVEE SVIVANT LES ROYAVMES, SOVVERAINETES, ET PRINCIPALES PARTIES QVY SY TROVVENT A PRESENT
[at the base of a decorative animal-skin screen]:
Avec Privilege du Roy | A Paris chez Pierre Mariette rue St. Iacques a l'Esperance.
- No scale bar.
- Paris : Pierre Mariette, [1647].
- From: separately published.
- 1 wall map : copperplate engraving ; 2 map sheets ; 61 x 92 cm in total.
- Latitude coordinates on all sides. Longitude coordinates along the Equatorial Line.
- North at the top; one compass rose within the South Atlantic.
- Title above the map with the first two lines of the title in Latin directly followed by the next two lines of the title in French.
- At the bottom left of the map, there is a text box in the design of an African animal skin stretched between two Africans standing on a platform. The text box begins 'Aux amateurs de Geographie' and contains twelve lines of text. At the top of the platform is 'Avec Privilege du Roy' and at the bottom, 'A Paris chez Pierre Mariette rue St. Iacques a l'Esperance'.
- There are two sailing ships within the Indian Ocean and three ships within the Atlantic Ocean.
- The animals within Africa appear to be taken directly from the Blaeu 1608 wall map.

Examples: *Paris BN*, Ge C 7682 (Cornelis Danckerts 1647) *and* Ge D 11393 [1-2] (Cornelis Danckerts c.1670).

Description

This fine map in two sheets was compiled and engraved by Cornelis Danckerts I (1603-1656). It was produced especially for the French market to meet a need for a French wall map of Africa. It would still be another twenty-two years before Alexis Hubert Jaillot would publish his own copy of the Blaeu wall map of Africa in Paris. The publisher for this map was Pierre Mariette, who was the most active publisher of cartographic material in Paris at this time. Danckerts had collaborated on many occasions in the past with the Paris map publishers, especially Melchior Tavernier, from whom Mariette purchased his business.

The Danckerts were a prominent print and map publishing family active in Amsterdam for almost 100 years. The founding member of the family business was Cornelis I (1603-1656) who started producing and publishing maps in the second part of the 1620s. His sons, Dancker (1634-1666) and Justus (1635-1701) were also active in the business. Justus issued a number of atlases with his sons. Justus' sons, Theodorus I (1663-c. 1727), Cornelis II (1664-1717), Johannes (?-1712), Eduard (?-after 1721), and his grandson, Theodorus II (1701-1727), continued the family business into the early eighteenth century. Because of the use of the same first names by various family members, there is some confusion in the literature over authorship of some of the maps produced by the Danckerts.

This map generally follows the 1608 Willem Blaeu wall map as its general model for the geography. The political divisions reflect other maps of the period as well. *Abassinie* (Abyssinia) continues to extend far south of the Equator. The river systems also follow Blaeu's 1608 wall map. However, there are some differences. For example, Danckerts removes 'Cas:[le] Portogal' and the other references to the Portuguese in Southern Africa that were so evident on the Blaeu map.

The Americas map has text panels surrounding the map, prepared by Pierre Duval. The examples of this map of Africa at the Bibliothèque Nationale de France do not have surrounding text panels. However, there is a reference to Duval's role in the preparation of this map in the text box at the bottom left of the Africa map. Roughly translated: 'This map was engraved in Holland for the use of persons who want to see the countries put together in a volume. I have requested Sr Du Val to add text where you can use the readings of the panels/tables [missing in these examples] as a treatise on modern geography... '.

Publication Information

It appears that this map of Africa was part of a set of the four continents prepared by Danckerts and published by Mariette. The Americas and Asia maps are reported in the literature (see Sotheby's auction of 7 December, 1993), though in only one known example of the Americas map. The date of 1647 is assigned based on the geography of the Americas and Africa maps and taking into consideration Danckerts' map-making history. This map provides an interesting link between the then dominant Dutch cartography and the emerging French cartography, which came to challenge Dutch map-making supremacy with its own innovative and ultimately more accurate view of Africa.

There are two known examples of this map, both at the Bibliothèque Nationale de France (BNF) in Paris. One example is dated by the Bibliothèque Nationale as 1647 and the other example is dated c.1670.
The author did not observe differences in the two Cornelis Danckerts wall maps at the BNF, though, if the date of c.1670 is correct, it is likely possible that one of Cornelis' sons re-issued this wall map, since Cornelis died in 1656.

A close derivative of this map was a two-sheet wall map published in Rome by Giovanni Giacomo de Rossi in 1670 with its title beginning TOTIVS AFRICAE NOVA ET EXACTA TABVLA... (See Map # 113). This map even uses Danckerts' decorative text cartouche, though without text, copying the African figures in reverse. Unlike the Danckert's map, the De Rossi derivative acknowledges Blaeu's wall map as the model for the geography.

References

Correspondence with György Danku, Curator, National Széchényi Library, Budapest, who has conducted extensive research on the Danckerts family, particularly Justus and his sons.
Burden (1996: map # 273) states that Sotheby's on 7 Dec. 1993 as lot 332 offered a Danckerts Americas wall map and an Asia wall map (Sotheby's 1993).

78   1650 Nicolas Sanson, Paris

AFRIQVE | Par N. Sanson d'Abbevile, Geog. du Roy | A Paris | Chez l'Auteur | Et chez Pierre Mariette, rue S. Iacques a l'Esperance | 1650 | Auec privilege du Roy pour vingt ans.
[imprint of engraver at bottom right of map]:
APeyrounin sculp.
- No scale bar.
- Paris : Pierre Mariette, 1650.
- From: separately published / *Les Cartes Generales de toutes les parties du Monde*.
- 1 map : copperplate engraving ; 39 x 56 cm.
- Latitude and longitude coordinates on all four sides of the map. Longitude coordinates across Equatorial Line. The Prime Meridian is to the west of *Isle de Fer* in the Canary Islands.
- North at the top; no compass rose.
- Title cartouche at the top right of the map.
- No decorative elements on the map, apart from title cartouche.

Examples: *Chicago NL*; *London BL*, Maps C.39.e.1 (in 1658 atlas entitled *Cartes generale de toutes les parties du monde*); *Washington LC*, 1659 (as dated by LC; as a separate map); *Author's collection*.

Description

This is a landmark map of Africa. It is the first French map to depict significantly new information on Africa as opposed to earlier French maps, which generally copied Dutch maps. Further, the publication of this map, along with Sanson's maps of the other continents and regions, signaled the beginning of a growing French competition with the Dutch mapmakers of the seventeenth century.

Nicolas Sanson (1600-1667) was born in Abbeville and moved to Paris in 1627, where he was made 'Geographe Ordinaire du Roi' in 1630. He later would become tutor to the future king, Louis XIV. Early in his career, Sanson worked with Melchior Tavernier and others. Sanson's future success was ensured by his partnership with the publisher Pierre Mariette, who had previously purchased the business of Tavernier. Mariette helped Sanson financially in the production of his maps for the planned *Les Cartes Generale* atlas, enabling Sanson to disseminate his separate maps from the Mariette address in Paris. In 1657 Mariette died and Sanson's atlas *Les Cartes Generales* was not published until 1658 in partnership with Mariette's son, also named Pierre. The *Les Cartes Generale* is the first folio-size French atlas. Sanson died in July 1667 and the atlas continued to be published by his son, Guillaume. In 1668, Guilaume prepared a new map of Africa that replaced this map by his father.

Austere in comparison to the Dutch folio maps of this period, Sanson's 1650 map of Africa is precise and scientific in its approach to presenting information. Sanson tends to be sparse in his information, omitting placenames, one may assume, because they were only based on hearsay. More information, however, is provided on those areas that were coming under increasing French control in Madagascar and in West Africa. The *R da Volta* (Volta River) is more precisely placed on the map and *Accara* (Accra) appears near its mouth on the Guinea coast in present-day Ghana.

For his depiction of much of the interior, Sanson relies on the still dominant Blaeu wall map of 1608. He does not appear to utilize the information from Blaeu's folio map of 1617; Sanson reverts to Blaeu's 1608 depiction of the *Cuama* and *R. de Spirito Santo* Rivers as both flowing from the unnamed Zambere which originates in the unnamed Sachaf lacus (Blaeu had changed the source of the Cuama to the southern side of the Mountains of the Moon in his 1617 folio map).

For his major divisions within Africa, Sanson tends to use even older conventions. He has the land of *Ethiopie* stretching far into Southern Africa, below the Nile source lakes. *Libye* dominates most of West Africa, and above that is *Afriqve*, an early reference to that part of Africa along the North African coast. He does provide geographic subdivisions that are more in line with the conventions of his day.

This map was engraved by Abraham Peyrounin (c.1620-c.1666), whose imprint of '*APeyrounin sculp.*' is at the bottom right of map.

Publication Information

This map of Africa is known in one state. There may be additional states of this map with geographical changes, based on changes to the Americas map, but the author has not located any even after an examination of numerous examples.

This map was prepared and issued separately in 1650 in anticipation of Sanson's atlas. It is known to first appear in an untitled collection of maps of 1652. Pastoureau cites seven separate editions of the *Les Carte Generales* in 1658, 1665, 1666, 1667, 1670, 1675, and 1676 (as *Les Carte Generale... ancienne et nouvelle*). The last three editions of the atlas were published by Nicolas' son, Guillaume, after Nicolas' death in 1667. What makes the listing and collation of Sanson's atlases particularly difficult is that often the contents do not agree with the printed list of maps, as on a regular basis extra maps were inserted into the atlases.

References

Burden 1996: map # 294 (Americas map); Pastoureau 1984: 345 (for Mariette) and 387-9 & 400-2 (for Sanson); Pastoureau 1980: 61-3; Tooley 1969: 98-100.

79   1650 Nicolas Sanson, Paris

AFRICA | VETVS | Autore N. Sanson Abbavilleio | Christianiss. Galliar. Regis Geographe | Parisiis | Apud Autorem | Et apud Petrum Mariette via Iacoboea | sub signo spei | 1650 | Cum Privilegio | Annorum Viginti.
[imprint of engraver at bottom right of map]:
APeyrounin sculp.
- No scale bar.
- Paris : Pierre Mariette, 1650.
- From: separately published / *Les Cartes Generales de toutes les parties du Monde.*
- 1 map : copperplate engraving ; 33 x 55.5 cm.
- Latitude coordinates and longitude coordinates on all four sides of the map. Longitude coordinates across Equatorial Line.
- North at the top; no compass rose.
- Title cartouche at the top right of the map.
- No decorative elements on the map, apart from title cartouche.

Examples: *Various locations.*

Description

This map of ancient Africa is one of a series of maps depicting parts of the ancient world. Sanson had a particular interest in subjects dealing with the past; his early training in Abbeville was in history.

The map generally shows Africa as it was prior to the European discoveries of the sixteenth century, but it does include a more modern shape for Africa. The geographical representation of the interior follows Ptolemy and other antiquarian sources. Placenames reflect knowledge of ancient traditional African groups as well as ancient geographical places. On the east coast, *Prassum Prom.*, the southern limit acknowledged by Ptolemy and others is placed far to the south on the Tropic of Capricorn.

This map was engraved by Abraham Peyrounin (c.1620-c.1666), whose imprint of '*APeyrounin sculp.*' is at the bottom right of the map.

Publication Information

This map was prepared and issued separately in 1650 in anticipation of Sanson's atlas. It appeared in the first edition of Nicolas Sanson's *Les Cartes Generales* in 1658 and in subsequent editions of the atlas until 1667, when it was replaced by a map prepared by Sanson's son, Guillaume.

Pastoureau cites seven separate editions of the Les Carte Generales in 1658, 1665, 1666, 1667, 1670, 1675, and 1676 (as *Les Carte Generale... ancienne et nouvelle*). It is not clear why the publication of the atlas was delayed from 1650, when the maps were prepared, until 1658. It may partly have been due to the death of Pierre Mariette in 1657, though this would not have been the sole reason for the delay. The 1658 *Les Carte Generales...* was co-published by Mariette's son, also named Pierre, in conjunction with Sanson.

The last three editions were published by Nicolas' son, Guillaume, after Nicolas' death in 1667. What makes the listing and collation of Sanson's atlases particularly difficult is that often the contents do not agree with the printed list of maps, as on a regular basis extra maps were inserted into the atlases.

References

Norwich 1997: map # 35, Tooley 1969: 99, Pastoureau 1980: 61-3, Pastoureau 1984: 345 (for Mariette) and 387-9 & 400-2 (for Sanson).

80   c. 1650 Pierre Mariette (attr.), Paris

AFRICÆ | Noua | TABVLA.
- No scale bar.
- [Paris : Pierre Mariette (attr.), c.1650].
- From: separately published.
- 1 map : copperplate engraving ; 36 x 46.5 cm.
- Latitude coordinates along left and right sides of the map. Longitude coordinates across Equatorial Line.
- North at the top; no compass rose.
- Simple drapery title cartouche at the top right of the map.
- There are two French sailing ships in the Atlantic with two flying fish above the upper ship. Numerous animals are within Africa.

Example: *Author's collection.*

Description

Little is known about this map. It has no imprint for either the cartographer or engraver and is thus difficult to identify. The map is not described in Tooley's *Collector's Guide to Maps of Africa* nor in Loeb-Larocque's more detailed article on the French derivatives of Dutch maps, but it is depicted in Tooley's 1972 *Sequence of Maps of Africa* as plate no. IX. Tooley does not provide an attribution for this map, but simply identifies it as being by an anonymous mapmaker.

Burden believes the similar Americas map is by Pierre Mariette, as he noted that the Americas map appeared on two occasions in *Theatre Geographique de France* which had editions published by Pierre Mariette in 1650 and 1653.

While this map appeared in Pierre Mariette's *Theatre Geographique de France*, it is possible that this map was actually made by a Dutch engraver, possibly even Cornelis Danckerts, who was known to work on occasion with Pierre Mariette during this period (See Danckert's Africa map, Map # 77). The engraving style and the decorative elements all suggest that it may be a Dutch-produced map. As this is purely conjecture, the map's attribution is still with Pierre Mariette.

This map is a derivative of Jodocus Hondius' 1619 map of Africa in its later states without the decorative borders. The geography is virtually identical and the animals within Africa are copied directly from the Hondius map, although they are shown in reverse on this map. A unique feature on this map is the square drapery cartouche and the absence of a reference to Hondius on the cartouche, compared to the other Hondius derivatives. The text on the map is in Latin. The verso of the map is blank.

Publication Information

This map was separately published, though Burden mentions the appearance of the similar Americas map in two editions of Mariette's *Theatre Geographique de France*. The only example of this map of Africa that the author has seen has evidence of having been bound into some type of atlas.

References

Burden 1996: map # 300 (Americas map); Loeb-Larocque 1993: 15-30; Tooley 1972: 7: plate IX.

81   1651 Nicolas Berey, Paris

Nouuelle description | DAFRIQVE.
- Scale bar at bottom left corner: 200 Lieues d'Alemag. = 2.2 cm.
- [Paris : Nicolas Berey, 1651].
- From: *Carte generalle de la geographie Royale... chez N. Berey proche les Augustins.*
- 1 map : copperplate engraving ; 13.5 x 19 cm.
- Latitude coordinates along left and right sides. Longitude coordinates across bottom of the map.
- North at the top; no compass rose.
- Simple box title cartouche at the bottom left, to right of scale bar.
- No decorative elements on the map.

Examples: 1651 edition: *Rennes BM*: 52 166; 1655 edition: *London BL*, Maps C.43.b.1.(1) (note: verso of 1655 edition is blank in the BL example); *Paris BN*: G 3151 1655; *London RGS*; *Chicago, Baskes Collection*.

Description

This map of Africa is from the publisher Nicolas Berey's *Carte generalle de la geographie Royale* of 1651. This is a small atlas, modeled on the Janssonius *Atlas Minor* of 1628, and contained maps of the world and continents. The atlas is often attributed to Christophe (Nicolas) Tassin, the author of the book, although he had sold all of his plates in 1644 to Nicolas Berey (c.1606-1665) and Antoine de Fer, Nicolas de Fer's father, prior to the publication of this edition.

This map of Africa by Berey is very similar to the Africa map in Jean Boisseau's *Tresor des cartes* of 1643. The Boisseau map in turn was taken from Janssonius' *Atlas Minor* of 1628. For information on the geography presented on this map, refer to the Janssonius 1628 map of Africa (Map # 64).

There is no imprint for the engraver on this map. The world map bears the imprint of H. Picart as the engraver, and as the engraving style is similar, it is likely that Huges Picart engraved this map of Africa as well.

Publication Information
The map first appeared in Nicolas Berey's *Carte generalle de la geographie Royale... chez N. Berey proche les Augustins* of 1651. This was a small volume with a total of 15 maps, including this one of Africa. Berey published one more edition of the *Carte generalle...* in 1655. Due to the popularity of this and Boisseau's work, another series of plates were produced by Nicolas Picart for his *Trésor des cartes geographiques* in c.1651 and then again in c.1659.

There were five similar maps produced around the same period of time, all using new copperplates, derived from the Janssonius 1628 map; the last four of which were produced in Paris. These maps are this Boisseau map of 1643, Berey of 1651, Picart (1) of c.1651, and Picart (2) of c.1659. To differentiate these maps, refer to Map # 64, the Janssonius map of 1628.

References
Burden 1996: map # 301 (Americas map); Pastoureau 1984: 438-9.

82   c. 1651 Nicolas Picart (1), Paris

82.1 First State c.1651

Nouuelle Description | D'AFRIQVE.
- Scale bar at the bottom left corner of the map: Lieues d'Allemag 200 = 2.3 cm.
- [Paris : Nicolas Picart, c. 1651].
- From: *Trésor des cartes géographiques.*
- 1 map : copperplate engraving ; 13.5 x 19.5 cm.
- Latitude and longitude coordinates along all four sides of the map.
- North at the top; no compass rose.
- Simple drapery title cartouche to right of the scale bar at the bottom left of the map.
- No decorative elements on the map.

Example: *London, Burden Collection* (undated atlas; The Burden copy is the only known example of the c. 1651 map of Africa; it is an unrecorded example. The title page in the Burden copy does not bear a date, though there are three maps in the atlas with the date of 1651. In the Burden example, the verso is blank; the Africa map is bound into the book on the left edge).

82.2 Second State 1657

Addition of number '7' within grid at bottom right corner of the map (the first state did not have a number).

Examples: *Paris BN*, Ge.FF.11314; *Chicago, Baskes Collection* (atlas dated 1657).

Description

Nicolas Picart is known as an engraver and mapmaker and produced a map of Africa with decorative panels in 1644. He is also known to have published editions of *Trésor des cartes géographiques* from c.1651 to 1662, with two different small maps of Africa. The Picart (1) map of Africa first appeared in an edition of c.1651 and the Picart (2) map first appeared in an edition of c.1659.

This map of Africa is derived from Jean Boisseau's *Trésor des cartes* of 1643 which in turn was modeled after Johannes Janssonius' *Atlas Minor* of 1628. For information on the geography presented on this map, refer to the Janssonius 1628 map of Africa (Map # 64).

It is a scarce map. Burden attributes the scarcity of maps from this atlas to the competition from the more popular *Carte Generalle* published by Nicolas Berey, also in 1651.

Publication Information

The map of Africa first appeared in an undated edition of Picart's *Trésor des cartes géographiques* of c.1651. Although the map is not dated, the date of c.1651 is derived from three maps in the undated atlas that bear the date of 1651. The Picart copperplates were later acquired by Antoine de Fer, who, with Pierre Duval as author, published the *Cartes de geographie* in 1657. There were later editions of the *Trésor des cartes géographiques* in which this map was used in 1659, 1661, and 1662.

A new series of copperplates were produced by Nicolas Picart in c.1659 under the same title (the Picart (2) map of Africa, Map # 92). These new plates were used later by Pierre Duval.

There were five similar maps produced around the same period of time, all using new copperplates, derived from the Janssonius 1628 map; the last four of which were produced in Paris. These maps are the Boisseau map of 1643, Berey of 1651, Picart (1) of c.1651, and Picart (2) of c.1659. To differentiate these maps, refer to Map # 64, the Janssonius map of 1628.

References

Burden 1996: maps # 306-307 (Americas and Arctic maps), and information provided to the author by Philip Burden; Phillips 1909-1992: Phillips entry # 3429.

83   1652 Henry Seile, London

Map 83.2.

83.1 First State 1652

AFRICÆ | Descriptio Nova | Impensis | Henrici Seile | 1652.
[bottom left of map]:
Will: Trevethen. Sculp.
- No scale bar.
- London : Henry Seile, 1652.
- From: *Cosmographie in foure Bookes Contayning the Chorographie & Historie of the whole World and all The Principall Kingdomes, Provinces, Seas, and Isles, Thereof.*
- 1 map : copperplate engraving ; 33.5 x 42 cm.
- Latitude coordinates along left and right sides. Longitude coordinates along the Equatorial Line.
- North at the top; no compass rose.
- Simple strapwork title cartouche at the top right of the map.
- In the Indian Ocean, there are one sailing ship and one sea monster. In the Atlantic Ocean, there are

four sailing ships, two sea monsters, and four flying fish. Animals within Africa, generally following those on the John Speed map of Africa.

Examples: 1652 edition: *Chicago NL*, C f F 09.4; *London RGS*, 133.j; 1657 edition: *London BL*, 210.i.10; 1660-62 Chetwind edition: *London BL*, L.35/82 (basically a re-issue of the 1657 edition).

83.2 Second State 1666
Seile's name replaced in the title cartouche with 'PHILIPPI CHETWIND', with the date changed to 1666.

Examples: 1666 edition: *London BL*, 505.R.R.1; 1670 edition: *London BL*, 10003.f.7; 1682 edition: *London RGS*, 133.j; *Author's collection* (separate map).

Description
This map appeared in the first edition of the *Cosmographie* published by Henry Seile (? – 1662) in 1652 along with maps of the world and the other three continents. The author of the *Cosmographie* was Peter Heylin, who sometimes has the attribution for this map. Peter Heylin or Heylyn (1599/1600-1662) was a noted seventeenth century English cosmographer and geographer who was also Chaplain to King Charles I of England. Heylin wrote *Microcosmus, or a Little Description of the World* in 1621. This book was expanded into his *Cosmographie*. The *Cosmographie* proved to be popular and was published in London from 1652 into the early eighteenth century. The last edition was published by Chiswell, Tooke, Hodgkin, and Bennet in 1703 with newly engraved maps.

The *Cosmographie* provided comprehensive information on all of the known world. The section on Africa is in the Fourth Book, Part I, with this map of the African continent. This map was engraved by William Trevethen who also engraved the other three continent maps.

This map is a close copy of John Speed's map of Africa of 1626. The placement of sailing ships, sea monsters, and even much of the lettering conforms to Speed's map. This is not especially surprising as Speed's map of Africa was the dominant and most readily available Africa map in the English language during this period. The geographical information for this map is also taken from John Speed's map. Seile's map does little to advance geographic understanding of Africa and was out of date by the time it appeared in Heylin's *Cosmographie*.

Publication Information
The map appeared in editions of the *Cosmographie* of 1652, 1657, 1660/2, 1666, 1670, and 1682. Sometime prior to Henry Seile's death in 1662, Philip Chetwind became involved in the book's publication. For the edition of 1666, Chetwind replaced Henry Seile's name on the map of Africa with his own name. This map, with the Chetwind name, also appeared in later editions of Heylin's *Cosmographie* in 1670 and 1682.

This set of continent maps is also found in examples of Lewes Robert's *The Merchants Map of Commerce*. Other English maps of the period, such as Robert Walton's map of Africa, are sometimes found bound into examples of Heylin's *Cosmographie*.

The Africa map is known in two states. Burden (1996: map # 308) refers to additional states of the Americas map dated 1653, 1674, and a final undated state based on correspondence. The author has not located any additional states of the Africa map beyond the ones listed here.

Following Henry Seile's death, his widow Anne published her own edition of the *Cosmographie* in 1663. For some reason, it appears she did not inherit her husband's copperplates. As a result, she was required to use a new set of plates, dated 1663 and engraved by Robert Vaughan (Map # 100).

A third and final set of maps, after the Henry Seile and Anna Seile maps, was prepared for a 1703 edition of Heylin's *Cosmographie* of 1703, published by Chiswell, Tooke, Hodgkin, and Bennet. The Africa map, titled 'AFRICA', is placed within an oval cartouche at the bottom left of the map.

References
Mayhew 2003; Norwich 1997: map # 36 (first state) and map # 39 (second state); Burden 1996: map # 308 (Americas map).

84 c. 1655 Jan Mathisz., Amsterdam

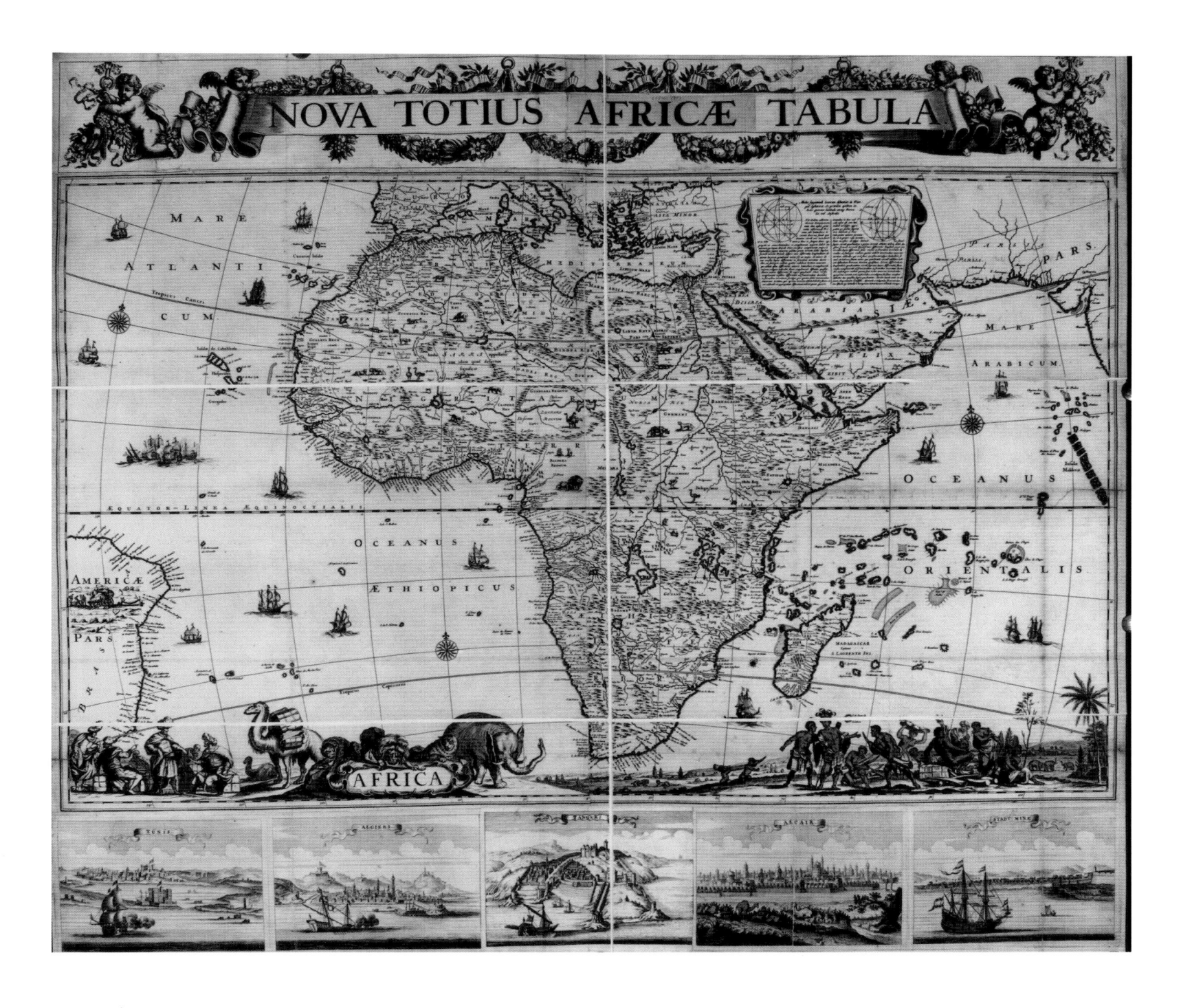

84.1 First State c.1655

AFRICA
(above the map):
NOVA TOTIUS AFRICÆ TABULA.
- No scale bar.
- [Amsterdam : Jan Mathisz., c.1655].
- From: separately published wall map.
- 1 wall map : copperplate engraving and etching ; 4 map sheets ; 59 x 95.5 cm (for the 4 map sheets); 82 x 95.5 cm (for map sheets, title strip, and town views [10.5 x 95 cm for title, 12.5 x 18.5 cm for each of five city views at bottom]) (N.B.: The only copy of this map available to examine at the British Library has been dissected and reassembled into 6 sheets).

- Latitude and longitude coordinates along all four sides of the map. Longitude coordinates across Equatorial Line.
- North at the top; one compass rose in the northern Indian Ocean and two compass roses in the Atlantic Ocean.
- Title cartouche at bottom left of the map within a scene of African animals. To the left of the cartouche are a group of Arab traders and along the right bottom side of the map is a scene of Africans engaged in various activities, including transporting and trading ivory tusks. A second title strip is above the map.
- Text box at upper right on determining geographical measurements.
- There are five sailing ships in the Indian Ocean, and 14 sailing ships in the Atlantic Ocean. Numerous animals are within Africa.
- There are five large views of important towns of Africa along the bottom of the map.

Example: *London BL*, *Maps 63510 (37).

84.2 Second State c.1696
With the addition of the imprint of Cornelis Danckerts ('Amsterdam by Cornelis Danckerts op den Nieuwen-dyk inden Atlas met Privilegie') at the bottom left of the third map sheet.

Example: Current location is not known. This state is known to exist based on the Sotheby's auction catalogue of 25 June 1992, and Christie's auction catalog of 27 April 1994.

84.3 Third State c.1705
With the addition of the imprint of De Wit ('F. de Wit Excudit Amstelodami Cum Priv.') at the bottom of the fourth map sheet.

Example: *Amsterdam HM*.

Description

This is a beautifully designed map that contains updated information on Africa. Little is known of this map or its maker, Jan Mathisz. Jan Mathisz. (1627-87), or Jan Mathes as he is sometimes identified, was mostly known as an Amsterdam copperplate printer and cutter (referred to as a 'Plaatsnyder', on the Americas map) as well as a leather merchant.

In c.1655, Mathisz. published a set of the four continents. Previously, the set had been given a date of c.1650, but a date of c.1655 is probably more accurate owing to the first appearance of the view of New Amsterdam on the companion Americas map (Burden 1996: map # 313).

The attribution to Mathisz. as the mapmaker for this map of Africa is based on the imprint on the companion map of the Americas, which reads: 'Gedruckt | T'AMSTERDAM. | BIJ | JAN MATHYSZ | PLAATSNYDER'.

The overall shape of Africa, many of the placenames, and the hydrography are based on Willem Blaeu's wall map of 1608. However, Mathisz. considerably updates his map. Of special note is the much greater detail along the Southern African coastline. The first Dutch settlement under Jan van Riebeeck at the Cape of Good Hope had occurred in 1652, and it is likely that Mathisz. wanted to show this information on his map. Mathisz. introduces a number of Dutch names interspersed with older Portuguese names. Among these names are *Tafel bay* (Table Bay), *Tafel berg* (Table Mountain), *Robben Eyl* (Robben Island), and *Schorre hoek* (inland from Cape Agulhas, the southern-most point in Africa).

The entire bottom of the map contains an especially decorative scene. From the left, these are Arab traders, and then various animals of Africa (two camels, two leopards and a lion on top of the cartouche, and an elephant). On the right side across the bottom of the map, there is a scene of Africans engaged in various activities, including transporting and trading ivory tusks. Various parts of this decorative scene would be re-used on later maps of Africa by Hugo Allard in c.1661 and others.

There are five large views of important towns of Africa along the bottom of the map beneath the decorative scene. These views are from the left: *Tunis, Algiers, Tangier, Alcair* (Cairo), and *Stadt Minæ* (the Fort at El Mina). These views have a much finer style than the views appearing on the various maps with carte-de-figures of this period by Blaeu, Hondius, and others. The source for these Mathisz. views is not known.

Publication Information

The map's history is not clear, largely due to its extreme rarity. Burden mentions three examples of the Americas map: at the British Library, the New York Public Library, and the Nederlands Scheepvaartmuseum in Amsterdam. The author is only aware of an example of the Africa map at the British Library. The New York Public Library does not have the map of Africa.

It is possible that the Mathisz.' plates were acquired by Visscher in c.1680, as Burden speculates, but this is impossible to prove as an example of this state is not known to exist. It is known that the Mathisz. plates re-appeared in c.1696 in the hands of Cornelis Danckerts II (1664-1717). In 1992, Sotheby's auctioned a set of wall maps of the continents bearing the imprint of Cornelis Danckerts, which had been found in a composite Danckerts atlas. The imprint on the Africa map bears the address of Danckert's wife's family, 'op den Nieuwen-dyk inden Atlas' from which he operated his business after his marriage in 1696. Whether the plates were acquired by Danckerts directly from Mathisz. or his heirs, or through another party is not known.

In about c.1705, Frederick de Wit used the plates to produce the map. The only known example of the third state is located at the Amsterdam Historisch Museum.

References

Burden 1996: map # 303 (Americas map); Sotheby's auction catalog of 25 June 1992; Christie's auction catalog of 27 April 1994; Koeman 1967-71, II: 88-89.

85   c. 1656 Pierre Duval, Paris

Map 85.3 Map sheet.

85.1 First State c.1656

L'AFRIQVE | Par le Sieur Du Val Geographe Ord. du Roy | A PARIS.
- No scale bar.
- Paris : Pierre Duval, [c.1656].
- From: separately published / French composite atlases.
- 1 map : copperplate engraving ; 40 x 49 cm for entire sheet with text above.
- No latitude and longitude coordinates.
- North at the top; no compass rose.
- Title above map.
- No decorative elements on the map.

Examples: *London BL; London, photo from Ashley Baynton-Williams.*

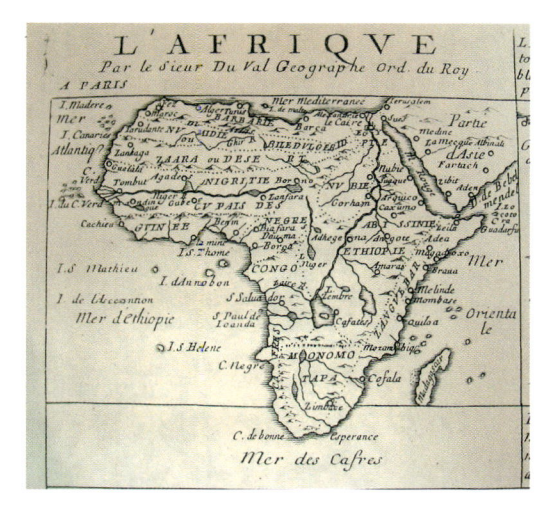

Map 85.1.

Map 85.2.

Map 85.3.

85.2 Second State 1661
Addition of date and publisher as follows: 'L'AFRIQVE | Par le Sieur Du Val Geographe Ord. du Roy of 1661. | A PARIS Ches I. Lagniet sur le quay de la Megisserie au fort leuseq. Auec priuil'. Longitude and latitude coordinates are added on all four sides of the map. An Equatorial Line and the Prime Meridian are also added to the Africa map.

Jacques Lagniet, whose imprint appears in the title in the second and third states, is known to have worked with both Duval and Antoine de Fer as a publisher.

Examples: *Paris BN; Bedburg-Hau, Antiquariat Gebr. Haas; USA, MacLean Collection.*

85.3 Third State after 1661
Evidence of erasure on plate to: 'Lagniet sur le quay de la Megisserie au fort leuseq.'

Example: *Amsterdam UB, 33-17-40; Bedburg-Hau, Antiquariat Gebr. Haas; USA, MacLean Collection.*

Description

This is an unusual map sheet by Pierre Duval depicting a map of the entire African continent and 14 regional maps of Africa. On the sheet, the map of Africa is placed at the top left side. The verso is blank.

This map of Africa is part of a set of four continent maps with regional maps produced by Duval, all of which are scarce. The map is undated. Based on the nomenclature, Burden (1996: map # 322) dates the map of the Americas c.1656 as the information on the map pre-dates Sanson's 1657 map showing Mexico. Likely the Africa map was published around the same date.

Considering the small size of the map, Duval was able to supply a surprisingly high degree of detail to it. Duval used his uncle Nicolas Sanson's 1650 map of Africa as his model, though in this map, Duval omits the lake in the Abyssinian highlands that was to eventually become Lake Tana, the source for the Blue Nile river.

Publication Information

There is no text on the verso of this map. While the map was issued separately, it is known to appear in French composite atlases. The author has only seen this map as a separate sheet, but Burden (1996: map # 322) mentions that a set of the four continents in its second state was seen in a Jollain composite atlas. The map of Africa is known in three states.

References

Burden 1996: map # 322 (Americas map); Pastoureau 1984: 136; Magna Gallery catalogue 1984: item 53.

86   1656 Nicolas Sanson, Paris

86.1 First State 1656

AFRIQVE. | Par le S^r. Sanson d'Abbeville, | Geographe du Roy. | Avec Privilege pour vingt ans. | A Paris chez l'Auteur. | 1656.
[at bottom right]:
APeyrounin sculp.
- No scale bar.
- Paris : Nicolas Sanson, 1656.
- From: *L'Affrique en plusieurs cartes nouvelles, et exactes... .*
- 1 map : copperplate engraving ; 20 x 27.5 cm.
- Latitude and longitude coordinates on all four sides of the map. Longitude coordinates along the Equatorial Line.
- North at the top; no compass rose.
- Title cartouche, in form of an oval wreath, at bottom left of the map.
- No decorative elements on the map.

Examples: 1656 edition: *Greenwich NMM*, 912.44(100):094; *London BL*, Maps C.21.b.1; *New York*

PL, BKC Sanson; 1662 edition: *New York PL*, *KB 1662 (The first state of the map appeared in the editions of 1656 and 1662).

86.2 Second State 1667
Addition of an engraved '1. Vol. page 1.' in upper right corner between the map gridline and the neatline. No other changes on the map.

Examples: *London BL*, 145.b.14-16 *and* 982.g.16-18.

Description
This map of Africa appeared in a quarto-size atlas, first published in 1656 by Nicolas Sanson. The atlas was devoted solely to describing Africa and contained eighteen maps of Africa and its regions, starting with this map of the continent and ending with a map of Malta.

The atlas was produced to provide the French market with a more affordable and conveniently-sized atlas. Up to this point, Dutch *Atlantes Minores* had dominated the market for smaller-sized atlases. Companion atlases, separately covering Europe, the Americas, and Asia, were also produced around this time by Sanson.

Sanson's son Nicolas II had started the series of quarto atlases with one on Europe in 1648, but he died prematurely of gunshot wounds resulting from his involvement in the 'Barricades' insurrection in that year. His father continued the work preparing *Atlantes Minores* of the other continents. These were Asia, first issued in 1652, Africa in 1656, and America in 1657.

This map of Africa is a reduced version of Sanson's 1650 Africa map with some loss of details owing to the smaller format of this atlas. The major divisions are the same as on the 1650 Sanson map with *Ethiopie* (Ethiopia) extending south of the Equator to include the Nile source lakes. The lake in the Congo region to the west of the Nile source lake, *L. de Zaire/Zembre*, is not clearly shown on this map, unlike the 1650 map (though this lack of properly engraving the map might be an oversight by the engraver who omitted the hachuring for this Congo source lake). As a result, *L. de Zaire/Zembre* appears to be shown as the only source for the various Congo rivers.

Publication Information
There were editions of Sanson's *Atlas Minor* of Africa in 1656, 1662, c.1667, and 1699 with this map of Africa. There was also an edition in c.1706, published by Jacques van Wesel, that contained this map.

The Africa map is known in two states. It appeared in a second state in a French translation of Luis del Marmol Carvajal's book with the new title of *L'Afrique de Marmol, de la Traduction de Nicolas Perrot* in 1667 (Marmol's original work on Africa, *Descripion general de Affrica*, appeared in Granada in 1573). The book with this second state of the Sanson map of Africa was published by Louis Billaine (*London BL*, 145.b.14-16) in 1667, and also by Thomas Jolly in Paris in 1667 (*London BL*, 982.g.16-18).

Derivatives
The Sanson atlas proved especially popular and the maps were imitated by others into the eighteenth century using new copperplates.

In Frankfurt in 1679, Johann David Zunner published his rare four volume *Die Gantze Erd-Kugel Bestehend in... Europa, Asia, Africa und America* with maps containing titles in Latin and including a new map of Africa. See Map # 123 for this map.

Zunner title cartouche (map # 123).

In Amsterdam in 1683, an atlas was published with the title: *L'Europe | Dediée | a Monseigneur | Monseigneur le Tellier... Par N. Sanson le fils...* Simultaneously, in Utrecht in 1683, Joannes Ribbius published a copy of the Sanson atlas, *L'Afrique en plusieurs cartes nouvelles, et exactes....* . Both atlases contain the same new map of Africa, engraved by Antoine de Winter ('A d'Winter schu:'). See map # 129 for this map.

De Winter imprint (map # 129).

In Paris in 1716, another newly engraved copy of the Sanson map appeared in *Methode Pour Etudier La Geographie* by Abbé Nicolas Lenglet Dufresnoy (1674-1755) (Paris: Chez Charles Estienne Hochereau.). This was followed by an edition in Amsterdam in 1718 (Amsterdam, Aux depens de la Compagnie, 1718), using this same map.

This Dufresnoy map is smaller in size than the 1656 Sanson map at 14.5 x 16.5 cm. The map is further distinguished by having no imprint for the engraver. Both the 1716 and 1718 editions have 'Tom: 3 Pag: 1.' in letterpress at the top right above the map (1716 edition: *Paris BN*, G 10525-28; 1718 edition: *London BL*, 1001.aaa.35.). There were subsequent editions of the atlas in Paris by P. Gandouin in 1729 and 1735, and in Amsterdam 'Aux depens de la Compagnie' in 1737. The author has a separate Dufresnoy map with "Tom: I Pag: 454.' in letterpress above the map at the upper right.

Dufresnoy map of Africa.

References
The author is grateful to Jason Hubbard for sharing the results of his research on this series of atlases and works of geography.
Shirley 2004: Vol. 1: 886; Norwich 1997: map # 37 (second state); Burden 1996: map # 324 (Americas map); Pastoureau 1984: 387-9.

87 c. 1658 Nicolaas Visscher I, Amsterdam

87.1 First State, c. 1658

AFRICÆ ACCURATA TABULA | ex officina | NIC. VISSCHER.
[at bottom left within cartouche]:
Nob^mo Spectat^mo Prudent VIRO | D. GERARDO SCHAEP, I.V.D.... D.D. | N. Visscher.
- No scale bar.
- Amsterdam : Nicolaas Visscher I, [c. 1658].
- From: separately published / *Atlas Contractus.*
- 1 map : copperplate engraving ; 43.5 x 54.5 cm.
- Latitude and longitude coordinates on all four sides of the map.
- North at the top; one compass rose in South Atlantic Ocean.
- Title cartouche is at the top right, surrounded by two Africans and at the bottom by two putti.
- At the bottom left, there is a large decorative scene with an elaborate dedication to Gerard Schaep, with his coat of arms at the top.

- In the Indian Ocean, there are three sailing ships. In the Atlantic Ocean, there are eight sailing ships, with one sea monster and four flying fish just below the Equatorial Line. Within Africa, various animals are depicted.

Examples: *Amsterdam UB,* 1805 A 2 (dated est. 1656-1677); *London BL,* Maps C.39.f.1; *New York NYPL,* [catalogued as] Map Collection in Visscher *Atlas Minor.*

87.2 Second State c.1677

With the privilege added at the bottom right of the map: 'cum Privilegio Ordin: General: Belgii Fæderati.'

Examples: *Amsterdam UB,* 33-17-43 (color) and 33-17-42 (b&w); *London BL,* Maps C.39.f.1. Numerous other locations.

Description

This is a beautifully designed and finely engraved map by Nicolaas Visscher I (1618-1679). From about 1631 and prior to the issuance of this map, the Visscher publishing family and its founder, Claes Jansz. Visscher, used the 1614 Van den Keere map of Africa in its second, third, and fourth states of 1631, 1636, and 1652 respectively. These earlier maps were based on the even older geography of the Jodocus Hondius map of 1606.

To prepare this map, Visscher turned to the common model of this period which was Willem Blaeu's 1608 wall map, likely in a later state, but not the more recent 1617 folio map by Blaeu. As in Blaeu's 1608 wall map, but not in Blaeu's 1617 map, Visscher shows a common source for the *Cuama* and *Spirito Santo* Rivers of the Zambere River originating in *Sacaf Lacus* in Southern Africa. The general outline of Africa is surprisingly modern in appearance.

Of special note is the much greater detail in this folio map that Visscher provides along the South African coastline. The first Dutch settlement under Jan van Riebeeck at the Cape of Good Hope had occurred in 1652, and it is likely that Visscher wanted to show this information on his map. Visscher introduces a number of Dutch names interspersed with older Portuguese names. Among these names are *Tafel bay* (Table Bay), *Tafel berg* (Table Mountain), *Robben Eyl* (Robben Island), and *Schorre hoek* (inland from *C. das Anguillas* (Cape Agulhas), the southern-most point in Africa). Much of this information appears to be taken from Jan Mathisz.' wall map of c. 1655.

Publication Information

The Visscher publishing firm was comprised of a prolific Dutch family of mapmakers, engravers, and publishers. The founder of the firm was Claes Jansz. Visscher (1587-1652) who established himself in Amsterdam likely working for Van den Keere and Hondius. He issued several of his maps under the Latinized form of his name, Piscator. From about 1620, he designed a number of individual maps, but his first atlas consisted of maps printed from plates purchased from Van den Keere. His son Nicolaas Visscher I (1618-1679) and grandson Nicolaas Visscher II (1649-1702) issued a considerable number of atlases. Among these were the *Atlas Contractus* and the *Atlas Minor* (actually in folio size). Many of their maps also appeared in atlases produced by Janssonius. The widow of Nicolaas Visscher II carried on the business until it finally passed to Pieter Schenk around 1717.

Various dates have been attributed to this map from as early as 1656 to 1677 (Koeman 1967-71, III: 163) and as late as 1690 (Tooley 1969:119-120). What makes the process of dating Visscher's maps so difficult is that the map is contained within composite atlases, often undated and often with manuscript title and contents pages.

Burden notes that a first state of Visscher's Americas map, without the 'privilege', appears in some examples of Volume III of Johannes Janssonius' *Novus Atlas* of 1658 (Burden 1996: map # 332). The author has not been able to locate a copy of Janssonius' *Novus Atlas* that contains the first state of the Visscher Africa map. The example the author has examined at the British Library of the 1658 *Novus Atlas* has the later Visscher map with the 'privilege'. Therefore, no firm date can be derived from its appearance in some examples of the *Novus Atlas.* However, upon the death of Nicolaas' father Claes Jansz. Visscher in 1652 the family was still using the old 1614 Van den Keere plates of the continents with the last state being from 1652. Burden postulates that it would be logical for Nicolaas Visscher to prepare new copperplates of the continents and world. As the world map bears features in China introduced in 1655, it is possible that the set of accompanying continent maps would likely have a date of c.1658 (Burden 1996: map # 322).

Koeman (1967-71, III: 163) mentions a date of 1656-1677 for this map using, as examples, the *Atlas Contractus Orbis Terrarum* in the Amsterdam Universiteitsbibliotheek (Amsterdam UB: 1805 A 2) and in the Rotterdam Maritiem Museum in which the atlases have no map dated later than 1657.

Besides possibly appearing in Janssonius' *Novus Atlas*, dated 1658, the map also appeared in Visscher's undated *Atlas Contractus*. There is an example of the first state of this map (without the privilege) bound within Visscher's *Atlas Contractus* in the British Library. This example of the Africa map has been dated c.1670 by the British Library, but the latest date on a map in this atlas (map no. 50 of East Asia) is dated 1662. The New York Public Library has an example of the first state bound into Visscher's *Atlas Minor*, dated 1655-80 in pencil on the printed title page. As a result of this body of information, the author has dated the map of Africa as c.1658.

In 1677, Nicolaas Visscher I obtained a publishing 'privilege' or copyright by the States of Holland and West Friesland. This privilege was added to the plate on the bottom right. Thus, the author has dated the second state of this map as c.1677. Upon Nicolaas' death in 1679, his son Nicolaas II continued the business, issuing his own *Atlas Minor* in which this map is found. The second state of the map has been found in atlases dated as late as c.1696.

Both states of the map have an elaborate dedication in the lower left of the map to Gerard Schaep. Gerard Schaep (1581-1655) was sent by the States General to London on 27 Dec. 1651 as part of an embassy. The intention of the embassy was to convince the English to withdraw the Navigation Act and to negotiate a new commerce treaty. They were not successful and the First Anglo-Dutch War resulted. It can be assumed Visscher's dedication was to acknowledge this statesman upon his death.

Derivatives
There were a number of derivatives of this Visscher map using new copperplates.

The map of Africa by Gerrit Lucasz. van Schagen (Map # 115), an Amsterdam engraver and art dealer, is a derivative of the Visscher map of Africa. The map has a decorative scene in the lower left of Neptune and Venus on a shell being pulled by horses and surrounded by sea horsemen, in place of the large dedicatory cartouche in the Visscher map. There is no privilege on this map. The Americas map by Van Schagen was published in 1671 (Burden 1996: map # 332). Accordingly, the map of Africa was likely published in 1671 as well. It is readily identified with the imprint of *Gerardi A Schagen* in the title cartouche.

Jacob van Meurs (c.1620-1680), an Amsterdam publisher, used the Visscher map with a blank dedicatory cartouche as his model for Olfert Dapper's *Naukeurige Beschrijvinge der Afrikaensche gewesten...*, which he published in 1668 (Map # 108). The Van Meurs map also appeared in John Ogilby's *An Accurate and Complete History of Africa* of 1670, based on the Dutch work of Dapper.

References
Gay 1998: 219; Norwich, 1997: 68: map # 55 (second state); Burden 1996: map # 332 (Americas map); Mendelssohn 1993: II: 119; Koeman 1967-71, II: Me 119, and III: 150-184; Tooley 1969: 119-120, with image of the first state as plate 88; Tiele 1867: 298.

88 1658 Robert Walton, London

Map 88.2

88.1 First State 1658

New, Plaine, & Exact Mapp of | AFRICA, described by: N:I: Vischer | and done into English, enlarged and | corrected, according to: I:Blaeu: with | the habits of the countries and | manner of the cheife Cities; | the like never before. 1658. | Printed, colered and Are to be | Sould by Robert Walton at the | Globe and Compass in St. Paules Church yard between | the two north doores.

- No scale bar.
- [London] : Robert Walton, 1658.
- From: separately published / *Cosmographie* (in second state).
- 1 map : copperplate engraving ; 42 x 53 cm (34 x 45 cm for map only).
- Latitude coordinates along left and right sides. Longitude coordinates along Equatorial Line.
- North at the top; no compass rose.
- An elaborate title cartouche is at the bottom left with an allegorical depiction of the continent above the title: an African woman sitting on a crocodile.

- In the Indian Ocean, there are two sea monsters. In the Atlantic Ocean, there are two sailing ships, one flying a Dutch flag. No animals are within Africa. Various notations in English within Africa denoting information such as Amara and the legend of Prester John.
- Decorative borders on all four sides as described below.

Example: *Boston PL* (only one example of the first state is known at The Boston Public Library. *The Map Collector* 17 (December 1981): 48-49; Schilder 2000: 425).

88.2 Second State 1660-62
Date removed with evidence of erasure to the date in the seventh line of the title cartouche.

Examples: *London BL*, L.35/82; *Author's collection* (as a separate map).

Description
Robert Walton (1618-1688) was a noted London map publisher, one of a small number working in London in the 1650s. In 1656, Walton completed a world map and after that a set of continent maps.

This scarce map was closely copied from Pieter van den Keere's 1614 map of Africa in its fifth state of 1652 as issued by Visscher. Within the map, the placenames have been changed from Latin to English. The general outline of Africa and the hydrography follow the Van den Keere map (for example the unnamed Cuama and Spirito Santo Rivers flow from a common river, the unnamed Zambere River which rises from Sachaf Lake). The map is such a close copy that the decorative elements on the map (ships, sea monsters, town views, figures, etc.) are taken directly from the Visscher state of the Van den Keere map.

Across the top and bottom borders of the map are views of African towns. At the top from left to right are MINA (The Mine at St. George in Guinea), TANGER (Tangiers), TUNES, AMARA (the home of Prester John), and ALGAR. At the bottom from left to right are CEUTA, TSAFFIN, CEFALA, I. MOZAMBIQUE, and CANARIE (in the Canary Islands). These town views are interspersed by faces of various kings of Africa: at the top, KING OF ABISSINES and KING OF CONGA; and at the bottom, KING OF GUINEA, KING OF MAROCCA, KING OF MADAGASCAR, and KING OF MOZAMBIQUE.

The left and right borders shows various costumed African figures. On the left side from the top are A WOMAN OF FEZ, A CONGENSIAN, A MADAGASCAR WOMAN, and A GUINEAN. On the right side from the top are A MOROCCO, ABISSINE WOMAN, AZENGENSIAN, and MOZAMBIQUE WOMAN.

The town views at the top and bottom borders are mostly taken from Volume I and II (1572 and 1575) of Georg Braun and Frans Hogenberg's *Civitates Orbis Terrarum*, a collection of views of famous towns of the world. The information for these views was mainly from unknown Italian and Portuguese sources.

Publication Information
This separately issued map is known in two states. The second state can be found on thin paper with evidence of multiple folds. The second state is dated 1660-62 since this state appears in an edition of Peter Heylin's *Cosmographie* printed in London for Philip Chetwind in 1660-62. In this edition, the title page bears the date of 1660 and the appendix bears the date of 1662.

References
Christies London auction catalog June 2000; Schilder 2000: 425; Burden 1996: map # 330 (Americas map); *The Map Collector* 15 (June 1981): 48-50; *The Map Collector* 17 (December 1981): 48-49; Tyacke 1978: 145-146; Tooley 1973: 304.

89   1658 Nicolas Berey, Paris

89.1 First State 1658

Carte de | L'AFRIQVE | Corrigeé et augmenteé, deßus toutes | les aultres cy deuant faictes.
- No scale bar.
- [Paris : Nicolas Berey, 1658].
- From: separately published.
- 1 map : copperplate engraving ; 39.5 x 51 cm.
- Latitude coordinates on left side of map and south of Equatorial Line on right side of the map. Longitude coordinates along Equatorial Line.
- North at the top; no compass rose.
- Title cartouche is at the top right.
- In the Indian Ocean, there is one sailing ship and one sea monster. In the Atlantic Ocean, there are five sailing ships, two sea monsters, and flying fish above and below the Equatorial Line. Some animals are within Africa.
- Text borders, describing Africa, on all four sides of the map.

Example: *Private French Collection.*

89.2 Second State 1671
With addition of the date of 1671 below the title in the title cartouche.

Example: *London BL*, *Maps 63510.(30) (N.B.: The British Library lists this map under Jean Michael Picart).

Description
This folio-size map by Nicolas Berey is one of a series derived from the Petrus Bertius map of 1624 which in turn was taken from the Jodocus Hondius Jr. map of 1619. Berey even copies some of the same sailing ships and sea monsters that appear on the Hondius map.

The map is quickly identified by the unusual small panels of text describing Africa that surround all four sides of the map. As with most of the separately issued maps from this period, it is rare.

This map of Africa is one in a set of the four continents. As the Europe and Asia maps in the set bear the imprint of Berey and the date of 1658, the Africa map, though unsigned and undated, is also assigned to Berey with a date of 1658.

Publication Information
Nicolas Berey (c.1606-1665), who was the father-in-law of Alexis-Hubert Jaillot, appears to have been active as a mapmaker and publisher working from 'au bout du Pont Neuf proche les Augustins aux Deux Globes' in Paris from about 1645 to his death. Berey is known to have published an edition of Tassin's *Carte generalle* in 1651. Berey's son, Nicolas II (1640-1667) continued the business for a short time after his father's death. Upon the death of Nicolas II, the business was taken over by Berey's son-in-law, Alexis-Hubert Jaillot. Jaillot issued a second state of this map in 1671.

To aid in the identification of this Berey map and a number of similar maps using different copper-plates by Jodocus Hondius Jr. (Map # 58), Bertius (Map # 61), Bertius-Tavernier (Map # 63), and Boisseau (Map # 66), please refer to the chart under Map # 58, the Jodocus Hondius map of 1619. All of these similar maps are derivatives of the Jodocus Hondius Jr. map of 1619. All lack decorative borders as did the third and later states of the Hondius 1619 map.

References
Burden 1996: map # 328 (Americas map), Loeb-Larocque 1989: 23-24, Pastoureau 1984: 229, Tooley 1999-2004, A-D: 122.

90 1658 Gabriel Bucelin, Ulm

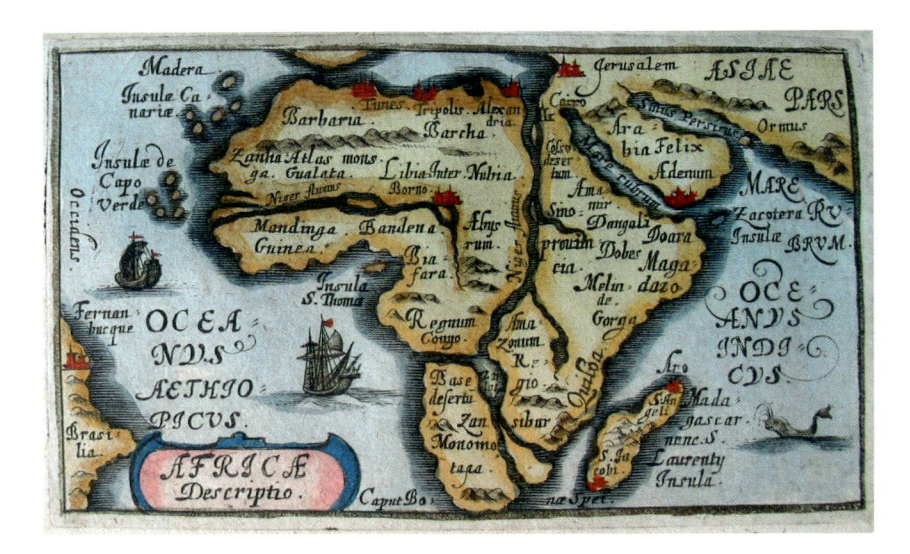

AFRICÆ | Descriptio.
- No scale bar.
- [Ulm : Johannes Görlin, 1658].
- From: *Praecipuarum Universi Terrarum Orbis*; Volume Two of Bucelin's *Historiae Universalis*.
- 1 map : copperplate engraving ; 6.5 x 10.7 cm.
- No latitude or longitude coordinates. No Equatorial Line.
- North at the top; no compass rose.
- Simple oval title cartouche is at the bottom left.
- In the Indian Ocean, there is one sea monster. In the Atlantic Ocean, there are two sailing ships. No other decorative elements.
- At the top of the page in letterpress is 'III.', and at the bottom of the page on the right side is a l etterpress 'AME-'.

Example: *New York PL*: BAC Bucelinus (1658 edition); *Various other locations*.

Description
This attractive little map is by Gabriel Bucelin (1599-1681). Bucelin was a Swiss genealogist and an author of several books and a number of manuscript maps of Germany and Austria.

Bucelin's map of Africa is modeled after the Ortelius-Galle (2) map of Africa which first appeared in 1593. The overly-large river systems on the map are the most obvious feature. Some mountains are also shown. The placenames on the map are generally derived from Ortelius (for example *Zansibar* denoting much of Southern Africa) with some omissions such as the Portuguese mine at *Mina* in Guinea.

Publication Information
Bucelin's *Praecipuarum Universi Terrarum Orbis*, which contained this map, was published in 1658, 1659, and 1664 (Tooley 1999-2004, A-D: 204). This book can sometimes be found bound with Bucelin's three volume *Historiae Universalis Auctorium sive Nuclei Historici* which can show a publication city of Augsburg. The first part of the *Historiae Universalis* contains woodcut maps.

The book *Praecipuarum Universi Terrarum Orbis*, containing thirty copperplate maps of the world, shows the publication city of Ulm and was published by Johannes Görlin. King (2003: 127) believes the maps were engraved by Melchior Küsel, whose signature appears faintly on the engraved title page.

The map is only known in one state from Bucelin's *Praecipuarum Universi*.

References
King 2003; Tooley 1999-2004, A-D: 204; Burden 1996: map # 329 (Americas map).

91   1659 Joan Blaeu, Amsterdam

NOVISSIMA | AFRICÆ | DESCRIPTIO. | AMSTELÆDAMI, Apud I. BLAEV. 1659.
- No scale bar.
- Amsterdam : Joan Blaeu, 1659.
- From: separately published.
- 1 wall map : copperplate engraving ; 6 map sheets: 104 x 127 cm; 120 x 156 cm (including letterpress text panels in London BL example).
- Latitude and longitude coordinates along all four sides of the map. Longitude coordinates along Equatorial Line.
- North at the top; five compass roses on the map; two in the Indian Ocean and three in the Atlantic Ocean.
- Title cartouche at the bottom right, surrounded by African and Arab figures and African animals. Above the cartouche, is an African holding an ivory tusk and an elephant.
- Text box at upper right detailing how to determine positions. Informational banner at lower left.
- In the Indian Ocean, there are eight sailing ships. In the Atlantic Ocean, there are six sailing ships. There are two ships with lateen sails in the Mediterranean.
- Various animals are within Africa.
- Text panels on each side and at bottom describing Africa in Latin, Dutch and French.

Examples:
This map is known in four examples and in only one state as follows:

91.A. *London BL*, K.A.R. Klenck Atlas of 1660 (map no. 6 is the Africa wall map). The British Library example is bound into the Klenck Atlas. The Klenck Atlas, nearly six feet tall, was given to King Charles II by a group of Dutch merchants upon Charles' ascension to the throne in 1660.

91.B. *Berlin SB*, in The Atlas of the Great Elector (bound into a large atlas).

91.C. *Rostock UB*, bound into a large atlas.

91.D. *Paris BN*, Ge DD 5079.

Description

By the late 1640s, Joan Blaeu had assumed complete control of the Blaeu family publishing business after the death of his father Willem in 1638, and his younger brother Cornelis in 1644. As his father's wall map had first been issued some 50 years before and as others were re-publishing this wall map without geographical changes, it seems logical that Joan would want to issue his own set of up-to-date wall maps of the world and the four continents. This wall map of Africa of 1659 is part of that effort; a twenty-one sheet world map preceded it in 1648 and a wall map of the Americas complemented the Africa map also in 1659.

This map of Africa follows the general outline of Joan Blaeu's father's wall map of Africa of 1608 as well as his own world wall map of 1648. Blaeu updates his father's information and provides further placenames in those areas where increasing European political and economic expansion was underway: in Guinea with the British and Dutch, in Senegal with the French, in Southern Africa with the Dutch, and in East Africa with the Portuguese and the Jesuit missionary activities.

In the Congo region, he includes newer Dutch placenames; e.g. *t'Rood Steilandt, Rode Bay, Melk Bay*, and *t'Fort Mols*, and replaces many of the much earlier Portuguese names. In Southern Africa, partly owing to the large format available to him, Blaeu adds further detail to the area of the Cape of Good Hope which the Dutch had settled in 1652. Besides using the nomenclature of the Mathisz. map of c.1655 and the Visscher map of c. 1658, he adds a further name of *Schorre hoeck met een Laegland* inland from *C. das Anguillas. Struys Bay* is now to the west of *C das Vacas*. In South Central Africa, he

NOVA AFRICÆ DESCRIPTIO.

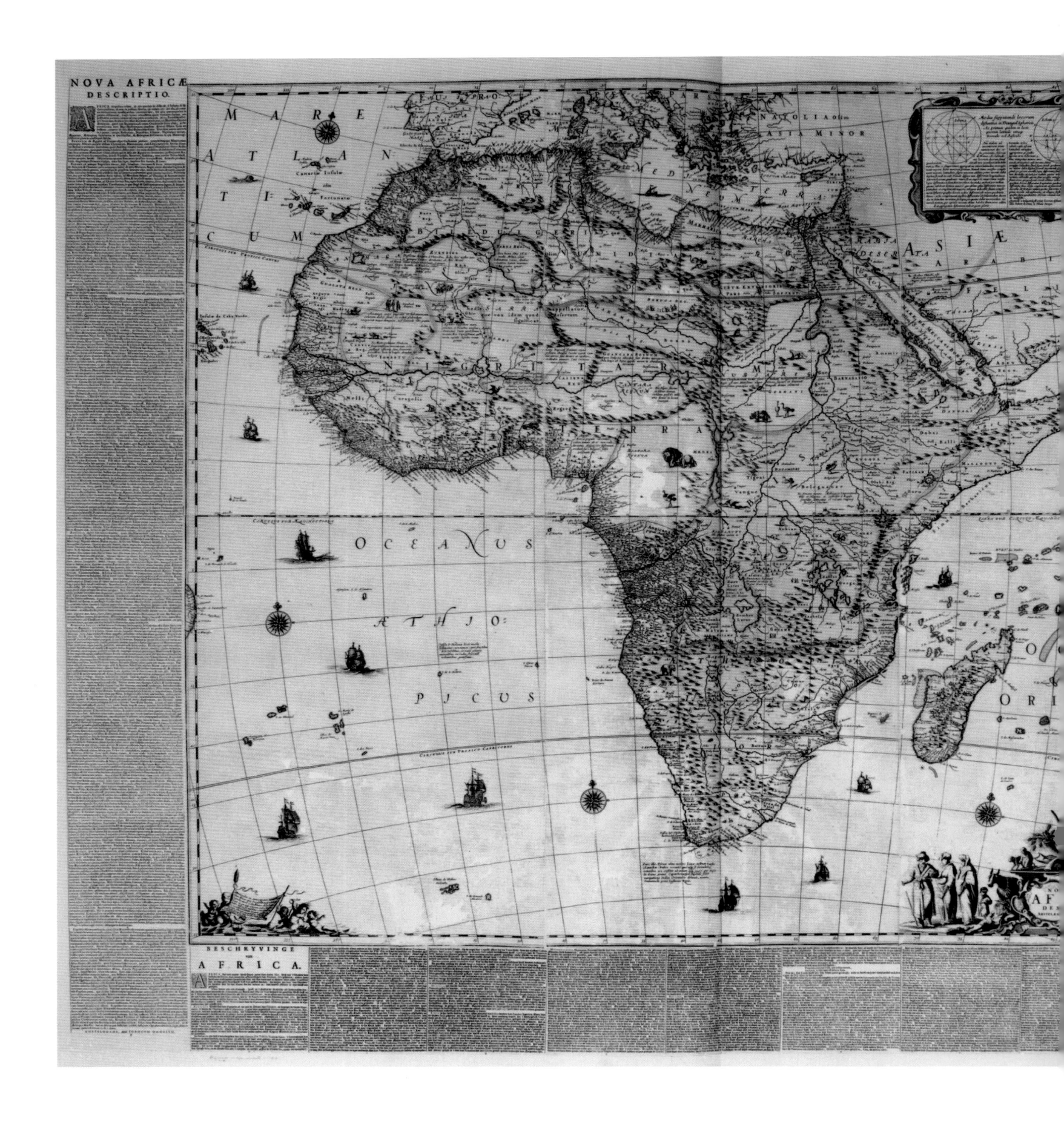

BESCHRYVINGE VAN AFRICA.

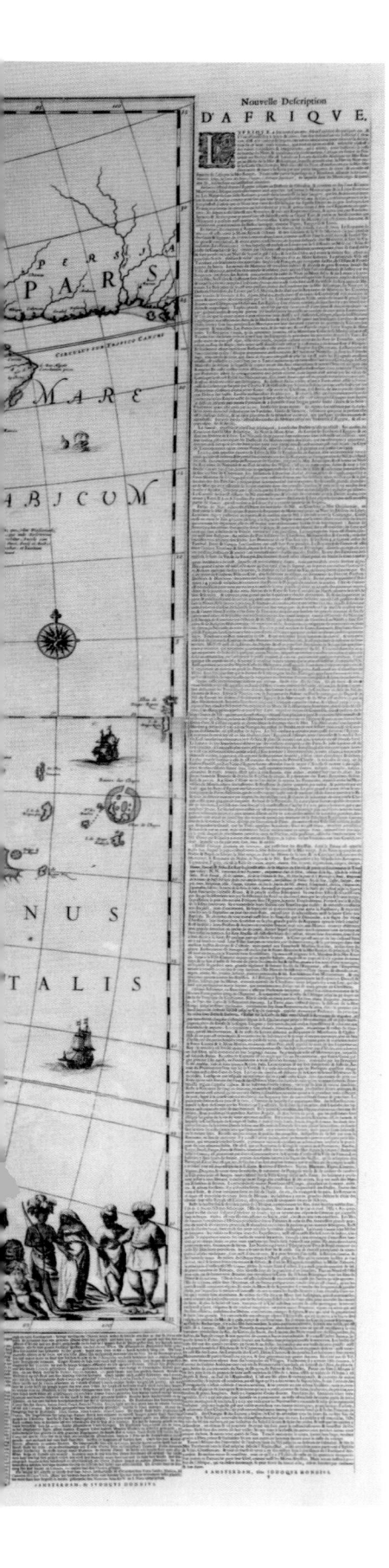

still shows much of the earlier representation of the region. *Lunæ Montes* are still on the map, but with no connection to the two Ptolemaic lakes to their north. He continues the Mercator tradition of a Portuguese fort, *Cast Portugal*, at the intersection of where the *Cuana* and *Rio Spirito Santo* join the *Zambere flu* upriver.

In East Africa, Blaeu introduces another lake, which, from its placement, may be Lake Tana, the lake referred to by Jesuit missionaries as the source of the Blue Nile River in the Abyssinian highlands. On Madagascar, further Dutch influence is shown by names such as *Hollands kerckhof*.

The engraver for this finely detailed map is not known. Josua van den Ende (c.1584-1634), who engraved many of the maps for the Blaeus, including the 1608 wall map of Africa, had died by this time.

Unlike many other wall maps from the seventeenth century, this map is comparatively devoid of decorative features. There are small sailing ships in the sea and an attractive cartouche, but the emphasis of this map is on its geographic content.

An oddity in the British Library example is the imprint of Jodocus Hondius at the end of each text panel. Evidently some previous owner in this map's long history had the text panels from an earlier Hondius work added to the map sheets for the Joan Blaeu wall map. In the British Library example, the text panels are as follows: text panel on left side is in Latin with the title of 'NOVA AFRICÆ DESCRIPTIO.'; text panel at bottom is in Dutch with the title of 'BESCHRYVINGE van AFRICA.'; and text panel on right is in French with the title of 'Nouvelle Description D'AFRIQVE.'.

Publication Information
This map is known in four examples referenced above, and in only one state. The examples at the British Library, Staatsbibliothek Berlin, and Universitätsbibliothek Rostock have surrounding text panels. The example at the Bibliothèque Nationale de France in Paris is without text panels.

References
Schilder 1990 (covering the two states of the Blaeu World wall map, 1619 & 1645-6; Burden 1996: map # 334 (Americas map); Klemp 1971.

92 C. 1659 Nicolas Picart (2), Paris

92.1 First State c.1659

Nouuelle description | DAFRIQVE
- Scale bar: Lieues d'Allemag 200 = 2.3 cm.
- [Paris : Nicolas Picart, c. 1659].
- From: *Trésor des cartes géographiques des principaux états de l'univers.*
- 1 map : copperplate engraving ; 13.5 x 19 cm.
- Latitude coordinates along left and right sides. Longitude coordinates along top and bottom of the map.
- North at the top; no compass rose.
- Title cartouche at the bottom left.
- No decorative elements on the map.

Examples: *Paris BN*: Ge.FF.3941; *Chicago, Baskes Collection*.

92.2 Second State c.1660

Addition of a plate number. The Africa map bears plate number at lower right of '32'. Other changes to the map.

Shortly after 1659, Gerard Jollain (1638-c.1724) acquired the Picart (2) copperplates. He published later editions of *Trésor des cartes géographiques* in c.1660 (an undated edition), and in 1667. The 1667 example of the atlas has the title page imprint of 'chez Jollain, rue st. Jaque a la ville de Cologne, 1667'. The map of Africa in the 1667 edition is bound into an atlas on the left edge. Jollain's imprint as the engraver is on the Prester John/East Africa map in the edition of 1667. Many of the maps, but not the Africa map, bear the date of 1667.

Example: *Paris BN*, Ge FF 4473.

92.3 Third State 1667
With the scale bar erased and showing erasure of all but 'L' in 'Lieuses d'Allemag' bottom left.

Examples: 1667 Jollain edition: *Paris BN*: Ge FF 18767; *London, Burden Collection*; *Author's collection* (as separate map).

Map 92.3.

92.4 Fourth State after 1667
With the imprint of 'par P. Duval | Geographe' at lower left in place of scale bar.

Sometime after 1667, Pierre Duval acquired the copperplates and published an edition of *Trésor des cartes géographiques*. The author has seen an undated edition of *Tresor des cartes geographiques* by Pierre Duval in the possession of Phillip Burden. The publisher is Huot. The date is sometime after the Jollain 1667 map. All evidence of the scale bar and the 'L' above it have been replaced with 'par P. Duval | Geographe.' in bottom left corner. The verso of the map is blank.

Example: *London, Burden Collection*.

Map 92.4.

Description
This is Nicolas Picart's second miniature map of Africa. The map appears in an undated edition of *Trésor des cartes géographiques* published by Nicolas Picart in c.1659.

Not much is known about Nicolas Picart. He is known as an engraver and a mapmaker who produced a map of Africa with decorative panels in 1644. He is also known to have published editions of *Trésor des cartes géographiques* from 1651 to 1662 with two different small maps of Africa (Picart 1 and Picart 2).

This atlas by Nicolas Picart is the last of the French derivatives of Joannnes Janssonius' *Atlas Minor* of 1628. The map is most directly derived from the Boisseau map of 1643 and the Picart (1) map of c.1651. For details on the geography, please refer to the Janssonius map of 1628 (Map # 64).

Publication Information

There are two examples of the first edition of *Trésor des cartes géographiques* (and the map of Africa) known to the author. The first is in the collection of Roger Baskes. The other is at the Bibliothèque Nationale de France in Paris. Neither example contains a date on the title page or in the table of contents of the atlas. The *Paris BN* has dated this book as c.1659, as the map of Russia in the atlas bears that date. The only other map in the atlas with a date is the map of France with a date of 1657. As a result, the author has applied a date to this atlas (and the map of Africa) of c.1659.

The map is known in four states as described above. Significant changes were made to the copperplate that produced this map over the course of its existence as the plate passed from Nicolas Picart to Gerard Jollain, and finally to Pierre Duval.

There were five similar maps produced around the same period of time, all using new copperplates, derived from the Janssonius 1628 map; the last four of which were produced in Paris. These maps are the Boisseau of 1643, Berey of 1651, Picart (1) of c.1651, and this Picart (2) map of c.1659. To differentiate these maps, refer to Map # 64, the Janssonius map of 1628.

References

Burden 1996: maps # 342-343 (Americas and Arctic maps); Pastoureau 1984: 499-500; Sotheby's 2000: Lot 73 (for Jollain); National Maritime Museum 1969-1976: Vol. III: Part I: entry 241; Bagrow 1964: 253; Phillips: 1909-1992: Phillips entry # 477.

93 1659 Philipp Cluver – Elzevier, Amsterdam

AFRICA.
- No scale bar.
- [Amsterdam : Elzevier publishing house, 1659].
- From: *Introductionis in Universam Geographicam.*
- 1 map : copperplate engraving ; 12 x 12.5 cm.
- Latitude coordinates along left and right sides. Longitude coordinates along Equatorial Line.
- North at the top; no compass rose.
- Title cartouche at the bottom left, set on a platform.
- No decorative elements on the map.

Examples: 1659 Elzevier Amsterdam edition: *Amsterdam UB*; *Washington LC*, G1015 .C56 1659 Vault; 1661 Elzevier Amsterdam edition: *London BL*, 10004.a.42 (1661); *Washington LC*, G120.C62 1661; 1677 Elzevier Amsterdam Edition: *Cambridge Harv RB,* *GC6 C6275I 1677.

Description

Philipp Cluver (1580-1623), originally from Gdansk and then Leiden, wrote many works on geography, the most famous being his *Introductionis in Universam Geographicam* which appeared over a number of years into the eighteenth century and which had a number of imitators. The first edition in Leiden of 1624 did not contain maps.

This is the second Cluver map of the African continent. The 1641 Cluver-Buno map of Africa was used as its model. Broadly across the middle of the map in large capital letters is *ÆTHIOPIA* with the last '*A*' in the word far into Southern Africa below the Ptolemaic Nile source lakes. Other divisions within the continent follow most of the conventions of the period. The map only shows a few common placenames such as *Monomotapa* in Southern Africa and some of the more important trading ports on the east coast. The Mountains of the Moon are shown as *Montes Lunæ* below the two Ptolemaic lakes.

Publication Information

This map of Africa appeared for the first time in a 1659 edition of Philipp Cluver's *Introductionis in Universam Geographicam*, published by the Elzevier publishing house. According to King (2003: 128), the cousins Louis and Daniel Elzevier were responsible for the issue of this edition with new maps of the world. Their relative Isaac Elzevier (1596-1651) was one of the most famous members of the firm that bears his name. He published the first edition of Cluver's *Introductio* in 1624 in Leiden (Tooley 1999-2004, E-J: 21), and the Elzevier family was responsible for numerous subsequent editions. This 1659 edition was edited by Joseph Vorstius.

This map continued to appear in the Elzevier Amsterdam editions of Cluver's *Introductio* in 1661, 1670, 1672, and 1677. In an example of the 1661 edition at the British Library, Cluver's *Introductio* is found bound with Petrus Bertius' *Breviarum Orbis Terrarum*. This map was used in a German edition of Cluver's *Introductio* published in 1678 in Nuremberg. There is also a 1686 edition with this map published by Wetstenium. There may be other editions in which this map appears. The map also appeared in the G. Broedelet editions of Cluver's *Introductio* of 1702 and 1717, published in Utrecht (Shirley 2004: 346).

There were numerous other editions of Cluver's *Introductio* that were published using different maps of Africa. Shirley (2004: 345) states that over 45 editions of the *Introductio* appeared over a 100 year period. Not all editions contained maps. Those editions that did contain a map of Africa used one of the following: Cluver-Conrad Buno from 1641; Cluver Elsevier from 1659; Cluver-Johann Buno from 1661; Cluver-Buno-Mosting, with the Herman Mosting imprint from 1686; Johannes Janssonius in 1661, using his 1628 Goos map; Janssonius van Waesberge in 1676 using the 1630 Cloppenburch map; and Cluver-Wolters from 1697. All are generally derived from the Cluver-Conrad Buno map of 1641, which in turn was based on the Blaeu folio map of 1617. Refer to the chart under the 1641 Cluver-Conrad Buno map, Map # 71, for a list of the different maps of Africa that were used for different editions of Cluver's *Introductio* and the distinguishing characteristics of these maps.

References

Shirley 2004: 345-350; King 2003; Tooley 1999-2004, E-J: 21; Burden 1996: map # 335 (Americas map); Shirley 1993: 379-380; Phillips 1909-1992: no. 4261.

94 1660 Giovanni Battista Nicolosi, Rome

94.1 First State 1660

AFRICA | IOANNE BAPTISTA NICOLOSIO S.T.D. | Sic Describente
- No scale bar.
- [Rome : V. Mascardi, 1660].
- From: *Dell' Hercole e studio geografico di Gio. Battista Nicolosi….*
- 1 map : copperplate engraving ; 4 map sheets ; 79 x 92 cm (when the 4 sheets are joined). Each sheet has the following characteristics: Southeast Africa sheet: 40 x 45.5 cm (on top left of map a '1.' with a small raised 'a' and the title cartouche); Southwest Africa sheet: 40 x 46 cm (on top right of map a '2.' with a small raised 'a' and with a large inset map of Galliam (France) in upper 2/3 of the map); Northeast Africa sheet: 40 x 46 cm (on bottom left of map a '3.' with a small raised 'a'); West Africa sheet: 40 x 46 cm (on bottom right of map a '4.' with a small raised 'a').
- Latitude and longitude coordinates along all four sides of all four map sheets.
- South at the top; one compass rose on Northeast sheet of Africa.
- Simple box title cartouche on the sheet of Southeast Africa (that is, sheet 1 at top left).
- No decorative elements on the map.
- Only a total of four coastal placenames in an area within the surrounding sea off the southeast coast, south of 30° South.

Examples: Edition of 1660: *New York PL*, *KB+ 1660 (in G.B. Nicolosi, *Dell' Hercole e studio geografico*); *Paris BN*; *Washington LC*; *Author's collection*.

94.2 Second State 1670

This state has an additional four coastal placenames within an area between *C. S. Francisco* and *C. Primeiro* off the coast of Southeast Africa. These are: *Golfo d'Anguilhas, I. de Cruz, P^e Camsial,* and *I.S. Christoual*.

Other additional placenames are also along the southeast coast south of 30° south (The first state of this map only had a total of four coastal placenames with the names engraved in an area of the surrounding sea south of 30° south).

Example: Edition of 1670-1: *London BL*, 568.l.5.

Description

Giovanni Battista Nicolosi (1610-1670) was a priest and cartographer for the Vatican's Propaganda Fide in Rome. At the request of the Propaganda Fide in 1652, Nicolosi produced an atlas, *Dell' Hercole e studio geografico di Gio. Battista Nicolosi* in 1660 which contained a map of the world and four-sheet maps of the four continents.

Like the Italian maps of Africa from the previous century by Ramusio and Bertelli, this map has Southern Africa aligned at the top.
The main source for this map appears to be Nicolas Sanson's map of Africa of 1650. However, Nicolosi's map of Africa is generally conservative, particularly in the first state, in how it presents geographic information. Nicolosi does show a lake in Abyssinia which may be a pre-cursor to Lake Tana, the source for the Blue Nile River, likely from the Portuguese travel accounts. What is unusual is that Nicolosi does not generally fill the map with placenames and geographic features based on other various travel accounts on Africa and from the prevailing information of his time. This information surely was readily available to him, yet he chose to not include it on his map. However more details and placenames are added to the second state of this map.

This map follows Ptolemaic conventions for its depiction of much of the interior. The source of the Nile River is shown as *Ld Zaflan* and *Zembro L./Zaire L.* Madagascar, the Canary Islands, and Madeira are shown, but not the Cape Verde Islands.

CARTOBIBLIOGRAPHY

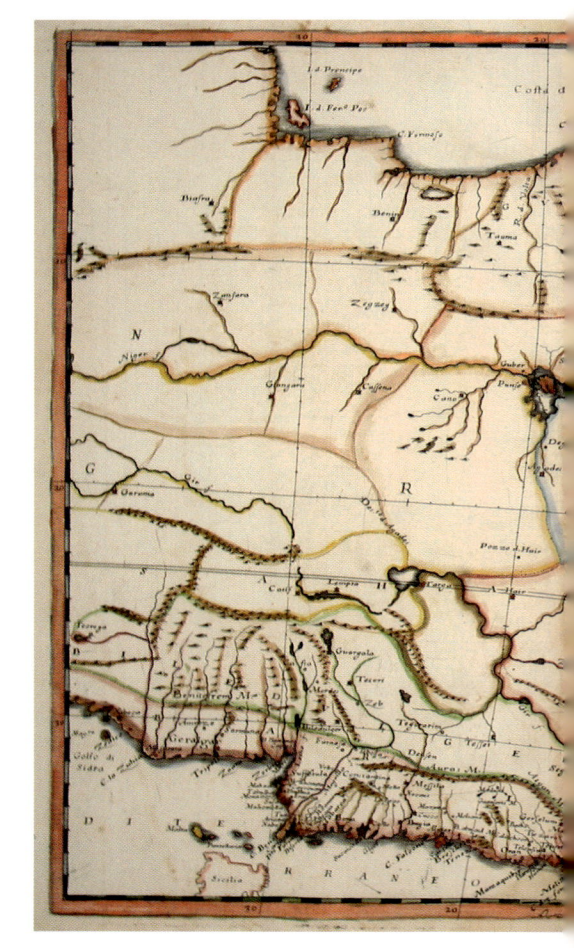

Publication Information

This four-sheet map of Africa was published by V. Mascardi in *Dell' Hercole e studio geografico di Gio. Battista Nicolosi....* in two volumes in 1660. The maps are in Volume II of the book. The verso of all four map sheets of Africa are blank.

There was one further edition of the atlas in Latin entitled *Hercules, Siculus sive studium geographicum* in 1670-71. It was published shortly after Giovanni's death by Joanne Baptista Nicolosi. Though both editions are considered scarce, the first edition is the rarer of the two. The map of Africa is known in two states.

References

Reiss & Sohn 1999 (auction catalog); Norwich 1997: map # 45 (second state); Burden 1996: map # 354 (Americas map); Shirley 1984: 440; Phillips 1909-1992: entry # 467; Sabin 1868: entry # 55258.

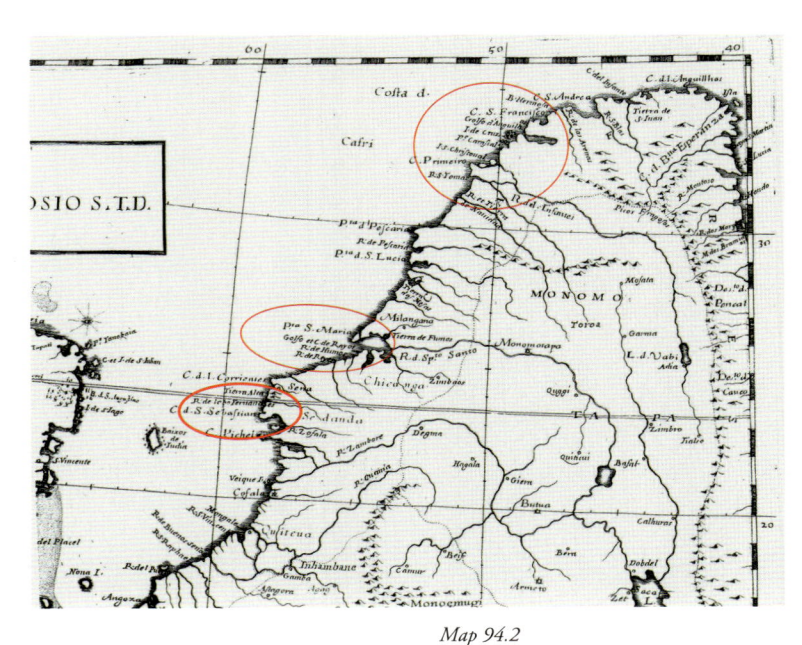

Map 94.2
Southeast sheet with additional coastal placenames.

95 c. 1660 Pierre Duval, Paris

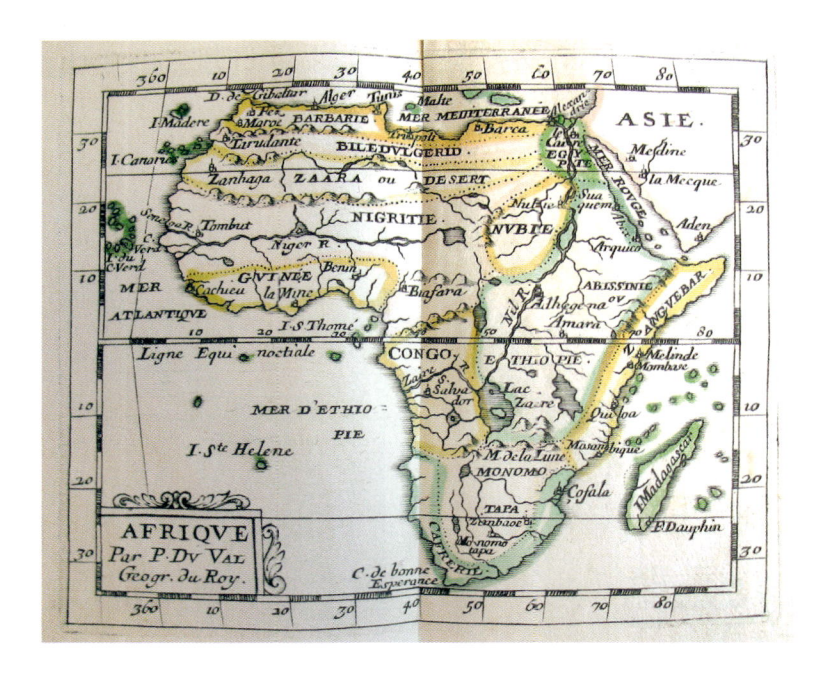

95.1 First State c. 1660

AFRIQVE | Par P. DV VAL | Geogr. du Roy.
- No scale bar.
- [Paris] : Pierre Duval, [c. 1660].
- From: *Le Monde ou La Geographie Universelle En plusieurs Cartes.*
- 1 map : copperplate engraving ; 10 x 12.5 cm.
- Latitude and longitude coordinates along all four sides. Longitude coordinates along Equatorial Line. Prime Meridian on map to west of Ferro Island.
- North at the top; no compass rose.
- Title cartouche at bottom left corner.
- No decorative elements on the map.

Example: *New York PL*, Map collection Atlas Case (in Pierre Duval, *Le Monde Terrestre*, Paris, 1661, with 76 maps. This atlas is catalogued with the title of *Le Monde Terrestre* because the title page is missing and the world map was thought at one time to be an illustrated title page).
[Insert Map Image: Map 95b en 95c]

95.2 Second State 1663
Numerous additional placenames such as: *I. de l'Ascension, Tropique de Capricorne,* MER DES INDES.

Examples: *Milwaukee AGS*, (in atlas; this example includes an unaltered copy of the 1660 title page.). *London, Burden Collection*; *Author's collection* (as separate map).

95.3 Third State 1672
Inclusion of additional placenames including *S. Helene la nouuelle. Gonthar* (in Central Africa by Equator).

Examples: *Cambridge Harv MC*, MA 17.82 (edition of 1682); *Washington LC*, G1015.D95 1682 (two volumes with Volume One as Europe only) *and* G1015.D95 1682b; *Paris, Libraire Le Bail-Weissert* (edition of 1682); *Author's collection* (as separate map).

Map 95.2.

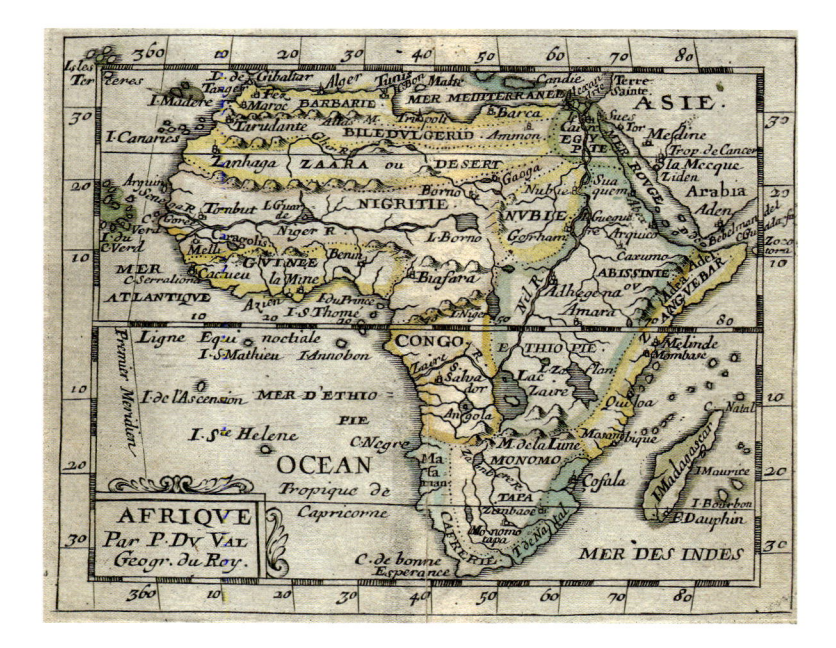

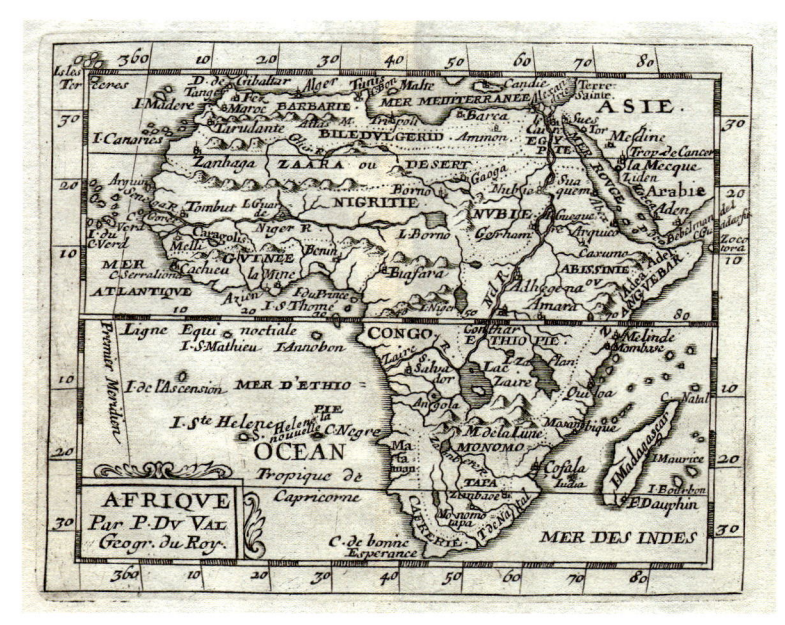

Map 95.3.

There may be a fourth state of this map of c.1687 or 1688, with a plate number in the lower right corner, based on the fourth state of the Americas map of c.1687 as per Burden. This possible fourth state has not yet been located.

Description

Following his folio-size atlas *Cartes de geographie* in 1657, Pierre Duval published a pocket-size atlas *Le Monde ou La Geographie Universelle En plusieurs Cartes* which contained this map of Africa and 75 other maps. This was a much larger number of maps than in his earlier folio atlas. This atlas minor proved to be popular, appearing in numerous editions, in part because it was of a convenient size and likely more affordable to an emerging European middle class than the larger, more expensive folio-size atlas.

As he was the nephew and pupil of Nicolas Sanson, Duval had an early access to mapmaking techniques and information. Duval (1618-1683), born in Abbeville in France, moved to Paris along with other intellectuals at the urging of the French king where he eventually became Geographe Ordinaire du Roy. He died in Paris and was succeeded by his wife, Marie Desmarests, and his two daughters, Marie-Angélique and Michèle. One of the daughters inherited the business at their father's last business address at Rue St. Jacques au Dauphin.

The primary models that Duval used for this map were Nicolas Sanson's folio map of 1650 and Sanson's quarto-size map of 1656. However, this map is lacking a number of the placenames that appear in the Sanson 1656 map.

Publication Information

The earliest known edition of this atlas appears to have been in 1660. According to Burden, this date is found on the title page of the example of a 1663 edition found in the American Geographical Society. However, no complete example of the 1660 edition has been found.

The earliest surviving example of the atlas containing the Africa map is at the New York Public Library. Though this example lacks an engraved title page, a number of the maps within it are dated no later than 1661; some maps have a date as early as 1659 (the La Virginie map). There is also an example with a dated title page of 1661, *Le Monde | ou | La Geographie Universelle | En plusieurs Cartes | Par P. Duval geographe ord[inaire] du Roy. | A Paris*, within a private collection in the United States.

Based on information supplied by Geoffrey King, Jason Hubbard, and Philip Burden, this Africa map appeared in the following editions: *Le Monde ou La Geographie Universelle* (1660?, 1661, 1663. Paris, Pierre Duval); *Le Monde ou la géographie universelle* (1663, 1670, and 1672. Paris, Pierre Duval &

Nicolas Pepingue; and 1676, 1677, 1682. Paris, Pierre Duval); 1682 (Pierre Duval & Nicolas Langlois); 1688 (Mlle. M. Duval [Duval's daughter]). The last three editions of the atlas had an extra title of *La geographie du temps*.

In 1678, a new copperplate of the Africa map was engraved for a Nuremberg edition of Duval's *La Geographie universalis...* translated into German by Johan Christoph Beer and published by Johann Hoffmann (*New York PL*) (Map # 121). Duval's *La Geographie universalis...* was also published in 1688 and 1712 by Jean Certe in Lyon using new copperplates (Map # 143). These other two maps of Africa are close copies of the Duval map and are sometimes confused with this map. The Duval, Johann Hoffmann, and Jean Certe maps of Africa are differentiated by the following:

Map # 95	c.1660, Duval	Map title at lower left is 'AFRIQVE	Par P. DV VAL	Geogr. du Roy.'
Map # 121	1678, Hoffmann	Map title at lower left is 'AFRICA.' Toponyms are in Latin.
Map # 143	1688, Certe	Map title at lower left is 'AFRIQVE	Par P. DV VAL	Geogr. du Roy.' Imprint of the engraver, 'M. Ogier fecit', below neatline at the bottom right corner.

In London, there were further editions of Duval's *La Geographie universalis* as *La Geographia Universalis: The Present State of the Whole World...Written... by Sieur Duval...And now made English*, translated by Ferrand Spence and printed by J. Rawlins and H. Clark in 1685 (*London BL*, 570.c.45) This 1685 edition does not contain maps, and there are no maps in the 1691 edition (*London BL*, 569.e.3).

References
King 2003; Burden 1996: map # 341 (Americas map); Pastoureau 1984: 162-165.
Also, Geoffrey King notes on Duval to the author; Jason Hubbard notes on Duval to the author.

96 1660 Frederick de Wit, Amsterdam

Map 96.2.

96.1 First State 1660

NOVA | AFRICA | DESCRIPTIO | Auct. F. de Wit.
[with publisher's imprint in decorative cartouche at bottom left]:
T'AMSTERDAM | By Frederick de Wit inde Calverstraet inde Witte Paskaert 1660.
- No scale bar.
- Amsterdam : Frederick de Wit, 1660.
- From: separately published / Dutch composite atlases.
- 1 map : copperplate engraving ; 38.5 x 46.5 cm (map only) ; 43.5 x 55 cm (including decorative borders).
- Latitude coordinates along left and right sides. Longitude coordinates along Equatorial Line.
- North at the top; no compass rose.
- Title cartouche at top right corner with a common allegorical representation for Africa of an African woman, holding a balsam branch in her right hand, atop a crocodile. Large vignette at the bottom

left showing Neptune and numerous cherubs and seahorses above a shield containing the imprint of De Wit.
- There are no animals within Africa, or sailing ships or sea monsters in the oceans.
- The decorative borders of six town views at the top, and four figures on left and right sides are described below.
- In some cases, one folio number '3' has been added in the upper right in manuscript ink (Schilder 2000: 126).

Examples: *Amsterdam UB*; *Leiden UB*, COLLBN Atlas 23/24 [5] (in composite atlas); *Author's collection*; *Private Foundation, formerly collection Stopp*.

96.2 Second State 1660
'f° 3' has been added outside the upper right corner of the decorative border frame.

Examples: *Amsterdam UB*, 33-17-44; *Leiden UB*, COLLBN Port 179 N 14; *Private Foundation, formerly collection Stopp*.

96.3 Third State Undated
Date of 1660 has been removed. Addition of woman and child within the shield on the right side of the title cartouche.

Examples: *Wien ÖNB*, 273-3 [75]; *Private Foundation, formerly collection Stopp*.

96.4 Fourth State Undated
'f° 3' has been removed and replaced with '4' in the upper right corner between the two lines of the outer frame of the map.

Examples: *Amsterdam UB*, 33-17-45; *Author's collection* (with a manuscript '3' over the printed '4'); *Private Foundation, formerly collection Stopp*.

Description

In 1660 De Wit prepared and published a set of maps of the four continents with decorative figures on each side border and town views across the top. These are among the earliest known maps by De Wit. Interestingly, they are also the last folio-size maps with decorative borders issued in Amsterdam.

Frederick de Wit (1629-1706) was known as a map publisher, engraver, and seller. He worked in Amsterdam from the Kalverstraat ('by den Dam inde dry Crabben') to 1654 and then from 'in de Witte Paskaert' from 1655 to 1706. He produced a prodigious amount of cartographic material including separate terrestrial maps and sea charts, sea and terrestrial atlases, wall maps, and town views. Many of his maps were superbly colored by master Dutch colorists of his day such as Dirck van Santen.

De Wit also acquired numerous copperplates from those auctioned by the Blaeu and Janssonius publishing houses sometime after 1674 and republished these works, often with his imprint added. On his death in 1706, the business was continued for a time by his widow Maria until 1709. In 1710, the De Wit plates were sold to Mortier, Van der Aa, and Renard. It is not believed that this map was re-issued as it had already been replaced by De Wit's second map of Africa of c. 1670 (Map # 114).

In designing this map of Africa, De Wit was influenced by the Visscher map states (1631, 1636, and 1652) of Van den Keere's map of 1614. De Wit used Visscher's allegorical depiction of the continent of an African woman sitting on a crocodile for his title cartouche.

Across the top border of the map are views of African towns. From left to right, these are *Alcair* (Cairo), *Alexandria*, *Alger* (Algiers), *Tunis*, *Tanger* (Tangiers) and *Ceuta*. These town views are close copies of the 1623 Jodocus Hondius Africa map. Unlike the Van den Keere 1614 (and later states by Visscher) carte-a-figure map of Africa, this De Wit Africa map does not have portraits of African kings interspersed between the town views.

The left and right borders show various costumed African figures. On the left side from the top are FEZZAICA MULIER, CONGENSIS, MADAGASCARICA MULIER, GUINEUS. On the right side from the top are MAROCUS, MULIER ABISSINEA, AZANAGENSIS, AND MOZAMBIQUE MULIER. These figures are the same as on the Van den Keere 1614 (and later states by Visscher) carte-a-figure map of Africa except that the top figures on each side border are completely new drawings.

The geography for the map is closely based on the Jodocus Hondius map of 1623 which in turn is based on the Blaeu map of 1617. The geography presents no new information on Africa. This is somewhat surprising, given the new information, particularly in Southern Africa, that must have been available to De Wit in Amsterdam at this time.

Publication Information

This map of Africa was issued separately and, like all separately issued maps, is not common. Besides appearing separately, the author has seen this map of Africa in composite Dutch atlases.

In its later map states, the map of Africa (along with the other continent maps) has numbers in the upper right corner, indicating an intention to include the map in regular atlases. In some cases, these numbers are in manuscript ink. According to Schilder (2000: 126-127), there are no known copies of standard atlases that included these maps, although Burden (1996: map # 356) states that the Americas map sometimes appears in the Zee Atlas of Hendrick Doncker. The map is known in four states, based on Schilder (2000).

References

Schilder 2000: 126-127 and facsimile 12; Tooley 1999-2004, Q-Z: 402; Burden 1996: map # 356 (Americas map); Koeman 1967-71, III: 191-192.

97 c. 1661 Hugo Allard, Amsterdam

NOVA | AFRICA | Hugo Allardt | Excudit.
[at bottom right side within map gridline]:
Velde Schulp.
[on a line of text above the map]:
NOVA AFRICÆ GEOGRAPHICA ET HYDROGRAPHICA DESCRIPTIO, auct: H; Allaert.
- No scale bar.
- [Amsterdam] : Hugo Allard, [c.1661].
- From: separately published / Dutch composite atlases.
- 1 map : copperplate engraving ; 45 x 56.5 cm.
- Latitude and longitude coordinates along all four sides of the map.
- North at the top; no compass rose.
- Title cartouche at top right corner. Second title on a line of text above the map.
- There is a large vignette across the bottom of the map. On the left side are three Arab traders and three Africans, one a chief. On the right side are various African animals: rhinoceros, elephant, two

lions, a leopard, and a baboon, all following earlier conventions for depicting these animals. There is one sailing ship in the Atlantic Ocean. There are no animals within Africa.

Examples: *London BL*, Maps C.39.f.3 (in a composite Hugo Allard atlas at the BL which the BL dates as c. 1665); *Portland USM*, 1675-01; *Author's collection*.

Description

This is a rare, separately issued map of the continent of Africa. It is part of a set of the four continents prepared and published by Hugo Allard. The map is undated, but is given a date of c.1661, based on its close similarity to the De Wit map of 1660.

Hugo Allard or Allardt (1625-1691) was a map engraver and publisher working from the 'Calverstraet bij den Dam' in Amsterdam and was the founder of a Dutch map publishing firm with his son Carel. Allard's name appears as 'Allardt' and 'Allaert' on the map. Allard is known for reprinting, in about 1650, a noteworthy map of Guinea, *Regni auriferae Guinae*, first published by Cornelis Claesz. and engraved by Baptista van Doetecum in 1602. This map was based on the travel information of the Dutchman, Pieter de Marees, on the Guinea coast.

The geography of this map of Africa is a close copy of Frederick De Wit's 1660 map of Africa. The common geographic antecedent for both maps is the Jodocus Hondius' map of 1623. It is not known if this Allard map was produced with De Wit's approval. It has been given a wide range of dates, but, as it is clearly a copy of the De Wit map of 1660, this map can be no earlier than 1660 and was probably produced shortly after De Wit's map was issued.

Of particular note is the domineering scene across the bottom of the map of African and Arab traders on the left and African animals on the right. The three figures of Arab traders are modeled after three of the figures in the larger scene at the bottom left of the Jan Mathisz. wall map of c. 1655. The three Africans are from part of the large scene at the bottom right of the wall map. Both groups of figures are shown in reverse on this Allard map. The map is engraved by Jan van den Velde, whose imprint, *Velde Schulp.*, is at the bottom right side of the map within the gridline.

Publication Information

While this map was published separately, Burden notes that the set of continent maps appeared in two examples of Doncker's *Zee Atlas* (Burden 1996: map # 370). The author has seen this map in a Hugo Allard composite atlas at the British Library. There is only one known state of this map.

References

Shirley 2004: 156-157; Schilder 2003: 298-302 (for the Allard second state of the Claesz. map of West Africa); Burden 1996: map # 370 (Americas map); Koeman 1967-71, I: 31-39.

98   1661 Philipp Cluver – Johann Buno, Wolfenbüttel

AFRICA | Antiqua | et | Nova.
- No scale bar.
- [Wolfenbüttel : Conrad Buno, 1661].
- From: *Introductionis in Universam Geographicam.*
- 1 map : copperplate engraving ; 21.5 x 26 cm.
- Latitude and longitude coordinates along all four sides of the map. Longitude coordinates along Equatorial Line.
- North at the top; no compass rose.
- Title cartouche at top right corner.
- No other decorative elements on the map.

Examples: 1661 edition: *Wolfenbüttel HAB*, 10,5:5; 1667 edition: *München SB, 4 Geo.u.31;* 1672 edition: *Wolfenbüttel HAB, Ca 173;* 1678 edition: *München SB*, Res 4 Geo.u.31.h; *Washington LC*, G1015.C56 1667 (title page in manuscript with a manuscript date of 1667). *Author's collection* (separate map).

Description

This map is from a German edition of 1661 of Philipp Cluver's famous *Introductionis in Universam Geographicam*, published in Wolfenbüttel or, Guelpherbytum, as it appears on the title page of the atlas in its Latinized form. This edition was edited by Johann Buno and contained a different set of maps than Conrad Buno's 1641 Brunswick (Braunschweig) edition, or the Elzevier 1659 Amsterdam edition.

The two men principally involved in the German editions were Conrad Buno (c.1613-1671), and his brother, Johann Buno (1617-1697). Conrad was an engraver and publisher at the court of Wolfenbüttel. Johann was a theologian and pedagogue in Lüneburg. The engraver for this map is not identified though it could have been Conrad Buno, who engraved the 1641 Cluver maps.

This map borrows from the 1641 Cluver-Conrad Buno map; however, as this map is larger, numerous placenames have been added to it. Much of the nomenclature is decidedly old; for example, *Aethiopia Inferiors Pars* is placed along the Southern African coast. For the interior, this map follows the 1623 Hondius and the 1617 Blaeu folio maps.

Publication Information

Besides the 1661 edition, this map appeared in later editions of 1667, 1672, and 1678. The 1678 edition was published with the imprint of 'heredum Conradi Bunonis' since Conrad had died in 1671. There may also be other editions in which this Philipp Cluver – Johann Buno map appeared, but none of these has been located. The map is only known in one state.

There were numerous other editions of Cluver's *Introductio* that were published using different maps of Africa. Shirley (2004: 345) states that over 45 editions of the *Introductio* appeared over a 100 year period. Not all editions contained maps. Those editions that did contain a map of Africa used one of the following: Cluver-Conrad Buno from 1641; Cluver Elsevier from 1659; Cluver-Johann Buno from 1661; Cluver-Buno-Mosting, with the imprint of Herman Mosting as the engraver as from 1686; Johannes Janssonius in 1661, using his 1628 Goos map; Janssonius van Waesberge in 1676 using the 1630 Cloppenburch map; and Cluver-Wolters from 1697. All are generally derived from the Cluver-Conrad Buno map of 1641, which in turn was based on the Blaeu folio map of 1617. Refer to the chart under the 1641 Cluver-Conrad Buno map, Map # 71, for a list of the different maps of Africa that were used for different editions of Cluver's *Introductio* and the distinguishing characteristics of these maps.

References

Peter Meurer 2006: notes to author; Shirley 2004: 345-350; Burden 1996: map # 360 (Americas map); Shirley 1993: 447; Phillips 1909-1992: entry # 3432; Sabin 1868: entry # 13805.

99　1661 Dancker Danckerts, Amsterdam

99.1 First State 1661

AFRICÆ | nova | discriptio.
(at the bottom left of the map): AMSTELODAMI, Apud Danckerum Danckerts. A° 1661
- No scale bar.
- Amsterdam : Dancker Danckerts, 1661.
- From: separately published / Dutch composite atlas.
- 1 map : copperplate engraving ; 43 x 54.5 cm.
- Latitude coordinates along left and right sides of the map. Longitude coordinates along Equatorial Line.
- North at the top; no compass rose.
- Title cartouche at top right corner. Above the title cartouche, there is a common allegorical representation for Africa showing an African woman, holding a balsam branch in her right hand, atop a croc-

odile. Large vignette at the bottom left showing Neptune and numerous cherubs and seahorses above a shield containing the imprint of Danckerts.
- No decorative elements on oceans or within Africa.

Example: *Paris BN*, Ge D 11379.

99.2 Second State c.1664
with change in the imprint to 'Apud Clement de Ionghe. A° 1661'.

Example: *Washington LC*, G1015.J7 1675.

The second state of this map is in a Library of Congress composite atlas by Clement de Jonghe with a handwritten title of *Tabula Atlantis...* and a date of 'anno 1675'. This atlas contains 52 maps with the latest date on one of the maps of 1664. The Library of Congress example is one of two copies of the de Jonghe atlas known to exist. The second example was once in the collection of Loeb-Larocque; the current location is not known.

Map 99.3 Cartouche bottom left.

99.3 Third State 1679
with change in the imprint to 'Apud Carolum Allard. A° 1679'. Evidence of erasure of the de Jonghe imprint still visible.

Example: *Leiden UB*, COLLBN Port 179 N 16.

It is possible that there is a fourth state of this map as follows: dated 1695 with imprint of 'Apud Carolum Allard 1695', based on a fourth state of the Americas map (Burden, 1996: map # 362). The author has not yet been able to locate an example of this state, if it exists.

Description
Dancker Danckerts (1634-1666) published this map of Africa. He was the eldest son of Cornelis and a member of an active seventeenth and early eighteenth century Dutch publishing family.

The map of Africa is largely based on the De Wit map of 1660, but omits the decorative borders. No changes were made to the body of the map during its publishing life.

Publication Information
In 1666, Dancker Danckerts died leaving no immediate heirs. Clement de Jonghe (1624 or 1625 – 1677) acquired the Danckerts plates, including this map of Africa, in 1661 or shortly before this date. De Jonghe was an art and bookseller and sold composite atlases of maps acquired from others. De Jonghe produced a world map in 1664 and it is likely that he would have desired maps of the four continents as well.

Upon De Jonghe's death in 1677, his son Jacob continued the business to 1687. Sometime between 1677 and 1679, the copperplates came into the possession of Carel Allard who re-published the maps in 1679, including this map of Africa, after altering the plates.

The map did not appear in standard atlases from this period. The map is known in three states.

References
Burden 1996: map # 362 (for Danckerts Americas map); Nebenzahl issue 6: item 29 (fourth state of Danckerts' Americas map); Löwenhardt 1983: 10-12 (for Danckerts world map); Koeman 1967-71, II: 88-90 (for Danckerts) and 213 (for De Jonghe).

100 1663 Anne Seile, London

AFRICAE | NOVA DESCRIPTIO | Impensis | Annæ Seile 1663.

[at bottom right of the map]:

Rob: Vaughan Sculp:

- No scale bar.
- [London] : Anne Seile, 1663.
- From: *Cosmographie in foure Bookes Contayning the Chorographie & Historie of the whole World and all The Principall Kingdomes, Provinces, Seas, and Isles, Thereof.*
- 1 map : copperplate engraving ; 33.5 x 42 cm.
- Latitude coordinates on left and right sides of the map. Longitude coordinates along the Equatorial Line.
- North at the top; no compass rose.
- Simple strapwork title cartouche at the top right of the map.
- In the Atlantic Ocean, there are two sailing ships, one sea monster and four flying fish. Animals within Africa generally follow those on the John Speed map of Africa.

Examples: 1663 edition: *New Haven Yale*; *New York PL*, *KC+ 1669 Heylyn; *Paris BN*; 1666 edition: *Edinburgh NLS*, *P.17.18; *London BL*, 505.kk.1 *and* Maps.570.i.27 (as a separate map).

Description
Upon Henry Seile's death in 1662, his widow Anne issued another edition of Peter Heylin's *Cosmographie*. She employed Robert Vaughan to engrave a set of maps of the four continents and the world that were similar to those her husband had used for his edition of Heylin's *Cosmographie*. For some reason, Anne did not inherit her husband's copperplates. These passed to Henry Seile's business partner, Philip Chetwind.

Peter Heylin or Heylyn (1599/1600-1662) was a noted seventeenth century English cosmographer and geographer who was also Chaplain to King Charles I of England. Heylin wrote *Microcosmus, or a Little Description of the World* in 1621. This book was expanded into his *Cosmographie*. The *Cosmographie* provided comprehensive information on all of the known world. It proved to be popular and was published in London from 1652 into the early eighteenth century.

This map closely follows Henry Seile's map of 1652 (Map # 83), except for a few changes in the decorative elements. For example, there are no ships or sea monsters in the Indian Ocean in the Anne Seile map. Like the 1652 Henry Seile map, this map derives its geography from John Speed's map of Africa of 1626. This is not especially surprising as Speed's map of Africa was the dominant and most readily available map of Africa in the English language during this period. Like the Henry Seile map of Africa, the map does little to advance geographic understanding of Africa and was out-of-date by the time it appeared in Heylin's *Cosmographie*.

Publication Information
Anne Seile's first edition of Heylin's *Cosmographie* was published in 1665-66. This was followed by an edition in 1669 and her last edition in 1677. The map is only known in one state.

A third and final set of maps, after the Henry Seile and Anna Seile maps, was prepared for a 1703 edition of Heylin's *Cosmographie* of 1703, published by Chiswell, Tooke, Hodgkin, and Bennet. For this Africa map, the title 'AFRICA' is placed within an oval cartouche at the bottom left of the map.

References
Burden 1996: map # 379 (Americas map); Sabin 1868: entry # 31655-31656.

101   1664 Pierre Duval, Paris

L'AFRIQVE | Par P. Du Val | Geographe Ordinaire | du Roy. | A PARIS. | Chez l'Auteur prés le Palais. | Avec privilege du Roy | pour Vingt Ans. | 1664.
(within bottom gridline): Somer Sculp.
- No scale bar.
- Paris : Pierre Duval, 1664.
- From: *Cartes de Géographie les plus nouvelles et les plus fideles.*
- 1 map : copperplate engraving ; 40.5 x 53 cm.
- Latitude coordinates and longitude coordinates along all four sides of the map. Longitude coordinates along Equatorial Line.
- North at the top; no compass rose. The Prime Meridian is drawn to the west of *I. de Fer* (Ferro Island) in the Canaries as *Ligne du Premier Meridien.*
- Title cartouche at the bottom left showing an African holding an out-stretched drape.
- Legend at upper right corner on a simple drapery cartouche providing information on symbols for a royal kingdom, bishopric (or diocese), seaport, a cape ('C'), Isle ('I'), Port ('P'), and River ('R').

Map 101.2 Cartouche.

- No decorative elements on the map apart from title cartouche and drapery legend.

Example: *Bourges BM*, E 96 (for edition of *Cartes de géographie* of 1667); *Various other locations*.

101.2 Second State 1676
Text in cartouche changed to: L'AFRIQVE | Reveue et augmentée, | Par P. Du VAL | Geographe Ordinaire | du Roy. | A PARIS. | Chez l'Auteur, prés le Palais, | sur le Quay de l'Orloge. | Auec privilege du Roy | pour Vingt Ans. | 1676.

Example: *Bedburg-Hau, Antiquariat Gebr. Haas*.

101.3 Third State 1684
Text in cartouche changed to: L'AFRIQVE | Reveue et augmentée, | Par P. Duval | Geographe Ordinaire | du Roy. | A PARIS. | Chez Mlle Du-Val Fille de l'Auteur. | sur le Quay de l'Orloge. | Auec priuilege du Roy | pour Vingt Ans. | 1684.

The map re-inserts *B. das Vacas* and *C. Talhado* from the Sanson 1650 map. *B. da Lagoa* and *C. du Recif* in South Africa are added. The map was prepared by Duval's daughter.

Examples: *Bedburg-Hau, Antiquariat Gebr. Haas*. Map image in Tooley 1968: *Map Collectors' Circle* no. 47, plate no. 38.

101.4 Fourth State 1698
Text in cartouche changed to: L'Afrique. | Par P. Du-Val, | Geographe Ordinaire | du Roy. | A Paris. | Chez Melle DV-VAL, Fille de l'Auteur | 1698.

The map was prepared by Duval's daughter. Verso of the map is blank.

Example: *Amsterdam UB*; *Bedburg-Hau, Antiquariat Gebr. Haas*.

Map 101.4 Cartouche.

Description
Pierre Duval's first edition of 1654 of his folio-size atlas *Cartes Géographiques methodiquement divisées* used the 1650 map of Africa by his uncle, Nicolas Sanson. By 1655, Duval began the process of preparing new maps for inclusion in this folio-size atlas, now called *Cartes de géographie...*, including a new map of the Americas. Based on Pastoureau, it was not until the 1667 edition of *Cartes de géographie* that this Duval map of Africa first appeared.

This map was still modeled on Nicolas Sanson's earlier work of 1650. Pierre Duval did add some placenames to this map. In Southern Africa and the Cape of Good Hope, these new names were *Tafel Bay, I. Da Fera, B. de Saldagne, C. Falso, C. de i'Infante,* and *Beauport*, reflecting the increased commercial attention in the area of the Cape of Good Hope and the recent Dutch settlements in Southern Africa.

While the first state of this map of 1664 presents a generally conventional view of Central Africa, especially with the two Ptolemaic lakes for the source of the Nile River, it is interesting that by the fourth state of 1698, this map breaks the two Ptolemaic lakes' connection to the Nile River. It appears that Duval and his successors were determined to utilize new information in their depiction of Africa.

Publication
This map appeared in the numerous editions of Duval's *Cartes de géographie* until Duval's death in September 1683. The business was continued by one of his daughters, either Marie-Angelique or Michele Duval, though it is not known which one. This daughter altered the imprint on the map and continued the business into the early 1700s, though the last known appearance of this map was in 1698.

References
Norwich 1997: map # 38 (first state); Burden 1996: map # 311 (Americas map); Pastoureau 1984: 135-137 and 139-151; Tooley 1968: 44.

102 c. 1665 Antonio Francesco Lucini, Florence (?)

AFRICA.
- No scale bar, however below the title is the wording 'Longezza 2000 leghe in circa'.
- [Florence(?) : Lucini, c.1685].
- From: separately published (as part of a gameboard).
- 1 map : copperplate engraving ; 9.0 x 8.5 cm (map of Africa only).
- No latitude and longitude coordinates.
- North at the top; no compass rose.
- Simple box title cartouche at the bottom left of the map.
- No decorative elements on the map.

Example: *London, Burden collection* (only known example).

Description

This is an unusual map of Africa. It is part of what appears to be a game printed on one sheet of paper. The other continents are also represented on the game along with regions of the world. Only one partial example of this game, missing the top ? of the game, is known to exist with the Burden Collection in England.

The game has an imprint on the top right of 'luci', signifying the likely authorship of Lucini. Antonio Francesco Lucini (1605-?) was an engraver and map printer, active primarily in Florence, from about 1631 until about 1665. Lucini was responsible for the engraving of the copperplates for Robert Dudley's Arcano di Mare in 1646, with a second edition in 1661. He also is responsible for a reduced version of Dudley's charts of c.1665. Immediately above the map is the wording of *GIOVANE NOBILTA VENETIA*. It is not known to what this refers.

CARTOBIBLIOGRAPHY

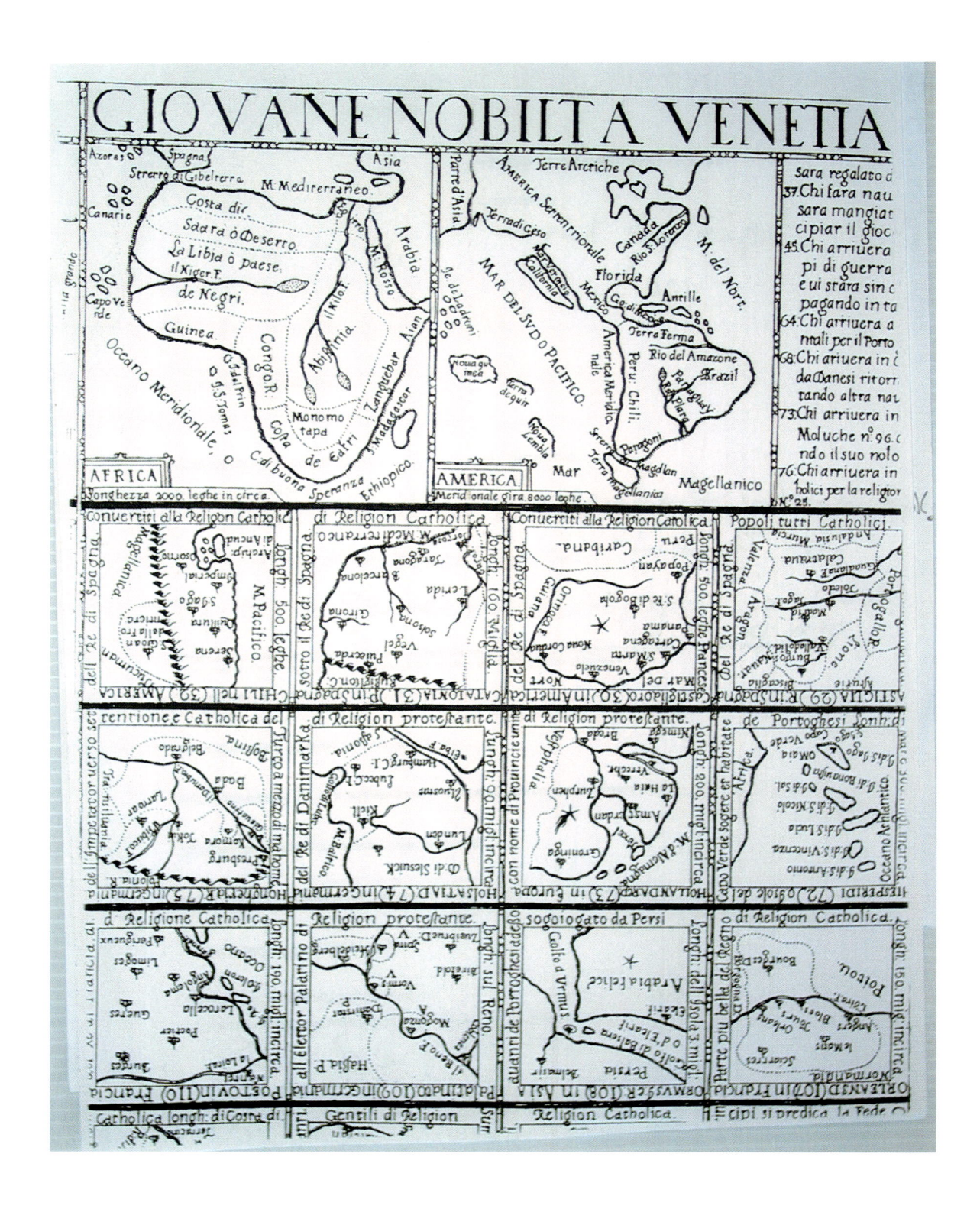

The map presents a simplified understanding of seventeenth century Africa, partly owing to its small size. Geographically, the Niger and Nile Rivers with their source lakes are shown. The map is divided into ten regions plus the surrounding islands of the Canaries, Cape Verde, and Madagascar.

Publication Information

The map contains no date. The date assigned to this map board game is purely based on conjecture and reflects the period when Lucini was cartographically active. Similarly, there is nothing on the map to indicate that this map was produced in Florence, except for the fact that Lucini was known to have mainly published his work in that city.

References

Discussion with Philip Burden.

103   1665/1666 Thomas Jenner - Wenceslaus Hollar, London

A New and Exact Map | of | AFRICA | and the Ilands thereun to | belonging | Anno 1666.
[at bottom right of the map]:
W: Hollar fecit. 1665.
- No scale bar.
- [London : Thomas Jenner], 1666.
- From: separately published / Composite Atlas / *A Map of the Whole World*(?).
- 1 map : copperplate engraving ; 37 x 48 cm.
- Latitude and longitude coordinates on all four sides of the map. Longitude coordinates on Equatorial Line.
- North at the top; one compass rose in the Indian Ocean and one in the Atlantic Ocean.
- Title cartouche at the bottom left of the map.
- There are numerous decorative elements on the map. In the Indian Ocean, there are nine sailing ships and three sea monsters. In the Atlantic Ocean, there are 19 sailing ships and nine sea monsters. In the Mediterranean Sea, there are six various sailing ships, one sea battle, and three sea monsters. Numerous animals are within Africa.

Examples: *Chicago NL*; *London BL*, K.Top.117.5 (as a separate map) *and* Maps C.39.f.5 (map # 19 within composite atlas).

Description

This map is equally attributed to Wenceslaus Hollar and Thomas Jenner. It is known that Hollar and Jenner collaborated on a set of separately published maps of the four continents, including this map of Africa. Wenceslaus Hollar (1607-1677) was a noted London artist, engraver, and etcher. He was recognized as being one of the foremost engravers of his day and was appointed Iconographer to the King in 1660. Hollar was born in Prague and received his training as an engraver under Matthäus Merian in Frankfurt before immigrating to London in 1636.

The map is also attributed to Thomas Jenner as his imprint as the map's publisher and seller appears on the companion map of the Americas. Jenner (1618-1673) was a bookseller, engraver and publisher in London.

The map is based on the geography in Jodocus Hondius Jr.'s map of 1619, though using the third state of 1631 by Henricus Hondius with no decorative borders. Many of the placenames on the Jodocus Hondius Jr. map are placed in the same positions on this map.

Publication Information

This map of Africa was initially published as a separate map. It is assigned a date of 1665/1666 as the map was engraved by Hollar in 1665 (W: Hollar fecit. 1665) and published in 1666 with that date on the title.

The Americas map appeared in Thomas Jenner's *A Map of the Whole World* in 1668; the only known example is at the Huntington Library (Burden 1996: map # 393). The author has not yet been able to confirm that this map of Africa also appears in this book, though it is possible that this is the case.

Shortly after Jenner's death in 1673, his business was acquired by John Garrett. Garrett published Donald Lupton's *A Most exact and accurate Map of the whole world* in 1676. This book contained a second state of the Americas map, with Jenner's name replaced by John Garrett and the date erased. Based on an examination of the British Library example (BL: MAPS C.21.b.24), the Africa map did not appear in this book.

While the Americas map by Jenner is known to be in three states, to date, the author has only identified one state of the Jenner-Hollar map of Africa.

Map 103A

Hollar engraved a prodigious amount of print material including subjects as diverse as religious and historical prints, maps, portraits, costumes, and natural history. The University of Toronto Libraries has a print in their Wenceslaus Hollar Digital Collection (plate number P644) showing a small map of Africa on what appears to be an outstretched lion skin with an African man and an elephant on the left side and an African woman and a rhinoceros on the right side of the map. The map has *AFRICA* as a title across North Africa with only a few placenames and no physical features (mountains, rivers, etc.) on it. The general shape of Africa is quite modern in appearance. The print is 5 x 20 cm. See the title page (p. 4) for an image of this map.

References

Shirley 2004: T.COM-33a; Tooley 1999-2004, E–J; Burden 1996: map # 393 (Americas map); Pennington 1982; Tyacke 1978: 114-116 (for Garrett) and 118 (for Jenner).

104 1666 Nicolas Sanson, Paris

L'AFRICQUE, | divisée en ses principales parties, | et en ses ESTATS, ROYAUMES, &c. | Tirée de Livio Sanuto, de Marmol, et autres | Par le Sr SANSON d'Abbeville Geogr ord^{ie} de Sa Majesté | A PARIS. | Chez Pierre Mariette, Rue S Iaques a l'Esperance | Avec Privilege du Roy pour Vingt Ans . | 1666.
[at bottom left within gridline below scale bar]:
Johannes Somer Pruthenus Sculp.
- Scale bar:
    Eschelle
    Degrés de Latitude, ou de | Longitude sur l'Equateur 10 = 10.6 cm.
    Mille Pas Geometriques 600 = 10.6 cm.
    Lieües comm de France 250 = 10.6 cm.
    Lieües comm de d'Allemagne 150 = 10.6 cm.
    Lieües comm de Mer 125 = 10.6 cm.
- Paris : Pierre Mariette, 1666.
- From: separately published.

- 1 wall map : copperplate engraving ; 4 map sheets; 60 x 46 cm each sheet.
- Latitude and longitude coordinates within gridline on all four sides of the map; Longitude coordinates along Equatorial Line.
- North at the top; no compass rose.
- Title is set within a decorative cartouche on the upper right sheet. At the top of the cartouche are a cheetah and a lion with two water gods below. At the bottom of the cartouche is an oval view of several sailing ships surrounded by a crocodile and a dragon-creature.
- There are six ships of various sizes and types within the Indian Ocean to the east of Southern Africa. There are 13 ships of various sizes and types within the Atlantic Ocean to the west of Southern Africa. No decorative elements are within Africa.

Example: *Paris BN*, Ge D 11380 (1-4).

Description

Nicolas Sanson prepared and issued this attractive wall map of Africa along with maps of the other continents in 1666, one year before his death. It is not clear why Sanson waited so long to issue a large format wall map of Africa. Possibly he felt he could not compete in this area with the commercially dominant Blaeu wall maps of the continents produced in The Netherlands and its derivatives produced in France as well as Italy at this time. It is also possible that Sanson produced an earlier issue of this wall map of Africa. The companion wall map of North America, also dated 1666, has text panels with a date on them of 1665, suggesting an earlier issue of the North America map.

In the title, Sanson acknowledges Livio Sanuto's 1588 continent and regional maps of Africa as a source for his map, though by this time Sanuto's maps were out-of-date. Sanson also acknowledges Marmol-Caravajal's work on Africa, *Descripcion general de Affrica...*, which originally appeared in Granada in parts in 1573 to 1599. This work describes the account of a prisoner-of-war captured after the fall of the Spanish-held fortress of La Goulette and the surrounding area in North Africa. Much of the geography in Marmol-Caravajal's work borrowed heavily from the much earlier work of Leo Africanus.

The basic model for this map is Sanson's 1650 folio map of Africa with the interior river systems generally following the Blaeu 1608 wall map. Sanson does update this map with new information on *ou de St. Laurent et Isle Daufine* (Madagascar) with the placement of *Fort Dauphin* at the southern end of the island. He changes the name of the Cuama River to the *Zambere River*, though *Zefala* (Sofala) is still placed far to the north of the river. Sanson also introduces a fictitious second *Nouuelle I. de St. Helene* (St. Helena Island) to the east of the island of St. Helena. Sanson's map of the coast of Southern Africa does not appear to utilize new Dutch information; in fact, Sanson does not even use the information from his 1650 map and incorrectly places the Cape of Good Hope at generally the same latitude to the south as Cape Agulhas. This information from Nicolas Sanson' 1666 map was to re-appear in his son, Guillaume's, 1668 folio map of Africa.

Publication Information

This wall map is exceedingly rare. It is only known to the author in one example at the Bibliothèque Nationale de France in Paris. Until recently, Sanson's companion wall map of North America was also thought to exist in only one example with a date of 1667 as part of the title. Burden (1996: map # 396) states that another example of the Sanson North America wall map dated 1666 was discovered in 1992.

References

Burden 1996: map # 396 (Americas map).

105 1666 Giovanni Giacomo de Rossi, Rome

Map 105.2.

105.1 First State 1666 (see ill. p. 343)

NOVA | TOTIVS | AFRICAE | TABVLA | AVCTORE IOANNES BLAEV | AMSTELÆDAMI | Jo. Jacobus de Rubeis | Formis Romæ.
[within oval cartouche at the lower right corner of the 4th map sheet]:
Joannes Jacobus de Rubeis | Formis Romæ ad | Templum Pacis cum Privileg. | Summi Pontificis | Anno 1666.
- No scale bar.
- Rome : Giovanni Giacomo de Rossi, 1666.
- From: separately published.
- 1 wall map : copperplate engraving ; 4 map sheets ; In first state: 2 sheets with costumed figures on each side and twelve town views at the bottom; Letterpress text with illustrations at the bottom and sides. The entire map is surrounded on all four sides by decorative scroll strips ; 120 x 180 cm (overall size).

- Longitude and latitude coordinates on all four borders of the map. Longitude coordinates across Equatorial Line.
- North at the top; there are five compass roses on the map: two in the Atlantic Ocean, two in the Indian Ocean, and one in the Mediterranean Sea ; one windrose in the South Atlantic.
- Title at upper right of the map is set within a decorative oval cartouche with three figures at the top.
- In the lower left corner, there are two circular diagrams at the top of a cartouche with text; in the upper left corner, there is a cartouche with text. Letterpress text panels on bottom and both sides of the map.
- In Indian Ocean, one ship and five sea monsters ; In Atlantic, eight ships, three sea monsters, vignette of Neptune and mermaid. In Mediterranean, one smaller ship and one sea monster.
- Numerous animals and figures of Africans are found on the map of Africa.
- Eight groups of costumed figures on each side of the map. Twelve town views across the bottom of the map.
- Various other decorative elements on the map as on the Blaeu 1608 wall map.

Example: Only known example of the first state is at Museo dell'Opera del Duomo di Orvieto (the Cathedral), Orvieto, Italy.

105.2 Second State 1686
Date changed to 1686 on imprint within cartouche at bottom of fourth map sheet.

Example: *Rotterdam MM*, K 248 (this example is without the text panels and the decorative borders).

Description
This map was published by Giovanni Giacomo de Rossi (1627-1691) who was a printer and publisher in Rome. De Rossi was part of an active publishing family, founded by De Rossi's father Giuseppe and continued by Giovanni's sons Domenico and Lorenzo to 1738 when the De Rossi plates were sold to the government of the Papal States (Tooley 1999-2004, Q-Z: 75-76). This map shows the imprint of Joannes Jacobus de Rubeis, the Latinized name for Giovanni Giacomo De Rossi.

This is the third Italian copy of the Willem Blaeu wall map of Africa of 1608, after the Scaicki wall map of c.1620-35 and the Scolari wall map of 1646. There are no geographic improvements to this map. De Rossi produced it simply to take advantage of the tremendously popular and financially successful Blaeu 1608 wall map and its derivatives.

The letterpress text on the borders begins with 'NOVA | AFRICÆ | DESCRIPTIO' and ends on the bottom right with the imprint 'ROMÆ, | Apud Antonium Mariam Gioiosum. M.DC.LIV. | SVPERIORVM PERMISSV. | Venundantur Romæ prope Ecclesiam B. Mariæ de Pace. | Typis Io. Iacobi de Rubeis ad signum Lutetiæ'. It is important to note that the letterpress type ends with an imprint date of 1654 (M.DC.LIV). It is therefore very possible that De Rossi acquired these plates from an unknown mapmaker who previous published this map using these plates. Schilder believes this person may have been Godefridus de Scaicki who was originally from Utrecht and whose original name was probably Van Schayck.

The map is bordered on the two sides by representative, costumed figures of Africa. Along the bottom are twelve town views. Outside of this is text in Latin with various illustrations. The entire map and text are framed on all four sides by decorative strips (See Map # 54, the entry for the Willem Blaeu 1608 wall map of Africa for a discussion of these decorative borders.).

References:
Tooley 1999-2004, Q-Z: 75-76; Schilder 1996: 199-208; Palma 14-19 June 1991.

Map 105.1.

106 1667 Guillaume Sanson, Paris

106.1 First State 1667

AFRICA | VETUS | NICOLAI SANSON Christianiss. Galliar. Regis | Geographi. | Recognita Emendata et | Multis in locis Mutata, | Conatibus Geographicis | GULIELMI SANSON N. FILII. | Lutetiae Parisiorum | Apud Petrum Mariette Via Jacobaea | Sub Signo Spei. | CIƆ IƆ C LXVII | Cum Privilegio ad Viginti Annos.
- No scale bar.
- Paris : Pierre Mariette, 1667.
- From: *Carte Generale de toutes les parties du Monde.*
- 1 map : copperplate engraving ; 40 x 55 cm.
- Longitude and latitude coordinates on all four sides of the map. Longitude coordinates across Equatorial Line.
- North at the top; no compass rose.
- Title cartouche at the upper right.
- There are no decorative elements on the map, apart from the attractively engraved title cartouche.

Examples: *London RGS*, 7.H.12; *Paris BN*, Ge DD 2687 Pl 55; *Author's collection.*

106.2 Second State 1690
New date of 1690 added at the bottom of the title cartouche.

Example: *Arizona, Old World Auctions.*

Description
This map was produced by the noted Sanson publishing firm. Nicolas I (1600-1667) founded the firm and was later appointed Geographer to the King (Geographe Ordinaire du Roy). His son and intended heir Nicolas II died in 1648. After Nicolas I's death in July 1667, the business was succeeded by his second son Guillaume (1633-1703) who also received the title of 'Geographe du Roy'. Guillaume Sanson prepared this Africa map for inclusion in the atlas *Carte Generale de toutes les parties du Monde*. This map replaces a similar one by Nicolas I of 1650.

The map of ancient Africa is one of a series of maps depicting parts of the ancient world. Guillaume followed his father's particular interest in subjects dealing with the past. Geographically, this map is similar to the 1650 map of ancient Africa, *AFRICA VETVS*, by Nicolas Sanson (Map # 79) but it is distinguished from it by a date within the title of 1667 (N.B.: 'CIƆ' represents M, and 'IƆ' represents D, thus the full date is MDCLXVII or 1667).

The map generally shows Africa as it was prior to the European discoveries of the sixteenth century, but it does include a more modern shape for Africa. The geographical representation of the interior follows Ptolemy and other antiquarian sources, notably the Greeks and Romans. Placenames reflect knowledge of ancient traditional African groups as well as ancient geographical places. There are a number of 'strange', misplaced mountains ranges; the Nile River shows a Ptolemaic origin; and Southern Africa is devoid of all names except a few. The inhabitants of Southern Africa are called *Anthropophagi Aethiopes*.

Publication Information
Pastoureau cites seven separate editions of the *Les Carte Generales* in 1658, 1665, 1666, 1667, 1670, 1675, and 1676 (as *Les Carte Generale... ancienne et nouvelle*). The last three were published by Nicolas' son, Guillaume, after Nicolas' death in 1667. What makes the listing and collation of Sanson's atlases particular difficult is that often the contents do not agree with the printed list of maps, as on a regular basis extra maps were inserted into the atlases. As noted on the title, this map was issued at the establishment of the printer Pierre Mariette.

Pierre Mortier is known to have re-produced this map in 1705 with the following information: *AFRICA | VETVS | Autore N. SANSON.* and AMSTELODAMI : Apud P. MORTIER., [1705]. - 39 x 54.5 cm. - North up. - [Cart. top right] Cum Privilegio. It was issued in a second state by Covens and Mortier in c.1737 with the addition of: Apud I. CÓVENS et C. MORTIER.

References
Norwich 1997: map # 41 (first state); Pastoureau 1984; Tooley 1969: 98-99.
Discussion with Marco van Egmond, Universiteit Utrecht, who wrote his doctoral dissertation on Covens and Mortier and who kindly shared his information on Pieter Mortier.

107 1668 Guillaume Sanson, Paris

107.1 First State 1668

AFRIQUE, | Par N. SANSON Geographe Ordinaire du-Roy | corrigée et changée en plusieurs endroits | suivant les Relations les plus recentes; | Par le Sr SANSON le Fils. | A PARIS; | Chez PIERRE MARIETTE rue S. Iacques à l'Esperance | Avec privilege de S. Mai.té po.r [pour] 20 ans. | 1668.

- No scale bar.
- Paris : Pierre Mariette, 1668.
- From: *Carte Generale de toutes les parties du Monde*.
- 1 map : copperplate engraving ; 40 x 55 cm.
- Longitude and latitude coordinates on all four sides of the map. Longitude coordinates across Equatorial Line.
- North at the top; no compass rose.
- Title cartouche at the upper right of the map.
- No decorative elements on the map except for the title cartouche.

Examples: *London BL*, Maps.C.39.e.2; *Author's collection*.

107.2 Second State 1669
Dated 1669 at the end of the title cartouche. No other changes on the map.

Examples: *London BL*, Maps C.36.f.6; *London RGS*, 7.H.12; *Paris BN*, Ge DD 2687; *Washington LC*, G1015.S25 1670.

There may possibly be a third state of this map as follows, but the author has not located an example of this state.

107.3 Third State? 1690
Dated 1690. Following the fifth line of title, insertion of 'augmentee et corrigee en cette seconde edition'.

Example: Not yet located; possible existence of this state is based on the third state of the Americas map (see Burden 1996: map # 405).

Description
Guillaume Sanson (1633-1703) assumed control of the Sanson publishing business upon the death of his father Nicolas in July 1667. Shortly after this, Guillaume began work on a new Africa map for inclusion in the *Carte Generale de toutes les parties du Monde*. This new map was intended to replace his father's map of 1650 which was considered out-of-date by this time.

The map was produced in partnership with Pierre Mariette the younger (1634-1716). Their partnership lasted until about 1673, when Sanson began to work instead with Alexis-Hubert Jaillot. it is unclear as to why Sanson and Mariette ended their business relationship. The 1668 Africa map was possibly engraved by Mariette, although there is no indication of this on the map itself.

Guillaume's map has additional placenames in those areas of Africa that were coming under increasing French influence, particularly in West Africa and on Madagascar. Madagascar has the additional title of *Isle Dauphine*, replacing his father's title of S. Laurens. *Fort Dauphin*, which was not on his father's 1650 map, is shown at the southern end of Madagascar.

The map shows *St. Helena* in the South Atlantic Ocean and to the east *N. I. de S.ta Helena* (New St. Helena Island), a fictitious island which did not appear on his father's map. However, on this 1668 map Sanson does not appear to utilize new Dutch information on the coast of Southern Africa. He incorrectly places the Cape of Good Hope at generally the same latitude to the south as Cape Agulhas, the true southern-most extent of the continent.

The map has a traditional view of the Niger River flowing westward into the Atlantic. The *R. de Spiritu Santo* and *Zambere* (replacing the Cuama) Rivers originate from one river that emerges from an unnamed lake (Sachaf Lacus) below the two Ptolemaic Lakes of Central Africa.

Publication Information
Guillaume Sanson sold the business to his nephew Pierre Moullart-Sanson in 1692 or 1693. It is not known if the copperplate for the Africa map was included in this sale. As this map was considerably out-of-date by this time, due to the numerous cartographic advances in the mapping of Africa through the French Académie Royale des Sciences, it is likely that this copperplate was not reused.

References
Shirley 2004: 883-884; Burden 1996: map # 405 (Americas map); Pastoureau 1984: 416.

108 1668 Jacob van Meurs, Amsterdam

AFRICÆ | ACCURATA TABULA | ex officina | IACOBUM MEURSIUM.
- No scale bar.
- [Amsterdam] : Jacob van Meurs, [1668].
- From: *Naukeurige beschrijvinge der Afrikaensche gewesten van Egypten, Barbaryen, Libyen, Biledulgerid, Negroslant, Guinea, Ethiopiën, Abyssinie … .*
- 1 map : copperplate engraving ; 43.5 x 54.5 cm.
- Longitude and latitude coordinates on all four sides of the map.
- North at the top; one compass rose in the South Atlantic Ocean.
- Title cartouche at upper right.
- There are two sailing ships in the Indian Ocean; eight sailing ships, one sea monster, and four flying fish are in the Atlantic Ocean. Animals are within Africa.

Examples: *London BL*, 146.i.1 (in John Ogilby's *From Africa, Being An Accurate*

Description of the Regions... London, 1670 with an imprint of 'Printed by Tho. Johnson for the Author and are to be had at his House in White Fryers, MDCLXX.'); *Author's collection.*

Description

Jacob van Meurs (c.1620-1680), an Amsterdam publisher, used the c.1658 Visscher map of Africa as his model for this map in Olfert Dapper's *Naukeurige Beschrijvinge der Afrikaensche gewesten...* which he published in 1668. Dapper's book was one of the more famous seventeenth century books on Africa. As well as describing places in Africa, it provided detailed information on all aspects of the people of Africa, including their customs, costumes, languages, religions, and governments. Also in the book were descriptions of the flora and fauna of Africa.

This map is a close copy of the Nicolaas Visscher I map of Africa of c.1658 in its first state, without the publisher's privilege. Geographically, there are no changes in this map from the Visscher model. Besides the new title cartouche, Van Meurs' version of the map has a blank dedicatory cartouche and only two ships in the Indian Ocean versus three on the Visscher map.

Publication Information

This map first appeared in Olfert Dapper's *Naukeurige Beschrijvinge der Afrikaensche...* of 1668 in Dutch with a second Dutch edition of 1676. The map also appears in the German (in 1670, published by Johann Hoffmann), English (in 1670), and French (in 1686) versions of Dapper's work.

The English version was John Ogilby's *Africa, Being An Accurate Description of the Regions...with all Adjacent Islands...their Coasts, Harbour, Creeks, River, Lakes, Cities...Customs, Modes and Manners, Languages...Plants, Beasts, Birds and Serpents* published in 1670 in London. Ogilby's version is considered the most authentic and comprehensive work on Africa published in the English language in the seventeenth century. Besides the map of Africa, the book contained regional maps and numerous views. The map is only known in one state.

References

Norwich 1997: Map # 43; Gay 1998: 219; Burden 1996: 431; Mendelssohn 1993: Vol. I: 413-414; Tiele 1867: 298.

109 1668 John Overton, London

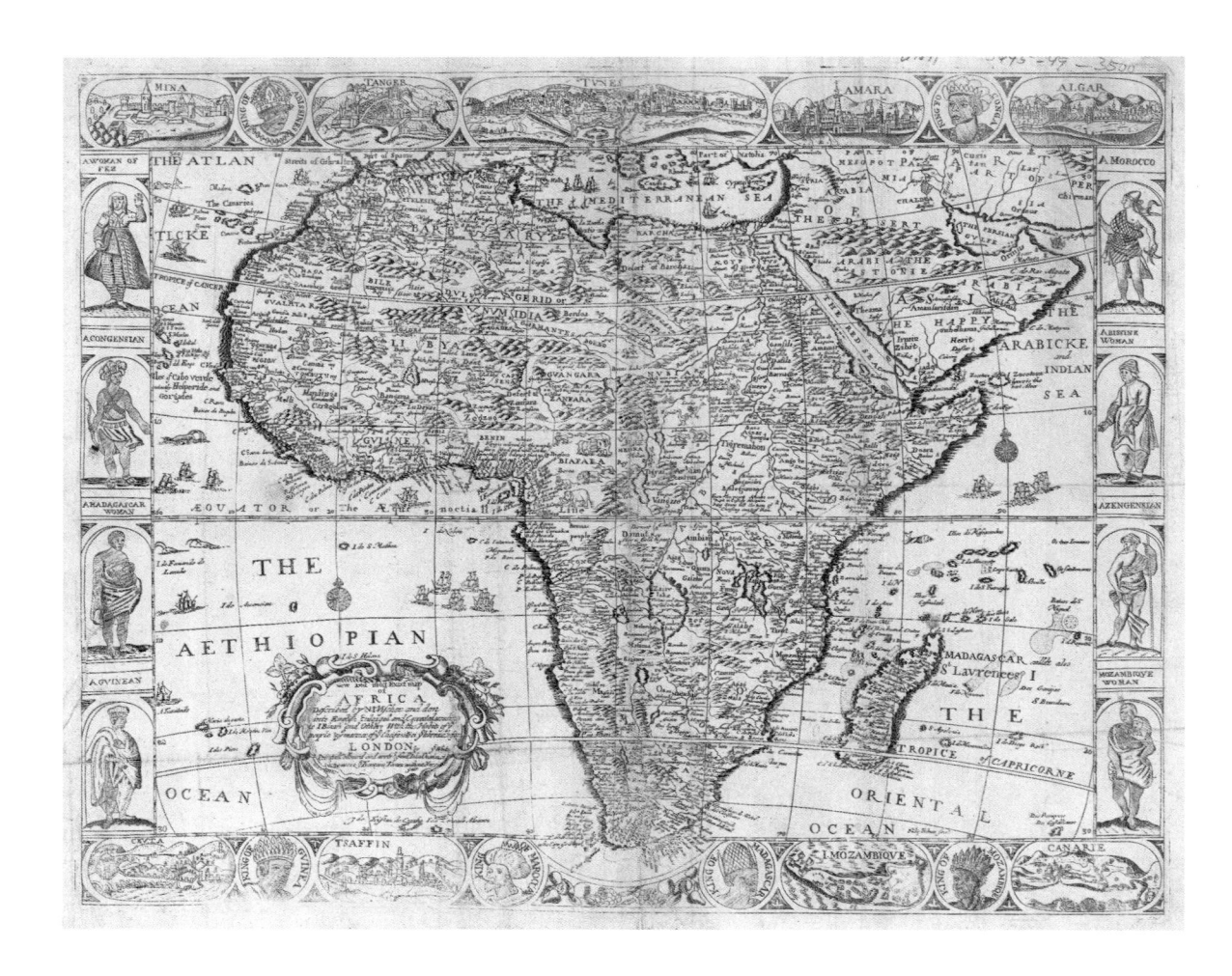

109.1 First State 1668

A | new and most Exact map | of | AFRICA | Described by N:I: Vischer and don | into English Enlarged and Corrected acording | to I Bleau and Others With the Habits of y^e | people & y^e manner of y^e Cheife sitties: ye like neuer before | LONDON | Printed Colloured and are to be sould by Iohn Ouerton at ye White | horse in Little Brittaine neare the Hospitall | 1668.
[at bottom right]:
Phillip Holmes fecit.
- No scale bar.
- London : John Overton, 1668.
- From: separately published.
- 1 map : copperplate engraving ; 42 x 52 cm including borders.
- Longitude and latitude coordinates on all four sides of the map. Longitudinal coordinates across Equatorial Line.
- North at the top; one compass rose in the Indian Ocean and one in the Atlantic Ocean.
- Title cartouche at bottom left.
- There are five sailing ships in the Indian Ocean, 14 ships and three sea monsters in the Atlantic, and six sailing ships in the Mediterranean Sea. The decorative borders on all four sides are similar to the Visscher states of the Van den Keere map of 1614. Animals within Africa unlike on the Van den Keere-Visscher map.

Example: *London RGS*, 264.H.14, plate 68.

109.2 Second State c.1670

Evidence of erasure to the date. Last line of text on the title cartouche changed to 'horse neere ye Fountaine Tavern without New | gate'.

Example: *London BL*, Maps C.39.e.3 (dated 1670 at BL); *Various other locations.*

Description

John Overton (1640-1713) was a mapseller who worked from a number of premises around London. This map was produced at the *Whitehorse in Little Britain near the Hospital* where he worked after 1666 and before 1669. Overton lost much of his map stock in the Great Fire of London of 1666; this map of Africa was published shortly after the Great Fire.

The title in the Overton map gives an acknowledgement to Visscher (and Visscher's 1652 fifth state of the Van den Keere map of 1614) in the production of this map. This is most evident in Overton's use of Van den Keere's town views and figures on the decorative borders, although the titles are in English rather than in Latin.

The left and right borders show various costumed African figures. On the left side from the top are *A WOMAN OF FEZ, A CONGENSIAN, A MADAGASCAR WOMAN, A GVINEAN*. On the right side from the top are *A MOROCCO, ABISSINE WOMAN, AZENGENSIAN,* and *MOZAMBIQVE WOMAN.*

Across the top and bottom borders of the map are views of African towns. At the top from left to right are *MINA* (The Mine at St. George in Guinea), *TANGER* (Tangiers), *TVNES, AMARA* (the home of Prester John), and *ALGAR*. At the bottom from left to right are *CEVTA, TSAFFIN, I. MOZAMBIQVE,* and *CANARIE* (in the Canary Islands). These town views are interspersed by faces of various kings of Africa; at the top, *KING OF ABISSINES* and *KING OF CONGA*; and at the bottom, *KING OF GVINEA, KING OF MAROCCA, KING OF MADAGASCAR,* and *KING OF MOZAMBIQVE*. Overton used all of the Van den Keere borders except for the bottom middle town view of Cefala, which was removed as the southern tip of Africa extended into the bottom border to accommodate newer placename information.

Overton also uses the Van den Keere-Visscher fifth state for some of his depiction of the interior of Africa. However, considerable information is updated on this map compared to the Van den Keere-Visscher map. This information may have come from Visscher's c.1658 folio map of Africa, particularly with new Dutch placenames in South Africa: *Robben Eyl., Tafel berg* (Table Mountain), etc.

For the general outline of Africa and Madagascar, Overton follows the Blaeu model, which he also acknowledges in his title. As well, Overton has included additional placenames along the coast, particularly in Southern Africa, when compared to the Blaeu map.

Overton also borrows from the two prominent English maps of Africa of this period. These are the Walton map of 1656 (based on the Van den Keere map) and the Jenner-Hollar map of 1666 (based on the Blaeu model and specifically the Jodocus Hondius Jr. map of 1619). Oddly, there is no clear evidence of the Dutch *Mossel Bay* on the Overton map, unlike on the Jenner-Hollar map.

Publication Information

After Peter Stent died in 1665, Overton acquired his map stock but it did not include a set of continent maps. To compete with the two other English maps of this period, those by Robert Walton and Thomas Jenner-Wencesalus Hollar, Overton produced this map of Africa and the other continent maps as separate maps in 1668. From about 1670, Overton began compiling and selling atlases. This map appeared in these atlases as well. According to Burden (1996: map 401), the map sometimes appeared in Lewes Roberts' *The Merchants Map of Commerce* and Robert Fage's *Cosmography*, both in 1671.

References

Schilder 2000: 425-426 (p. 425 for an image of the first state of the map); Tooley 1999-2004, K–P: 363-365; Norwich 1997: map # 40 (second state); Burden 1996: map # 401 (Americas map).

110 1669 Pierre Duval, Paris

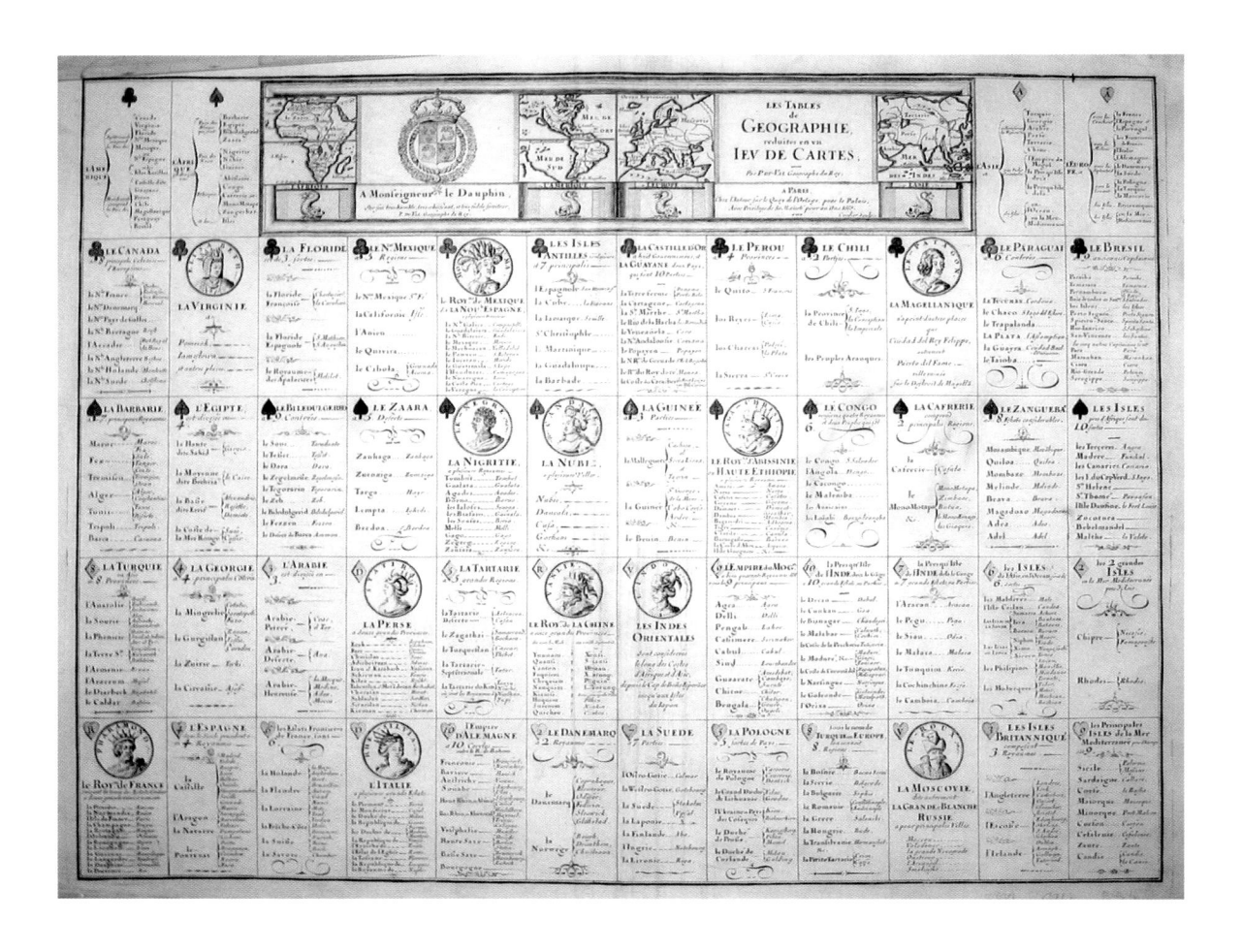

L'AFRIQUE.
[to the right of the map of Africa, there is a dedication to the Dauphin]:
A Monseigneur le Dauphin, | Par son tres-humble, tres-obeissant, et tres-fidele serviteur, | P. Dv-Val Geographe du Roy.
[to the right of the map of Europe, there is the title for the sheet]:
LES TABLES | de | GEOGRAPHIE, | reduites en vn | IEV DE CARTES; | Par P. DV-VAL Geographe du Roy.
[just below the title for the sheet]:
A PARIS, | Chez l'Auteur sur le Quay de l'Orloge, pres le Palais; | Avec Privilege de Sa. Maïesté, pour 20. Ans. 1669. | Cordier Sculp.
- No scale bar.
- Paris : Pierre Duval, 1669.
- From: separately published / *Cartes de Géographie les plus nouvelles et les plus fideles.*
- 1 map : copperplate engraving ; 41 x 54 cm (entire sheet) ; 4.5 x 4.7 cm (map of Africa only).
- No longitude and latitude coordinates.
- North at the top; no compass rose.
- Title below the map of Africa at top of a decorative box with a fish.
- No decorative elements on the map.

Example: *New York, Richard D. Arkway Inc. Rare Books* (January 2005 internet catalogue).

Description
Pierre Duval was a prolific Parisian publisher of numerous maps and map games. This map of Africa is on a sheet intended to be used as playing cards.

At the top center of the map sheet are small maps of the four continents as part of the title cartouche and dedication; from left to right these are Africa, the Americas, Europe, and Asia.

The twelve playing cards of Africa are the spades suit: *LA BARBARIÉ, L'EGIPTE, LE BILEDULGERID, LE ZAARA, LA NIGRITIE, LA NUBIE, LA GUINEÉ, LE ROY DE ABISSINIE ou HAUTE ETHIOPIE, LE CONGO, LA CAFRERIE, LE ZANGUEBA,* and *LES ISLES*. The ace of spades for the Africa cards, *L'AFRIQUE*, is to the left of the map of Africa at the top and represents all of Africa. Each card provides geographical information to the card player. For example, the card *LA CAFRERIE* states that there are two principal regions: *La Cafrerie* with its important port of *Cefala* (Sofala) and *La Mono-Motapa Etc.* with its important areas of *MonoMotapa, Zimbaoe, Butua, le MonoEmugi,* and *les Giaques*.

The Africa map, at the top of the sheet, is demarcated into the same twelve major geographical areas which comprise the twelve playing cards of Africa. The two Ptolemaic Nile lakes are shown in the interior along with the Nile river. Neither the Blue Nile in Ethiopia nor its source lake are shown. The Niger is conventionally shown in West Africa. Other than the unnamed Mountains of the Moon to the south of the Nile Lakes, there are no other geographical features on the map. *C. de Bonne Esperāce* (the Cape of Good Hope) is shown as are numerous islands off the African coast. Madagascar is shown as *I. Dauphine*.

Publication Information
The sheet was issued separately to be used as playing cards. It also appeared in some editions of Duval's *Cartes de Géographie les plus nouvelles*. It is known to appear in the edition of 1679 (Pastoureau 1984: 146: map no. 5). The sheet and the map of Africa are only known in one state.

References
Pastoureau 1984: 146; D'Allemagne 1950: 221/2.

111   1669 Richard Blome, London

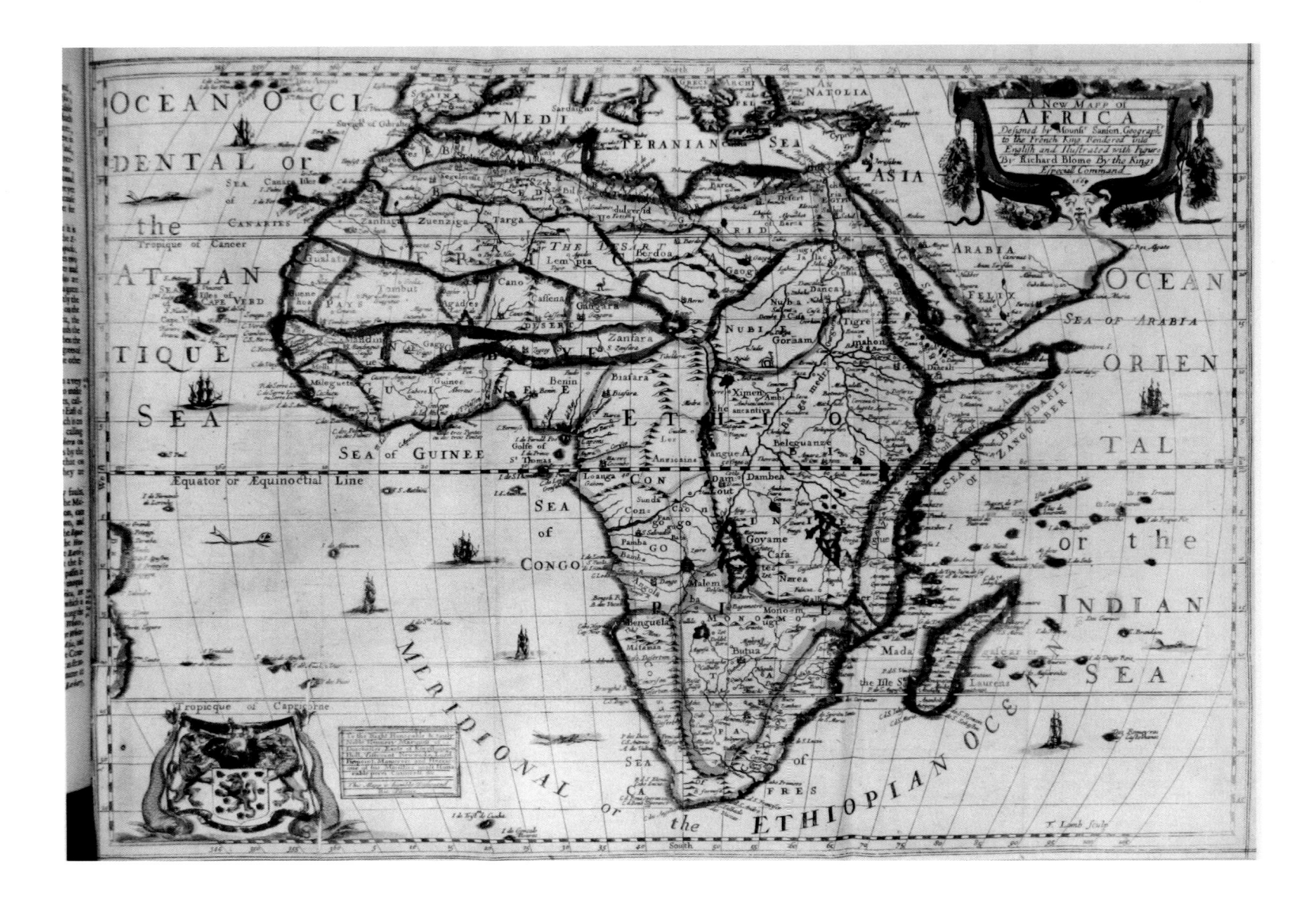

111.1 First State 1669

A New MAPP of | AFRICA | Designed by Mounsi[r] Sanson, Geograph[r] | to the French King. Rendered into | English and Ilustrated with Figurs | By Richard Blome By the Kings | Especiall Command | 1669.

[at bottom right of the map]:

F. Lamb sculp.

[at the bottom left of the map within a text box]:

Right Honorable & truly | Noble HENNERY Marquis of | Dorchester...This Mapp is humbly Dedicated | Ric Blome.

- No scale bar.
- [London] : Richard Blome, 1669.
- From: *A Geographical Description of the Four Parts of the World, which is the 2nd Part....*
- 1 map : copperplate engraving ; 38 x 54 cm.
- Longitude and latitude coordinates on all four sides of the map. Longitude coordinates across Equatorial Line.
- North at the top; no compass rose.
- Title cartouche is at the top right of the map.

- A text box is at the bottom left of the map with a heraldic shield to the left of the text box.
- There are two sailing ships in the Indian Ocean, and five ships and three sea monsters in the Atlantic Ocean. An ostrich and a camel are within Africa.

Example: *London BL*, Maps C.39.d.2 (1670 edition of *A Geographical Description of the Four Parts of the World*).

111.2 Second State 1670
Text box to the right of the heraldic shield has been erased. New dedication to the 'Right Honorable John Edgerton, Earle of Bridgewater...'. The title cartouche is the same.

Example: *London BL*, 1003.T.3 (1670 edition of *A Geographical Description of the Four Parts of the World*).

111.3 Third State 1682
Evidence of erasure of the date in the title cartouche. New dedication to 'Charles Howard, Earl of Carlisle ... by R.B.' below the heraldic shield.

Examples: 1682 edition: *New York PL*; *Washington LC*, GA7.V3 1682; [There is no map of Africa in the 1682 edition in *London BL*,1481.f.14]. 1683 edition: *Washington LC*, GA7.V3 (in 1683 atlas, same map of Africa as 1682 with same dedication). 1693 edition: *Washington LC*, GA7.V295 1693 (in 1693 atlas, same map of Africa as in 1682 with the same dedication).

Description
Richard Blome (1641-1705) was a prolific and successful English publisher of geography books and maps. His book *A Geographical Description of the Four Parts of the World* containing maps of the continents, is noteworthy as being the second folio atlas of the world to be published in England after John Speed's atlas of 1627.

As stated on the title, the Africa map is derived from the Nicolas Sanson Africa map of 1650 which, by this time, was generally assumed to be geographically the most up-to-date map of Africa. The Africa map was engraved by Francis Lamb with his imprint at the bottom right of the map as 'F. Lamb sculp'.

Publication Information
According to Burden (1996: map # 397), Blome announced the publication of his atlas on November 15, 1669, but it was not actually printed until 1670 when he likely received sufficient funding for his publication. Blome proved to be an astute business person. Rather than risk his own money in the publication of this geography book, he actively sought sponsorship for his maps and also sold subscriptions for his book to pay for its production. Wealthy patrons could thus ensure that their coats of arms and dedications to them would appear on a map in the book. This may also account for the numerous changes to the dedications over the life of the publication, as newer patrons provided funds for subsequent editions of the book.

After the edition of 1670, there were further editions of the atlas in 1682, 1683, and 1693. This map also appeared in Blome's edition of Bernhard Varenius' *Cosmography and Geography* of 1682. The Africa map is known in three states.

References
Shirley 2004; Norwich 1997: map # 42 (third state); Tooley 1999-2004, A-D: 151; Burden 1996: map # 397 (Americas map); Tyacke 1978: 110-111.

112   1669 Alexis-Hubert Jaillot, Paris

112.1 First State 1669

CARTE | DE | L'AFRIQVE | Nouuellement | Dressée sur les Memoires des | Meilleurs Geographes de nostre | temps et distinguée suiuant les | Royaumes, souuerainetés et | principales parties, qui | se trouuent iusques | apresent | 1669.
[in a cartouche in the upper left corner of the map]:
A PARIS | CHEZ H. IALIOT | PROCHE LES | GRAND | AUGVSTIN | AV BOV DV PONT | NEVF | 1669.
[the woodcut title strip, in Latin, is along the top edge of the map]:
NOVA AFRICÆ GEOGRAPHICA ET HYDROGRAPHICA DESCRIPTIO, auct: G: Blaeu 1669.
[at bottom right on surrounding text panels]:
A PARIS, | Chez Nicolas Berey, Enlumineur du Roy, proche les Augustins, | aux deux Globes.
- No scale bar.
- Paris : Alexis-Hubert Jaillot, 1669.
- From: separately published.
- 1 wall map : copperplate engraving ; 4 map sheets ; 2 sheets with figures. 83 x 110 cm (map only). 123 x 168.5 cm (map, figures, town views, title in woodcut, and outer letterpress text). 16 sheets in total.

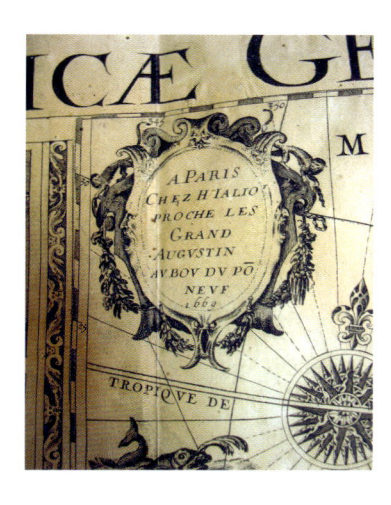

- Longitude and latitude coordinates on all four sides of the map. Longitude coordinates across Equatorial Line.
- North at the top; there are five compass roses on the map: two in the Atlantic Ocean, two in the Indian Ocean and one in the Mediterranean Sea; one windrose in the South Atlantic.
- Title, in French, is within a cartouche in the upper right corner of the map, replacing the Latin in the Blaeu map. Woodcut title strip is above the map.
- The text in the panels begins in French at the upper left: 'DISCOVRS GEOGRAPHIQVE DE L'AFRIQVE...'. This is followed by a description in Latin: 'NOVA AFRICÆ DESCRIPTIO'. The text ends with the imprint at the bottom right: 'A PARIS, | Chez Nicolas Berey, Enlumineur du Roy, proche les Augustins, | aux deux Globes'.
- Decorative elements on the map follow the Blaeu wall map of 1608. The map has borders with costumed figures on the left and right sides, and twelve town views along the lower edge. These correspond with the positioning on the original Blaeu wall map.

Examples: *Firenze IGM*; *Washington LC* (missing woodcut title strip in Latin and the text panels in French and Latin); *New York, Martayan Lan, Fine Antique Maps*, 2005 catalogue (as described above).

112.1.A Variant A 1669
At bottom right on surrounding text panels, the publisher's imprint is changed to Jaillot from Nicolas Berey as follows: 'A Paris | Chez | Hubert Jaillot proche les grand Augustins, | aux deux Globes.'. The map is still dated 1669.
Example: *Bologna BUN*.

112.2 Second State 1678
The date on the map is now changed to 1678.
Example: *Paris, Patrick Dudragne Book Dealer* (for example of 1678 wall map; from advertisement in *The Map Collector* 1994: 68: 9; N.B.: It is not clear if the imprint remains the same as on the earlier states.).

112.3 Third State 1685
The date on the map is now changed to 1685.

Example: *USA, MacLean Collection* (through correspondence, Ashley Baynton-Williams notes that there is a later state of this map with the collector Barry MacLean which has the date on it of 1685).

Description

Reflecting the importance of Willem Blaeu's 1608 wall map of Africa, it was copied in France by Alexis-Hubert Jaillot. Geographically, the map is identical to Blaeu's wall map of 1608. One major change is that the Latin nomenclature on the map has been translated into French by Jaillot, thus making the map more accessible to Jaillot's reading audience. Another difference is that this map does not have the cartouche containing the privilege on the bottom right map sheet which was on the Blaeu 1608 map.

Alexis-Hubert Jaillot (c.1632-1712) was a noted geographer, print and mapseller, and publisher. With his brother, Simon, Jaillot, came to Paris in 1657 to take advantage of King Louis XIV's encouragement to artists and scientists from throughout France to settle in Paris. In 1664, Jaillot married Jeanne Berey, the daughter of the established publisher and mapseller, Nicolas Berey I (c.1606-1665).

This wall map was one of Alexis-Hubert Jaillot's first works, published shortly after he took over the business from his brother-in-law Nicolas Berey II (1640-1667) upon Berey's death in 1667. Nicolas Berey I, the father, had already died in 1665.

Publication Information

Schilder (1996: 208-209) theorizes that Jaillot was likely involved with Berey in the preparation of this

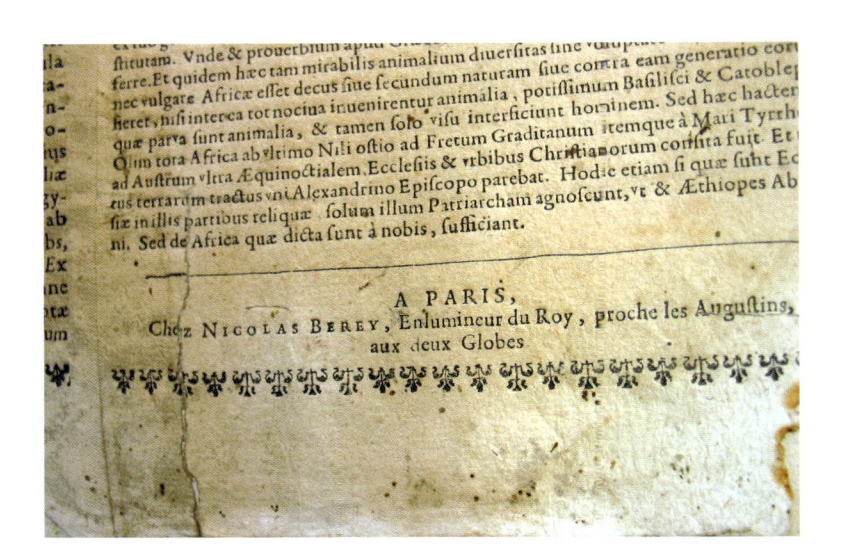

Map 112 Berey's imprint on text panel.

map prior to 1669 (the date on the map), due to the Berey imprint on the surrounding text panel of 'Nicolas Berey, Enlumineur du Roy'. The author believes this refers to Berey, the father, and not to Berey, the son, who would have been only 27 at his death and possibly too young to have used that designation alone. If this is correct, then Jaillot may have been working on this map with both Bereys from 1664 when Jaillot joined the business upon his marriage to Berey's daughter. There is no record of an earlier wall map of Africa produced only by the Bereys.

The author has identified three states of this map. The Jaillot map appears to have been re-published on a number of occasions up to 1712 when Jaillot died. There may therefore be other dates for this map.

Another example of the Jaillot wall map was listed in the Zisska & Kistner auction catalog of 8-11 May 2001. According to their catalog, this example has the imprint of Alexis Hubert Jaillot and the date of 1669 as follows: 'Nova Africae Geographica et Hydrographica Descriptio, auct: G: Blaeu 1669'. The author was not able to examine this example so was not able to determine the particular state of this map. Further, the Zisska example did not contain the text panels. The size reported for this example is 93.5 x 137 cm.

References
Zisska, F. & R. Kistner auction 2001: Catalog 37: lot 3518; Schilder 1996: 208-213; Burden 1996: map # 403 (Americas map); *The Map Collector* 68 (1994): 9 (for image of 1678 wall map); Schilder 1993; Sotheby's 25 June 1987: lot 339 (offering all four continents); Pastoureau 1984: 229-33; Keuning 1973; Ristow 1972; Kraus 1970: Catalog 124: 10-11; Ristow 1967; Wieder 1925-33: Vol. III: 70.

113   1670 Giovanni Giacomo de Rossi, Rome

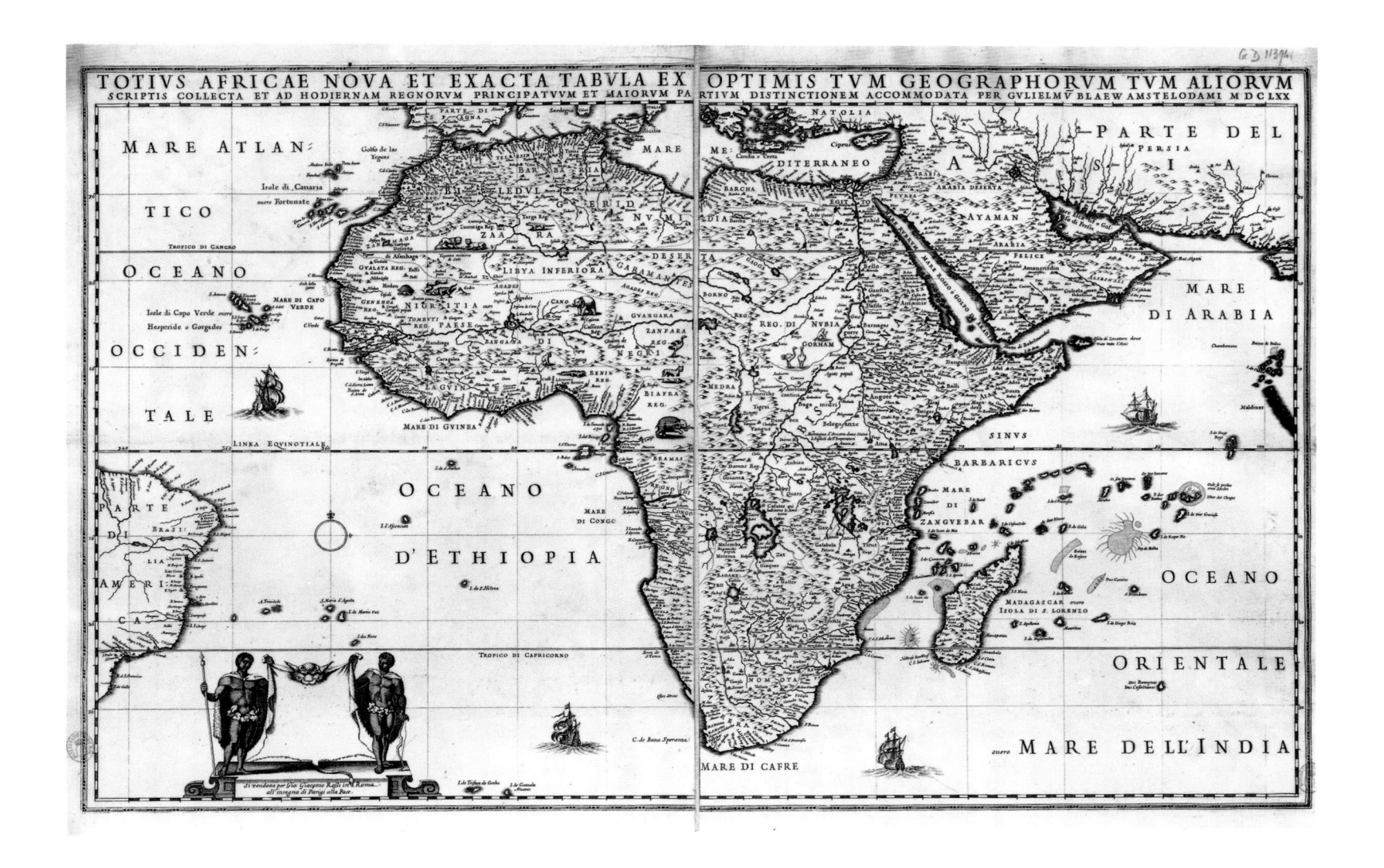

113.1 First State 1670

TOTIVS AFRICAE NOVA ET EXACTA TABVLA EX OPTIMIS TVM GEOGRAPHORVM TVM ALIORVM | SCRIPTIS COLLECTA ET AD HODIERNAM REGNORVM PRINCIPATVVM ET MAIORVM PARTIVM DISTINCTIONEM ACCOMMODATA PER GVLIELMV BLAEW AMSTELODAMI M D C LXX.

[at bottom left, at the base of the vignette]:

Si vendono per Gio: Giacomo Rossi in Roma | all'insegna di Parigi alla Pace.

- No scale bar.
- Rome : Giovanni Giacomo de Rossi, 1670.
- From: separately published.
- 1 wall map : copperplate engraving ; 2 sheets ; 58 x 90 cm (total of two sheets).
- Latitude coordinates on left and right sides of the map. Longitude coordinates across Equatorial Line.
- North at the top; compass rose below Equator in Atlantic Ocean.
- Title on two lines of text above the map.
- Two ships each in Indian and Atlantic Oceans. Various animals within Africa. There is a cartouche at the bottom left of the map of two Africans holding an 'animal skin' scroll with no text on the scroll.

Examples: *Paris BN*, Ge D 11394 (1 & 2); *Roma BVat*.

113.2 Second State 1679
Date altered on map title to 1679 as: 'M D C LXXIX'.

Example: *Berlin, Nicolas Struck Book and Map Dealer*, Catalog no. 77, item 95.

Description

This attractive, finely engraved map is a reduction to two sheets of Willem Blaeu's 1608 wall map of Africa. It is one of a set of four continent maps published by Giovanni Giacomo de Rossi.

The title acknowledges Willem Blaeu as the source for this map and the map generally follows Blaeu's wall map. The hydrography is a close copy of Blaeu. There are areas on the map where newer information is presented. For example, in Southern Africa, De Rossi shows evidence of the Dutch settlement with placenames such as *Tafel berg*, etc.

The cartographic elements on the map are taken from Cornelis Danckerts' map of 1647 (Map # 77). This map even uses Danckerts' decorative text cartouche, though without text, copying the African figures in reverse. Unlike Danckerts' map, the De Rossi derivative acknowledges Blaeu's wall map as the model for the geography.

Giovanni Giacomo de Rossi (1627-1691) was a member of a printing and publishing family, founded by his father Giuseppe. Working from 'all'insegna di Parigi alla Pace' in Rome, De Rossi produced a number of separate maps and atlas maps over a considerable period of time. Besides this map and a reproduction of the large Blaeu wall map in 1666, De Rossi published an atlas in 1677, *Mercurio Geografico*, derived from the maps of Guillaume Sanson.

According to Schilder (1996: 202), the two copperplates for this map were engraved well before 1670 by Daniel Vidman, possibly as early as 1653. Vidman was contracted by De Rossi's brother, Giovanni Domenico, to engrave reduced versions of Blaeu's 1608 wall maps of the continents which were finally produced by 1670.

Publication Information

This separately published Africa map by De Rossi is known to exist in two states.

De Rossi published an atlas in 1677, *Mercurio Geografico*, using maps derived from those of Guillaume Sanson. The map of Africa, L'Africa Nuovamente corretta, et accreseiuta secondo le Relationi piu moderne da Guglielmo Sansone..., is based on Sanson's 1668 map of Africa with the engraving by Giorgio Widman. See Map # 120 for a discussion of this map.

References

Tooley 1999-2004, Q-Z: 76; Schilder 1996: 199-207.

114 c. 1670 Frederick de Wit, Amsterdam

Map 114.3.

114.1 First State c.1670

TOTIUS | AFRICÆ | Accuratissima | TABULA | Authore | FREDERICO DE WIT | Amstelodami.
- No scale bar.
- Amsterdam : Frederick de Wit, [c. 1670].
- From: separately published / Dutch composite atlases.
- 1 map : copperplate engraving ; 49 x 58 cm.
- Latitude and longitude coordinates on all four sides of the map. The longitudinal gridline at the top and the bottom of the map begins 350...360...10... .
- North at the top; one compass rose below Equatorial Line in Atlantic Ocean.
- Title cartouche within vignette at the bottom left of the map.

CARTOBIBLIOGRAPHY

- There are three ships in the Indian Ocean, four ships in the Atlantic Ocean, and five small ships to the right of Sicily in the Mediterranean Sea. Various animals are within Africa.
- Large decorative vignette, with Africans and some animals (two lions above title cartouche).
- Without a publisher's privilege.

Example: *London BL*, C.39.f.5.

114.1

114.3

114.5

114.6

Map 114 Title cartouches of first, third, fifth and sixth states of Map 114.

114.2 Second State c.1675
Map is now with an engraved number "3" in the top right corner.

114.3 Third State c.1680
The longitudinal gridline at the top and bottom is changed to begin 348...358...8... . Evidence of re-engraving to the copperplate. For example, in this state, the title cartouche has lines added to represent cracking on the right side of the stone block.

114.4 Fourth State c.1685
Additional details (new mountains, forests, placenames, and borders within countries) have been added to the map. For example, the *BIAFARÆ | REGN.*, above the Equator in west-central Africa, now has mountain ranges to the left and right of the Lon where none existed in previous states.

Examples (state 1-4): *Amsterdam UB*, 33-17-47; *Leiden UB*, COLLBN Port 179 N 17; *Washington LC*.

114.5 Fifth State 1689
With the privilege and with a new title: TOTIUS | AFRICÆ | Accuratissima Tabula, | denuo correcté revisa, | Multis locis aucta, in Partes | tam maiores quam minores | divisa. Per F. de Witt Amstel. | Cum Priv. D.D. Ord. Holl. et West-fr. Also, with the addition of letters on the outer gridlines.

Examples: *Amsterdam UB*, 33-17-48; *London BL*, Maps C.45.f.3 (in De Wit composite atlas containing De Wit imprint with the Privilege; dated 1690 by BL); *New York PL*, Map Division Atlas Case, De Wit *Atlas Maior*, Vol. 2; *Washington LC. Author's collection.*

114.5.A Fifth State Variant A 1689
There is a variant of the fifth state of this map as follows: On either side of the map are tables listing the cities, keyed to the map. Lower right hand corner 't' Amsterdam, By Frederick de Wit in de Kalverstraat, in de witte Paskaart, werden deze Kaarten gedrukt en verkocht met Privilegie.' The size is 49 x 84 cm including text surround.
Examples: *Various locations.*

114.6 Sixth State 1710?
With the imprint on the last line of title cartouche for Pieter Mortier, as follows: 'ex Officina P. Mortier.'

The copperplate for this map, like many of the De Wit plates, were sold in 1710 shortly before the death of De Wit's widow Maria in 1711. The copperplate of the De Wit map of Africa was sold to Pieter Mortier.

Example: *Königstein, Reiss & Sohn Buch- und Kunstantiquariat Auktionen* (as observed by the author, April 2005 auction catalog).

Description

This map by Frederick de Wit is finely and attractively engraved with a title cartouche that shows 'typical' figures of North Africa and sub-Saharan Africa, along with trade goods and elephants and lions. (see also table on page 364). It is a classic late seventeenth century map of Africa by probably the foremost mapmaker and publisher working in The Netherlands after the demise of the Blaeu and Hondius-Janssonius publishing houses.

Frederick de Wit (1629/30-1706) was a noted and prolific map publisher with a wide-ranging cartographic output including wall maps, separately issued maps, terrestrial and sea atlases, and town views. He worked in Amsterdam from the Kalverstraat ('by den Dam inde dry Crabben') to 1654 and then from 'in de Witte Paskaert' from 1655 to 1706. Late in 1689, De Wit applied for and obtained a privilege from the States of Holland and West Friesland to published his maps (Koeman 1967-71, III: 191).

For much of the geography on this map, De Wit follows his earlier map of 1660 with its decorative borders which in turn was based on the geography of the Jodocus Hondius Jr. map of 1623 and the Willem Blaeu map of 1617. However, De Wit seems to incorporate some of the details of the Visscher c.1658 map. For example, in the De Wit map, the Cuama (here named the *Zambere* river) and the *R. de Spiritu Santo* Rivers join upriver with the *Zambere* River flowing from *Zachaf Lacus*. There are numerous European placenames along the Guinea coast: *Acara, Fort Nassau*, etc. However, De Wit does not use the newer information from the Dutch settlements at the Cape of Good Hope, which was so apparent on the Visscher map of c.1658.

By the time of this map, major advances and innovations in the representation of Africa were passing to the French, notably Sanson, Duval, De Fer, and others. One noticeable feature that De Wit does borrow from the French is the depiction of the fictitious island in the South Atlantic of *N.I.de Sta. Helena* (New St. Helena).

Publication Information

As many of De Wit's maps were prepared separately for inclusion in numerous undated composite atlases, as well as his *Atlas Maior* in various editions from c.1670-1707 which sometimes had manuscript title pages, it is especially difficult to precisely date his maps. The conventional date used by Tooley (1969: 123), and Norwich (1997: 59) for this map is 1680, though without a specific rationale for that date. There is a possibility that this map may have been prepared then, but it seems more likely that De Wit would have prepared this map earlier for use in his first atlas from c.1670. Also, of greater importance, the information on this map does not utilize the information available to De Wit in his wall map of 1672. Finally, Shirley (1993: 468-469) also concurs with a date of c.1670 for the companion world map in his atlas, based on the out-dated geography represented on the map. In fact, Shirley states that the world map may have been engraved as early as the mid-1660s. The map of Africa is known in six states.

Maps 114, 114A, 131, 168, and a later map by Johannes Covens & Cornelis Mortier (with the same title as # 114.5, but with the addition of "ex Officina Covens et Mortier."; published after 1721 when their partnership was formed) are similar in appearance, but are from different copperplates. They are easily distinquished from one another by changes to the placement for the words for the island of *Martin Vaz*, located on the bottom left of the map above the left lion's head of the title cartouche. For comparison purposes, the differences are as follows:

Map # 114	De Wit	c.1670	*I. Martin Vaz* through the parallel of 20° South.
Map # 114A	De Wit-Anon.	c.1680	*I Martin Vaz* above and in line with the parallel of 20° South.
Map # 131	J. Danckerts	1683	*I. Martin Vaz* below the parallel of 20° South.
Map # 168	J. Danckerts	c.1699-1700	*Martin Vaz* (not preceded by "I") above the parallel of 20° South.
- - -	Covens & Mortier	after 1721	No *I. Martin Vaz*, with parallel of 20° South just touching a few locks of the left lion's head.

References

The author is grateful to George Carhart, Osher Map Library, Portland, Maine, for sharing the results of his unpublished doctoral dissertation research on Frederick de Wit. Norwich 1997: map # 48 (second state?); Shirley 1993: 468-469; Werner, 1994; Koeman 1967-71, III: 85-87 and 191-205.

114A c. 1680 Frederick de Wit - Anonymous, Amsterdam(?)

TOTIUS | AFRICÆ | Accuratissima | TABULA | Authore | FRIDERICO DE WIT | Amstelodami.
- No scale bar.
- Amsterdam(?) : [Unknown, c. 1680].
- From: Pasted into an undated De Wit atlas.
- 1 map : copperplate engraving ; 49 x 58 cm.
- Latitude and longitude coordinates on all four sides of the map. The longitudinal gridline at the top and the bottom of the map begins 348...358...8... .
- North at the top; one compass rose below Equatorial Line in Atlantic Ocean.
- Title cartouche within vignette at the bottom left of the map.
- There are three ships in the Indian Ocean, and four ships in the Atlantic Ocean. An elephant and rhinoceros, and two ostriches are in Central Africa.

- Large decorative vignette, with Africans and some animals (two lions above title cartouche).
- Without a publisher's privilege.

Example: The sole known example is at the *Österreichische Akademie der Wissenschaften*, K-V: WE 98.

Description

Just before the final printing of this cartobibliography, this map of Africa was recently discovered in Vienna at the Österreichische Akademie der Wissenschaften through the research work of George Carhart of the Osher Map Library. Though this map appears to be virtually identical to Map # 114, it has numerous differences in the placement of lettering, the exclusion of the five ships to the right of Sicily in the Mediterranean Sea, the engraving of the ships, etc. It is clearly prepared from a new copperplate.

The geography for this map is taken directly from Map # 114 with no noted changes or additions. See Map # 114 for information on the geography.

There is a possibility that this map could have been prepared by De Wit, using a different copperplate, prior to the issuance of the first state of Map # 114. However, the author does not believe this to be the case. The longitudinal grid at the top & bottom of Map # 114A begins 348...358...8... . The first state of Map # 114 has a longitudinal grid that begins 350..360..10... . This seems to confirm that # 114A was produced by an unknown copyist using the third state of the De Wit map (with the identical gridline) as the model.

Publication Information

The single known copy of Map # 114A is in an undated De Wit atlas. However, the map was printed from different paper and is pasted onto a page in the atlas, indicating that it was likely added at a later date.

References

The author is grateful to George Carhart, Osher Map Library, Portland, Maine, for bringing this map to his attention.

115 c. 1671 Gerrit Lucasz. van Schagen, Amsterdam

AFRICÆ | ACCURATA TABULA | ex officina | GERARDI A SCHAGEN.
- No scale bar.
- [Amsterdam] : Gerrit Lucasz. van Schagen, [c.1671].
- From: separately published / Dutch composite atlases.
- 1 map : copperplate engraving ; 43.5 x 54 cm.
- Latitude and longitude coordinates on all four sides of the map.
- North at the top; compass rose below Equator in Atlantic Ocean.
- Title cartouche at top right is the same as on the Nicolaas Visscher c. 1658 map.
- A new vignette with Neptune and Venus at the bottom left, unlike that on the c.1658 Nicolaas Visscher map. Animals within Africa. Ships in oceans as described below.

Examples: *Amsterdam UB*, 1806 A 19 (in a composite atlas); *Charleston CL*, Mss 91 (in composite Dutch atlas, referred to as the Mitchell King Atlas); *London BL*, Maps 63510.(3) (as a separate map); *Amersfoort, Leen Helmink* (as a separate map).

Description

This map is a close copy of the Nicolaas Visscher I (1618-1679) map of Africa of c.1658 in its first state, without the privilege. There is also no privilege on this Van Schagen map. There are a few differences in the decorative elements of the two maps. The Van Schagen map has five ships in the Indian Ocean and twelve ships in the Atlantic Ocean, while the Visscher map has only three ships in the Indian Ocean and eight ships in the Atlantic Ocean. The decorative scene at the bottom left on the Van Schagen map replaces the dedicatory cartouche on the Visscher map. This decorative scene shows Neptune and Venus on a seashell, being pulled by horses and surrounded by sea horsemen.

Gerrit Lucasz. van Schagen (c.1642-c.1690) was an Amsterdam publisher, engraver and art dealer. Little has been written about this relatively unknown map and its author. It is supposed that Van Schagen produced maps of the four continents at the same time for the Dutch market. Whether he received permission to copy Visscher's map is not known.

Publication Information

This map is often found in Dutch composite atlases, usually without title pages or other identifying features. Burden (1996: 431) dates the Americas map by Van Schagen as being published in 1671. Van Schagen would have been about 29 years old when this map was published. The Africa map was likely published at or around the same time.

References

Discussion with Jan Werner, Map Curator, Universiteitsbibliotheek Amsterdam.
Burden 1996: 431.

116   1672 Frederick de Wit, Amsterdam

116.1 First State 1672

AFRICA.
[on the right side of the fourth map sheet, that is the bottom right side of the map]:
Gedruckt tot Amsterdam : bij | FREDERICK DE WIT | inde Kalverstraet bij den Dam | inde Witte Pascaert, A° 1672.
[title strip above the map]:
NOVA TOTIUS | AFRICÆ TABULA | Emendata a F. de Wit.
- No scale bar.
- Amsterdam : Frederick de Wit, 1672.
- From: separately published.
- 1 wall map : copperplate engraving ; 6 map sheets ; 99.5 x 125 cm (map only). 150 x 196 cm (including map, text panels, title strip [21 cm for height of top title strip], and borders in *Greenwich NMM* example) ; 121 x 168 cm (for map and borders in *Amsterdam UB* example).
- Latitude and longitude coordinates on all four sides of the map. Longitude coordinates across Equatorial Line.
- North at the top; one compass rose in Indian Ocean and three compass roses in Atlantic Ocean.
- Title within oval cartouche at bottom center of the map. Title strip above the map, set within floral decorations in *Greenwich NMM* example.
- At the upper right of the map, there is a text box describing how to determine distances.
- There are three ships and one sea monster in the Indian Ocean, four ships and one sea monster in the Atlantic Ocean, and two ships in the Mediterranean. No animals are within Africa.
- Ten town views at left border and ten town views at right border as described below in *Greenwich NMM* example.
- Across the entire bottom of the map, there are decorative scenes of Africa.

Examples: *Greenwich NMM*, De Wit 1672; A.7823 (as described above); *Enkhuizen MB*.

On the example at the Greenwich National Maritime Museum, inside of the town views on left side there is a text strip in English beginning 'A Description of AFRICA'. This example also has, inside of the town views on the right side, a text strip in French beginning 'Nouvelle D'AFRIQUE'. Across the bottom, there is descriptive text on Africa in Latin and Dutch beginning 'NOVA AFRICA' and then 'Beschryvinge van Africa'.

116.2 Second State 1700
Dated changed to 1700 within the imprint on right side of the fourth sheet of the map, below the sailing ship.

Example: *Amsterdam UB*, W.X. 002.

The example at Universiteitsbibliotheek Amsterdam, in the map's second state, has a different title strip above the map of NOVA ET ACCURATA TOTIUS AFRICÆ TABULA. emendata a F. de Wit. Also, there are no text boxes on the left and right side borders inside of the town views. There are only nine town views on the left and right sides in a different order, rather than ten as in 1672 example at the Greenwich National Maritime Museum. On the left side, these are *ALEXANDRIA IN EGIPT*; *ALGIERS*; *TUNIS*; *ORAN*; *CASTRUM MINAE*; *STADT MINAE*; *LOANDA S. PAULI*; *ALGER*; *AMARA*. On right side, these are *ALCAIR*; *MALTA*; *CEUTA*; *TRIPOLI IN BARBARIA*; *ARX NASAUII*; *CANARIEN*; *INSULA S. THOMAE*; *TANGER*; *ARX DELPHINI FORT DAUPHIN*. Across the bottom, the descriptive text on Africa is only in Latin, beginning 'NOVA AFRICÆ | DESCRIPTIO...'.

116.3 Third State c.1730
The imprint is now changed to 'R & J Ottens.'

Example: *London, Burden Collection.*

CARTOBIBLIOGRAPHY

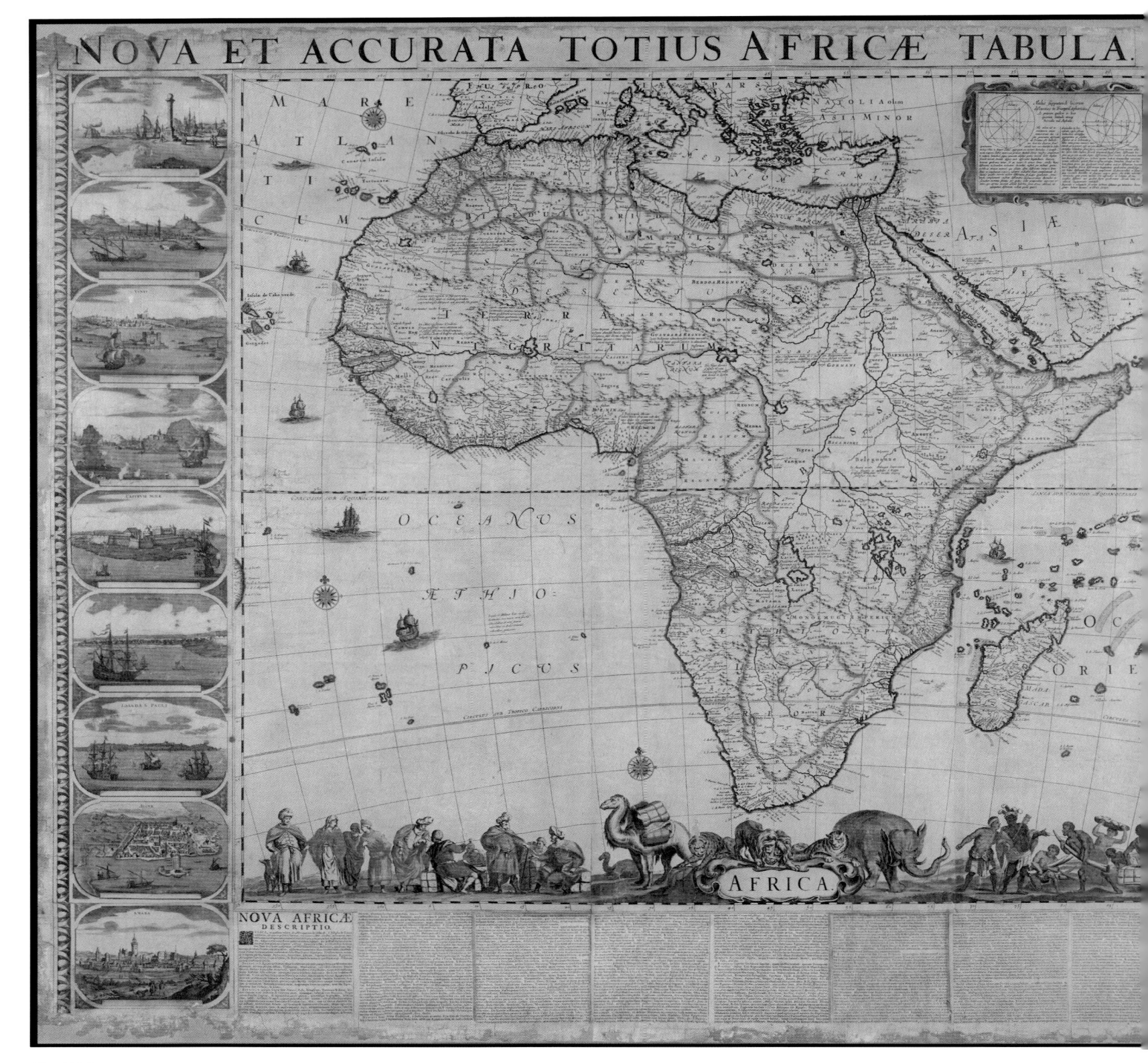

Map 116.2

Description

This wall map of Africa is one of a set of the world and the four continents that Frederick de Wit produced in the early 1670s. De Wit used Willem Blaeu's wall map of 1608 for his model for this map owing to its continuing popularity, although it was out-of-date by this time. De Wit does up-date his map with newer information available to him, notably of the Dutch in Southern Africa.

In the first state, the map has ten town views on each side border. On the left border from the top are *ALEXANDRIA IN EGIPT*; *ALGIERS*; *TUNIS*; *ORAN*; *CASTRUM MINAE* ; *COSA…MINAE*; *LOANDA S. PAULI*; *ALGER*; *ANGRA*; *CANARIEN*[?]). On the right border from the top are *ALCAIR* (Cairo); *MALTA*; *CEUTA*; *TRIPOLI IN BARBARIA*; *ARX NASAUII* (Fort Nassau on the Guinea Coast); *CANARIEN*;

INSULA S. THOMAE; *TANGER*; *MALTA*; *LE FORT DAUPHIN* (The French fort on the southern end of Madagascar). In the Amsterdam Universiteitsbibliotheek example of the second state, there are nine town views on each side border.

Much of the attractive decorative scene across the bottom of the map has been copied from the Jan Mathisz. wall map of Africa of c. 1655: many of the Arab figures on the left; the entire title cartouche with the surrounding camel, elephant and lions; and the African figures, particularly those on the left side.

Frederick de Wit (1629-1706) was known as a map publisher, engraver, and seller. This map carries De Wit's publication location of 'Kalverstraet bij den Dam inde Witte Pascaert' where he worked from 1655 to 1706. He produced a prodigious amount of cartographic material including separate terrestrial maps and sea charts, sea and terrestrial atlases, wall maps, and town views. Many of his maps were superbly colored by master colorists of his day such as Van Santen.

Publication Information

De Wit is known to have produced a wall map of the world in about 1660 with an example at the Deutsche Staatsbibliothek (Shirley 1993: 446). It is not known if De Wit published a companion map of Africa at this time. De Wit did publish a wall map of the world with new plates in 1670. This map of Africa of 1672, was intended to accompany that wall map of the world.

Sometime after De Wit's death in 1706, the copperplates for this wall map were acquired by Reiner and Josua Ottens, who published it in its third state in c. 1730.

References

Schuckman, Veldman, and Scheemaker 1999: 31: image as 85/11; Werner 1994; Shirley 1993; Frabetti 1959.

117   1673 Pietro Todeschi, Bologna

AFRICA.
[woodcut title above map]:
NOVA ET ACVRATA TOTIVS AFRICÆ TABVLA auct: G.I: Blaeu.
- No scale bar.
- [Bologna : Giuseppe Longhi(?), 1673].
- From : separately published.
- 1 wall map : copperplate engraving ; 4 map sheets ; 82 x 107.5 cm; 115.5 x 164 cm including 4 map sheets, decorative borders, letterpress text, and title strip.
- Latitude and longitude coordinates on all four borders of the map. Longitude coordinates across Equatorial Line.
- North at the top; there are five compass roses on the map: two in the Atlantic Ocean, two in the Indian Ocean, and one in the Mediterranean Sea ; one windrose in the South Atlantic.
- Title is within a cartouche in the upper right corner of the map. Woodcut title strip is above the map.
- Decorative elements on the map are similar to the Blaeu 1608 wall map.

Examples: This map is only known in one state. There are six known examples of this wall map, with variations as noted in the titles, uses of decorative borders, and text panels (Schilder 1996: 195-199). One example, Map # 117.A, is new to the literature.

117.A. *New York, Graham Arader Galleries.* Title strip in woodcut above the map: NOVA ET ACVRATA TOTIVS AFRICÆ TABVLA auct: G.I.: Blaeu. 4 map sheets and 2 sheets with figures, and town views. Text borders surrounding the map starting in upper left in Italian (NOVA DESCRIZIONE | DELL AFFRICA), and then in Latin (NOVA AFRICÆ | DESCRIPTIO).

117.B. *Roma BVat*. Title strip in woodcut above the map: NOVA ET ACCVRATA TOTIVS AFRICA TABVLA auct: G.I. Blaeu. 4 map sheets and 2 sheets with figures, and town views. No text borders. As loose maps.

117.C. *London RGS*. Similar to 117.B, Title strip in woodcut above the map: NOVA ET ACCVRATA TOTIVS AFRICA TABVLA auct: G.I. Blaeu. 4 map sheets and 2 sheets with figures, and town views. No text borders. As a wall map.

117.D. *Roma SGI*. 4 map sheets and decorative borders along sides and bottom. No text borders. As a wall map.

117.E. *Amsterdam, Koninklijke Hollandsche Lloyd Collection* (a private collection). The Americas map in this set has the date on it of 1673 which is used to date all of the other, undated examples (referred to by Wieder 1925-33: Vol. III: 112-15). Wieder states that this example of the Africa map has text surrounding the map in Italian and in Latin.

117.F. *Lüneburg RB*. 4 map sheets, decorative borders and a title strip. The title strip has been trimmed so that it ends with 'auct.', and not with 'G.I. Blaeu'. With no text borders. In good condition as a wall map.

The University of Texas at Austin has Todeschi wall maps of Asia and the Americas, but not one of Africa.

Description

This is the Bologna derivative of Willem Blaeu's 1608 wall map of Africa. The immediate model for this map was likely the Venetian example of the Blaeu wall map, produced in 1646 by Stefano Scolari (1598-1650).

Though there is no date on this map of Africa, the map is likely from 1673 as that date appears on the companion wall map of the Americas in a set of the four continents by Todeschi at the Koninklijke Hollandsche Collection in Amsterdam. In this set, the Americas wall map has text ending with a date of M.D.C.L XX III (1673). The publisher for the maps is likely Giuseppe Longhi as he is known to have published Blaeu's wall map of Italy.

The cartouche in the lower right, which on Blaeu's original map contained the text of the privilege, is blank in this example as in the earlier Scolari example. Unlike in the Scolari example, the name of the original map engraver for Blaeu, 'I. vanden Ende sculp', does not appear on this map. It is possible that the copyist for the Todeschi map could have used the original Blaeu 1608 map as a model, rather than Blaeu's second state of 1612, which for the first time contained the imprint of the engraver, Joshua van den Ende. More likely, Todeschi directly used the Scolari model and simply did not copy the Van den Ende imprint onto his map.

Like the Scolari example, this is a close, well-executed copy of the Blaeu map, duplicating Blaeu's geography, placenames, and decorative borders. Differences are noted in the long legends in the cartouches which are not as finely done as in the Blaeu map. Also, in some cases, there are minor deviations from the Blaeu map in the text in the long legends (Schilder 1996: 195). For a detailed discussion of the geography depicted on this map, please refer to the entry for Willem Blaeu's 1608 wall map of Africa (Map # 54).

References
Schilder 1996: 195-199.

118 1674 Alexis-Hubert Jaillot, Paris

118.1 First State 1674

L'AFRIQUE | divisée suivant l'estendɩe de ses principales parties | ou sont distingués les vns des autres | LES EMPIRES, MONARCHIES, ROYAUMES, ESTATS, ET PEUPLES, | qui partagent aujourd'huy L'AFRIQUE. | sur les Relations les plus Nouvelles. | Par le Sr.SANSON, Geographe Ordinaire du Roy. | PRESENTÉE | A MONSEIGNEUR LE DAUPHIN | Par son tres-humble, tres-obeissant, et tres-fidele Serviteur | Hubert Iaillot.
[above the map on two lines of text]:

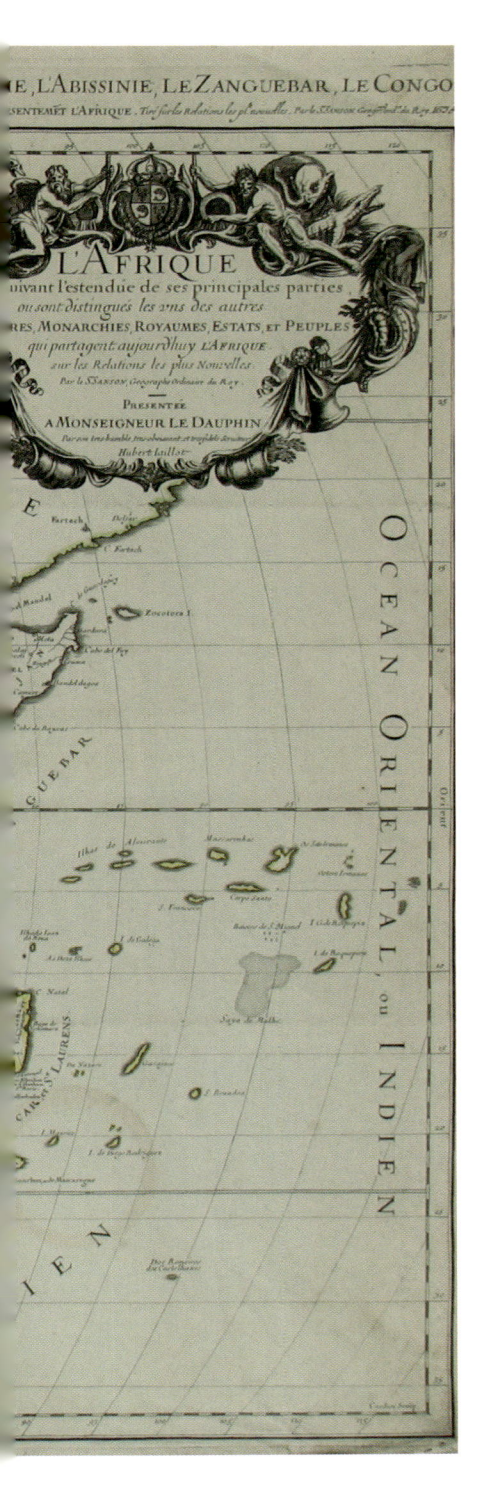

L'AFRIQUE DISTINGUÉE EN SES PRINCIPALES PARTIES SÇAVOIR LA BARBARIE, LE BILEDUL-GERID, L'EGYPTE, LE SAARA ou LE DESERT, LE PAYS DES NEGRES, LA GUINÉE, LA NUBIE, L'ABISSINIE, LE ZANGUEBAR, LE CONGO, | LE MONOMOTAPA, LES CAFRES, LES ISLES DE CANARIES, DU CAP-VERD, DE ST THOMAS, L'ISLE DAUPHINE autrement MADAGASCAR OU SONT REMARQUÉS LES EMPIRES, MONARCHIES, ROYAUMES, ESTATS, et PEUPLES, QUI PARTAGENT PRESENTEMENT L'AFRIQUE. Tire sur les Relations les pl. nouuelles. Par le Sr. Sanson, Geogr. ordre. du Roy. 1674.

[at the bottom of the scale bar, there is a publisher imprint as follows]:
A Paris, | Chez H. IAILLOT joignant les grands Augustins, aux 2. Globes | Auec Privilege du Roy pour 20 Ans. | 1674.

[at bottom right corner of the map]:
Cordier Sculp.

- Scale bar on 1674 map set within drapery cartouche at upper left of the map:
  Eschelle
  Milles Pas Geometriques ou Milles d'Italie 600 = 6.9 cm.
  Lieues Communes de France 250 = 6.9 cm.
  Lieues Communes d'Allemagne 150 = 6.9 cm.
  Lieues Communes d'Espagne 178 = 6.9 cm.
  Lieues d'vne Heure de Chemin 200 = 6.9 cm.
  Dietes ou Iournees de 40 M.P.G. Chacune 15 = 6.9 cm.
- Paris : Alexis-Hubert Jaillot, 1674 [1681].
- From: separately published / *Atlas Nouveau.*
- 1 map : copperplate engraving ; 2 map sheets ; 54.5 x 87.5 cm
  (not including text above the map).
- Latitude and longitude coordinates on all four sides of the map. Longitude coordinates across Equatorial Line.
- North at the top; no compass rose.
- Title cartouche is at the upper right of the map in the form of a drapery surrounded by water gods, animals, and decorative elements. Dedication within the title cartouche is to the Dauphin, the son of King Louis XIV, with his coat-of-arms at the top center. A second title is above the map.
- No decorative elements on the map except for a large decorative title cartouche at the top right and a scale bar at top left on a drapery cartouche.

Example: *Amsterdam UB*, 34-31-02 (as a separate map).

118.2 Second State 1685
Map dated 1685 (Appeared in the 1689 edition of *Atlas Nouveau*).

Examples: *Paris BN*, Ge.CC.1002 *and* Ge.DD.4796.

Description

As the title indicates, this map is based on the geography of Guillaume Sanson (1633-1703), in this case his map of Africa of 1668. Guillaume Sanson's 1668 map was produced in partnership with Pierre Mariette II. They continued to produce further works together until about 1673. At this point, Sanson began to work instead with Alexis-Hubert Jaillot (c.1632-1712). This map is one of the first results of that new collaboration. The imprint of the engraver Cordier (Cordier Sculp) is at the bottom right corner of the map.

Following the 1650 and 1668 Sanson maps of Africa, this map continues to present the *Empire des Abissins* (Abyssinia) as extending far to the south of the Equator. The name of *S. Laurens* is re-introduced for the Island of Madagascar as on the 1650 Sanson map. The fictitious island of *St. Helena Nova* (New St. Helena) is still on the map in the South Atlantic. Compared to the 1668 map, the African coastline, especially south of the Equator, takes on a more accurate appearance in this map. However, little new information is shown on the South African coast, which is somewhat unusual since the Dutch had established a permanent supply base at the Cape of Good Hope by this time.

There are a number of departures from Sanson's earlier work. Of most significance, this map introduces a new depiction of the rivers south of the Ptolemaic lakes. Attempting to correct the misplaced modern-day Zambezi River which appeared too far to the south on the earlier maps, Jaillot places a new major river, the *Zambeze*, above the *Zambere R.* and *Rio de Spiritu Santo* which exits into the Indian Ocean by the trade port of Quiloane. This new alignment correctly places *Sofala* to the south of the Zambezi. The new *Zambeze* River originates in a new, unnamed lake to the east of *Zachaf Lac*, the traditional source for the Zambere/Cuama and Spirito Santo Rivers.

Publication Information

The map is dated 1674 and was issued initially as a separate map. It is found in Alexis-Hubert Jaillot's publication of Sanson's *Atlas Nouveau* of 1631 with reprints in 1684 and 1689. Partially based on Pastoureau, this map is known in two states.

Due to the importance and popularity of this 1674 Jaillot map, it was copied by Johann Hoffmann in c.1682 in Nürnberg (see map # 127). The Jaillot map was also copied in Italy by an unknown publisher. Only one set of the maps of the four continents, including Africa, by this anonymous Italian publisher is known to exist. This set is with an American collector. The engraving style is somewhat rough, though the map of Africa is clearly modeled after the Jaillot map. The maps in the set are dated 1674, but they were obviously published some time after that date. Burden (2007: 286) dates the Americas map in the set as c.1685.

The Jaillot map was also copied in Amsterdam by Pieter Mortier. There is some confusion with what appear to be later states of Jaillot's map of Africa, dated 1692 and later. In fact, these maps are not by Jaillot, but Pieter Mortier.

In 1692, Pieter Mortier and Marc Huguetan published their own edition of Jaillot's *Atlas Nouveau* in Amsterdam with the title and address of Jaillot in Paris, but no mention of their own names or location. It is unclear if this edition had Jaillot's approval but, as Koeman states (1967-71, III: 8), this use of the Jaillot name was done to promote the sale of Mortier and Huguetan's version of the popular Sanson atlas, which had been published by Jaillot in 1674 and 1685.

Mortier's map of Africa (See Map # 156) is a close copy of the 1674 Jaillot map with the date on the map of 1692, but without the name of Jaillot's engraver, Cordier. Mortier and Huguetan published a second edition in 1696 with the date on the map of 1696, again without Cordier's name.

A third edition of the Pieter Mortier copy of the *Atlas Nouveau* was published in Amsterdam with no date on the title page of the atlas or on the map of Africa. A copy of this undated Mortier atlas with this map of Africa is at the Royal Geographical Society (*London RGS*, MR 14.C.132). This example has evidence of an erasure of the date of 1696. This edition was issued by Mortier sometime after 1696.

Based on Blonk's analysis in *Hollandia Comitatus* (Blonk 2000), a fourth edition was published without a date on the title page and no date on the map of Africa. However the map of Africa in this edition does have Mortier's name and address as: 'A Amsterdam | chez Pierre Mortier & Compagnie | avec privilege' (also see Pastoureau, 1984: 257: no. 4). This edition was likely published between 1696 and 1708.

There is a final, also undated, copy at the Royal Geographical Society (*London RGS*, MR 14.C.131) with the imprint of Johannes Covens & Cornelis Mortier (Pieter's son). As Covens and Mortier entered into partnership in 1721 upon the marriage of Mortier and Coven's sister, this map is after that date.

See Pastoureau (1998: 246-262) for further information on the Amsterdam copies of the Jaillot map of 1674.

References

Shirley 2004: 579-582 and 717-724 (for Mortier); Blonk 2000; Norwich 1997: map # 46 (first state); Pastoureau 1984: 229-262 (Jaillot 1674 map: 235: no. 4); Koeman 1967-71, III: 4-10.
The author is grateful to Marco van Egmond, Universiteit Utrecht, who wrote his doctoral dissertation on Covens and Mortier and who kindly shared his information on Pieter Mortier.

119   c. 1676 John Seller, London

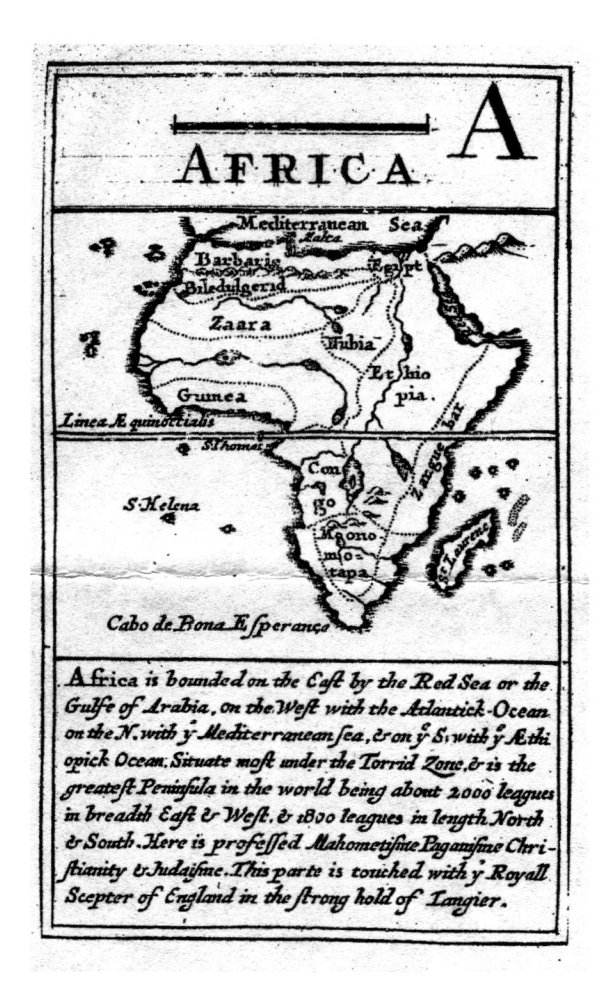

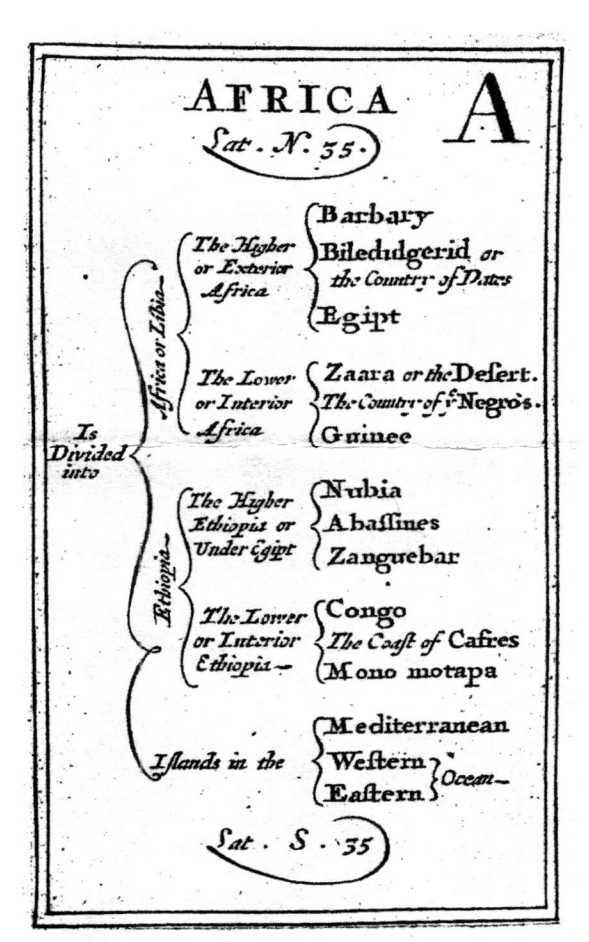

119.1 First State c.1676

AFRICA.
- No scale bar.
- [London : John Seller, c.1676].
- From: separately published as part of a set of playing cards / *Atlas Minimus or a Book of Geography...* .
- 1 map : copperplate engraving ; 6.5 x 5.5 cm (map only) ; 9.5 x 5.5 cm (map including text below the map).
- No latitude and longitude coordinates on the map. Equatorial Line across Africa.
- North at the top; no compass rose.
- Title above the map, with a large capital 'A' to the top right of the title.
- No decorative elements on the map. The map is at the top of a single leaf with nine lines of text in English describing Africa below the map.

Examples: First state intended as playing cards: *London BL*, Maps C.7.a.18 (bound into a book); *London BM*, Schreiber Collection 1901: 4: no. 44.

119.2 Second State 1678
Removal of the large capital 'A' at top right of the title.

Examples: *Atlas Minimus* of 1678/9: *Washington LC*, G1015.S5 1679 (with undated title page) *and* QB41.S5 1678 (with undated title page); *Atlas Minimus* of 1680: *Washington LC*, QB41.85 1680;

Undated *Atlas Minimus: London BL*, Maps C.7.a.16 *and* Maps C.7.a.17; *A Pocket Book containing severall Choice Collections* [n.d. but c.1680]: *London BL*, Maps C.21.a.7.

Description

John Seller initially issued this map of Africa, with the 'A' to right of the title, as the ace of spades for a set of playing cards of the four continents and regions of the world. Other playing cards in the set depict regions of Africa.

John Seller (c.1630-1697) was an author of books on navigation, a map and chart seller, and an instrument maker. He was active in London, primarily at the 'Mariner's Compass near Hermitage Staires in Wapping'. In 1671, he was appointed Hydrographer to the King. His numerous and ambitious projects and resulting financial difficulties required that he seek the funding and involvement of others in his projects. Though he later involved William Fisher, John Thornton, and others, Seller seems to have produced this map and the others in the set of playing cards without support from them.

The map of Africa, owing in part to its small size, is generally devoid of much geographic information. The Cape of Good Hope is prominently labeled as *Cabo de Bona Esperança*. The map does show the major geographic divisions of Africa at that time: *Barbarie, Biledugerid, Zaara, Nubia, Ethiopia, Guinea, Congo, Monomotapa, Zanguebar*, and the island of *St Laurens* (Madagascar). Hydrographically, Seller presents a typical view of Africa with the Nile source lakes in Central Africa though the Blue Nile is not shown.

The map was likely engraved by James Clark as the engraved title page for the set of cards is signed 'Ja. Clark sculp'. Clark was an engraver and printseller in London and is known to have worked for Seller on the engraving of the plates in the *English Pilot* sea atlas of 1671.

Publication Information

Shirley (1993: 497) believes that the original set of Seller playing cards was prepared as early as 1676 or a year or so later. The British Museum dates their set as c.1661 (O'Donoghue 1901: 4: no. 44). The British Museum date is likely too early. It would place these playing cards among the earliest of Sellers' work, and there is no indication in the literature to support an earlier date.

Shortly after the map playing cards were issued, Seller issued his maps in a small pocket-size atlas in 1678, the *Atlas Minimus*, or as it was also titled, *a Book of Geography*. Though these atlases are often not dated, the first advertisement for the sale of the *Atlas Minimus* is in the London Gazette of August 1678, so the publication of this atlas took place in 1678 or shortly thereafter (Shirley 1993: 497). There was also another edition of the *Atlas Minimus* in 1680.

The *Atlas Minimus* was followed by his *A New System of Geography* in c.1684 with a new map of Africa which is slightly larger than this map.

In about c.1705-1706, John Senex and Charles Price published their edition of the *Atlas Minimus*, having acquired these Seller copperplates. This is the assumed date since the title page for the Senex-Price atlas mentions the publication location as 'next to the Fleece Tavern in Cornhill'. It is known that Senex and Price were working from this location around 1705-1706 before they moved to 'Whites Alley, Coleman Street' in about 1707. The map of Africa is unchanged from the second state although the world map has the imprint of Senex and Price added to it, based on the British Library example (Senex & Price edition: *London BL*, Maps C.7.a.19).

References

Tooley 1999-2004, Q-Z: 143-145; Shirley 1993: 497; Tyacke 1978: 136-138 (for Price), 139-140 (for Seller), and 142 (for Senex); O'Donoghue 1901.

120 1677 Giovanni Giacomo de Rossi, Rome

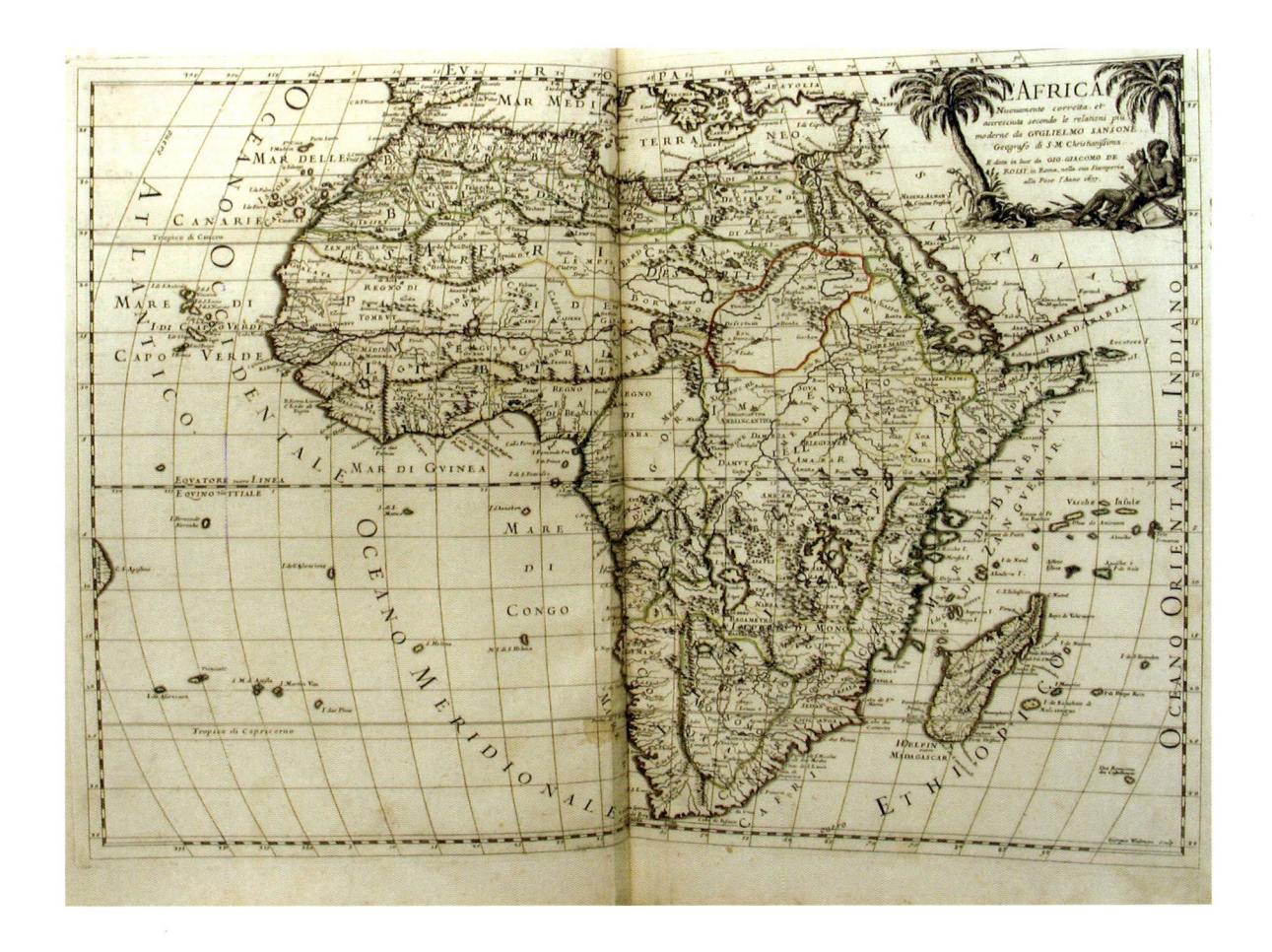

120.1 First State 1677

L'AFRICA | Nuouamente corretta, et | accresciuta secondo le relationi piu | moderne da GVGLIELMO SANSONE, | Geografo di S.M. Christianissima. | E data in luce da GIO · GIACOMO DE | ROSSI, in Roma, nella sua Stamperia | alla Pace l'Anno 1677.
[at the bottom right edge of the map]:
Giorgio Widman sculp.
- No scale bar.
- Rome : Giovanni Giacomo de Rossi, 1677.
- From: *Mercurio Geografico Overo Guida Geografica.*
- 1 map : copperplate engraving ; 40 x 55 cm.
- Latitude and longitude coordinates on all four sides of the map. Longitude coordinates across Equatorial Line.
- North at the top; no compass rose.
- Title cartouche at top right of the map.
- No decorative elements on the map except for a large decorative title cartouche at the top right.

Examples: *London BL*, Maps C.39.f.8 (undated but listed in catalog as 1684); *Washington LC*, G1015.R7 1685 (manuscript date on book of 1685) *and* G1015.R7 1694 (manuscript date on book of 1694).

120.2 Second State 1687
With date in the title cartouche changed to 1687.

Example: *London BL*, Maps 4.TAB.19,20 (in Vol. II) (The British Library example of the atlas, containing the 1687 map of Africa, is dated 1692-1723).

Description

Giovanni Giacomo de Rossi (1627-1691) was a member of a Roman printing and publishing family, founded by his father, Giuseppe. De Rossi, working from 'alla Pace all'insegna di Parigi' in Rome, produced a number of separate and atlas maps over a considerable period of time. Besides this map, De Rossi first issued a copy of the large Blaeu 1608 wall map in 1666 and a reduced two-sheet version of Blaeu's wall map in 1670. The map was engraved by Giorgio Widman with an imprint of 'Giorgio Widman sculp', at the bottom right edge of the map.

This map is directly modeled after Guillaume Sanson's map of Africa of 1668, as cited in the title cartouche. De Rossi's 1677 map is generally a faithful reproduction of the Sanson map, substituting Italian text for French text. It does not add any new geographic details.

Publication Information

It seems that De Rossi's *Mercurio Geografico* was developed over a number of years and is often found with no date, or only a manuscript date, on the title page. The first edition was begun in 1677. There were later editions of the *Mercurio Geografico* in 1685 and 1694. The map is known in two states.

References

Reiss & Sohn 2001: Auction 80.

121   1678 Pierre Duval - Johann Hoffmann, Nürnberg

121.1 First State 1678

AFRICA.
- No scale bar.
- [Nürnberg : Johann Hoffmann, 1678].
- From: *Geographie universalis pars prior. Das ist: Der allgemeinen Erd-Beschreibung erster Theil,....*
- 1 map : copperplate engraving ; 10 x 12 cm.
- Latitude and longitude coordinates on all four sides of the map. Longitude coordinates across Equatorial Line.
- North at the top; no compass rose.
- Title cartouche at the bottom left of the map.
- No decorative elements on the map except for the title cartouche.
- With an engraved number '150' at the upper left corner of the map (not visible on the illustration above because of tight binding..

Example: *New York PL*, *KB 1678 (1678 edition of *Geographie universalis*).

121.2 Second State 1681

Removal of engraved number '150' at the upper left corner of the map. The number is now '216'.

Example: *Bedburg-Hau, Antiquariat Gebr. Haas* (1694 edition of *Geographie universalis*).

Description

In 1678, Pierre Duval's *La Geographie universalis...* was translated into German by Johann Christoph Beer and published by Johann Hoffmann (1629-1698) in Nuremberg. Hoffmann began his career as a printseller in Nuremberg in 1655. Shortly thereafter, he started to publish numerous books. According to Meurer, Hoffmann's importance in the history of German cartography has been overlooked. He should be regarded as the most prolific German map publisher before Johann Baptiste Homann.

This map of Africa is similar in appearance and is the same size as the Duval map of c.1660 (for a discussion of the sources for Duval's depiction of Africa see Map # 95). The toponyms on this map are in Latin. In Duval's map they were in French.

Publication Information

Hoffmann's *Geographie universalis* appeared in two parts with the first part containing a description of Africa and this map of the continent. After the 1678 edition, there were other editions published by Hoffmann using this map in 1679, 1681, 1685, 1690, and 1694. The Duval-Hoffmann map is known in two states.

Duval's *La Geographie universalis...* was also published in 1688 and 1712 by Jean Certe in Lyon using new copperplates (Map # 143). The Duval-Hoffmann map of Africa is similar in appearance to the Duval (Map # 95) and the Duval-Certe maps and is sometimes confused with them. Refer to the chart under the Duval map of Africa (Map # 95) for differences in the three maps.

Johann Hoffmann also produced geographically-based playing cards. Hoffman produced a set of playing cards of Europe in Nuremberg, dated 1678 (*Europäische Geographische Spiel Charte, J.H. Seyfrid delineavit, Wilhelm Pfann sculpit.*). Hoffmann is also known to have produced a set of playing cards of the world, *Geographisches carten-spiel von Asia, Africa, und America* (O'Donoghue 1901: 81, G. 267).

The author has not been able to locate an example of this set of cards. The set of world cards at the British Museum has not been located and a supposed second set is no longer accessible at the Playing Card Museum in Cincinnati, Ohio, USA as that museum has closed. In appearance, it has been suggested by Shirley that the set is similar to the miniature maps prepared by Hoffmann for Beer's translation of Duval's *Le Monde* of 1678 (Map # 121).

This set of playing cards of the world has been dated as early as c.1678 by Shirley (1993: 496) and King (2003: 146). Meurer has located one copy of the booklet to accompany the cards, dated 1696. According to the text in the booklet, the general map of Africa is shown on the card 'Pique 6'. Meurer has found that the example in Cincinnati forms a second edition with the booklet dated 1710. Based on this evidence, the author has to conclude that the playing cards of the world including the map of Africa were first produced in 1696 and then reissued in 1710.

References

King 2003; Shirley 1993; O'Donoghue 1901; Information from Peter Meurer.

122   1678 Pierre Duval, Paris

Map 122 as wallmap.

122.1 First State 1678

L' AFRIQUE | Où sont exactement decrites | toutes les Costes de la Mer, | suivant les Routiers et Portolans de divers Pilotes | et le Dedans du Pais | Selon les Relations les plus nouvelles. | Par P. DU-VAL Geographe du Roy | A PARIS | Chez l'Auteur, en l'Isle du Palais, Sur le quay de l'Orloge, | proche le coin de la rüe de Harlay | Avec Priuilege de sa Majesté | 1678.
[on bottom right corner of map sheet 2 and map sheet 4, and at the bottom left corner of map sheet 1 and map sheet 2]:
R. Michault scripsit.
- No scale bar.
- Paris : Pierre Duval, 1678.
- From: separately published / *Cartes de geographie les plus nouvelle et les plus fideles.*
- 1 wall map : copperplate engraving ; 4 map sheets ; the map in total measures 80 x 102 cm (in 4 sheets), with each sheet as 40 x 51 cm including text at the sides (each sheet measures 40 x 41 cm without text) ; without the secondary titles, the map measures 72 x 101 cm.
- Latitude and longitude coordinates on gridline along all four sides of the map and longitude coordinates across Equatorial gridline. *Ligne du Premier Meridien* (Prime Meridian) to the west of the Canary Islands.
- North at the top; no compass rose.
- Title cartouche is at the bottom left of sheet 3 with an Arab at the left holding a book, and an African at the right holding a map showing the source of the Nile at Lake Tana.
- Secondary titles on each of the 4 map sheets: On the upper left sheet (sheet 1), the secondary title is below the Equatorial gridline and begins: *LA BARBARIE...* with the publisher's imprint to its left and a key to its right identifying various symbols on the map. On the upper right sheet (sheet 2), the secondary title is below the Equatorial gridline and begins: *L'EGIPTE...* with the publisher's imprint to its right and a key identifying symbols to its left. On the bottom left sheet (sheet 3), the secondary

CARTOBIBLIOGRAPHY

title is above the Equatorial gridline and begins: *L'OCEAN...* with a key to its left. This third sheet also contains the primary title. The bottom right sheet (sheet 4) has a secondary title that begins: *L'ETHIOPIE...* with the publisher's imprint to its right.
- There are three sailing ships in the southern Indian Ocean. No animals within Africa. Text on each side of the map describing Africa with a list of its parts. A key, identifying various symbols on the

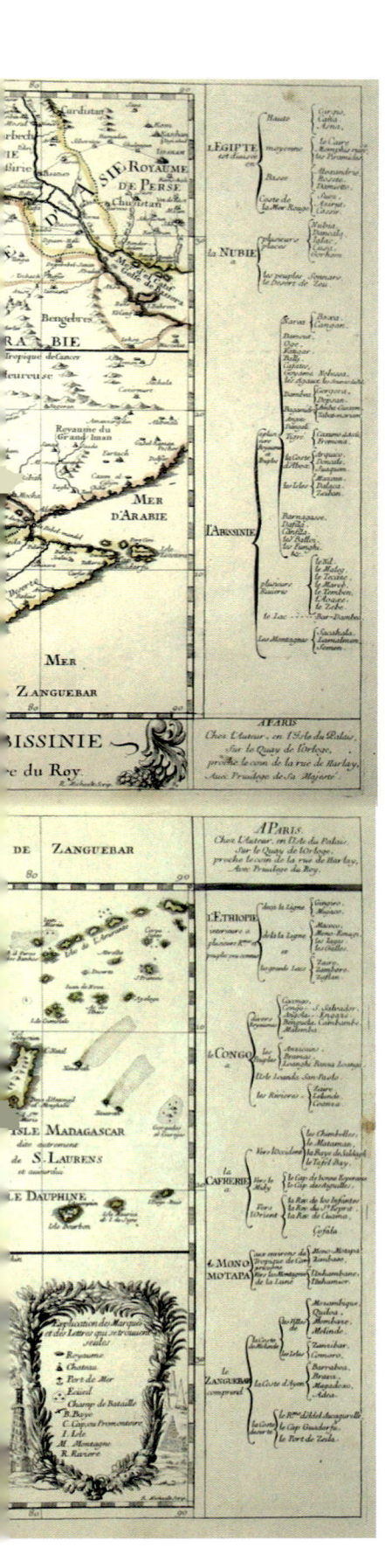

map (*Royaume, Chateau, Port de Mer*, etc.), is within a decorative wreath at the bottom right of the map.

Example: *New York, Martayan Lan Fine Antique Maps.*

122.1.A First State Variant A 1678
Joined to form one wall map. Dated 1678 on title cartouche. Without the secondary titles on each map sheet, which appear to have been trimmed.

Example: *København KB.*

122.2 Second State 1684
Now dated 1684 on title cartouche and also within the secondary titles of sheets 1, 2 and 4.

Examples: *New York, Richard B. Arkway Rare Books Inc.* (for 1684 example); *New York Martayan Lan Fine Antique Maps,* Catalogue # 17.

Pastoureau (1984: 146) mentions the appearance of this map in an edition of the *Carte de geographie* of 1679, located at the Toulouse Municipal Bibliothèque (*Toulouse BM*, Res A. XVII.) with another example at the Quimper Municipal Bibliothèque (*Quimper BM*: Arm. H.). Pastoureau describes a further edition of the *Cartes de geographie* of 1688 with this map at the Bibliothèque Nationale de France (*Paris BN*, Richelieu-Cartes et plans, Ge.DD.1174). The map in these editions is dated 1684. From the literature, it is not known if these examples are the second or third state of this map.

122.3 Third State 1684
Change in the map's author within the cartouche from Pierre Duval to his daughter as follows: from 'Chez l'Auteur' changed to 'Chez Melle, Du-Val'. Also, the new publication location of 'a l'Ancien Buis' is added to the title cartouche immediately after 'proche le coin de la rüe de Harlay'.

Tooley (1968: no. 47: 38) reports this map with dimensions of 72 x 100 cm. This conforms with the dimensions of the first state variant A, but not the first or second states of this map, which are larger owing to the additional secondary title strips. Given this map's dimensions, it is likely that this state of the map did not have the secondary titles.

Examples: Locations not known. As reported by Tooley in *Map Collectors' Circle*, no. 47: 38.

Description

This is an important transitional map of Africa. For the first time on a map of Africa, the connection between the Nile River and the two Ptolemaic central African lakes is purposely removed, and they are not shown as a source for the Nile. The map also clearly shows a correct orientation of the Blue Nile in the Abyssinian highlands and its source as Lake Tana. While the Ptolemaic lakes do not connect to the Nile, the western Ptolemaic lake of *Lac Zaire/Lac Zembere ou Zambere* is the primary source for the *Zaire* (previously Congo) River.

Many other aspects of this map follow the Jaillot map of 1674. The third southern river, which Duval names the *Indires R. al Zambere* (Zambezi), exiting into the Indian Ocean at *Quiloane* and to the north of the Cuama River, is shown, though without its separate source lake. In West Africa, the Niger River still shows its much earlier representation with an east-to-west direction entering into the Atlantic through the Senegal and Gambia Rivers. Above the Niger River, there remains the even older representation of a river (the *Ghir* and *Riviere de Nubie*) flowing eastward into the Nile. French influence is depicted in West Africa with placenames for French settlements (*Petit Diepe*, etc.).

In Southern Africa, the map reflects the Dutch influence at the Cape with Dutch names such as *Tafel Bay, Beauport*, etc. In Madagascar, *Fort Dauphin* is clearly shown at the southern point of the island.

This attractive, well-engraved map is one of a set of the four continents prepared by Pierre Duval. The Africa map was engraved by Michault ('R. Michault scripsit' on bottom right corner of the second and fourth map sheets and at the bottom left corner of the first and third map sheets).

Map 122.2 Title cartouche.

Publication Information

The four-sheet wall map appears to have been initially prepared for insertion in Duval's *Cartes de Geographie*, probably for the edition of 1679. The author has not examined an example of this 1678 map within a Duval atlas, but has only seen examples as four separate map sheets. The map was likely prepared for use in Duval's atlas to show larger maps of separate regions of Africa (Duval already had a folio-size Africa map of 1664 with later states which appeared in his atlases).

The author has also examined an example of this map at Det Kongelige Bibliotek in Copenhagen, also dated 1678. In this example, the four map sheets have been joined to form one large wall map. This example does not have the secondary titles at the top or bottom of each map sheet. As the information in the 1678 map was more current than in Duval's 1664 folio map of the continent, Duval may have had a demand for a more up-to-date map showing the entire continent. Thus, to meet this demand, he trimmed the secondary titles to join the four separate maps to produce a wall map of the continent.

The author has also seen an example of the set of four map sheets with the secondary titles that is dated 1684 at the antique map dealer, Richard Arkway Inc., in New York City. Another example of the map dated 1684 appeared in the Martayan Lan catalogue no.17.

Tooley reports an example of this map with the four map sheets joined in the *Map Collectors' Circle* (1968: no. 47: 38). Although dated 1684, in this example, the map's author is now changed to Duval's daughter with the secondary titles trimmed (this is based on this map's reported dimensions of 72 x 100 cm, the size without the secondary titles, since the author does not know the location of the Tooley example and has not been able to examine it).

The map appeared in editions of Duval's *Cartes de Géographie* until Duval's death in September 1683. The business was continued by one of his daughters, either Marie-Angelique or Michele Duval, though it is not known which one. This daughter altered the imprint on the map and included it in a further edition of 1688, using the third state of this map.

References

Shirley 1993: 501; Arkway n.d.: catalog no. 54; Martayan Lan n.d.: catalogue no. 17; Pastoureau 1984: 146-151, map numbers 68-71; Tooley 1968: 47: 38.

The author is grateful to Martayan Lan Fine Antique Maps, and Richard Arkway, Inc. for sharing images and information on their maps.

123   1679 Nicolas Sanson - Johann David Zunner, Frankfurt

AFRICÆ | Tabula | per ~ sanson.
- No scale bar.
- [Frankfurt : Johann David Zunner, 1679].
- From: *Die Gantze Erd-Kugel Bestehend in... Europa, Asia, Africa und America.*
- 1 map : copperplate engraving ; 20 x 27.5 cm.
- Latitude and longitude coordinates on all four sides of the map.
- North at the top; no compass rose.
- Title cartouche, in the form of a scroll, at the upper right side of the map.
- There are two sailing ships and two sea monsters in the Atlantic Ocean.

Examples: *München SB*, 4° MAPP. 95-1 ~ 95-4; *Wien ÖNB*; 393.913-B.K.

Description

This map of Africa appeared in a rare quarto-size atlas, published in 1679 in Frankfurt, Germany. The Zunner atlas filled a need for a detailed German-language atlas on Africa.

The map is generally derived from the 1656 map of Africa that Nicolas Sanson produced for his own atlas (See Map # 86). In contrast to the more austere Sanson map, this map is filled with considerably more geographic information, much of it imagined, as well as the addition of decorative elements. Also, Zunner locates the Prime Meridian further west, crossing the Cape Verde Islands at *I. del Fuogo* rather than *I. de Fer* in the Canary Islands.

Publication Information

In Frankfurt in 1679, Johann David Zunner published his rare four volume *Die Gantze Erd-Kugel Bestehend in... Europa, Asia, Africa und America*. This atlas contained volumes covering the four continents. Zunner was an active Frankfurt publisher from 1653, when he succeed his father of the same name, until his own death in 1703. There are two editions of *Die Gantze Erd-Kugel* with the same map of Africa.

Besides the earlier Sanson map of Africa, this Zunner map was followed by editions using new maps of Africa by De Winter (Map # 129) and then by Dufresnoy from the early eighteenth century.

In Amsterdam in 1683, an atlas was published with the title: *L'Europe | Dediée | a Monseigneur | Monseigneur le Tellier... Par N. Sanson le fils...* Simultaneously, in Utrecht in 1683, Joannes Ribbius published a copy of the Sanson atlas, *L'Afrique en plusieurs cartes nouvelles, et exactes....* . Both atlases contain the same new map of Africa, engraved by Antoine de Winter ('A d'Winter schu:'). See map # 129 for this map.

In Paris in 1716 and in Amsterdam in 1718, another new map of Africa appeared in *Methode Pour Etudier La Geographie* by Abbé Nicolas Lenglet Dufresnoy (1674-1755), using the Sanson 1656 map as a model. See the Sanson 1656 map (Map # 86) for an image and description of the Dufresnoy map.

References

The author is grateful to Jason Hubbard for sharing the results of his research on this series of atlases and works of geography containing the derivatives of the Sanson 1656 map.
Shirley 2004: Vol. 1: 886; Norwich 1997: map # 37 (second state of Sanson map); Burden 1996: map # 324 (Americas map); Pastoureau 1984: 387-9.

124 1680 William Berry, London

AFRICA | Divided according to the extent of its Principall Parts | in which are distinguished one from the other | the EMPIRES, MONARCHIES, KINGDOMS, STATES, and PEOPLES, | which at this time inhabite AFRICA. | To the Most Serene and Most Sacred Majesty of | CHARLES II. | by the Grace of God KING of Great Brittain, France, and Ireland. | This Map of AFRICA, is humbly Dedicated, and Presented, By | your Majesties Loyal Subject, and Servant | William Berry.
[a secondary title above the map]:
AFRICA DISTINGUISHED into its PRINCIPALL PARTS viz. BARBARY, BILEDULGERID, EGYPT, ZAARA or The DESART, THE COUNTRY OF THE NEGROS, GUINEA, NUBIA, ABISSINEA, ZANGUEBAR, CONGO | MONOMOTAPA, CAFFARES, The ISLANDS of the CANARIES, CAPE VERD. St THOMAS, MADAGASCAR or St LAWRENCE in which are observed The EMPIRES, MONARCHIES, KINGDOMES, STATES, and PEOPLES, which at present inhabite AFRICA. Described by Sanson | Corrected and amended By William BERRY.
[within cartouche at the upper left corner, immediately below scale bar]:
LONDON | Sold by William Berry at the Sign of the Globe | between Charing-Cross and Whitehall. | 1680.
- Scale bar at upper left:
  SCALE
  Italian Miles 600 = 7.0 cm.
  Common Leagues of France 250 = 7.0 cm.
  Common Leagues of Germany 150 = 7.0 cm.

- Common Leagues of Spain 180 = 7.0 cm.
  English Miles 480 = 7.0 cm.
- London : William Berry, 1680.
- From: separately published / A Bound Collection of Maps of the World by Berry.
- 1 map : copperplate engraving ; 2 map sheets, 57 x 89 cm (to outer black line).
- Latitude and longitude coordinates on all four sides of the map. Longitude coordinates across Equatorial Line.
- North at the top; no compass rose.
- Title within decorative cartouche at the top right of the map. Second title above the map on two lines of text.
- No decorative elements on the map, except for a large decorative title cartouche at the top right of the map.

Examples: The author has been able to locate five examples of this map: *Chicago NL*, Ayer 135 B48 1680; *London BL*, Maps 1.TAB.4 (the Africa map is bound into a non-standard atlas along with 36 other maps of the world); *London RGS*, 1.c.45; *Stanford SC*, Norwich number: 0047 (a separate map, as part of the Oscar Norwich Collection); *Washington LC*, G1015.B35 1689 (atlas).

Description

William Berry (1639-1718) was an engraver, map and bookseller, globemaker, and publisher active in several locations in London from the late 1660s to about 1708. On occasion, he collaborated with Robert Morden and others.

By tradition, the atlas is sometimes referred to as the English Sanson atlas, although all of the maps are by Berry. This large format map of Africa is most closely modeled on Alexis-Hubert Jaillot's map of Africa, dated 1674 (see Map # 118). That map in turn was derived from one originally prepared by Guillaume Sanson in 1668.

Publication Information

This map is from a collection of English maps prepared by William Berry and is often found bound into a non-standard atlas. In all known examples of the atlas seen by the author, there is no printed title page. This is a rare map. Phillips (1909-1992: 3442) found three examples of the Berry atlas: one at the Library of Congress (bound with other Berry maps of the world); one at The British Library; and one that was in the possession of Henry Stevens in the early part of the twentieth century, current location unknown. The map is only known in one state, dated 1680.

References

Tooley 1999-2004, A-D: 128; Norwich 1997: map # 47; Tyacke 1978: 109-110; Phillips 1909-1992: 3442.

125   1680 Robert Morden, London

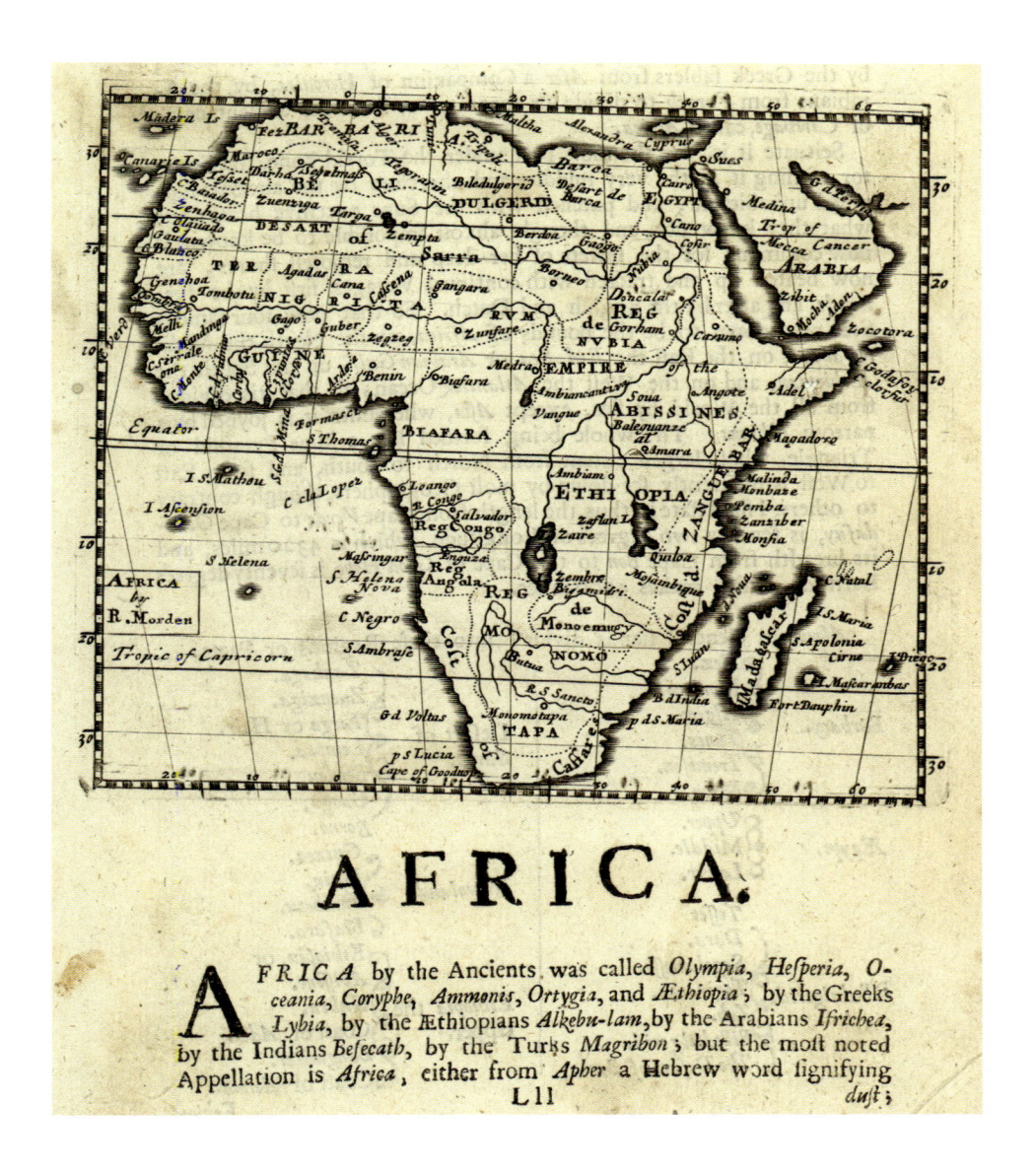

125.1 First State 1680

AFRICA | by | R. Morden.
- No scale bar.
- London : Robert Morden and Thomas Cockeril, 1680.
- From: *Geography Rectified: or, A Description of the World*.
- 1 map : copperplate engraving ; 11 x 13 cm (map only to outer gridlines; numbers extend beyond this; 20 x 15 cm (for entire page).
- Latitude and longitude coordinates on all four sides of the map. The Prime Meridian passes through West Africa. From the bottom left, longitudinal numbers of 20, 10, 0, 10…
- North at the top; no compass rose.
- Title is set within a simple box at bottom left just inside the gridline.
- No decorative elements on the map.
- Five lines of text describing Africa below the map (with text on the verso). ' (317)' in letterpress directly above the map).

Example: *Washington LC*, G1015.M68 1680 (from *Geography Rectified* of 1680).

125.2 Second State 1688

The Prime Meridian now passes through the Canary Islands. From the bottom left, longitudinal numbers are changed to 360, 8, 18, 28... . 'Of AFRICA' now appears in large letterpress directly above the map. Also, the number '441' appears in letterpress in upper right corner above 'Of AFRICA'.

Example: *Washington LC*, G120.M68 1688 (*Geography Rectified* of 1688).

125.3 Third State 1693

'Pag.93' etched into the top right corner of the map image. Also, the number '461' appears in letterpress in upper right corner above 'Of AFRICA'.

Examples: *Geography Rectified* of 1693: *London BL*, 570.d.4; *Washington LC*, G120.M83 1693 Vault.

This map also appeared in its third state (with 'Pag.93') in Morden's *Atlas terrestris*. In the *Atlas terrestris*, there is no text above or below the map.

Examples: *Atlas terrestris* of c.1700: *Washington LC*, G1015.M67 1700 (in *Atlas terrestris*, undated on title page of LC example).

125.4 Fourth State 1699/1700

'Pag. 293.' etched into the top right of the map image. In the *Geography Rectified* of 1700, the number '461' appears in letterpress in the upper right corner above 'Of AFRICA'.

Examples: 1699 edition of *Geography Anatomised* by Patrick Gordon: *London BL*, 570.c.13. 1700 edition of Morden's *Geography Rectified*: *Washington LC*, G120.M68 1700.

Morden's map of Africa also appeared in *Geography Anatomised* by Patrick Gordon, published in 1693, 1699, 1700, 1702, 1704, up to 1719. The 1699 edition of Gordon's *Geography Anatomised* used Morden's fourth state of the map of Africa ('Pag. 293.' is at top right within map image).

Description

Robert Morden (? – 1703) was a geographer, publisher, bookseller, and globemaker working in London, 'at the Atlas at Cornhill', in the later part of the seventeenth century. He is noted for his geography books, his separately published maps, and his publication of playing cards. He collaborated with William Berry, Philip Lea, John Overton, and Robert Walton among other London mapmakers of the period (Tooley 1999-2004, K-P: 278; Tyacke 1978: 123).

Like the William Berry map of 1680, Morden followed the French for his depiction of Africa. This map is most closely model on Alexis-Hubert Jaillot's map of Africa, dated 1674. This map in turn was derived from one originally prepared by Guillaume Sanson in 1668.

Publication Information

This map first appeared in *Geography Rectified* of 1680, written by Robert Morden and published jointly by Morden and Thomas Cockeril whose name also appears on the title page. The book was reprinted in 1688, 1693, and 1700. The atlas contains a world map and 75 maps of parts of the world including Africa and the other continents.

The map of Africa also appeared in Morden's undated *Atlas terrestris*. The exact publication date for this atlas is not known. The *Atlas terrestris* may have been published as early as 1690 or as late c.1700. It contains a world map and 75 maps of parts of the world including Africa and the other continents.

References

King 2003; Tooley 1999-2004, K-P: 278; Shirley 1993: 510; Tyacke 1978: 123; Phillips 1909-1992: entry # 498, # 4265, and # 3454.

126 1681 Jonas Moore – Herman Moll, London

AFRICA.
- No scale bar.
- [London : Robert Scott, 1681].
- From: *A New Geography with Maps to each country and Tables of Longitude and Latitude. The second part of the second volume of A New System of the Mathematics.*
- 1 map : copperplate engraving ; 16.5 x 21 cm.
- Latitude and longitude coordinates on all four sides of the map. Longitude coordinates across the Equatorial Line. The line of the Prime Meridian, identified as the *The first Meridian*, crosses through the western Canary Islands.
- North at the top; no compass rose.
- Title cartouche in simple box at bottom left.
- No decorative elements on the map.

Examples: *London BL*, 531.n.17,18; *New York PL*, *KC 1681, Moore.

Description

This is the earliest, printed map of Africa attributed in part to Herman Moll (c.1654-1732) as the map's engraver. Moll was a geographer, mapmaker and an engraver, originally from Germany. It is thought that Moll emigrated to England in 1678 via The Netherlands. Moll was originally employed as an engraver in London before undertaking his major atlas projects in the early part of the eighteenth century.

The geographic model for this map was the Duval map of Africa of 1660 and the later Duval – Hoffmann map of 1678. French settlements such as *F. Dauphin* at the southern end of Madagascar and *Goree* Island off the Senegal coast are shown. The map does not show recent Dutch or English settlements along the coasts. The interior is generally devoid of placenames, except for a few locations, such as *Zimbaoe, Amara*, etc., and the names of regional divisions. The east coast shows the traditional Arab-Portuguese placenames.

Publication Information

The map is from Jonas Moore's *A New Geography* of 1681. Jonas Moore died in 1679 before he could finalize the text of Part II of Volume II of *A New System of Mathematics* entitled *A New Geography*. It was this section of the book that contained maps. Based on an 'Advertisement to the Reader' at the beginning of this part, *A New Geography* was completed by the editors Vale and Fruere. Evidently, a portion of the text in this part of the book was also reworked by Edmund Halley.

The maps by unknown engravers, which were originally intended for this book, were evidently found to be unsuitable when the editors began to finalize the book for publication (Baker 1937: 16). Moll, who had recently moved to London from Germany in 1678, was selected to engrave new maps. At the time, he was about 27 years old. The maps of America and Europe have Moll's imprint as the engraver ('H. Mol schulp.', with the imprint of 'Londini' also added on the Europe map). Based on the similar engraving style, the map of Africa is believed to have also been engraved by Moll. The verso of the map is blank.

The book is scarce, and it is very rare to find it with the full set of world, continent, and regional maps. Not all editions of Moore's book contain the maps. The book was published by Robert Scott, 'Bookseller at the Princes Arms in Little Britain', and the printers were Godbid and Playford. This map is only known in one state.

References

Tyacke 1978: 122-123; Baker 1937: Vol. 2: 16; Sabin 1868: entry # 50415.

127   c. 1682 Alexis Hubert Jaillot – Johann Hoffmann, Nürnberg

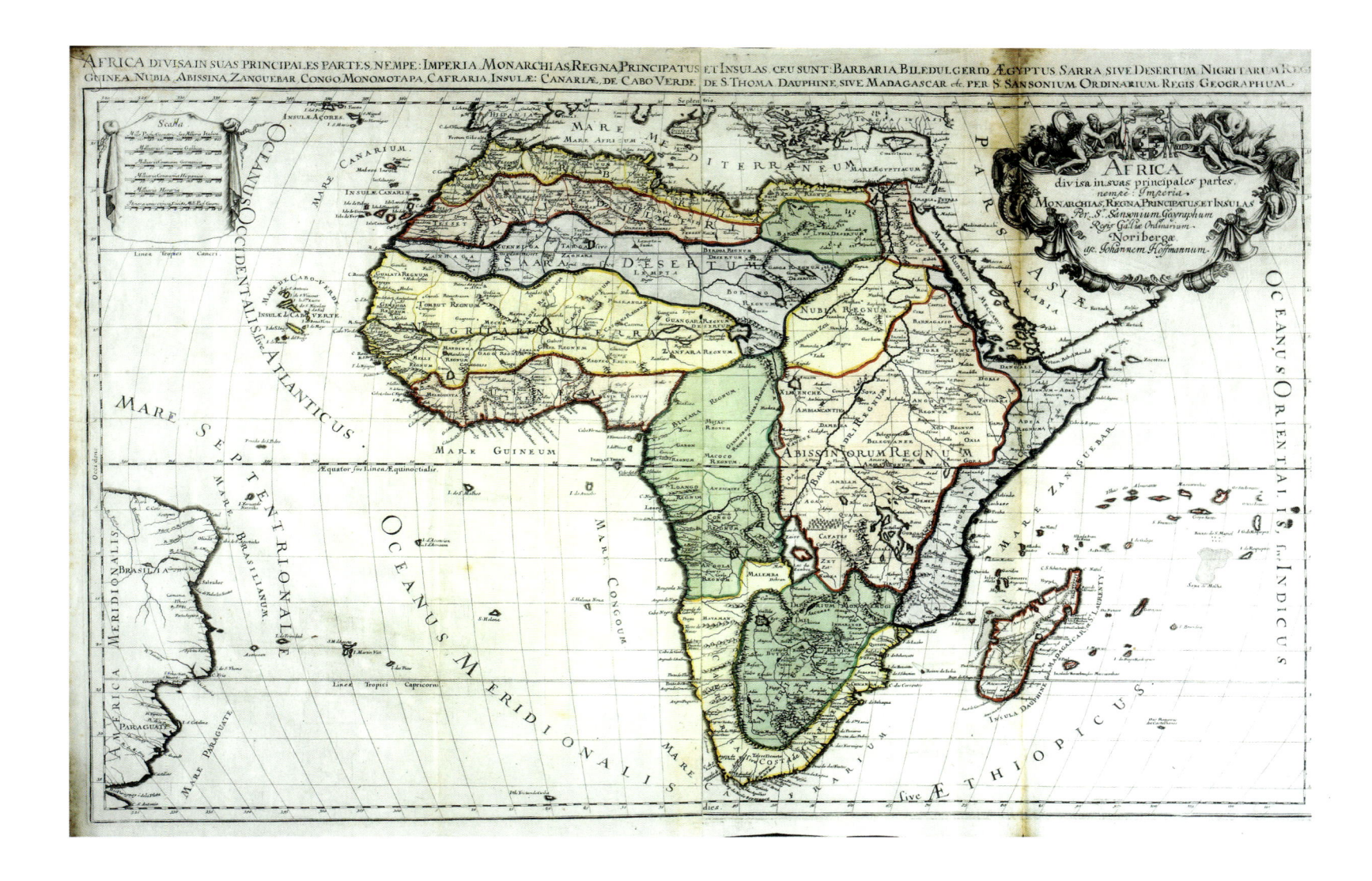

AFRICA | divisa in suas principales partes, | nempe: Imperia | MONARCHIAS, REGNA, PRINCIPATUS, ET INSULAS | Per .S.ʳ Sansonium, Geographum | Regis Galliæ Ordinarium. | Noribergæ, | ap. Johannem Hoffmannum.
[above the map on two lines of text]:
AFRICA DIVISA IN SUAS PRINCIPALES PARTES, NEMPE: IMPERIA, MONARCHIAS, REGNA, PRINCIPATUS ET INSULAS, CEU SUNT: BARBARIA, BILEDULGERID, AEGYPTUS, SARRA, SIVE DESERTUM, NIGRITARUM REGIO, | GUINEA, NUBIA, ABISSINA, ZANGUEBAR, CONGO, MONOMOTAPA, CAFRARIA, INSULÆ: CANARIÆ, DE CABO VERDE, DE S. THOMA, DAUPHINE, SIVE MADAGASCAR etc. PER S. SANSONIUM, ORDINARIUM REGIS GEOGRAPHUM.
[at the bottom of the title cartouche set within the decorative elements]:
Sigmund Gabriel Hipschman scu.
- Scale bar is set within drapery cartouche at upper left of the map on six lines:
  Mille Passus Geometrici sive Milliaria Italica 600 = 7.1 cm.
  Milliaria Communia Gallica 250 = 7.1 cm.
  Milliaria Communia Germanica 150 = 7.1 cm.
  Milliaria Communia Hispanica 178 = 7.1 cm.
  Milliaria Horaria 200 = 7.1 cm.
  Itinera unius cujusque Diei 40 Mill: Pass: Geom: 15 = 7.1 cm.
- Nürnberg : Johann Hoffmann, [c. 1682].
- From: separately published / *Prospect des ganzen Erdkreisses.*

- 1 map : copperplate engraving ; 2 map sheets ; 54.5 x 87.5 cm (an extra 2.5 cm for the title above the map).
- Latitude and longitude coordinates on all four sides of the map. Longitude coordinates across Equatorial Line.
- North at the top; no compass rose.
- Title cartouche at the upper right of the map is set within a decorative drapery surrounded by water gods and wild animals. The coat-of-arms at the top center of the title cartouche is of the Salzburg Prince Bishop, Max Gandolf, Count of Künburg, to whom this map is dedicated. A second title is above the map.

Examples: *Bern SUB,* Ryh 7601:29; *Dresden LB,* A 539; *Erlangen UB,* H61/Kat.C.37; *Göttingen SUB,* Mapp 2943.2 Ex; *Regensburg SB,* Lade 49.

Description

This large format map of Africa by Johann Hoffmann is a close copy of the Alexis Hubert Jaillot map of Africa, dated 1674. As the title indicates, it is based on the geography of Guillaume Sanson (1633-1703), in this case his map of Africa of 1668.

For a discussion of the geography represented on the map, please refer to Map # 118, the 1674 Jaillot map.

The Hoffmann map is most easily distinguished from the Jaillot map by this map's use of Latin for the titles instead of French and by the imprint for the publisher, Hoffmann, and the engraver, Hipschman, on this map instead of Jaillot and his engraver, Cordier.

Publication Information

Johann Hoffmann (1629–1698) was active in Nuremberg from 1655 starting as a printseller. Shortly after, he began publishing books. Hoffmann aggressively expanded his business. In some years, he issued nearly 40 new titles.

Hoffmann's importance in the history of cartography has been overlooked so far. Meurer states that Hoffmann must be regarded as the most prolific German map publisher before Johann Baptist Homann. Recent research indicates that he produced about 50 separate maps (many of them with accompanying booklets) and about 45 books, some illustrated with maps.

Shortly after 1670, Hoffmann began to prepare a series of reprints of the Jaillot maps with the intention of producing a large size folio atlas. In total he produced 24 maps until about 1686. He was not able to complete a large scale atlas. Hoffmann was producing a tremendous amount of cartographic material during this period and was probably not able to complete this project due to other priorities and also due to economic reasons.

The map of Africa appeared around 1682 as a separate publication. It has been dated c.1682 by Peter Meurer as it is known that the map was first mentioned in Hoffmann's sales catalogue of 1683 and the appearance of examples in 'plano' shows that the map also could be separately purchased.

The four maps of the continents and the world also served as illustrations in a highly rare book *Prospect des ganzen Erdkreisses* which was issued by Hoffmann in 1686. The map is known in one state.

Besides this Hoffmann map, there were later copies of the Jaillot 1674 map by Pieter Mortier, starting in 1692.

References

Meurer 2007: 9-19 with image on 17; Blonk 2000; Pastoureau 1984: 229-262 (Jaillot map of 1674 map: p. 235: no. 4); Koeman 1967-71, III: 4-10. The author is grateful to Peter Meurer for sharing his information on this map.

128 c. 1682-1685 Nicolaas Visscher II, Amsterdam

Map 128 with decorative and text panels surrounding the map (example A).

AFRICA.
[a second title above the map]:
EXACTA TOTIUS AFRICÆ TABULA.
[set within decorative scene at bottom left of the map]:
AMSTELODAMI | apud Nicolaum Vißcher | Cum Privil: Ordin: General: | Belgii Fœderati.
- No scale bar.
- Amsterdam : Nicolaas Visscher, [c.1682-1685].
- From: separately published wall map.
- 1 wall map : copperplate engraving ; 6 map sheets ; 100 x 123 cm for the map (with each sheet as 50 x 41 cm on the Visscher Paris example) ; 150 x 184 cm (including the map, all the borders and the title strip on the Visscher-Allard Stockholm example, Map 128.A).
- Latitude coordinates and longitude coordinates along all four sides of the map. Latitude coordinates across the Equatorial Line.

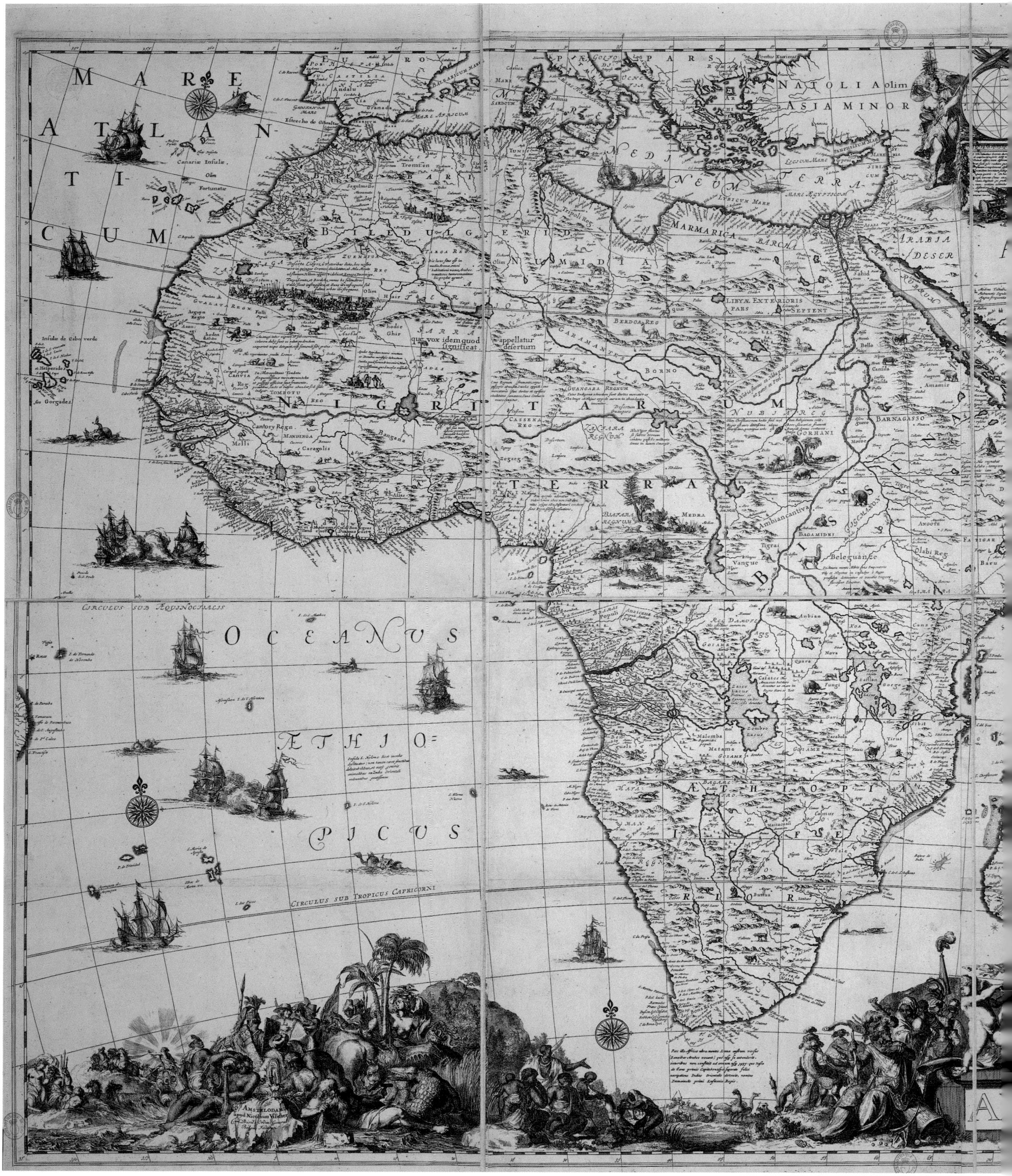

- North at the top; one compass rose in northern Indian Ocean, and three compass roses in Atlantic Ocean.
- The title of AFRICA set within a box at the bottom right of the map surrounded by an elaborate scene of African people, flora, and fauna across bottom of the map.
- A second title of EXACTA TOTIUS AFRICAE TABULA is set within a ribbon-shaped cartouche on five pasted strips above the map on the Visscher-Allard Stockholm example. In this example, the second title letters are in gold, set on a ribbon with a blue background, surrounded by a floral design with putti at each end.
- A large display is at the top right of the map with two geometrical spheres and text describing how to determine distance using triangulation.
- There are six sailing ships in the Indian Ocean, eleven sailing ships in the Atlantic Ocean, and three ships in the Mediterranean Sea.
- Within Africa, there are numerous legends; vignettes of African people and animals, including a large caravan beginning to cross the Sahara Desert; and many animals scattered throughout the map.
- There are various decorative elements on the Visscher-Allard Stockholm example. These are a title on five pasted strips above the map; six town views on each side of the map with each town view separately pasted and measuring 19 x 26.8 cm; and text panels below the map. The town views are described further below.

Examples:

128.A *Stockholm KB.*
The example at Kungliga Biblioteket contains decorative and text panels surrounding the map, with an imprint for Hugo Allard (1625-1691) at the end of the text panels.

128.B *Paris BN*, Ge DD 5082.
This Visscher example at the Bibliothèque Nationale de France includes only the 6 map sheets.

128.C *Karlsruhe LB.*
The separate title above the map in the Visscher Karlsruhe example is NOVA ET ACCURATA TOTIUS AFRICAE TABULA. The 6 map sheets have been joined in this example. There is no mention of side decorative borders nor of text panels below, as on the Stockholm example.

Description

This map is by Nicolaas Visscher II (1649-1702), son of Nicolaas I (1618-1679) and grandson of Claes Jansz. Visscher (1587-1652). Sometime before c. 1652, Claes Jansz. Visscher had issued the fourth state of Willem Blaeu's famous 1608 wall map of Africa. A fifth state of the Blaeu wall map was issued by his son, Nicolaas Visscher I, sometime after 1652, when Claes died, and possibly as early as 1656. By the time Nicolaas Visscher II assumed control of the business in 1679, when his father died, the Visscher-issued fifth state of the Blaeu wall map was clearly out of date. Nicolaas Visscher II produced this map to better compete, particularly with the French, in the market for large format maps of Africa.

In the lower left of the map, set within the decorative scene is the privilege granted to Nicolaas Visscher II in 1682 by the States of Holland and West Friesland (AMSTELO-DAMI | apud Nicolaum Vißcher | Cum Privil Ordin General | Belgii Fœderati). The privilege granted to his father in July 1677 had previously expired upon his father's death in 1679. As the map contains no date, the date of c. 1682-1685 is inferred through the appearance of this privilege on the map.

Map 128 Map sheets only (example B).

The general model for the geography of the interior of Africa, including the river systems, is Willem Blaeu's 1608 wall map of Africa in its later Visscher states, rather than the more up-to-date French maps from this period. However, there are numerous refinements to the map including the shape of Africa.

On this Visscher map, Southern Africa is labeled using the much older name of *Aethiopia Inferior* (Lower Ethiopia), rather than Monomotapa. However, along the coast, the map shows considerable up-to-date information, reflecting an increased Dutch presence at the Cape of Good Hope. *Tafel bay* (Table Bay), *Tafelberg* (Table Mountain), *C Falso* (False Bay), and *Mossel bay* (Mossel Bay) are readily identified. Off the coast, the map provides additional details related to water depth and shoals, which were a cause for numerous sailing mishaps. Saldanha Bay just to the north of the Cape of Good Hope is still not as accentuated with its inward sweep as it will become on later maps. Further north, Walfis Bay is not clearly shown. The Southern African river systems follow the Blaeu wall map with little change.

The general configurations of the lakes and rivers of other parts of the interior also appear unchanged. In East Africa, the placenames appear to be taken directly from the Blaeu wall map. On the Guinea coast, besides the more traditional *El Mina*, a number of additional placenames appear such as *Nassau* and *Acra*, reflecting an increase in permanent European coastal fortifications and trading settlements. Additional placenames for settlements continue north along the coast to the regions of the Senegal and Gambia Rivers.

In the Visscher-Allard Stockholm example, the six map sheets are surrounded by town views, six on each side, and text panels below the map. There are six town views on each side of the map. On the left side are *DEL MINA* (the mine on the Guinea coast), *MAROCO* (Marrakech), *SALEE* (in Morocco), *ALEXANDRIA*, *GIGIRI*, and *SABA* (?); on the right side are *TANGERE*, *ALCAIR* (Cairo), *TRIPOLI DE BARBARIA*, *THUNIS*, *ALGIERS*, *AMARA* (in Abyssinia).

Below the map in the Visscher-Allard Stockholm example, there is text describing Africa. The first section is in Latin (NOVA AFRICA DESCRIPTIO...) with the imprint at the end of the Latin text as follows: 'Amstelodami, Apud Hugo ALLARDT [in] platea vitulina, vulgo dicta de Kalverstraet, ad Cart[a Mundi]'. This is followed by text in French (NOUVELLE DESCRIPTION D'AFRIQUE...) with the imprint at the end of the French text as follows: 'Amsterdam, chez HUGO ALLARDT, demeurant a la Rue dicte Cal[verstraet, à la Carte du Monde]'.

Publication Information

This map is known to the author in three examples, though one example at the Badische Landesbibliothek Karlsruhe is based on the literature.

On all three examples, a title of *AFRICA* is set within a rectangular box at the bottom right of the map (on the bottom of the fifth and sixth map sheets) surrounded by an elaborate African scene across the entire bottom of the map of people with flora and fauna. Also within the scene are various putti and allegorical representations of Neptune and sea creatures.

References
Stopp 1974: 7: image C3.
Discussion with Günter Schilder, Universiteit Utrecht, The Netherlands.
Discussion with Göran Bäärnhielm, Map Curator, Kungliga Biblioteket, Stockholm, Sweden.

129 1683 Nicolas Sanson - Antoine de Winter, Amsterdam & Utrecht

129.1 First State 1683

AFRIQVE. | Par la S^r. Sanson d'Abbeville, | Geographe du Roy.
(at bottom right of the map):
A. d'Winter schu:
- No scale bar.
- [Amsterdam & Utrecht : Francois Halma & Joannes Ribbius, 1683].
- From: *L'Afrique en plusieurs cartes nouvelles, et exactes... .*
- 1 map : copperplate engraving ; 20 x 27.5 cm.
- Latitude and longitude coordinates on all four sides of the map. Longitude coordinates across the Equatorial Line.
- North at the top; no compass rose.
- Title cartouche, in the form of an oval wreath, at the lower left side of the map.
- No other decorative elements on the map.

Examples: Halma 1683 Amsterdam edition: *London, Hubbard Collection*; Ribbius 1683 Utrecht edition: *Amsterdam SMA*, A/I.43; *London BL*, Maps C.48.b.39; *Oxford Bodl*, 4° F.47.Art; *Chicago, Baskes Collection.*

129.2 Second State 1699
Addition of the wording for the Prime Meridian, Tropic of Cancer, and Tropic of Capricorn as follows: *Premier Meridien, Tropique du Cancer, Tropique du Capicorne.* There are also placename changes; for example, in Southern Africa the map now has *Monomegui* in place of *Monomo=*, as on first state.

Examples: *Dijon BM*, 16636 (in *Description de tout L'Univers...* , Amsterdam : Arnout van Ravestein, 1699); *London BL*, 568.d.7 (1700 edition of Halma's *Description de tout l'univers*).

Map 129.1 Original title.

Map 129.3 New title.

Map 129.5 Title identical to 129.1, but re-engraved.

129.3 Third State 1705
New title: AFRICA. | Door N. Sanson d'Abbeville, | Geographe du Roy.

Title is now in Dutch for an edition of *Algemeene Weereldbeschryving nae de recht verdeeling*, published by Halma in 1705.

Examples: *Leeuwarden Tr,* E2525; *Providence JCB,* E705-L147a; *Washington LC,* G114.L.14 Pre-1801 Coll.

129.4 Fourth State 1715-1734
Title cartouche is now again in French. Evidence of re-engraving of letters and letter spacing.

Example: *Washington LC,* G1015 .D87 1735 Vault (Du Sauzet's *Atlas Portatif*).

129.5 Fifth State 1734 (1738)
For the second printing of Du Sauzet's *Atlas Portatif,* the map now has an engraved '6' in the upper right corner within the map gridline.

Examples: *London BL,* Maps C.29.c.7; *Washington LC,* G1015.D87 1738 Vault (second printing of 1738 of Du Sauzet's *Atlas Portatif*).

Description
In Utrecht in 1683, Joannes Ribbius published a copy of the Sanson atlas, *L'Afrique en plusieurs cartes nouvelles, et exactes...*, with a new map of Africa. In the same year in Amsterdam, François Halma published an atlas with the title: *L'Europe...*, containing maps of all regions of the world including this same map.

This map is a close copy of the Nicolas Sanson 1656 map of Africa. The map is most easily distinquished from the 1656 Sanson map by the imprint of Antoine de Winter (ca. 1652/3 - 1707) as the engraver of the map of Africa ('A d'Winter schu:', at the bottom right corner of the map), replacing the imprint for the Sanson engraver, Peyrounin (A. Peyrounin sculp., at the bottom right corner of the map). For a discussion of the geography, refer to the Sanson 1656 map (Map # 86).

Publication Information

This map appeared in a number of publications in The Netherlands from 1683 to 1738. The publication history of these atlases using the De Winter plates is complicated.

In 1683, an atlas, containing maps of all regions of the world including this one of Africa, was published in Amsterdam with the title: *L'Europe | Dediée | a Monseigneur | Monseigneur le Tellier. . . Par N. Sanson le fils | Geographe du Roy | Sur la copie | A Paris chez l'Autheur | 1683*. The second title page again mentions: *. . .Sur la copie imprimée | A Paris, | Chez l'Autheur, dans le Cloître de Saint Germain de l'Auxer-| rois, joingnant la grande Porte du Cloître. | M.DC.LXXXIII.* [1683]. Pastoreau (1984) notes that this was a counterfeit edition, 'Édition hollandaise copiant l'édition française.'. Hubbard believes this atlas was published by François Halma. Simultaneously, in Utrecht in 1683, Joannes Ribbius published a copy of the Sanson atlas, *L'Afrique en plusieurs cartes nouvelles, et exactes...*, with the imprint of Antoine de Winter as the engraver of the map of Africa ('A d'Winter schu:') (Pastoureau 1984: 391-2).

In Amsterdam in 1692, François Halma (1653-1722) published an edition of Joannis Luyts' *Introductio ad Geographiam novam et veteren* containing an example of the De Winter copy. In Amsterdam, in 1699, Arnout van Ravestein published his rare *Description de tout l'Univers...* with this map. Also, in 1700, François Halma re-published an edition of *Description de tout l'univers en plusieurs cartes* with this De Winter map of Africa. This was followed by a Halma edition in Dutch In 1705 of *Allgemeene Weereld-Beschryving...* . This edition of the book also contained this map. The same De Winter copperplate was used in 1715 by N. Chamereau (also identified as Chemereau) for the publication, *Geographie Pratique*.

The last use of the De Winter map was by Henri du Sauzet in 1734 and 1738. In 1734, Du Sauzet published his *Atlas Portatif* in Amsterdam using the De Winter map. There was a second Du Sauzet printing of the *Atlas Portatif*, still dated 1734, with numbers engraved on the maps, and with some maps with dates of 1738, (e.g. Carte des Espagnes, etc.).

Also derived from the Sanson 1656 map of Africa, another new map of Africa appeared in *Methode Pour Etudier La Geographie* by Abbé Nicolas Lenglet Dufresnoy (1674-1755) in 1716 (Paris: Chez Charles Estienne Hochereau.) and again in 1718 (Amsterdam, Aux depens de la Compagnie, 1718). There were subsequent editions in 1729, 1735, and 1737. See the Sanson 1656 map (Map # 86) for an image and description of the Dufresnoy map.

References

The author is grateful to Jason Hubbard for sharing the results of his research on this series of atlases and works of geography containing the derivatives of the 1656 Sanson map.
Shirley 2004: Vol. 1: 886 (for Sanson), 890 (for Du Sauzet); Norwich 1997: map # 37 (second state of Sanson map); Burden 1996: map # 324 (Americas map); Pastoureau 1984: 387-9; Koeman 1967-71, II: Me 207-208, and Hal 1: 125-127; Sabin 1867: entry # 38504.

130   1683 Johannes de Ram, Amsterdam

130.1 First State 1683

AFRICÆ | ACCURATA | TABULA | EX OFFICINA | JOAN: DE RAM.
[across the bottom right of the map]:
Cum Privilegio Ordinum Hollandiæ et Westfrisæ.
- No scale bar.
- [Amsterdam : Johannes de Ram, 1683].
- From: separately published.
- 1 map : copperplate engraving ; 44 x 56 cm.
- Latitude and longitude coordinates on all four sides of the map. Longitude coordinates across the Equatorial Line.
- North at the top; one compass rose in the South Atlantic.
- The title cartouche at bottom left is similar to the earlier De Wit map of 1670.

- There is one sailing ship in the North Atlantic by Spain. No animals are within Africa.

Example: *Stanford SC,* MOA 0071 (part of Oscar I. Norwich collection of maps of Africa).

130.2 Second State after 1696 to 1711
With the addition in the title cartouche of: 'ex officina Jacobi de la Feuille'.

Example: Location not known. As reported by Tooley (1969: 123).

Description
Johannes de Ram (1648-1693) was known as an engraver, publisher, globemaker, and art dealer. De Ram copied the maps of other prominent mapmakers from this period, but also designed and issued his own maps. He is known to have produced an elaborate and highly decorative map of the world in 1683 (Shirley 1993: 518). This map of Africa was likely prepared and issued at the same time to accompany his world map. After De Ram's death, his widow, Maria van Zutphen, continued the business, marrying Jacques de la Feuille in 1696. La Feuille is known to have been active as a map publisher until 1711. He produced a second state of this map after 1696 (Koeman 1967-71, III: 98).

This map copies the same large, decorative title cartouche of the Frederick de Wit map of Africa of c.1670. However, the geography on this De Ram map closely follows the earlier De Wit map of 1660, which was based on the geography of the Jodocus Hondius Jr. map of 1623 and the Willem Blaeu map of 1617.

By the time of this map, major advances and innovations in the representation of Africa were passing to the French, notably Sanson, Duval, De Fer, and others. De Ram chose not to show any of these features on his map.

Publication Information
De Ram is known to have published an atlas in 1685 with a dated title page. This title page was engraved by Jacob Harrewyn whose signature appears on the page. It is possible that Harrewyn engraved this map of Africa. The map is known in two states.

The De Ram atlas is not common. Koeman reports an example of the De Ram atlas at the Bibliotech in Warsaw and at the Bayerische Staatsbibliothek, München. Tiele mentions a copy of the atlas published by De Ram's widow, though this might be the atlas produced with her new husband, Jacques de la Feuille.

References
Tooley 1999-2004, Q-Z: 8; Norwich 1997: map # 50; Shirley 1993: 518-519; Koeman 1967-71, III: Ram 1, 98; Tooley 1969: 123; Tiele 1969.

131   1683 Justus Danckerts, Amsterdam

Map 131.2.

131.1 First State 1683

TOTIUS | AFRICÆ | Accuratissima | TABULA | Authore | I. DANCKERTS | Amstelodami.
- No scale bar.
- Amsterdam : Justus Danckerts, [1683].
- From: separately published / Danckerts atlases starting from 1688-1689.
- 1 map : copperplate engraving ; 49 x 58 cm.
- Latitude and longitude coordinates on all four sides of the map.
- North at the top; one compass rose below the Equatorial Line in the Atlantic Ocean.
- Title cartouche at the bottom left of the map
- There are four ships in the Atlantic Ocean, three ships in the Indian Ocean, and five smaller ships

between Sicily and Greece. Animals are included as decorative elements within Africa as on the De Wit c.1670 map.

Examples: *Amsterdam UB*, 366 A 10 (atlas dated 1695); *Budapest NSL*, TA 225 /4; *Graz LB*, T 258003 VI /4; *Greenwich NMM*, B 6190 /4; *London BL*, Maps C.39.f.2 (the map is in a composite atlas dated '1650?'); *Wien ÖNB*, FKB 272-28 / 4.

Map 131.2 Title cartouche.

Map 131 Hachuring of the shadowy right side of the title 'stone': left early edition, right later edition (Graz LB).

Later prints (c. 1691/93) show extensive wear to the copperplate with some reworking of the map as follows: The ships are fairly worn, but some main features are slightly reworked and there are some new lines detectable. The cartouche is also worn and retouched a little. For example: the shadowy right side of the title 'stone' has new hachuring lines in //// direction - on the old worn lines. New /// lines are on the front/forehead of the elephant, and also on the bale at the left side.

131.2 Second State c. 1696/1698

The ships in the Atlantic and Indian Oceans have been removed, with some remnants still to be seen in some examples. The five smaller ships between Sicily and Greece are still on the map, though faint and worn. Other changes as follows: Animals are also detectable on some examples of this state, though in very bad condition with very shallow gray lines. The lines of hachuring on the seacoasts are renewed. The right sides of the mountains are also retouched. The cartouche is extensively reworked with many new lines.

Examples: *Amsterdam UB*, 33-17-38; *Wien UB*, IV 292877 /4.

Description

The Danckerts were a prominent print and map publishing family active in Amsterdam for almost 100 years. The founding member of the family business was Cornelis I (1603-1656) who started producing and publishing maps in the second part of the 1620s. His sons, Dancker (1634-1666) and Justus (1635-1701) were also active in the business. Justus issued a number of atlases with his sons. Justus' sons, Theodorus I (1663-c. 1727), Cornelis II (1664-1717), Johannes (?-1712), and Eduard (?-after 1721) and his grandson Theodorus II (1701-1727) continued the family business into the early eighteenth century. Because of the use of the same first names by various family members, there is some confusion in the literature over the authorship of some of the maps produced by the Danckerts.

The business was most active from about 1630 to 1727 when the stock of maps in the shop of Theodorus II, Cornelis I's great-grandson was sold. As Koeman states (1967-71, II: 88), 'Their cartographic work has, compared with that published by the Blaeus or Janssonius, attracted but little attention and has never received proper recognition.'.

This map of Africa was prepared by Theodorus I, probably in cooperation with Justus (1635-1701) and Cornelis II (1664-1717). Justus appears to have been active as a map publisher from 1664 when he was registered in the booksellers' guild of Amsterdam until 1701 when he died.

Justus' atlases, like those of other members of his family, are undated. This map is often dated c. 1680 based on Koeman. However, more recent research by Gyuri Danku, Map Librarian, Budapest National Széchényi Library, who has conducted extensive research on the history of the Danckerts atlases, suggests that a more proper publishing date for this map is 1683. This map of Africa was published into the late 1690s when it was replaced by a newer map by the Danckerts.

The map is a close copy of the Africa map of c. 1670 by Frederick de Wit, with the same decorative vignette and geography (see Map # 114). It was likely published without De Wit's permission. It appears that the Danckerts published this map without their publishers' privilege (which they obtained in 1684), in all issues of this map even after this date.

Publication Information

This map was issued separately and appears in undated Danckerts atlases starting in 1688-1689.

On Sept. 14th, 1684, Justus and his sons Theodorus I and Cornelis II obtained a privilege from the States of Holland and West Friesland for their prints and maps. Around 1698/1701, most probably in 1699/1700, the Danckerts issued a new map of Africa (see Map # 168), which replaced this map. It has a new title: *Novissima et perfectissima Africæ descriptio....* . Koeman dates this new map as after 1696, however, Gyuri Danku suggests the publishing date is 1699 or 1700.

References

The author is grateful for the correspondence and discussion with Gyuri Danku, Map Librarian, Budapest, National Széchényi Library, who has conducted extensive research on the production and publication history of the Danckerts atlases and who has freely shared this information.
Norwich 1997: map # 57; Koeman 1967-71, II: 88-97.

132   1683 Alain Manesson Mallet, Paris

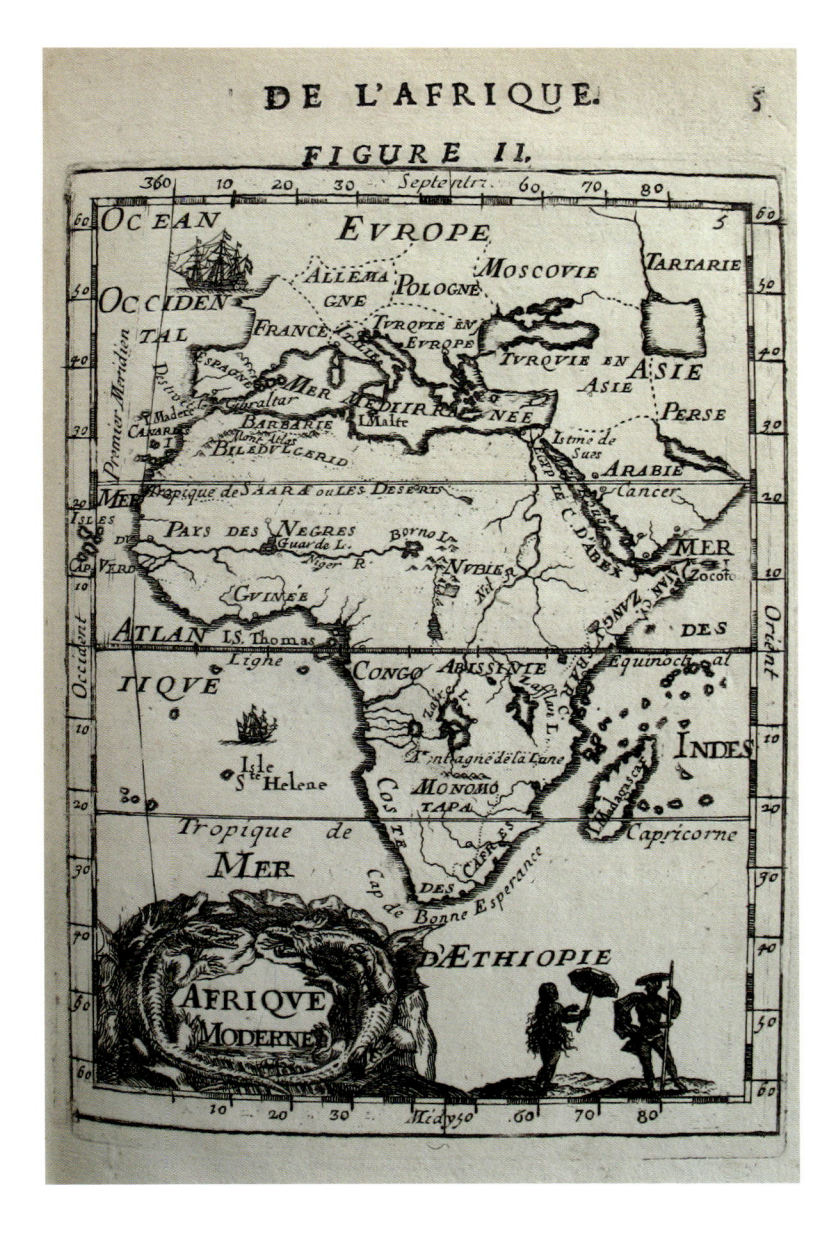

AFRIQVE | MODERNE.
- No scale bar.
- [Paris : Denys Thierry, 1683].
- From: *Description de l' Univers ... Par Allain Manesson Mallet...Tome troisième.*
- 1 map : copperplate engraving ; 14 x 10.5 cm.
- Latitude and longitude coordinates on all four sides of the map. The *Premier Meridien* (Prime Meridian) is shown crossing through the Canary Islands.
- North at the top; no compass rose.
- Title cartouche at bottom left.
- There are two sailing ships in the Atlantic Ocean.

The title cartouche is surrounded by two winged crocodiles. An African man and woman are to the right of the title cartouche.

Example: *Author's collection.*

Description
Alain Manesson Mallet (1630-1706) was a French engineer who, for a time, was in the employ of the Portuguese. Upon his return to Paris, he served under Louis XIV. He wrote on geometry, geography, mathematics, and particularly on military fortifications. This is Mallet's map of modern Africa. Partly owing to the map's size, little relevant geographic information is presented. It generally follows a much earlier Blaeu interpretation of Africa. However, in Southern Africa, Sachaf Lake is omitted and in Northern Africa the eastward-flowing Nubia River is also missing. The placenames are in French.

Publication Information
Alain Mallet's *Description de l' Univers* was published in Paris in 1683 by Denys Thierry at ruë S. Jacques, á l'Enseigne de la Ville de Paris. The book, covering all of the world, was published in five volumes. Volume III was a comprehensive history of ancient and modern Africa. The volume was extensively illustrated with maps, plans and views, including a map of ancient Africa and a map of modern Africa.

The book was only published in 1683 and the map is only known in one state as follows: Letterpress on two lines above the map of 'DE L'AFRIQUE. 5. | FIGURE II.' At the bottom right of the page 'A iij'.

Owing to the book's popularity, due in part to its convenient size and its extensive use of illustrations, the book (Volume III on Africa) was published in Frankfurt, Germany, by Johann David Zunner in 1685 with a new modern map of Africa (Map # 136). The book was published for a final time in Frankfurt in 1719 by J.A. Jung.

References
Pastoureau 1984: 309-322; Tooley 1969: 73; Phillips 1909-1992: entry # 3447.

133   1683 Alain Manesson Mallet, Paris

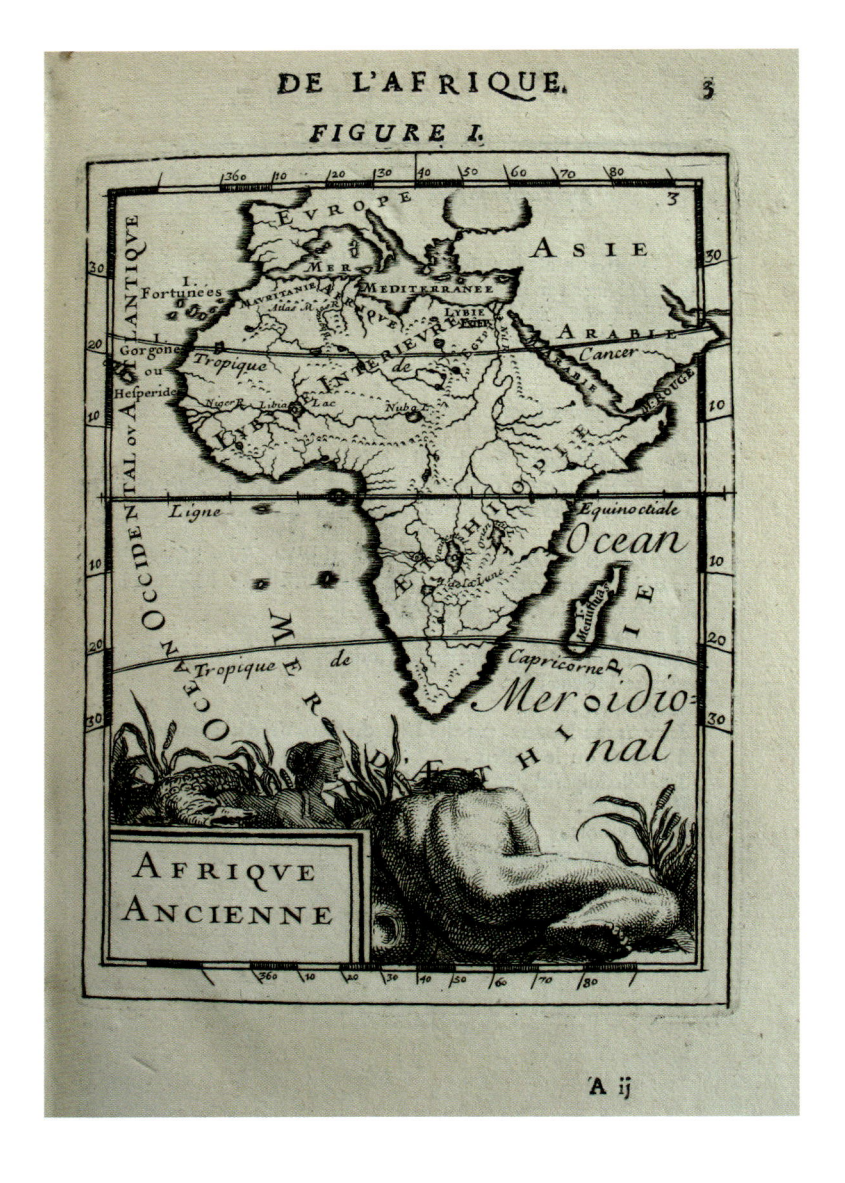

AFRIQVE | ANCIENNE.
- No scale bar.
- [Paris : Denys Thierry, 1683].
- From: *Description de l' Univers... Par Allain Manesson Mallet...Tome troisième.*
- 1 map : copperplate engraving ; 14.5 x 10.5 cm.
- Latitude and longitude coordinates on all four sides of the map.
- North at the top; no compass rose.
- Title at bottom left in simple, box-shaped cartouche.
- The cartouche is surrounded by a vignette of a crocodile, a woman, and a man with his back to the viewer.
- No decorative elements on the map, except for the vignette at the bottom.

Example: *Author's collection.*

Description
Alain Manesson Mallet (1630-1706) was a French engineer who, for a time, was in the employ of the Portuguese. Upon his return to Paris, he served under Louis XIV. He wrote on geometry, geography, mathematics, and particularly on military fortifications.

This is Mallet's map of ancient Africa. The map shows Africa as it was supposedly known to the classical Greek and Roman writers, although with a modern shape of Africa and with other current conventions. Surprisingly, unlike Mallet's 1683 modern map of Africa, this map does show Sachaf Lake in Southern Africa, and in Northern Africa, the unnamed eastward-flowing Nubia River, which empties into the Nile, is shown.

Publication Information
Alain Mallet's *Description de l' Univers* was published in Paris in 1683 by Denys Thierry, ruë S. Jacques, á l'Enseigne de la Ville de Paris. The book was published in five volumes. Volume III was a comprehensive history of ancient and modern Africa. The volume was extensively illustrated with maps, plans and views, including a map of ancient and modern Africa.

The book was only published in 1683 and the map is only known in one state as follows: Letterpress on two lines above the map of 'DE L'AFRIQUE. 3. | FIGURE I.' At right bottom of the page 'A ij'.

Owing to the book's popularity, due in part to its convenient size and its extensive use of illustrations, the book (Volume III on Africa) was published in Frankfurt, Germany, by Johann David Zunner in 1685 with a new map of ancient Africa (Map # 137). The book was published for a final time in Frankfurt in 1719 by J.A. Jung.

References
Pastoureau 1984: 309-322; Tooley 1969: 73; Phillips 1909-1992: entry # 3447.

134   c. 1684 John Seller, London

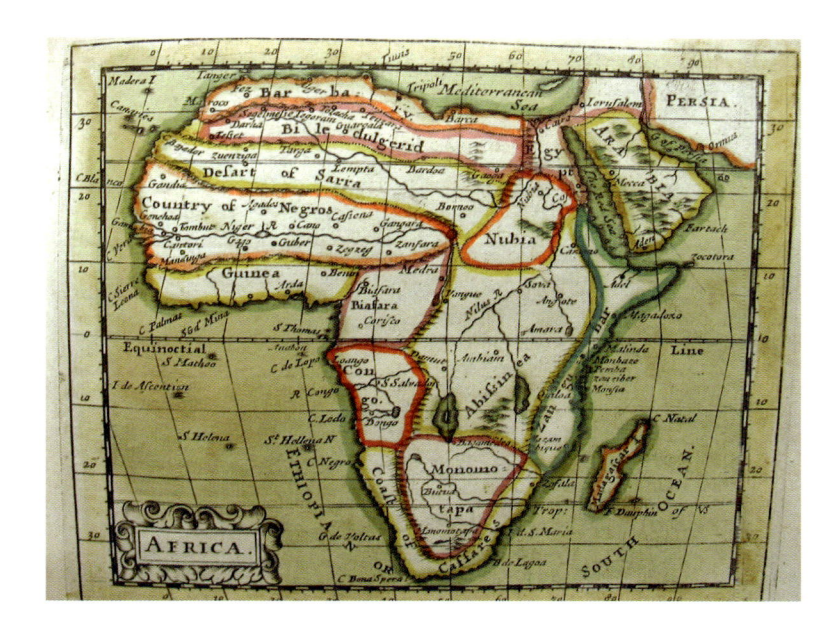

AFRICA.
- No scale bar.
- [London : John Seller, c.1684].
- From: *A New System of Geography…* .
- 1 map : copperplate engraving ; 12 x 14.5 cm.
- Latitude and longitude coordinates on all four sides of the map.
- North at the top; no compass rose.
- Title cartouche at bottom left of map.
- No decorative elements on the map other than the title cartouche.

Examples: *A New System of Geography* of c.1684: *Washington LC*, G1015.S53 1685 (with an undated title page) *and* G1015.S53 1690 (with a dated title page); *Atlas Terrestris* of c.1689: *London BL*, Maps C.39.a.12; *Atlas Terrestris* of c.1691: *London BL*, Maps C.29.f.3; *Atlas Terrestris* of c.1700: *Washington LC*, G1015.S52 1700 (undated title page).

Description
For his *A New System of Geography*, John Seller replaced his earlier map of Africa of c. 1676 with this slightly enlarged map. At this time, Seller was working from 'His shop on the West side of the Royal Exchange' (Tyacke 1978: 139).

This map has names partly in English (*Country of the Negros* for the region of *Nigirita*) and in Latin. Although it differentiates the various regions of Africa using the conventions which applied for most of the seventeenth century, the map adds a new region of *Biafara* in West-central Africa. The region of *Abyssinea* (Abyssinia) also includes an area far to the south of the Equator. The map is a composite of maps of the seventeenth century and, with its depiction of a third major river south of the Ptolemaic lakes, it is most closely modeled after the French maps of Africa by Guillaume Sanson in 1668 and by Alexis Hubert Jaillot in 1674.

Publication Information
This map appeared in Seller's undated *A New System of Geography*, which made its first appearance in c.1684. It is known that Seller's *The New System of Geography* was advertised for sale in the London Gazette of 29 September – 2 October 1684, thus the date for this map of c.1684. The map also appeared in Seller's undated *Atlas Terrestris*. Besides the map of the continent of Africa, these atlases contain numerous regional maps of Africa. There is only one known state of this map.

References
Shirley 1993: 497; Tyacke 1978.

135   1684 Nicolas de Fer - Jacques Robbe, Paris

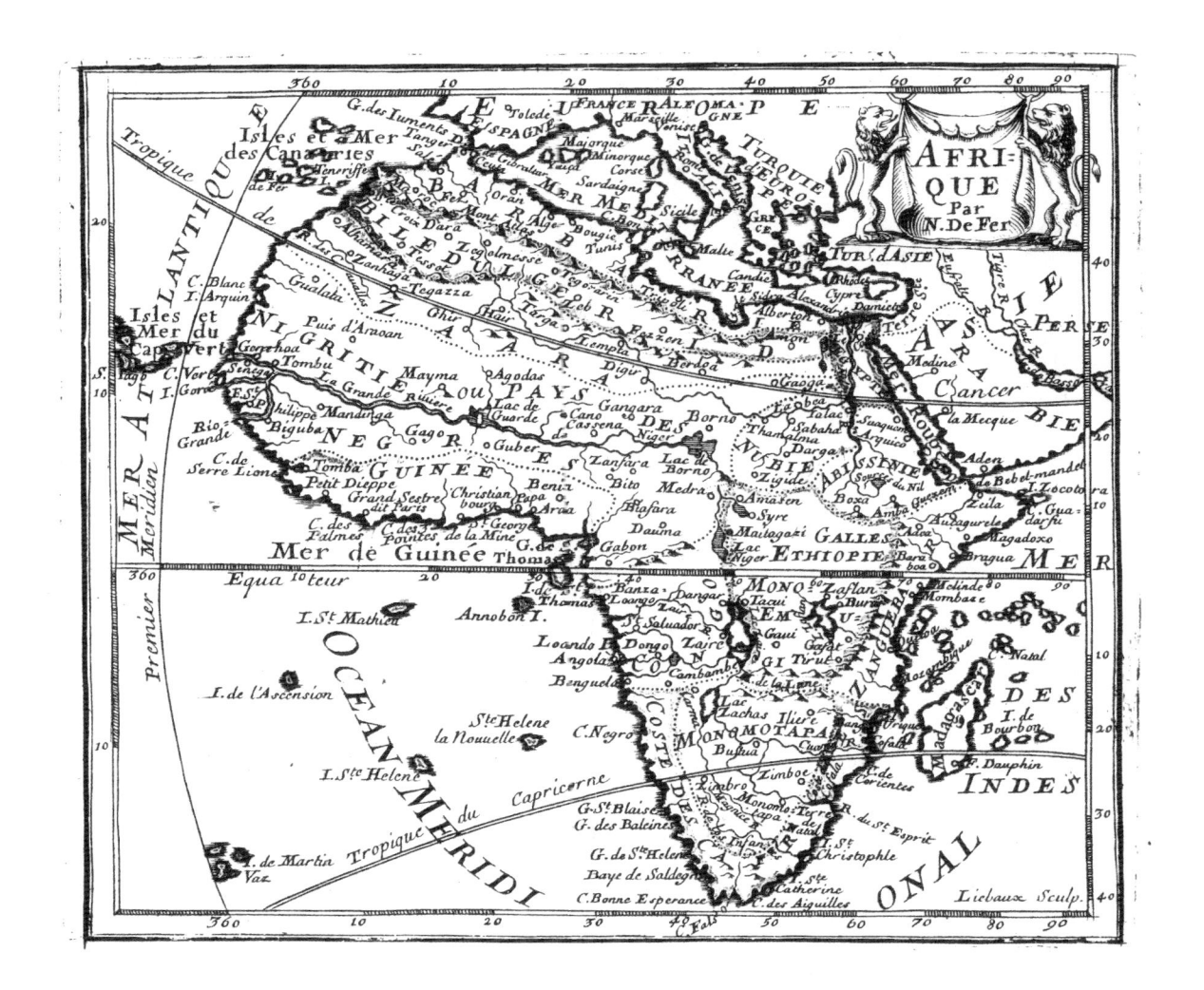

135.1 First State 1684

AFRI= | QUE | Par | N. De Fer.
[at the bottom right of the map]:
Liebaux Sculp.
- No scale bar.
- [Paris] : Nicolas de Fer, [1684].
- From: *Proof Atlas / Methode Pour Apprendre Facilement la Geographie*.
- 1 map : copperplate engraving ; 13.5 x 15.5 cm
- Latitude and longitude coordinates on all four sides of the map. The *Premier Meridien* (Prime Meridian) is shown crossing through the Canary Islands.
- North at the top; no compass rose.
- Title at top right within a cartouche with two lions on each side holding a drapery.
- '24' in manuscript in upper right corner of page.
- The Nile River from the two Ptolemaic lakes below the Equator to Abyssinia is missing.

Examples: 1684 'proof' copy: *London, Hubbard Collection;* Loose copy: *Paris BN*, Ge F 5053.

135.2 Second State 1685
Addition of the Nile River from the two Ptolemaic lakes to Abyssinia with inclusion of *Nils des anciens* with this and subsequent states. '24' in manuscript in upper right corner of page is not present.

Addition of an etched 'Tome 2. p. 156.' just inside upper <u>left</u> side of map within the gridline.

Examples: 1685 edition of *Methode*: *London BL*, 569.c.12 *and* 569.c.13; *New York PL*, *KB 1685 Robbe; London, Hubbard Collection.*

Map 135 Comparison of the central part with the sources of the Nile River, left state 1 and right state 2.

135.3 Third State 1689
Page number altered to: 'Tome 2. p. 161.' just inside upper <u>left</u> side of map within the gridline. Appears in editions from 1689 to 1739.

Examples: 1695 edition: *Paris BN*, G 32.380 *and* G 32.381; *London, Hubbard Collection.*

135.4 Fourth State 1746
Page number altered to: 'Tome 2, p. 173.'.

Example: *Paris BN*, Ge.FF.505.

Description

This map is modeled after Pierre Duval's four-sheet map of 1678. As on the Duval map, this map breaks the connection between the two Ptolemaic source lakes and the Nile River. It shows a generally correct view of the Blue Nile River with its source of Lake Tana in East Africa. Interestingly, this map inserts *Sources du Nile* just above the unnamed Lake Tana. Given the map's relatively small size, it still provides numerous up-to-date placenames, particularly in those areas under French influence. *Christianbourg* is placed on the map on the Guinea coast. The map still perpetuates the erroneous 'New' St. Helena island in the South Atlantic.

This map of Africa was prepared by Nicolas de Fer (1646-1720), whose name is on the title as the mapmaker. It represents one of the earliest works by De Fer. Upon the death of his father, Antoine de Fer, in 1673, his mother headed the family map business for which Nicolas worked. Eventually, the business was transferred completely to Nicolas by his mother in 1687. In 1690, Nicolas was appointed 'Géographe du grand Dauphin', and his business began to be successful.

Attempting to replicate the financially successful, small-format atlases by Duval and others that had been appearing in France, De Fer prepared a series of maps of the world, the four continents, and various regions. The engraver for the Africa map is Jean Baptist Liebaux, whose imprint 'Liebaux Sculp.' is at the bottom right of the map.

For what appears to be financial reasons, he did not publish his atlas, but instead sold the copperplates, including the one of Africa, for use in Jacques Robbe's general geography book *Methode Pour Apprendre Facilement la Geographie* published by Antoine Dezallier in 1685. Robbe's book was first issued in Paris in 1678 without the maps. It seems Robbe needed a set of maps to accompany his text so he purchased the De Fer copperplates.

For the 1685 edition of his book, Robbe had the maps revised with significant placename and numbering differences. For example, in the 1685 and subsequent examples of this map, Robbe erroneously reintroduced a connection between the Nile and the two Ptolemaic lakes, though with a caveat that this was the Nile as known to the ancients (*Nils des anciens*).

Publication Information

The map of Africa is known in four states. This is most easily identified by changes to the binder's instructions with the inclusion of printed page numbers, though there are placename changes as well:

To date, one example of a 'proof' copy of the atlas with the De Fer maps is known to exist prior to the sale of the copperplates to Robbe. The atlas is tentatively dated 1684, as the map of France bears that date. The collection of the Bibliothèque Nationale, Department des Cartes et Plans contains eighteen of the twenty-six maps in the first or 'proof' state, which were planned for the unpublished De Fer atlas,.

Robbe's *Methode Pour Apprendre Facilement la Geographie* of 1685 was evidently well received as it was published again in 1689, 1695, 1703, 1714, 1721, 1739, and finally in 1746.

In 1687, the *Methode* was copied and published in The Netherlands. For this edition a new set of copperplates was produced, still mentioning De Fer, but with the imprint of De Winter as the engraver ('A[ntoine] de Winter') on some of the maps and not Liebaux (note: Liebaux was only one of the engravers of maps in the De Fer-Robbe book along with Roussel and Lhuilier). The map of Africa does not have the imprint for De Winter. It seems that the publishing history for this edition is somewhat cloudy. However, it appears that this De Winter edition was a counterfeit one; that is, it was issued without the permission of the author or publisher.

Further editions were published in The Netherlands again in 1687, and in 1688, 1691, 1704, and 1743, using the De Winter plates. These editions were published by François Halma in The Hague, and Henry van Bulderen in Utrecht, working together or independently.

A new copperplate of this map was engraved, using the second state of the De Fer – Robbe map as a model, for use in De la Croix's *Relation Universelle de L'Afrique, Ancienne et Moderne*. This book was published by Thomas Amaulry in Lyon in 1688. This is a close copy of the De Fer – Robbe map of Africa, but without the name of 'Par N. De Fer' within the title cartouche or the engraver 'Liebaux Sculp.' at the bottom right. This map also has 'Tom.I. Pag.1.' at the top above the map grid. Most interestingly, this map does not show the eastern Ptolemaic lake of Zaflan connecting to the White Nile River but it does show it as almost connecting to the Blue Nile River in the Abyssinian Highlands. Whether this was simply an oversight, or if it was purposely done to project a more modern interpretation, based on the Duval 1678 map, is not known (See Map # 145).

The three similar maps by Nicolas de Fer, Antoine de Winter (Map # 142), and Thomas Amaulry (Map # 145) are distinguished from each other by the following:

Map # 135	1684, De Fer-Robbe	With the signature of 'Par N. De Fer' within the title cartouche. Imprint of the engraver 'Liebaux Sculp.', at the bottom right corner of the map.
Map # 142	1687, De Fer-Robbe-De Winter	With the signature of 'Par N. De Fer' within the title cartouche. No imprint for the engraver at the bottom right corner of the map.
Map # 145	1688, De Fer-Robbe-Amaulry	Without the signature of 'Par N. De Fer' within the title cartouche. No imprint for engraver.

References

The author is grateful to Jason Hubbard for sharing his notes on the history of the publication of these works of geography.
Hubbard Spring ed. 2005: 46-63, and Winter ed. 2004: 16-24; Norwich 1997: map # 53 (for Amaulry edition); Shirley 1993: 496.

136 1685 Alain Manesson Mallet - Johann David Zunner, Frankfurt

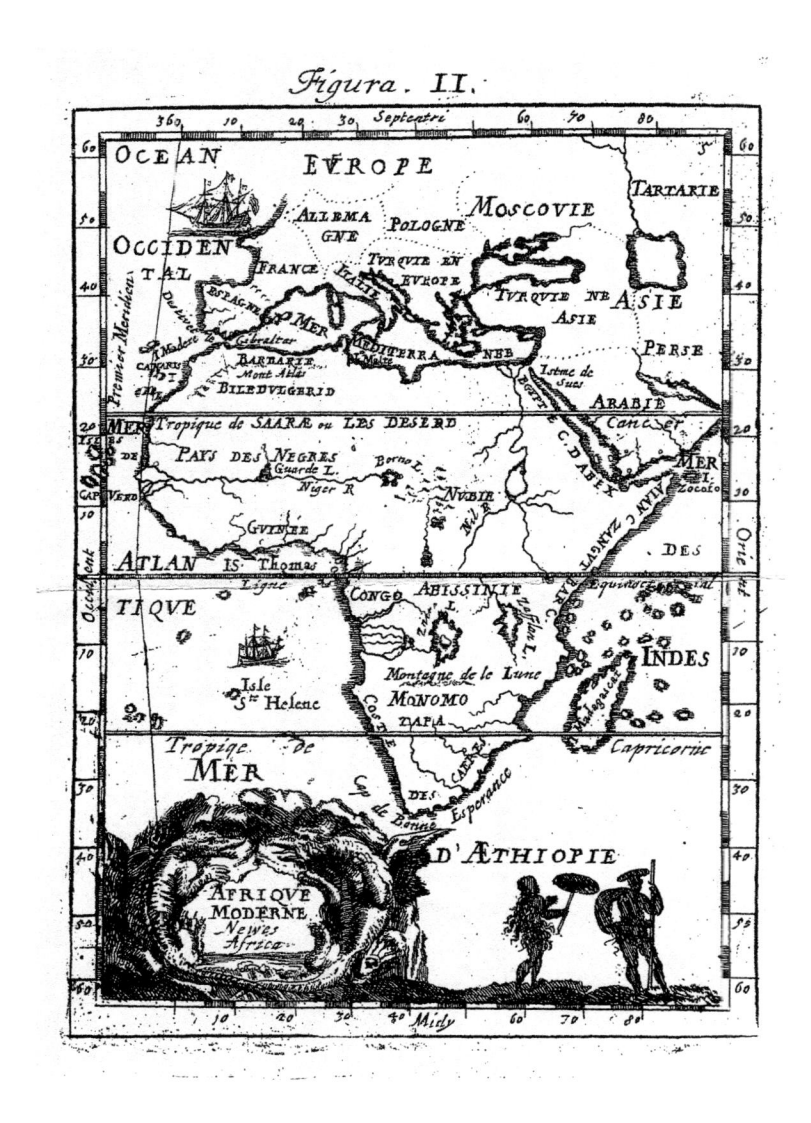

AFRIQVE | MODERNE | Newes | Africa.
- No scale bar.
- Frankfurt : Johann David Zunner, 1685.
- From: *Description de l' Univers' Contenant... Par Allain Manesson Mallet,...Tome Troisieme...Chez Jean David Zunner. M DC LXXXV.*
- 1 map : copperplate engraving ; 14 x 10.5 cm.
- Latitude and longitude coordinates on all four sides of the map. The *Premier Meridien* (Prime Meridian) is shown crossing through the Canary Islands.
- North at the top; no compass rose.
- Title cartouche at bottom left.
- There are two sailing ships in the Atlantic Ocean. The title cartouche is surrounded by two winged crocodiles. An African man and woman are to the right of the title cartouche.
- Engraved 'Figura. II'. on one line above the map.

Examples: 1685 edition: *London BL*, 10004.dd.8; 1719 edition: *Washington LC*.

Description

This map of Africa is from a reprint of Alain Manesson Mallet's *Description de l' Univers' Contenant* of 1683. Owing to the book's popularity, due in part to its extensive use of illustrations, the book was reprinted in Frankfurt by Zunner in 1685 shortly after its initial appearance in Paris. This edition used the same French text, but had new copperplates with German titles added to the original French titles.

This map closely follows Mallet's 1683 map of modern Africa. Except for 'Newes Africa' added to the title in German, the placenames are still in French as on the Mallet map. This map of Africa was likely engraved by J.J. Vogel as his imprint appears as the engraver on some other plates in the book.

Publication Information

The book, *Description de l' Univers' Contenant*, was published in five volumes; the first volume was published in 1684. Volume III, published in 1685, was a comprehensive history of ancient and modern Africa. The volume was extensively illustrated with maps, plans, and views, including a map of ancient and of modern Africa.

There was a later German edition published in Frankfurt by J.A. Jung under the title *Beschreibung des Gantzen Welt Kreises* in 1719. This edition was also in five volumes.

References
Pastoureau 1984: 322, Tooley 1969: 73, Phillips 1909-1992: entries no. 3447 and 4280.

137 1685 Alain Manesson Mallet - Johann David Zunner, Frankfurt

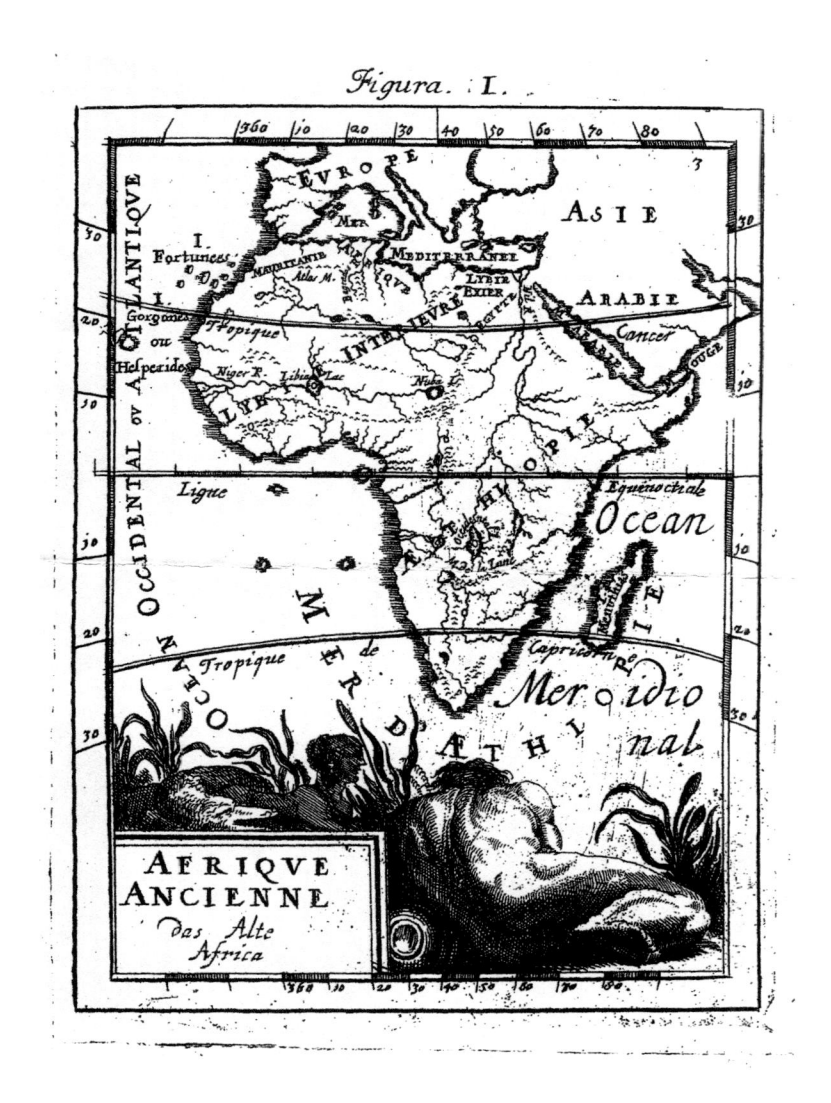

AFRIQVE | ANCIENNE | das Alte | Africa.
- No scale bar.
- Frankfurt : Johann David Zunner, 1685.
- From: *Description de l' Univers' Contenant... Par Allain Manesson Mallet,...Tome Troisieme...Chez Jean David Zunner. M DC LXXXV.*
- 1 map : copperplate engraving ; 14.5 x 10.5 cm.
- North at the top; no compass rose.
- Title at bottom left in simple box-shaped cartouche.
- The cartouche is surrounded by a vignette of a crocodile, a woman, and a man with his back to the viewer.
- No decorative elements on the map, except for the vignette at the bottom.
- Engraved 'Figura. I' on one line above the map.

Examples: 1685 edition: *London BL*, 10004.dd.8; 1719 edition: *Washington LC*.

This map of Africa is from a reprint of Alain Manesson Mallet's *Description de l' Univers' Contenant* of 1683. Owing to the book's popularity, due in part to its extensive use of illustrations, the book was reprinted in Frankfurt by Zunner in 1685 shortly after its initial appearance in Paris. This edition used the same French text, but had new copperplates with German titles added to the original French titles.

Description
The map shows Africa as known to the classical Greek and Roman writers. The map closely follows Mallet's 1683 map of ancient Africa. Except for 'das Alte Africa' added to the title in German, the placenames are still in French as on the Mallet map. This map of Africa was likely engraved by J.J. Vogel as his imprint appears as the engraver on some other plates in the book.

Publication Information
The book, *Description de l' Univers' Contenant*, was published in five volumes; the first volume was published in 1684. Volume III, published in 1685, was a comprehensive history of ancient and modern Africa. The volume was extensively illustrated with maps, plans, and views, including a map of ancient and of modern Africa.

There was a later German edition published in Frankfurt by J.A. Jung, under the title *Beschreibung des Gantzen Welt Kreises* in 1719. This edition was also in five volumes.

References
Pastoureau 1984: 322; Tooley 1969: 73; Phillips 1909-1992: entry # 3447 and # 4280.

138   1685 John Lawrence, London

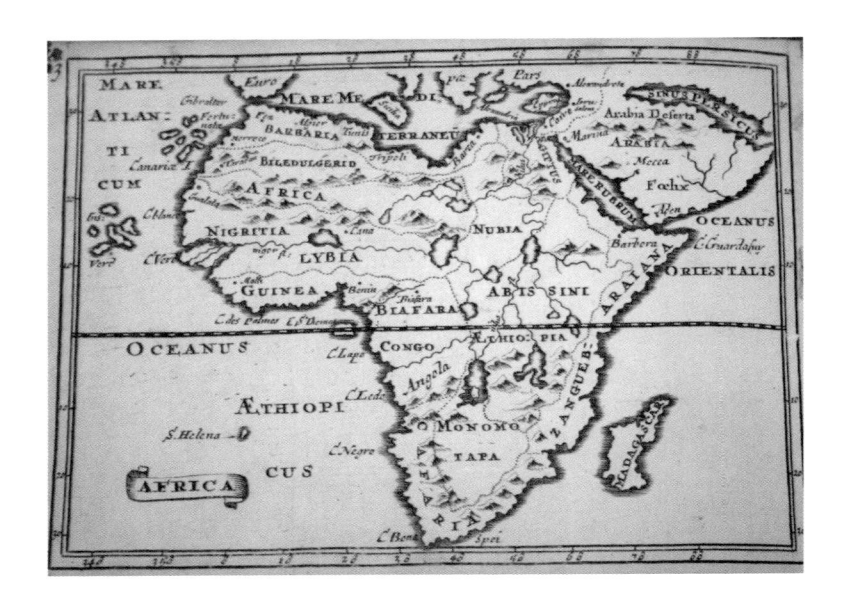

AFRICA.
- No scale bar.
- [London : John Lawrence, 1685].
- From: *Orbis imperantis tabellae geographico-historico-genealogico-chronologicae in quibus geographiae epitome, mappis quo fieri potuit exactioribus, descriptio historica imperiorum, regnorum, et rerumpublicarum, saeculorum series a Christo nato ad hunc usque annum 1685...* .
- 1 map : copperplate engraving ; 8.5 x 11.5 cm.
- Latitude and longitude coordinates on all four sides of the map. Equatorial Line across Africa.
- North at the top; no compass rose.
- Title at the bottom left set within a simple scroll cartouche.
- No decorative elements on the map.

Examples: *Chicago NL*, Ayer 135 .L42 1685; *Oxford Magd*, MAG Old Libr. L.3.13; *Washington LC*, D11 .L32; *Barry Lawrence Ruderman Antique Maps*;

Description
This uncommon map of Africa is found in John Lawrence's *Orbis imperantis tabellae*, a small (8vo) book that covered geography, history, and genealogy. The book also contains maps of the other continents, Europe, Asia, and America.

The map introduces no new cartographic information. It generally follows earlier English maps, especially the two Seller maps of c.1676 and c.1684 and the Moore-Moll map of 1681. There are some obvious exceptions. The Lawrence map does not depict the major river systems south of the Nile source lakes in Central Africa. Also, the usual two Nile source lakes in Central Africa flow north into an unusual lake which then continues its northward flow. While this might be a reference to a much earlier representation by Claesz.-Langenes of a third Nile source lake (See Map # 37), in this case it appears to be an engraver's mistake. Maps of the sixteenth and seventeenth centuries depicted a *Meroe Island* in the same location as the lake on this map; the engraver erroneously provided shading to this island to make it appear to be a lake.

Publication Information
Little is known of the book's publisher, John Lawrence (active 1681-1711). He is known as a bookseller in London at the *Angel in the Poultry, over against the Compter*. Wing (entry L670) attributes the authorship of the *Orbis imperantis tabellae...* to Lawrence.

Only one edition of this book was published.

References
Burden 2007: 271; Wing: entry L670.

139   C. 1685 Justus Danckerts, Amsterdam

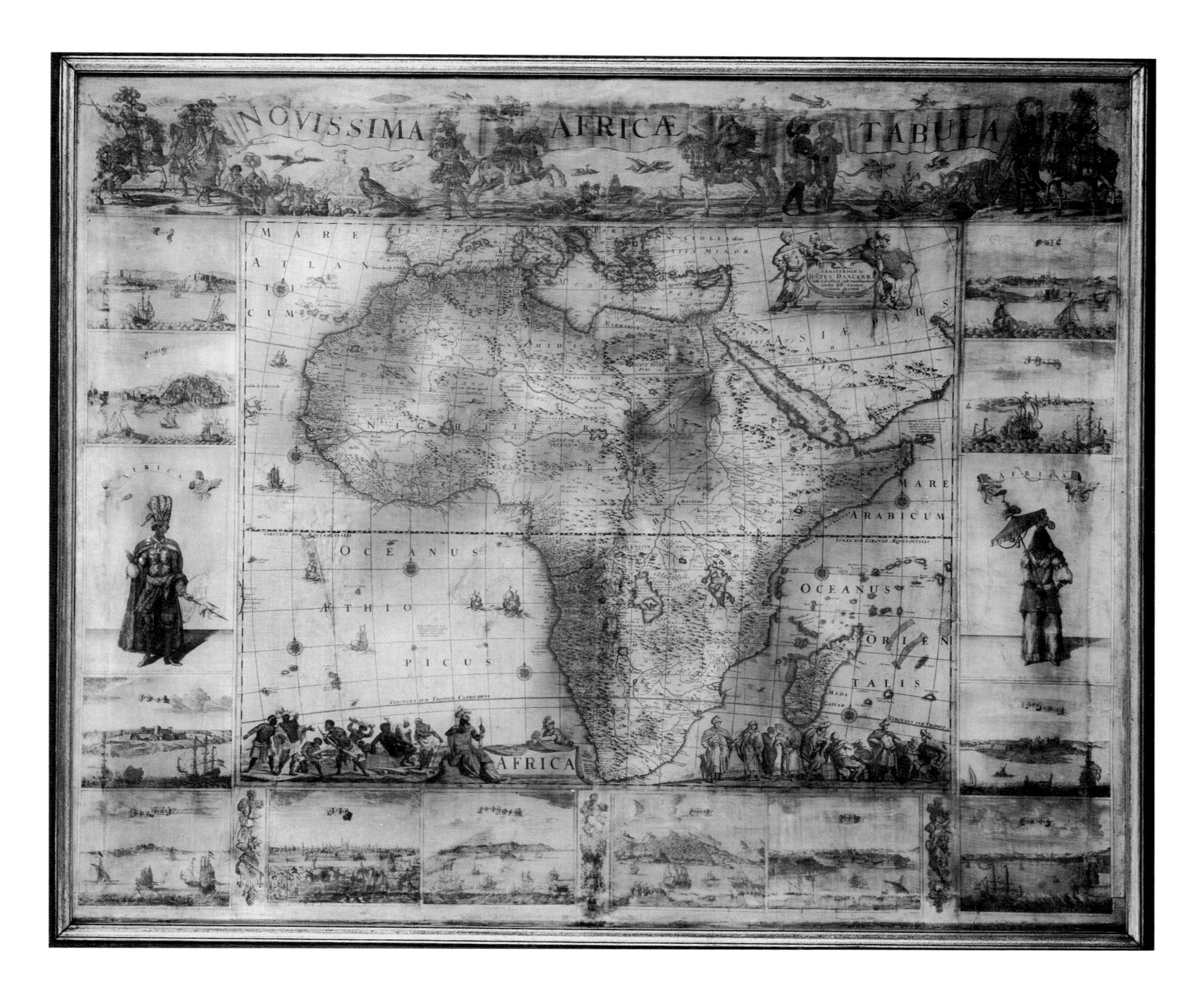

AFRICA.
[second title within a banner on title strip at top above the map]:
NOVISSIMA AFRICÆ TABULA.
[within a cartouche at the top right of the map with four figures to the side and behind]:
t' AMSTERDAM by | JUSTUS DANCKER | inde Calverstraet inde Danckbaerheyt | werden deze Caerten | gedruckt en vercoft.
[below the large figure of an African man on the left border, and again, below the African woman on the right border]:
Danckerts inventor et Fecit. Cum Gratia et Privilegio Ordinum Hollandiae et West-Frisiae. Justus Danckerts Excudit.
- No scale bar.
- Amsterdam : Justus Danckerts, [c.1685].

- From: separately published.
- 1 wall map : copperplate engraving ; 4 map sheets ; 88 x 107.5 cm (map only). 130 x 156 cm (total of map sheets, title strip at top, and side and bottom panels. Title strip at top is 22 x 156 cm. Town views are each 17 x 22 cm).
- Latitude and longitude coordinates on all four sides of the map. Longitude coordinates across Equatorial Line.
- North at the top; three compass roses within the Indian Ocean and five compass roses within the Atlantic Ocean.
- Title within a drapery cartouche at the bottom of the map with figures on either side.
- There are six sailing ships within the Atlantic.
- At the bottom of the map sheets is a decorative African scene, which is modeled after the scene in Jan Mathisz.' c.1655 map of Africa,
- There are various decorative elements surrounding the map sheets: a title strip within an elaborate scene above the map; three town views on each side of the map; one figure of an African man on the left side below the top two views, and one figure of an African woman on the right side below the top two views; six town views across the bottom of the map. The town views are described below.

Example: *Greenwich NMM*, G290:1/1 (only known example).

Description

Justus Danckerts I (1635-1701) produced this wall map of Africa as part of a complete set of wall maps of the world and continents. The geographic information, overall shape of Africa, and much of the decorative scene across the bottom of the map, is copied from the c.1655 map of Africa by Jan Mathisz. (See Map # 84).

The Danckerts were a prominent print and map publishing family active in Amsterdam for almost 100 years. The founding member of the family business was Cornelis I (1603-1656) who started producing and publishing maps in the second part of the 1620s. His sons Dancker (1634-1666) and Justus (1635-1701) were also active in the business. Justus issued a number of atlases with his sons. Justus' sons Theodorus I (1663-c. 1727), Cornelis II (1664-1717), Johannes (?-1712), and Eduard (?-after 1721) and his grandson Theodorus II (1701-1727) continued the family business into the early eighteenth century. Because of the use of the same first names by various family members, there is some confusion in the literature over the authorship of some of the maps produced by the Danckerts.

There are three town views on each side of the map and six town views across the bottom. On the left side *SALEE*; *TANGIER*; [African man]; and *ARX NASSOV*; on the right side are *TUNIS*; *TRIPOLI IN BARBARI*; [African woman]; and *CASTRVM MINA*. At the bottom are *INSULVA THOMAE*; *ALCAIR*; *LOANDA S. PAULI*; *ALGIER*; *MALTA*; and *STADT MINA*.

Publication Information

An example of the Danckert's wall map of the world is at the Greenwich National Maritime Museum, and other examples of the world map are known to exist (Shirley 1993: 503). The date assigned to this world map is c.1680. The only example known to the author of this wall map of Africa at the Greenwich National Maritime Museum has also been assigned a date of c.1680. It is noted that the example of this wall map of Africa contains the publisher's privilege for Justus Danckerts (Danckerts inventor et Fecit. Cum Gratia et Privilegio Ordinum Hollandiae et West-Frisiae.), just below the figure on the left border and also below the figure on the right border. However, Justus and his sons Theodorus I and Cornelis II did not obtain their privilege from the States of Holland and West Friesland for their prints and maps until September 14, 1684. As this map has the imprint of a privilege, it is likely that it was produced shortly after that date.

References

Shirley 1993: 502-503; Koeman 1967-71, II: 89.
Correspondence with Gyuri Danku, Map Librarian, Budapest Széchényi National Library, who has conducted extensive research on the Danckerts family, particularly Justus and his sons.

140 1686 Philipp Cluver – Johann Buno – Herman Mosting, Wolfenbüttel

AFRICA | Antiqua | et | Nova.
[at the bottom left]:
Herman Mosting Sculp: Luneburg.
- No scale bar.
- [Wolfenbüttel: Widow of Conrad Buno (Viduae Conradi Bunonis), 1686].
- From: *Introductio in Universam Geographicam.*
- 1 map : copperplate engraving ; 21 x 25.5 cm.
- Latitude and longitude coordinates along all four sides of the map.
- North at the top; no compass rose.
- Title cartouche at top right corner.
- No decorative elements on the map.

Examples: 1686 edition: *München SB*, Res 4 Geo.u.33; *Wolfenbüttel HAB*, 2º XVIII:109; 1694 edition: *München SB*, Res 4 Geo.u.34; *Wolfenbüttel HAB*, Ca 175; *Author's collection*.

Description

This is the third set of maps produced in Germany for Philipp Cluver's famous *Introductio in Universam Geographicam*. This edition, still using the Johann Buno text, was edited by Johann Friedrich Heckel (1640-1697) and contained a different set of maps than Conrad Buno's 1641 Brunswick (Braunschweig) edition (Map # 71), or Johann Buno's 1661 Wolfenbüttel edition (Map # 98). This edition was published in Wolfenbüttel, or Guelpherbytum as it appears on the title page of the atlas in its Latinized form.

The two men principally involved in the German editions were Conrad Buno (c.1613-1671), and his brother, Johann Buno (1617-1697). Conrad was an engraver and publisher at the court of Wolfenbüttel. Johann was a theologian and pedagogue in Lüneburg.

This map and the 1661 Cluver-Johann Buno maps are close copies of each other. They both are modeled after the 1641 Cluver-Conrad Buno map of Africa. Like the 1661 Cluver-Johann Buno map, this map is larger than the 1641 map and, as a result, has numerous placenames added to it. Much of the nomenclature is decidedly old; for example, *Aethiopia Inferiors Pars* is placed along the Southern African coast. For the interior, this map follows the 1623 Hondius and the 1617 Blaeu folio maps.

This map is readily distinguished from the 1661 Cluver-Johann Buno map by the inclusion of the imprint for the engraver for this map, 'Herman Mosting Sculp: Lüneburg', at the bottom left of the map. In addition to engraving this map of Africa, Herman Mosting of Lüneburg engraved other maps in the book. The remaining maps in the book were engraved by Martin Hailler whose imprint appears on some of the maps. It appears that with Conrad Buno's death in 1671, new engravers were needed for the new editions of the book.

Publication Information
There are two known editions containing this Philipp Cluver – Johann Buno – Herman Mosting map of Africa. The first was published in 1686 in Wolfenbüttel by the widow of Conrad Buno (Viduae Conradi Bunonis). The second edition was published in 1694 by the Heirs of Conrad Buno (Haeredum Buninianorum), also in Wolfenbüttel. This second edition was edited by Johann Reiske (1641-1701). Jäger (1982) states that the same maps were used for a German edition of 1671. While this is possible, since Conrad Buno had died in that year, the author has not been able to locate an example of the 1671 edition with the Mosting imprint.

There were numerous other editions of Cluver's *Introductio* that were published using different maps of Africa. Shirley (2004: 345) states that over 45 editions of the *Introductio* appeared over a 100 year period. Not all editions contained maps. Those editions that did contain a map of Africa used one of the following: Cluver-Conrad Buno from 1641; Cluver Elsevier from 1659; Cluver-Johann Buno from 1661; Cluver-Buno-Mosting, with the imprint of Herman Mosting as the engraver as from 1686; Johannes Janssonius in 1661, using his 1628 Goos map; Janssonius van Waesberge in 1676 using the 1630 Cloppenburch map; and Cluver-Wolters from 1697. All are generally derived from the Cluver-Conrad Buno map of 1641, which in turn was based on the Blaeu folio map of 1617. Refer to the chart under the 1641 Cluver-Conrad Buno map, Map # 71, for a list of the different maps of Africa that were used for different editions of Cluver's *Introductio* and the distinguishing characteristics of these maps.

References
Peter Meurer 2006: notes to author; Shirley 2004: 345-350; Burden 1996: map # 360 (Americas map); Shirley 1993: 447; Phillips 1909-1992: entry # 3432; Sabin 1868: entry # 13805.

141   1686 Philip Lea and John Overton, London

141.1 First State 1686

A New Mapp of | AFRICA | Divided into Kingdoms and | Provinces Sold by J Overton | at yᵉ White Horse with out New= | gate and by Philip Lea at yᵉ | Atlas and Hercules in | yᵉ Poul | try near Cheapside | LONDON.
- No scale bar.
- London : John Overton and Philip Lea, [1686].
- From: separately published / *An Atlas containing ye Best Maps of the Severall parts of the World Collected by Phil: Lea who selleth all sorts of Mathematicall Books & Instruments. Printed by H.C. for John Overton at the White-Horse without Newgate. Collected by Philip Lea, Globemaker, at the Atlas and Hercules, near the corner of Friday Street in Cheapside.*

- 1 map : copperplate engraving ; 49.5 x 58.5 cm.
- Latitude and longitude coordinates on all four sides of the map. Longitude coordinates across the Equatorial Line.
- North at the top; one compass rose in the South Atlantic Ocean.
- Title cartouche with a large decorative vignette surrounding it at the bottom left of the map.
- There are three sailing ships in the Indian Ocean, four sailing ships in the Atlantic Ocean, and five small ships in the Mediterranean Sea between Sicily and Greece. Animals are within Africa on the map. There is a lettered grid at the borders to find locations referenced in subsequent pages of the atlas.

Example: *London, Bernard Shapero Rare Books.*

141.2 Second State 1687
Removal from the title of the address for Lea of: '*y*ᶠ *Poultry near*'. Address is now ... *at y*ᶠ *| Atlas and Hercules in | Cheapside...* .

Examples: *London BL*, Maps C.45.f.4; *Paris BN*, Ge D 11428.

Description
Unlike other English mapmakers of this period such as Berry, who used French cartographic sources, Lea and Overton used the De Wit c.1670 map as their immediate model for their own map of Africa. Lea and Overton also closely copied the decorative cartouche and other elements that were on the De Wit map. For a discussion of the geography of this map, see Map # 114, the De Wit map of Africa.

Philip Lea was active as a globe and instrument maker, and map publisher from about 1683 until his death in 1700. Lea collaborated on a number of maps with John Overton (1640-1713), an active London publisher. The engraver for this map is not known, though it is known that James Moxon engraved 'A New Mapp of America' for Philip Lea and John Overton (Tyacke 1978: 127). Moxon may have engraved this map as well.

Publication Information
This map proves difficult to date precisely. As the geography on the map does not provide a clue to its date, the only other means to date this map is the publisher's location on the map. In 1686, Lea was working from 'yᵉ Poultry near Cheapside'. From here in 1686, he advertised the sale of this map. It was likely published in that year. It is known that, from 1687 until his death, Lea was primarily working from 'The Atlas and Hercules in Cheapside, London', which appears on the map title for the second state (Tyacke 1978: 120). As a result, this date of 1687 is assigned to the second state of this map, though it possibly could have been published slightly later to 1690.

The map was likely issued separately at first, and then included in atlases prepared by Lea and others. Tooley's Dictionary of Mapmakers (Tooley: 1999-2004: K-P: 102) also lists this map as first being issued as a separate map in 1687, along with those of the other continents.

Based on Moreland (1989: 157 & 162), the map first appeared in Lea's *An Atlas containing ye Best Maps of the Severall parts of the World...* in 1690 The British Library example of the atlas is only a collection of maps without text and is dated 1690 in their catalog, but it contains a map, number 9 showing the British Channel, with the date of 1695 on it. Lea's business was continued by his wife Anne until her death in 1730 whereupon the remaining copperplates were sold at auction.

References
Shirley 2004: 653-654; Tooley: 1999-2004: K-P: 102; Moreland 1989: 157 & 162; Tyacke 1978: 120-122 (for Lea) and 130-134 (for Overton); Tooley 1968: no. 48, under De Wit; Phillips 1909-1992: entry # 5694.

142 1687 Nicolas de Fer - Jacques Robbe - Antoine de Winter, Utrecht

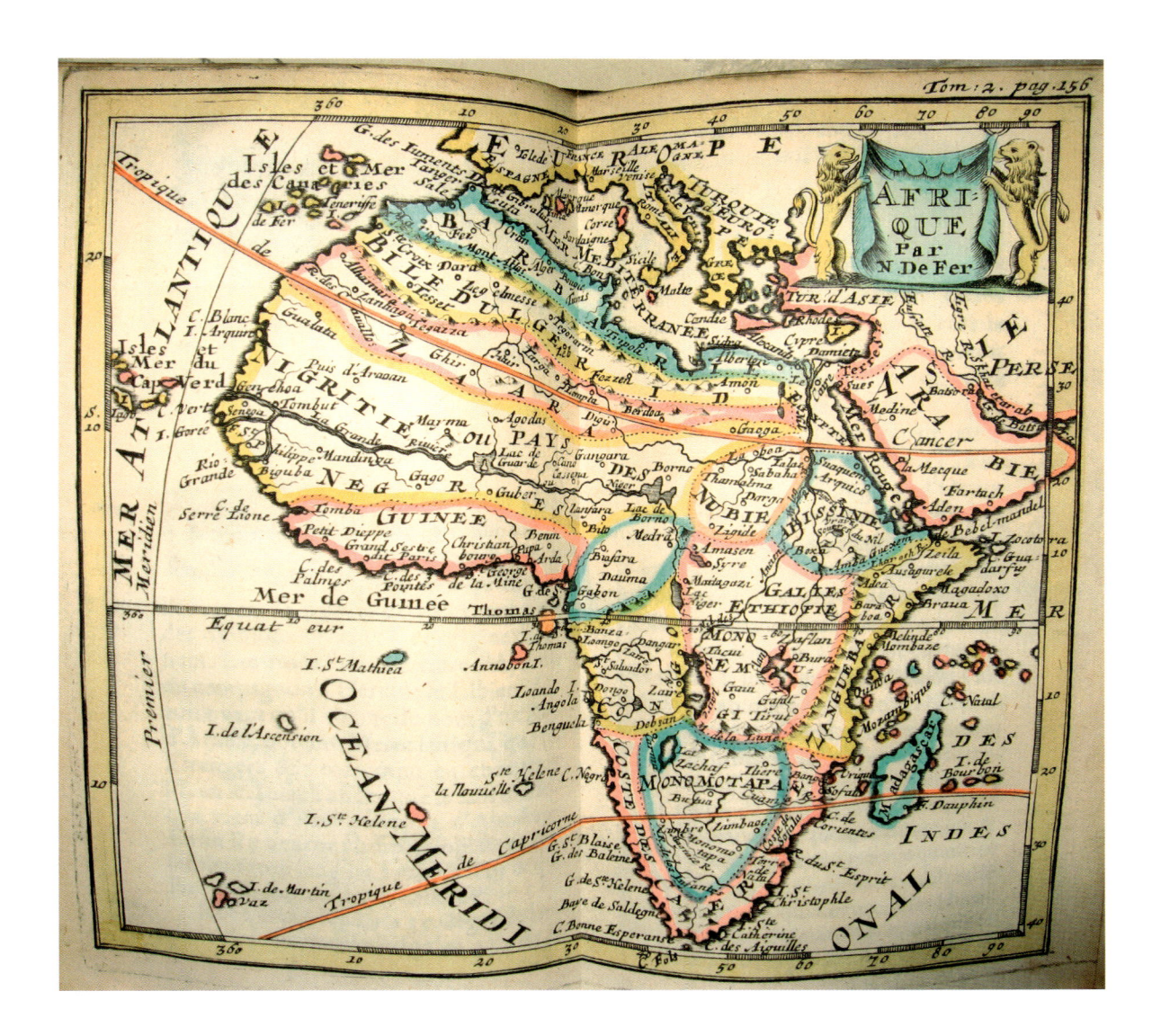

142.1 First State 1687

AFRI= | QUE | Par | N. De Fer.
- No scale bar.
- [Utrecht : François Halma, 1687].
- From: *Methode Pour Apprendre Facilement la Geographie.*
- 1 map : copperplate engraving ; 13.5 x 15.5 cm.
- Latitude and longitude coordinates on all four sides of the map. Longitude coordinates across the Equatorial Line. The *Premier Meridien* (Prime Meridian) is shown crossing through the Canary Islands to the west of *I de Fer.*
- North at the top; no compass rose.
- Title at top right within a cartouche with two lions, one on each side, holding a drapery.
- At top right corner above the map, the engraved printer's instructions of: 'Tome: 2. pag.156'.

Examples: *Cambridge Harv RB*, *58C-231 (Halma, 1687 edition, Vol. 2) *and* *58C-230 (Van Bulderen, 1688 edition, Vol. I); *Washington FSL* (1687 Van Bulderen and Halma editions); *London, Hubbard Collection* (Halma, 1687 edition, Vol. 2, 1688 edtion, Vol. 1).

142.2 Second State 1691-1743
At top right corner above the map, the printer's instructions are now: 'Tome: 2. pag.152'.

Example: 1691 edition: *Amsterdam SMA*, R 51 N2; *London, Hubbard Collection* (1704 van Bulderen, Den Haag edition, Vol. 2).

Description
This map is copied from the 1684 De Fer – Robbe map of Africa in its second state of 1685 (Map # 135). As in the earlier map, it shows the Nile River flowing from the two Ptolemaic source lakes in Central Africa. It adds text that states that this was the source of the Nile River as known to the ancients (*Nils des anciens*). The map correctly orients the Blue Nile River source of Lake Tana in the Abyssinian highlands (*Source du Nil*). A traditional view for the rest of the interior of Africa remains as on most maps of this period. The *La Grande Riviere ou Niger* (Niger River) shows its traditional source in *Lac Niger* just north of the Equator and flows due west into the Atlantic in West Africa. The unnamed Nubie River flows eastward into the Nile to the north of the Niger. Various French settlements are identified such as *Fort St. Philippe* and *Petit Dieppe* on the coast in West Africa, and *Fort Dauphin* on Madagascar.

Publication Information
In 1687, as a result of the success of Robbe's *Methode Pour Apprendre Facilement la Geographie*, the book and the maps were copied and published in The Netherlands.

For this edition a new set of copperplates was produced, still mentioning De Fer, but with the imprint of De Winter as the engraver (A[ntoine] de Winter) on some of the maps. The map of Africa does not have the imprint for De Winter. It seems that the publishing history for this edition is somewhat cloudy. However, it appears that this De Winter edition was a counterfeit edition; that is, it was an edition done without the permission of the author or publisher.
Further editions were published in The Netherlands again in 1687, and in 1688, 1691, 1704, and 1743 (published simultaneously in Frankfurt), using the De Winter plates. These editions were published by Francois Halma and Henry van Bulderen in Amsterdam, Den Haag and Utrecht, working together or independently. The map of Africa from the De Winter editions is known in two states.

A new copperplate of this map was engraved, using the second state of the De Fer – Robbe map as a model, for use in De la Croix's *Relation Universelle de L'Afrique, Ancienne et Moderne*. This book was published by Thomas Amaulry in Lyon in 1688 (See Map # 145). The Africa map is a close copy of the De Fer – Robbe map of Africa, but without the name of 'Par N. De Fer' within the title cartouche or the engraver 'Liebaux Sculp.' at the bottom right. This map also has 'Tom.I. Pag.1.' at the top above the map grid. Most interestingly, this map does not show the eastern Ptolemaic lake of Zaflan connecting to the White Nile River, but it does show it as almost connecting to the Blue Nile River in the Abyssinian Highlands. Whether this was simply an oversight, or if it was purposely done to project a more modern interpretation, based on the Duval 1678 map, is not known.

The three similar maps by Nicolas de Fer - Robbe (Map # 135), De Fer - Robbe - Antoine de Winter (Map # 142), and De Fer - Robbe - Thomas Amaulry (Map # 145), each with a different copperplate, are distinguished from each other by differences in the imprint for the author and the engraver. See the chart for Map # 135 to differentiate these maps.

References
The author is grateful to Jason Hubbard for sharing his notes on the history of the works of geography containing these maps.
Hubbard Spring ed. 2005: 46-63, and Winter ed. 2004: 16-24; Norwich 1997: map # 53 (for Amaulry edition); Shirley 1993: 496.

143   1688 Pierre Duval - Jean Certe, Lyon

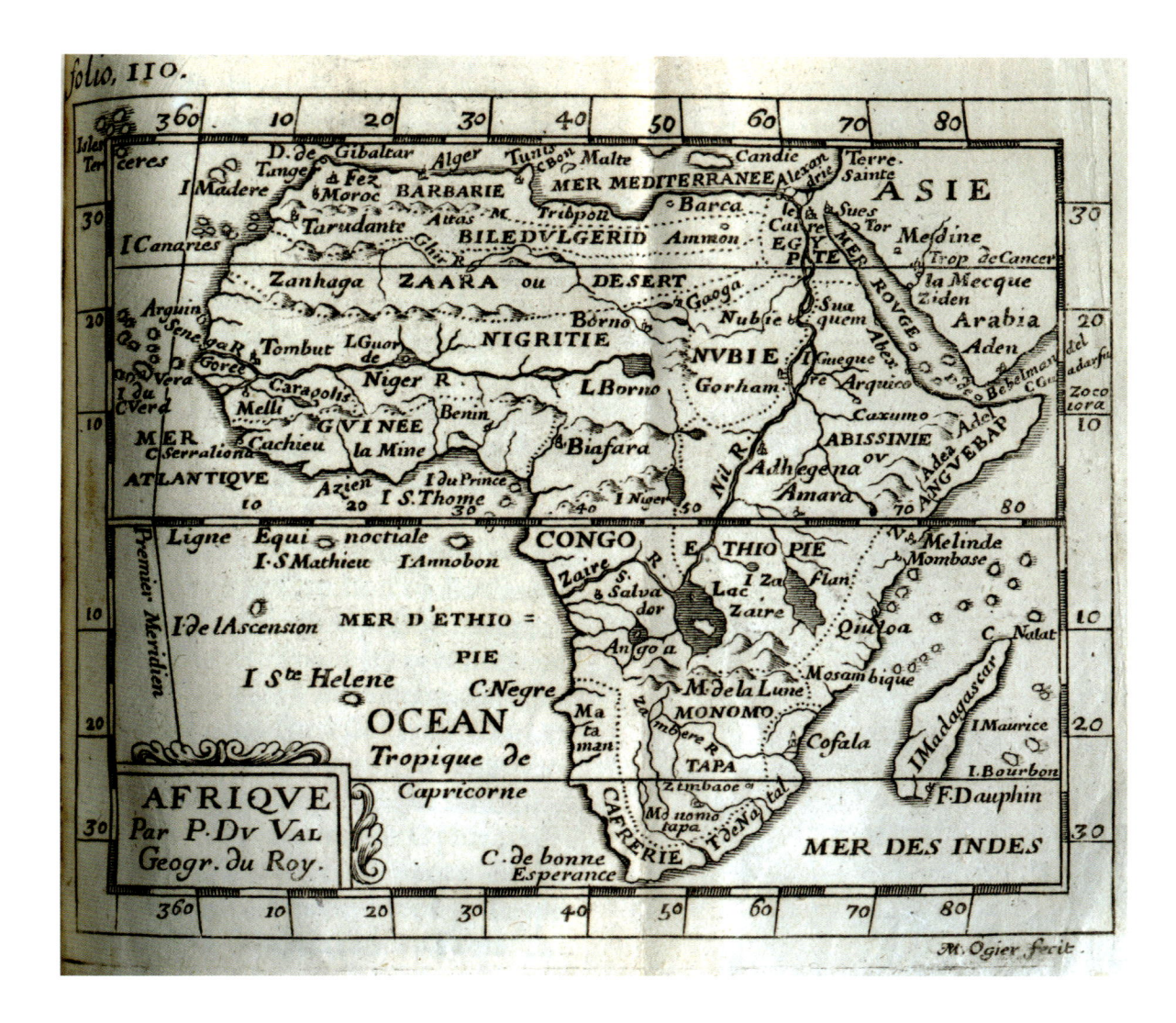

AFRIQVE | Par P. DV VAL | Geogr. du Roy.
[below the neatline at the bottom right corner]:
M. Ogier fecit.
- No scale bar.
- [Lyon : Jean Certe, 1688].
- From: *La Geographie Universelle Qui Fait Voir... Du Monde...* .
- 1 map : copperplate engraving ; 10 x 12.5 cm.
- Latitude and longitude coordinates along all four sides. Longitude coordinates along Equatorial Line. *Premier Meridien* (Prime Meridian) is to the west of *I. Canaries*.
- North at the top; no compass rose.
- Title cartouche at bottom left corner.
- No decorative elements on the map.
- 'folio, 110.' is above the map on the left.

Examples: 1688 edition: *Besançon BM*, 213854 and 213855 (the map of Africa is in volume 1, i.e. s/m ending in '4'); *Paris BArs*, 8° H 229. 1712 edition: *Bordeaux BM*, H 5668(1-2); *Various other locations*.

Description

Pierre Duval's *La Geographie universalis...* proved to be exceedingly popular. There were imitators in Germany and also in France. This map is from Duval's *La Geographie Universelle* as published by Jean Certe. For this edition, Certe had new copperplates engraved by Mathieu Ogier. Certe had approval to publish this edition of Duval's *La Geographie* as the title page of the atlas has the addition at the end of 'Avec permission'.

This map is modeled after the second state of Duval's c.1660 map of Africa from 1663. For this state, numerous placenames were added to the first state of Duval's 1660 map of Africa. Among the additions of placenames are: *Premier Meridien, I. de l'Ascension, Tropique de Capricorne,* and MER DES INDES. The geography on this map is unchanged from the first state of Duval's map. The primary models that Duval used for his own map were Nicolas Sanson's folio map of 1650 and Sanson's quarto-size map of 1656 (See Map # 95).

Publication Information

The Certe map of Africa was published in his edition of Duval's *La Geographie Universelle* in 1688 and then again in 1712. His map is most easily distinguished from the Duval map by the imprint of the engraver Mathieu Ogier on the Certe map ('M. Ogier fecit.') below the neatline at the bottom right corner. The Certe map also has letterpress type at the left corner above the map of: folio. 110.

The Duval-Certe map of Africa is similar in appearance to the Duval (Map # 95) and the Duval-Hoffmann (Map # 121) maps and is sometimes confused with them. Refer to the chart under the Duval map of Africa (Map # 95) to distinquish differences among the three maps.

References

Pastoureau 1984: 162-165.
The author is grateful to Jason Hubbard for bringing this map to his attention.

144 1688 Giuseppe Rosaccio – Giuseppe Moretti, Bologna

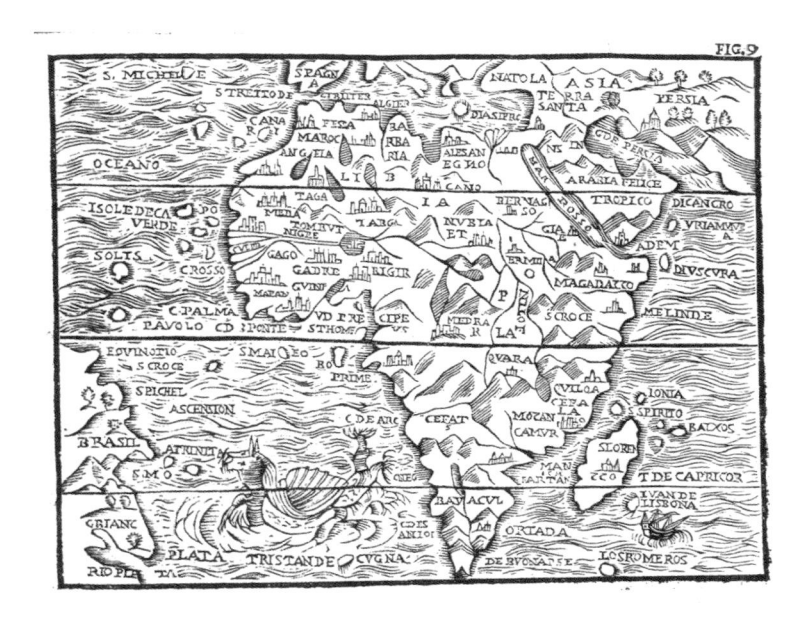

[Map of Africa].
- No scale bar.
- [Bologna : Antonio Pisarri, 1688].
- From: *Teatro del mondo e sue parti*.
- 1 map : woodcut ; 1 map sheet ; 13 x 17.5 cm.
- No latitude and longitude coordinates.
- North at the top; no compass rose.
- There is one large sea monster in the South Atlantic and one sailing ship in the Indian Ocean. The wood is cut to show the swirling seas as on the 1594 Rosaccio map of Africa.
- Above the map image 'FIG. 9'.

Example: *London BL*, 10007.a.15 (1688 edition of the *Teatro*); *Various other locations.*

Description
This is a later issue of Giuseppe Rosaccio's map of 1594 using a new woodblock. Rosaccio (1550-1620), from Venice, was a physician and an author of numerous books on geography including the *Teatro Del Cielo*, which first appeared in 1594, and the *Il Mondo e sue parti* . For this edition of Rosaccio's book, new woodblocks were cut by Giuseppe Moretti whose name replaces Rosaccio's name on the world map in the book.

This map shows all of Africa, the Arabian Peninsula, and part of Brazil. Numerous islands are shown in the South Atlantic, reflecting their importance to early shipping around the Cape of Good Hope. The coastal information is based on early Portuguese and Arab sources.

Geographically, the most significant feature on the map is a lack of the western Ptolemaic Nile source lake, which was evident on the earlier Rosaccio map (Map # 30). Other than this one aspect, the map is similar to the 1594 Rosaccio map. This map shows no new information and is an anomaly in an era increasingly concerned with a more scientific approach to the mapping of Africa.

Publication Information
This map of Africa is from the *Teatro del mondo e sue parti*, published by Antonio Pisarri in Bologna. It also appeared in an edition of Rosaccio's *Teatro Del Cielo*, published in Treviso, in 1693. It may also have appeared in other late seventeenth and early eighteenth century editions of *Teatro Del Cielo*. King (2003: 157) mentions an edition of 1724, published by Costantino Pisarri.

This map is distinguished from the earlier Rosaccio map of Africa by its larger size, its lack of text on the verso of the map, and by being printed from one woodblock rather than two.

References
King 2003.

145   1688 Nicolas de Fer - Jacques Robbe - Thomas Amaulry, Lyon

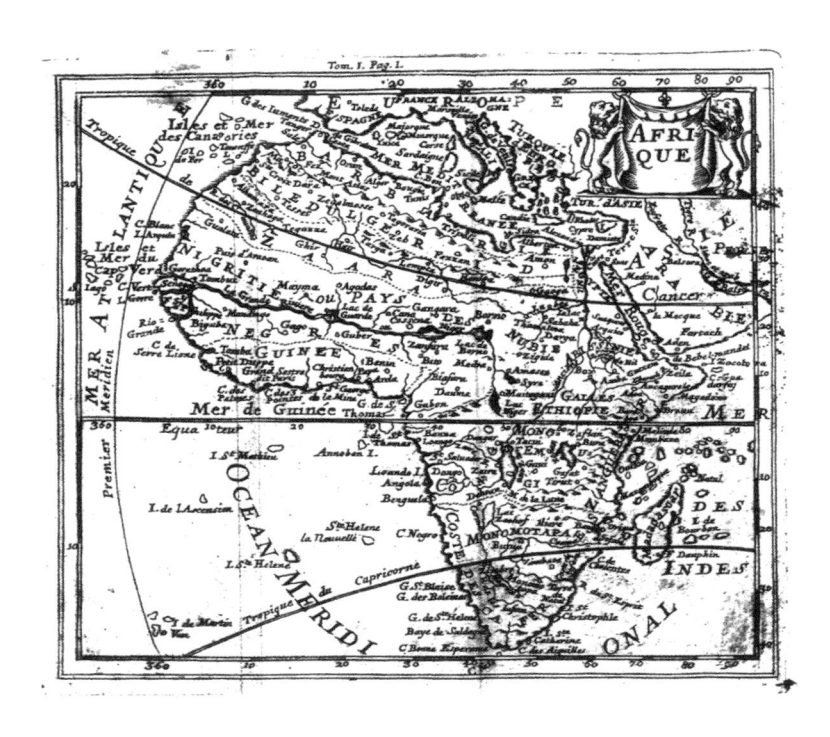

AFRI= | QUE.
- No scale bar.
- [Lyon : Thomas Amaulry, 1688].
- From: *Relation Universelle De L'Afrique Ancienne et Moderne… en Quatre Parties par le Sr. De La Croix.*
- 1 map : copperplate engraving ; 13.5 x 15.5 cm.
- Latitude and longitude coordinates on all four sides of the map. *Premier Meridian* (Prime Meridian) crosses to the west of the Canary Islands.
- North at the top; no compass rose.
- Title at top right within a cartouche with two lions on each side holding a drapery.

Examples: *London BL*, 978.b.16; *Various other locations.*

Description
A. Phérotée de la Croix (? – c.1715) was a teacher of geography, history, and mathematics in Lyon as well as an author of books on geography. Thomas Amaulry published an edition of De la Croix's *Relation Universelle De L'Afrique Ancienne et Moderne* in 1688.

For this edition, Amaulry used the 1684 De Fer-Robbe map in its second state of 1685 as the model for his own map of Africa, although there is a notable change. In the second state, the Nile River is shown flowing from the two Ptolemaic source lakes in Central Africa. The De Fer-Robbe map adds text that states that this was the source of the Nile River as known to the ancients (*Nils des anciens*), which Amaulry follows. However, for his map, Amaulry does not show the eastern Ptolemaic lake of Zaflan connecting to the Nile River. Whether this was simply an oversight, or if it was purposely done to project a more modern interpretation of the water systems of Central Africa based on the Duval 1678 map, is not known.

The Amaulry map correctly orients the Blue Nile River source of Lake Tana in the Abyssinian highlands. A traditional view for the rest of the interior of Africa remains as on most maps of this period. The *La Grande Riviere ou Niger* (Niger River) shows its traditional source in *Lac Niger* just north of the Equator and with its flow westward into the Atlantic in West Africa. The unnamed Nubia River flows eastward into the Nile to the north of the Niger. Various French settlements are identified on the coasts: in West Africa, *Fort St. Philippe* and *Petit Dieppe*; on Madagascar, *Fort Dauphin*, etc. The map still perpetuates the erroneous *Ste. Helene la Nouvelle* (New St. Helena island).

Publication Information
The Amaulry map is distinguished from the 1684 De Fer-Robbe map of Africa by the absence of 'Par N. De Fer' within the title cartouche below AFRI= | QUE. The imprint for the engraver, 'Liebaux Sculp.', at the bottom right on the De Fer-Robbe map is not on the Amaulry copy, and the engraver of this Amaulry map chose not to identify himself. Finally, this map also has 'Tom.I. Pag.1.' at the top above the map grid.

The three similar maps by Nicolas de Fer (Map # 135), Antoine de Winter (Map # 142), and Thomas Amaulry, each with a different copperplate, are distinguished from each other by differences in the imprint for the author and the engraver. See the chart for Map # 135 to differentiate these maps.

References
The author is grateful to Jason Hubbard for sharing his notes on the publication history and identification of the De Fer-Robbe maps.
Hubbard Spring ed. 2005: 46-63, and Winter ed. 2004: 16-24; Norwich 1997: map # 53 (for Amaulry edition).

146   1689 Jean Baptiste Nolin – Vincenzo Coronelli, Paris

Map 146.2.

146.1 First State 1689

AFRIQVE | selon les Relations les plus Nouvelles | Dressée et Dediée | Par le P. Coronelli Cosmographe de la Sere- | nissime Republique de VENISE. | A Monseigneur le Duc de BRISSAC | Pair de France. | A PARIS | Chez. I.B. Nolin. sur le Quay de l'Horloge du Palais, | proche la Rue de Harlay, à l'Enseigne de la | Place des Victoires. | Avec Privilege du Roy | 1689.
[at the bottom right of the map between the gridline and the neatline]:
H. van Loon Sculp.
- Scale bar in French within a box at the bottom right of the map:
  Échelle
  Milles d'Italie. 600 = 5.8 cm.
  Lieues Communes de France. 250 = 5.8 cm.
  Lieues d'Espagne. 180 = 5.8 cm.
  Lieues Communes d'Allemagne. 150 = 5.8 cm.
  Lieues d'un Heure de Chemin. 200 = 5.8 cm.
  Lieues de la Mer. 125 = 5.8 cm.
- Paris : Jean Baptiste Nolin, 1689.

- From: separately published.
- 1 map : copperplate engraving ; 45 x 60 cm.
- Latitude and longitude coordinates on all four sides of the map. Longitude coordinates across Equatorial Line.
- North at the top; no compass rose.
- Title cartouche at the top right of the map.
- No decorative elements on the map other than the title cartouche and a large text oval in Central Africa signed by Coronelli (Le P.[ère] Coronelli).

Example: See Tooley (1969) for an image of this map as plate 27.

146.2 Second State c.1693
Still dated 1689, but with the addition in title of: *Corrigée et augmentée Par le S*. Tillemon <u>and</u> with a new publication address as follows:
AFRIQVE | *selon les Relations les plus Nouvelles* | *Dresseé et Dediée* | *Par le* P. Coronelli *Cosmographe de la Sere* | *nissime Republique de* VENISE. | *Corrigeé et augmentée Par le S*. Tillemon. | *A Monseignor le Duc de* BRISSAC | *Pair de France.* | A PARIS | *Chez*. I.B. Nolin. *sur le Quay de l'Horloge du Palais,* | <u>Vers le Pont Neuf</u>, à *l'Enseigne de la* | Place des Victoires. | *Avec Privilege du Roy* | 1689.
Tooley's Dictionary (1999-2004: A-D: 404) states that Tillemon was Jean Nicolas Du Tralage (? - 1699), a French geographer who corrected and augmented ('Corrigee et augmentee') the maps of Coronelli for Nolin.

Examples: *Paris BN*, Ge CC 1272 (149); *Author's collection*.

146.3 Third State 1704
The reference to Coronelli in the title has been removed <u>and</u> a new date at the end of the title of 1704 as follows:
AFRIQVE | *selon les Relations les plus Nouvelles* | *Dressée* | *sur les Memoires du S*. de Tillemont | *Diviseé* | *en tousses Royaumes et grands Etats* | *avec* | *un discour sur la nouvelle discouverte* | *de la situation des sources du Nil.* | A PARIS | *Chez*. I.B. Nolin. *sur le Quay de l'Horloge du Palais,* | *Vers le Pont Neuf,* à *l'Enseigne de la* | Place des Victoires. | *Avec Privilege du Roy* | 1704.

Example: *Paris, Loeb-Larocque* (auction catalog of 4 November, 2005).

146.4 Fourth State 1742
New date of 1742. The map title is now:
AFRIQVE | *selon les Relations les plus Nouvelles* | *Dressée* | *sur les Memoires du S*. de Tillemont | *Divisée* | *en tousses Royaumes et grands Etats* | *avec* | *un discour sur la nouvelle discouverte* | *de la situation des sources du Nil.* | A PARIS | *Chez*. I.B. Nolin. *rue Saint Iacques* | à *l'Enseigne de la* | Place des Victoires. | *Avec Privilege du Roy* | 1742.
(and at the bottom right of the map between the gridline and the neatline):
/ H. van Loon Sculp.

Example: *London BL*, Maps 4.Tab.51-52 (The map is plate 47 in the second volume).

Description

Like the Duval map of 1678, this is another important transitional map of Africa. This map almost completely omits the two Ptolemaic lakes in Central Africa, except for the lower portion of the western Lake Zaire. Within this central area of Africa, there is a large text oval describing the sources for the interior of Africa. This text was purposely placed here to demonstrate that the two lakes were likely based on tradition and not on fact. The text describes the Nile River as known to the ancients and mentions the Jesuit explorations in Abyssinia of Pedro Paez, Manuel de Almeida, and others. It culminates by describing Hiob Ludolf's *Historia Aethiopia* of 1681, with his important map of Abyssinia which helped to disprove Ptolemy's belief that the Nile begins in the Mountains of the Moon. The text box is signed at the bottom with Coronelli's name (Le P.[ère] Coronelli).

The map does show the *Abawi le Nil, Nilus fl.* (Abay River or the Blue Nile) with its source in *Tzana Lac* (Lake Tana) in the Abyssinian highlands. It completely excludes the White Nile River with its true source further to the south in East Africa. Evidently, Coronelli believed that the Nile River had its primary source in the Abyssinian highlands as he labeled Tana as the source for the Nile on the map.

In West Africa, the Niger River still retains its east-to-west orientation, arising in *Lac Niger* in Central Africa and emptying into the Atlantic Ocean. This was a relatively consistent representation on maps since Leo Africanus' book, first published by Ramusio in 1550, described the Niger River. Above the Niger, the *Nubia* River continues its west-to-east flow into the Nile River by Egypt.

This map follows Jaillot's 1674 map in its presentation of the three major rivers in Southern Africa. From the north, they are the *Zambeze*, the *Cuame* (Cuama), and the *Spirito Santo*. Interestingly, in an attempt to clarify matters, Coronelli labels the common interior source river for both the Cuama and Spirito Santo as the *Tamoese* (and not the Zambere).

Besides attempting to correct the river systems of Africa, this map represents a tremendous step forward in its use of additional placenames reflecting up-to-date knowledge of European explorations and permanent settlements. This can be seen along the African coasts, as well as in the interior in West Africa, in Abyssinia, up the *Rio Zambeze* (the Zambezi River showing the town of *Tete*), and along the South African coast with the addition of *Bay Haut* (Hout Bay), etc. Some of the older representation of the interior remains however, such as *Castel Portugal* in the interior of Southern Africa.

This map was elegantly engraved by Hendrik van Loon (Van Loon's name as the engraver, 'H. van Loon Sculp.', is at the bottom right of the map between the gridline and the neatline). Van Loon was employed by various publishers in late seventeenth century Paris, including Nolin and De Fer, to engrave a variety of maps. To the left of the scale bar there is a map key identifying various types of places on the map.

Publication Information
Jean Baptiste Nolin (1657-1708) was known as a Parisian engraver and publisher. Nolin succeeded his father Jean Nolin who worked as an engraver in Paris on the Rue St. Jacques, a traditional area for engravers. It was on Rue St. Jacques that Jean Baptiste Nolin began his cartographic output: globe gores, a map of Italy and a map of Canada and North America, all produced in 1688.

From 1689, Nolin shows his publication location as: sur le Quay de l'Horloge du Palais, proche la Rue de Harlay, a l'Enseigne de la Place des Victoires (as it appears on the title of the first state of the Africa map). From about 1693, Nolin was working from: sur le Quay de l'Horloge du Palais, Vers le Pont Neuf, a l'Enseigne de la Place des Victoires (on the map title of the second state). Upon Nolin's death in 1708, Nolin's son Jean Baptiste II carried on the family business until 1762.

Nolin is probably best known for publishing the maps and globes of the noted Venetian Vincenzo Coronelli. From 1681 to 1683, Coronelli had worked in Paris to produce an enormous pair of terrestrial and celestial globes for King Louis XIV. It is not clear if Nolin had established a relationship with Coronelli, who was only seven years his senior, during this time. In any event, he began to publish a series of separately-issued maps of various areas of the world including this map of Africa which used much of the geographic information provided by Coronelli's detailed globe gores. It is possible that Nolin simply copied Coronelli's work without permission as he was later charged with copying the maps of Delisle in 1706 without permission (Pastoureau 1984: 357). However, it appears that Nolin had Coronelli's approval to publish his maps including this one of Africa. Helen Wallis (1982: 32-33) states that Coronelli had secured a 15-year privilege from King Louis XIV for the publication of his maps and had granted Nolin exclusive rights.

This map is not regularly found in atlases, but is usually found as a separate map. It is known in four states. In its later states, the map may appear in various composite and regular atlases published by Nolin and his son Jean Baptiste Nolin II.

The Nolins published an atlas *Le Theatre Du Monde* in the early 1700s. Pastoureau (1984: 363) mentions editions of 1701 to 1746.

References
Shirley 2004: 759-761; Tooley 1999-2004, A-D: 404; Norwich 1997: map # 85 (fourth state of the Nolin-Coronelli map); Pastoureau 1984: 363; Wallis 1982: 35: 30-34.

147 c. 1690 Johann Stridbeck, Jr., Augsburg

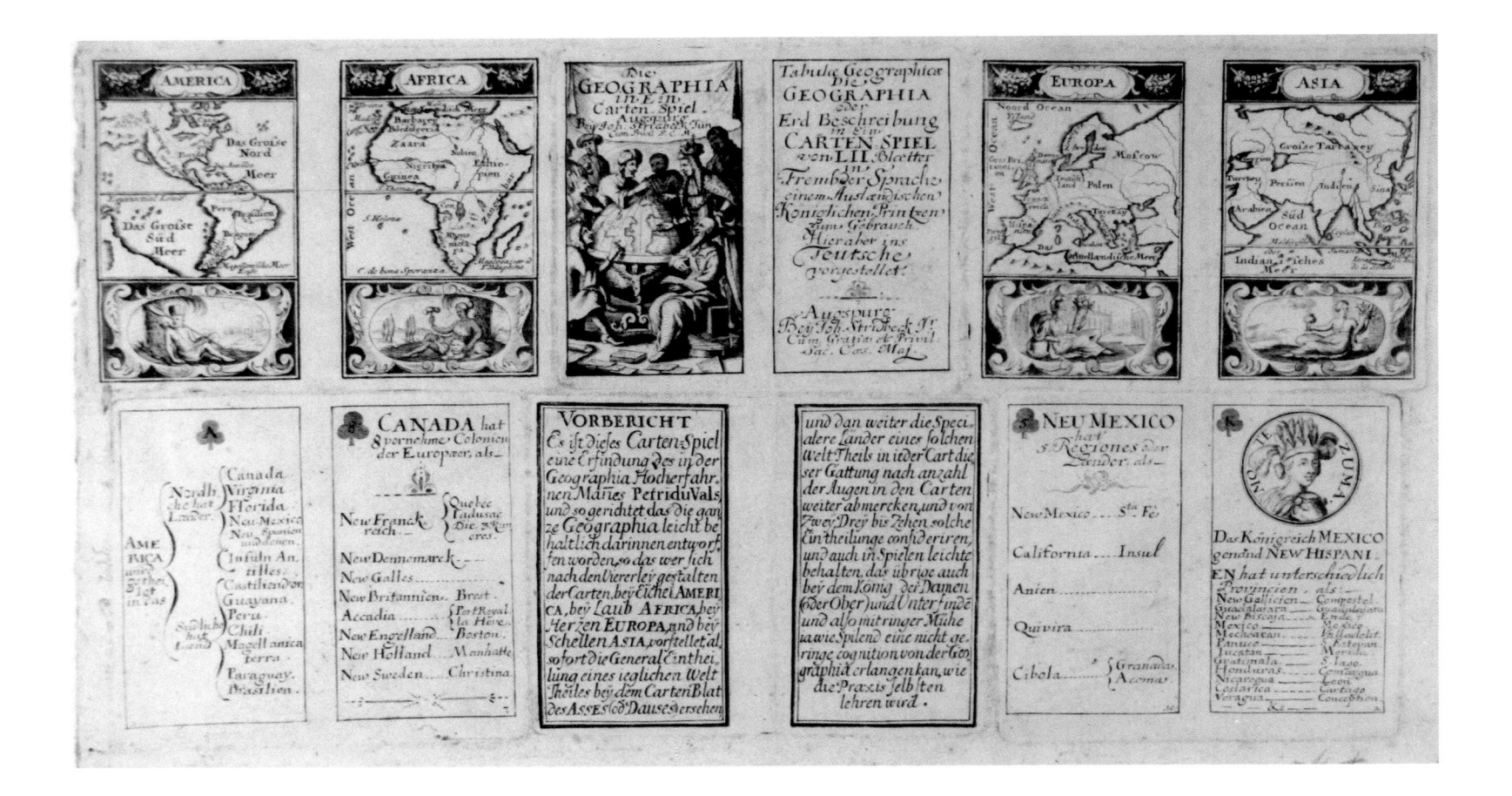

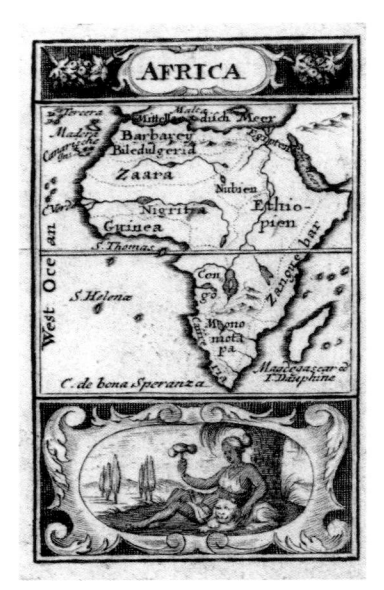

AFRICA.
- No scale bar.
- Augsburg ; Johann Stridbeck Jr., c.1690.
- From: *Die Geographia oder Erd Beschreibung in ein Carten Spiel... bey Joh. Stridbeck, Jr.*
- 1 map : copperplate engraving ; 8.0 x 4.9 cm (for map of Africa).
- No latitude and longitude coordinates on the map. Equatorial Line shown on the map.
- North at the top; no compass rose.
- Title is above the map set between two decorative 'fruit garlands'.
- Below the map is an allegorical representation of Africa as a woman with a lion at her side. No other decorative elements on the map.

Example: *London BM*, 1896-5-1-1111 (1) (also identified as *London BM*, Schreiber G 51, p. 208).

Description

This rare map is from a collection of playing cards. The example of the set at the British Museum is uncut and is on a total of five sheets. The first sheet contains: maps of the four continents, including Africa; the frontispiece; and the title page. The maps of the four continents at the front of the set are not intended for use as playing cards. The four sheets that follow contain a complete set of 52 playing cards of regions of the world.

The map of Africa has relatively limited information. Identified are the major regions of Africa: *Barbarey, Biledulger, Zaara, Nigritia*, and *Guinea* in the west; *Egipten*, and *Nubien* in the northeast; *Ethiopien, Zanguebar*, in the east; and *Congo, Monomotapa*, and *Caffreria* in the south. No towns are identified on the map. The shape of Africa and geography of the river systems generally follow Ortelius, although there is a Ptolemaic river configuration above the Niger.

Publication Information

Similar sets of playing cards with maps of Africa and the other continents were published in Germany (this set), France, and Italy. The Schreiber reference book (O'Donoghue 1901) mentions that these cards were first published in France and are under Schreiber France-123, though the author was not able to locate this example in the British Museum. There is also a similar collection in the British Museum as Schreiber Italy I-107, but this collection does not contain the separate sheet showing the continents. The author also observed a similar set of Italian playing cards on exhibit at the Correr Museum in Venice with the map of Africa.

The author has attributed this map to Johann Stridbeck Junior as no other name appears on any of the sheets, except for the publisher, *Joh. Stridbeck Jr.*. Tooley's Dictionary of Mapmakers (1999-2004: Q-Z: 226) mentions a Johann Stridbeck (1640-1716) and his son, also named Johann Stridbeck Junior (1666-1714), editors, engravers, and publishers from Augsburg.

There is no date on the playing cards. The Schreiber catalog (O'Donoghue 1901) at the British Museum dates this collection as c.1670. As the publisher of these playing cards was Johann Stridbeck, Junior and accepting the dates of 1666-1714 for the son, the date assigned to this collection of c.1670 is obviously too early. The British Library has an atlas of Italy produced by the son in 1692. Meurer (1986) suggests a date of c.1687 for Stridbeck's atlas of Italy. These dates seem to be more in line with the maps that were produced by the son. Accordingly, the collection and this map of Africa are dated c.1690.

References

Shirley 2004: 974-975; King 2003; Tooley 1999-2004, Q–Z: 226; Shirley 1993; Meurer 1986 (entry for Stridbeck in *Lexikon zur Geschichte der Kartographie*); O'Donoghue 1901.

148 c. 1690 Gerard Valk, Amsterdam

AFRICA.
[at the bottom of sheet 6 below title]:
cum Privilegio Ordinum Hollandiæ et West Frisiæ.
[and at side of sheet 6]:
Tot Amsterdam Gedruckt by GERARD VALK op den Dam inde Wackeren hont.
- No scale bar.
- Amsterdam : Gerard Valk, [c.1690].
- From: separately published.
- 1 wall map : copperplate engraving ; 6 map sheets ; 103 x 124.5 cm (in total, with sheet 1 (upper left) as 48.5 x 41.5 cm, sheet 2 as 48.5 x 41.5 cm, sheet 3 as 48.5 x 41.5 cm, sheet 4 as 54.5 x 41.5 cm, sheet 5 as 54.5 x 41.5 cm, and sheet 6 as 54.5 x 41.5 cm)
- Latitude and longitude coordinates on all four sides of the map. Longitude coordinates within Africa across Equatorial Line on sheets 1-3.
- North at the top; one compass rose in the Indian Ocean and two compass roses within the Atlantic Ocean.
- Title surrounded by a large decorative vignette on sheets 5 and 6.
- There are three sailing ships and one sea monster in the Indian Ocean, and fifteen sailing ships, a group of ships in a sea battle, and one sea monster in the Atlantic Ocean. A decorative vignette across the entire bottom of the map is an amalgam of different scenes of Africa, part of which shows the Dutch at Cape Town. There is another decorative vignette at the top of sheet 3.

Example: *London BL*, Maps C.5.a.1.(4) (only known example).

Description

Gerard Valk's six-sheet wall map of Africa is especially attractive with an elaborate vignette across the entire bottom of the map. It generally follows the Blaeu model, especially the 1659 wall map of Joan Blaeu in its depiction of Africa, and does not attempt to present information depicted on any of the new models by Sanson, Jaillot, Duval, and others. Not following the French, the map relies on a Ptolemaic image of the interior with the twin Nile source lakes. However, it does update placenames. This is particularly apparent along the coasts, although surprisingly not around the Cape of Good Hope which the Dutch had thoroughly settled by this time.

The *Lunae Montes* (Mountains of the Moon) divide the rivers of Southern Africa from the Nile source lakes. Flowing from *Sachaf Lac.* is the *Zambera flu*, which splits into the *Rio de Lagoa* (Spirito Santo River) and *Rio de S. Maria* (Cuama River). Considerable detail is shown in the Congo region.

Gerard Valk (1652 – 1726) produced a prodigious amount of cartographic material during his professional career and was a noted Amsterdam publisher, engraver, and globemaker. For a time Valk worked in London in association with Christopher Browne among others.

After Valk's return to Amsterdam from London in the late 1670s, he collaborated with Pieter Schenk, a map engraver and seller from Germany, from about 1680. Together they acquired the copperplates of Janssonius' *Novus Atlas* from Janssonius' heirs and other copperplates. Valk did continue to produce maps on his own through this period, including this wall map of Africa. In 1687, Schenk, who had been a pupil of Valk married Valk's sister and entered into a long, more formal relationship with Valk and Valk's son Leonard (1675-1746).

In 1687, Valk moved into a house on the Kalverstraat, near the Dam Square in Amsterdam. Some time after 1687, Valk moved into a house on the Dam, where the Hondius family had lived: 'op den Dam in de Wakkere Hond' (Koeman 1967-71, III: 136).

Publication Information

Gerard Valk is known to have produced a wall map of the world in c.1686 and wall maps of the four continents at some time after this date. None of these wall maps are known to contain a date, so the dating of this map of Africa is difficult. The author has dated this map as c.1690 since Valk moved to 'op den Dam inde Wackeren Hond' after 1687 (the publication location cited on the map). Shirley states (1993: 545) that there is a wall map of the world in eight sheets of c.1690 prepared by Gerard Valk and engraved by Leonard Schenk, Pieter's son. This wall map of Africa may have been prepared to accompany that world map.

References

Tooley 1999-2004, Q-Z: 307-308; Shirley 1993: 529; Koeman 1967-71, III: 136-140.

149   1690   A. Phérotée de la Croix, Lyon

Map 149.2.

149.1 First State 1690

L'AFRIQVE | Selon les | Autheurs les plus | Modernes.
- No scale bar.
- [Lyon : Jean Baptiste Barbier, 1690].
- From: *Nouvelle Methode pour apprendre...La Geographie Universelle...Chez Jean Baptiste Barbier...M.D.L.XXXX* (The map is in Vol. 4 of the BL example).
- 1 map : copperplate engraving ; 14.5 x 20.5 cm (including coats of arms along the two sides of the map).
- Latitude and longitude coordinates on all four sides of the map. Longitude coordinates across the Equatorial Line. *Ligne du Premier Meridian* (The Prime Meridian) is to the west of the Canary Islands.
- North at the top; no compass rose.
- Title within a cartouche at the upper right of the map.
- No decorative elements on the map. To each side of the map are coats-of-arms. From the top left are: *Barbarie*, *Maroch*, and *Egypte* and five blank coats-of-arms. From the top right are: one unidentified coat of arms, *Malthe*, *Congo* and five blank coats-of-arms.

Example: *London BL*, 795.d.31; *Various other locations.*

149.2 Second State 1693

Addition of engraved 'Tom 4. pag 225' at top right above the map, in the Grenoble example. Though this book is cataloged as 1690, the map is evidently later than the 1690 edition of the book due to the addition of the printer's instructions for the insertion of this map into Volume 4, page 225.

Example: *Grenoble BM*: E 28421.

Description

A. Phérotée de la Croix was a teacher of geography, history and mathematics in Lyon as well as an author of books on geography. In 1690, De la Croix's book of geography *Nouvelle Methode pour apprendre...La Geographie Universelle* which contained this map of Africa, was published by Jean Baptiste Barbier.

This map of Africa is quite up-to-date and reflects current French thinking regarding the interior of Africa. The general model for this map is Jaillot-Duval and especially the Duval map of 1678. Lake Tana is shown as the source for the Blue Nile in East Africa. A third river in Southern Africa is shown, which was to become a more accurately placed Zambezi River. Following tradition, De la Croix inserts *Amara*, the residence of Prester John, on the Equator. He also shows the *Mont. de la Lune* (Mountains of the Moon) south of the Equator.

Publication Information

This map was first printed in Lyon in 1690, again in 1705 and finally in 1717. It also appeared in a Paris edition in 1693. Burden (2007: 399) mentions the use of the Americas map in Alexis Hubert Jaillot's rare *Tablettes Geographiques ou Recueil des Principales Cartes Generales et Particulieres*, published in Paris in 1695, without text (the only known examples are at the Biblioteka Gdanska and mentioned in a Walter Reuben catalog of c.1980s). The author has not examined a copy of the Jaillot book to ascertain if it contains a map of Africa.

A derivative of the map was published in 1697 in a Leipzig edition of De la Croix's *La Geographie Universelle* by J.L. Gleditsch and the Heirs of M.G. Weidmann. Another derivative of the De la Croix map was published by Daniel de la Feuille in Amsterdam in 1702 in *Atlas Portatif*. In the De la Feuille copy, the map is distinguished by six coats of arms on each side of the map (in the 1690 map, there were only three coats of arms, with the remaining five coats of arms on each side blank).

References

Burden 2007: 399; Shirley 2004: T.LACR-1a: 630; Hoppen 1975: no. 108.

150 c. 1690 Joachim Bormeester, Amsterdam

150.1 First State c.1690

AFRICAE | NOVISS:CATERES | TABULA | AUCT: I. BORMEESTER.
[and at bottom right of the map]:
Spectat: Ampl: et Prudent: D.D. | IAC: JACOBI HINLOPEN | CONS: SENAT: ETC. | Amstelodamensium | Devovit. I. Bormeester.
- No scale bar.
- Amsterdam : Joachim Bormeester, [c.1690].
- From: separately published.
- 1 map : copperplate engraving ; 43 x 52.5 cm.
- Latitude and longitude coordinates on all four sides of the map.
- North at the top; one compass rose below Equatorial Line in Atlantic Ocean.
- Title within decorative cartouche at top right.

- There are two sailing ships in the Indian Ocean, and five sailing ships in the Atlantic Ocean. No animals are within Africa. Elaborate and attractive decorative scenes across the bottom of the map including a vignette surrounding the dedication and Bormeester's imprint at the bottom right.

Example: *Utrecht UB,* KAART: Ackersdijck 759.

150.2 Second State c. 1700+
Erasure of Bormeester's name within title cartouche (though some evidence of the Bormeester name still remains) and within the dedicatory cartouche. There is a new name within the title cartouche of 'T. Danckerts'. Also, there is a new name within the dedicatory cartouche of 'T. Danckerts'.

Example: *Private Belgian collection.*

Map 150.1 Title cartouche.

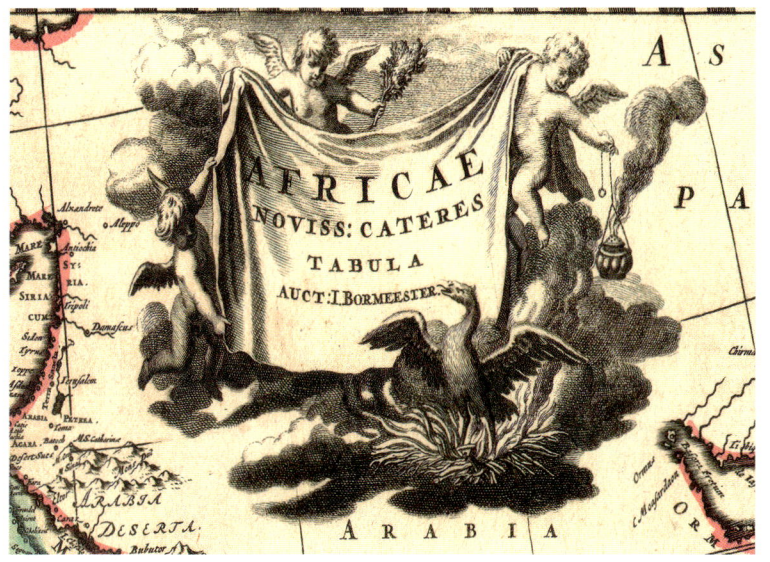

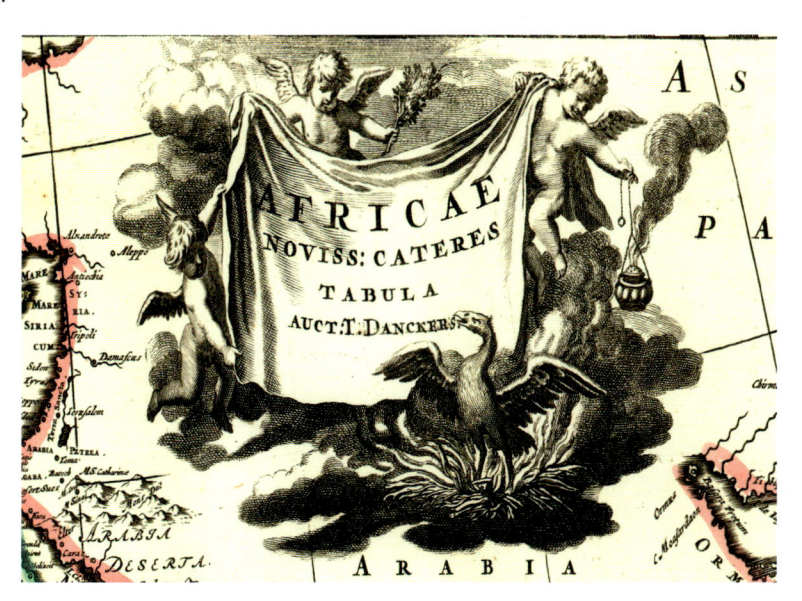

Map 150.2 Title cartouche.

Description
Joachim Bormeester was an engraver, printer, and publisher working from Warmoes Straadt, Amsterdam from about 1670 to 1690 according to Koeman (1967-71, II: 1). Little else is known of Bormeester.

Directly modeled after the Visscher map of c.1658, this map of Africa was clearly out-of-date by the time of its publication. Three separate vignettes decorate the bottom of the map: a scene on the left of traders with an elephant and ivory tusks; an elaborate coat-of-arms with a crown on top which is surrounded by classical figures; and a dedicatory cartouche on the right in the foreground with ships sailing behind it. A unique title cartouche shows a drapery held by three cherubs and a phoenix rising from a fire.

Publication Information
This map of Africa is not dated. Bormeester produced three maps of the world (Shirley 1993: 525-6), which Shirley has dated c.1685. He is also supposed to have produced a map of the Americas in c.1700 (Tooley 1999-2004, A-D: 167). As Koeman (1967-71, II: 1) states that Bormeester was active as a mapmaker from the period of c.1670 to 1690, the author has dated this map as c.1690.

At some point likely later in the 1690s or early 1700s, the copperplate that produced this map was acquired by the Danckerts publishing family. The imprint of 'T. Danckerts' probably refers to Theodorus I (1663-c.1727). Theodorus I was the son of Justus Danckerts. Theodorus I was succeeded by his son, Theodorus II (1701-1727). The Danckerts family business ended upon Theodorus II's death in 1727 when all the stock was sold. Besides appearing as a separate map, this map probably appeared in various Danckerts atlases. The verso of the map is blank.

References
Tooley 1999-2004, A-D: 167; Shirley 1993: 525-6; Koeman 1967-71, II: 1.

151 1690 Sebastián Fernández de Medrano - Jacques Peeters, Brussels

AFRICÆ.
[below the title within vignette]:
I.P.
[bottom left corner below the cartouche and vignette]:
Harrewyn f:
- No scale bar.
- [Brussels : Juan (Jean) Leonard, 1690].
- From: *Nueva Descripcion | del Mundo y sus Partes... En Bruselas | En Casa de Juan Leonard 1690.* with an engraved title page of *Atlas | Par Jacques Peeters a Anvers sur le Marche... .*
- 1 map : copperplate engraving ; 15 x 18.5 cm.
- Latitude and longitude coordinates on all four sides of the map.
- North at the top; no compass rose.
- Title set within a large decorative vignette at the bottom left of the map.
- There are four sailing ships in the Atlantic Ocean. A lion, zebra(?), and an ostrich are in West and Central Africa.

Examples: 1690 De Medrano edition: *München SB*, 8° Mapp.68 i (Map # 26 in the atlas); 1692 Peeters edition: *Den Haag KB*, 509 L 25; *London BL*, Maps C.43.b.2 (Map # 10 in the atlas); *London, Hubbard Collection*; 1696 Van Aefferden edition: *New York HSA*; *New York PL*, *KB 1696 Aefferden; *London, Hubbard Collection*; 1709 Verdussen edition: *London BL*. 1295.b.9; 1711 Pedro Ponton edition: *New York HSA*, 83224; 1725 Verdussen edition: *Leiden UB*, COLLBN Atlas 715; 1755 Lyon edition: *London, Hubbard Collection*.

Description

This map first appeared in the very rare book of 1690, *Nueva Descripcion del Mundo*, by Sebastián Fernández de Medrano. De Medrano (1646-1705), born in Spain, was a noted military figure serving in Flanders among other places. He is most known for his various books on the theory and practice of warfare. He also issued books of geography including this one.

The Antwerp publisher, Jacques Peeters (1637-1695), designed this map for De Medrano. The decorative vignette surrounding the title includes the signature of Peeters below the title with his initials as 'I.P.' to show that he was the designer for this map. The name of the engraver Jacob Harrewijn is indicated on the bottom left corner below the cartouche, 'Harrewyn f:'.

For some reason, this De Medrano-Peeters map of Africa does not incorporate the more recent cartographic information provided by the French mapmakers. Instead, the immediate model for this map is the c.1670 De Wit map.

Publication Information

This map has a long, complicated history. Its appearance in De Medrano's book, *Nueva Descripcion del Mundo*, is only recently known. It is most known for its appearance in Peeter's *L'ATLAS en abregé*, which was only published once in Antwerp in 1692. It appears that the copperplates for the maps were acquired by Jan (Juan) Duren shortly after Peeters death in 1695. In 1696, Franciscus van Aefferden (or Van Afferden as his name sometimes appears) (1653-1709) prepared a Spanish language version of the atlas called *El Atlas Abreviado* using these plates. This 1696 edition was also published in Antwerp in 1697 by Jan Duren.

In 1709, a third edition was published by Henricus (Henry) Cornelius Verdussen in Antwerp. There were two final editions, using this same map of Africa, one by Verdussen and one by the widow of Verdussen, both dated 1725.

The De Medrano-Peeters map, with the imprint of 'Harrewyn f:', was also used for a second edition of Sebastián Fernández de Medrano's *Nueva Descripcion del Mundo*, published by Jean Leonard in Brussels with Spanish text, in 1701. The map also appears in a 1709 edition of De Medrano's *Geographia o Moderna Descripcio del Mondo*, published by Verdussen in Antwerp.

The De Medrano-Peeters map of Africa appeared unchanged from its original issue in 1690 to its last publication in 1725 in Antwerp. All of the issues of this map have the name of the engraver in the lower left corner 'Harrewyn f:'

Sometime before 1709, a new set of copperplates was cut for a publication in Spain of Van Aefferden's *El Atlas Abreviado*. This version was published in Madrid in 1709 and 1711 by Pedro Ponton (*New York HSA*, 83224, 1711 edition). Many of the maps were redone with Spanish titles and placenames. The maps can be identified as well by the large crude page numbers. It has the title of *AFRICA* within a decorative scene at the bottom left similar to the De Medrano-Peeters map. The Africa map has an engraved 'N° 113' within the top right gridline. This map is also distinguished from the earlier De-Medrano-Peeters map by the removal of 'I.P' and 'Harrewyn f:' from within the vignette surrounding the title. These examples are scarce.

Finally, a third set of copperplates was prepared for use in a new edition of *El Atlas Abreviado,* published in 1739 and again in 1755 in Lyon by Jean Certe, still with text in Spanish. In these editions, the geography for the map of Africa is generally the same as presented on the 1690 De Medrano-Peeters map with a few exceptions (the 1755 map has removed the fictitious *N.I d.S Helena*). The 1739 and 1755 examples are easily distinguished from the earlier De Medrano-Peeters map of Africa by the removal of the ships from the Atlantic Ocean and the animals within Africa. Also, the cartouche at the bottom left has been greatly enlarged and fills much of the South Atlantic in the 1739 and 1755 examples.

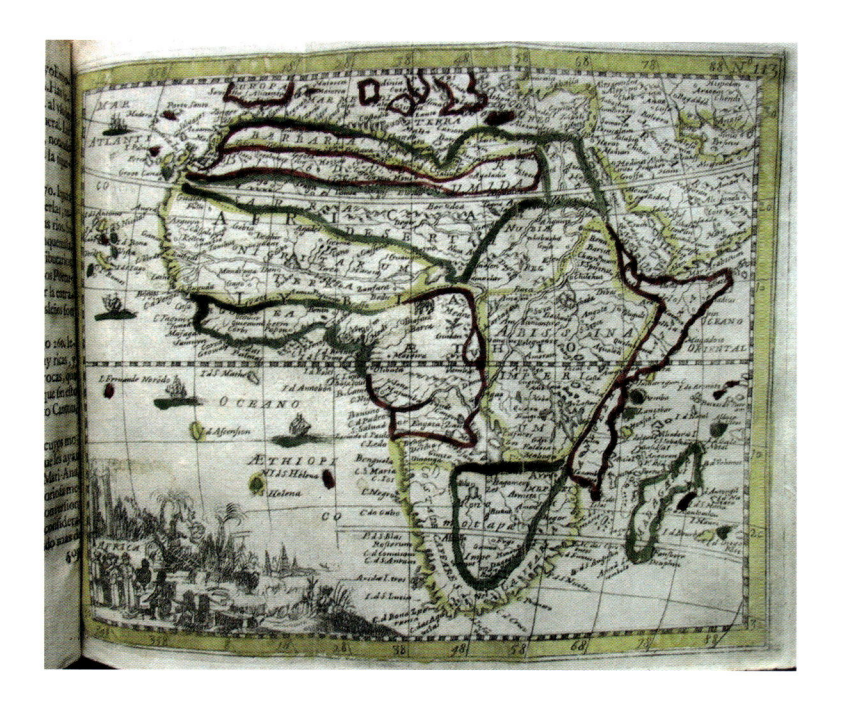

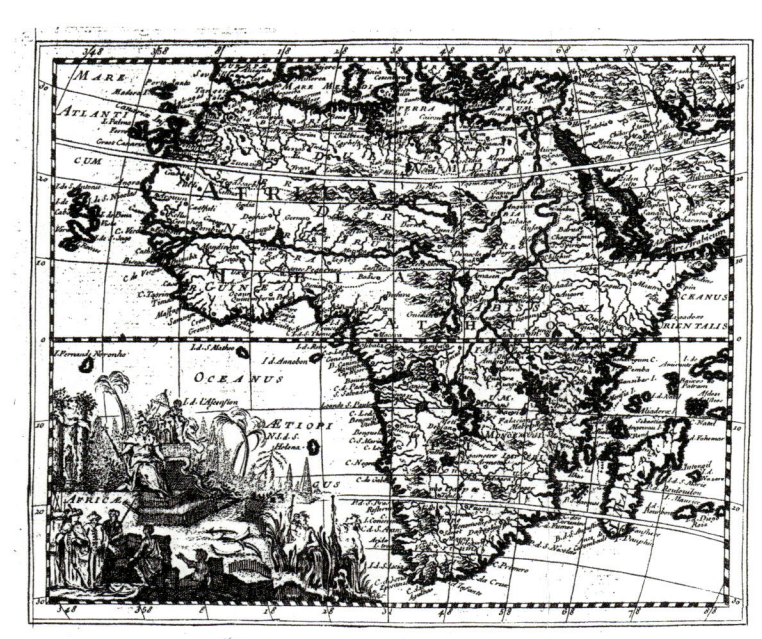

Copy of Map 151, Map of Africa from the Ponton edition, Madrid 1709-11.

Map of Africa from the Certe edition, Lyon 1739

References
Hubbard Spring 2005: 46-63, and Winter 2004: 16-24; Shirley 2004: T.MED-1a: 674; Norwich 1997: map # 65 (issue of 1709 of the De Medrano-Peeter's map); Shirley 1993: 553; Koeman 1967-71, I: Aff 1-4, and III: Pee 1: 94; Peeters-Fontainas 1933.

An acknowledgement is due Cyrus Alai of London, who initially discovered the rare De Medrano book of 1690 containing this map of Africa at the Bayerische Staatsbibliothek, München. The author is grateful to Jason Hubbard for bringing the 1690 De Medrano book to his attention, and for sharing his research notes.

152 1691 James Moxon, London

AFRICA.
- No scale bar.
- [London : James Moxon, 1691].
- From: separately published as part of a set of geographical playing cards.
- 1 map : copperplate engraving ; 5.6 x 9.0 cm (the size for the card of Africa).
- Latitude coordinates on two sides of the map. Longitude coordinates across the Equatorial Line.
- North at the top; one compass rose in South Atlantic Ocean.
- Title cartouche as a banner to the left of the map.
- Symbol for the ace of spades within Indian Ocean. No other decorative elements on the map.

Example: *London, Burden Collection*, (only one example is known; the map of Africa is part of a set of 52 playing cards).

Description
This rare map of Africa is shown on a playing card representing the ace of spades. Africa is divided conventionally into *Barbaria, Libya, Libya Interi*[or], *Nubia, Guyne, Biafara, Congo, Abissines, Aiana, AEthiopia* (for South Africa), and *Madagascar*.

A prominent geographical feature is the western Nile source lake as the only source for the White Nile River and the eastern Nile source lake as the source for the Blue Nile River, though far to the south of the true source as Lake Tana in the Abyssinian highlands. Apart from the geographical divisions of Africa and depictions of some of the major rivers, the map is generally devoid of geographic characteristics.

Publication Information
This set of cards, including the Africa map, is by James Moxon (?-1708) who was a London publisher and engraver. This map of Africa, along with the other playing cards in the set, was initially published in 1677. Moxon's *Term Catalogues* for February 1677 advertised a pack of Geographical Cards, '*describing the whole world; a particular Map in each Card, with Longitude and Latitude: newly engraven in Copper plates by J. Moxon*'. Moxon's playing cards from his first edition of 1677 are not known to exist. The 1691 set of playing cards is from a second edition.

References
The author is grateful to Philip Burden for bringing this map to his attention and for providing information on it.
King 2003; Tooley 1999-2004, K-P: 290; Tyacke 1978: 126-128.

153 1691 Vincenzo Maria Coronelli, Venice

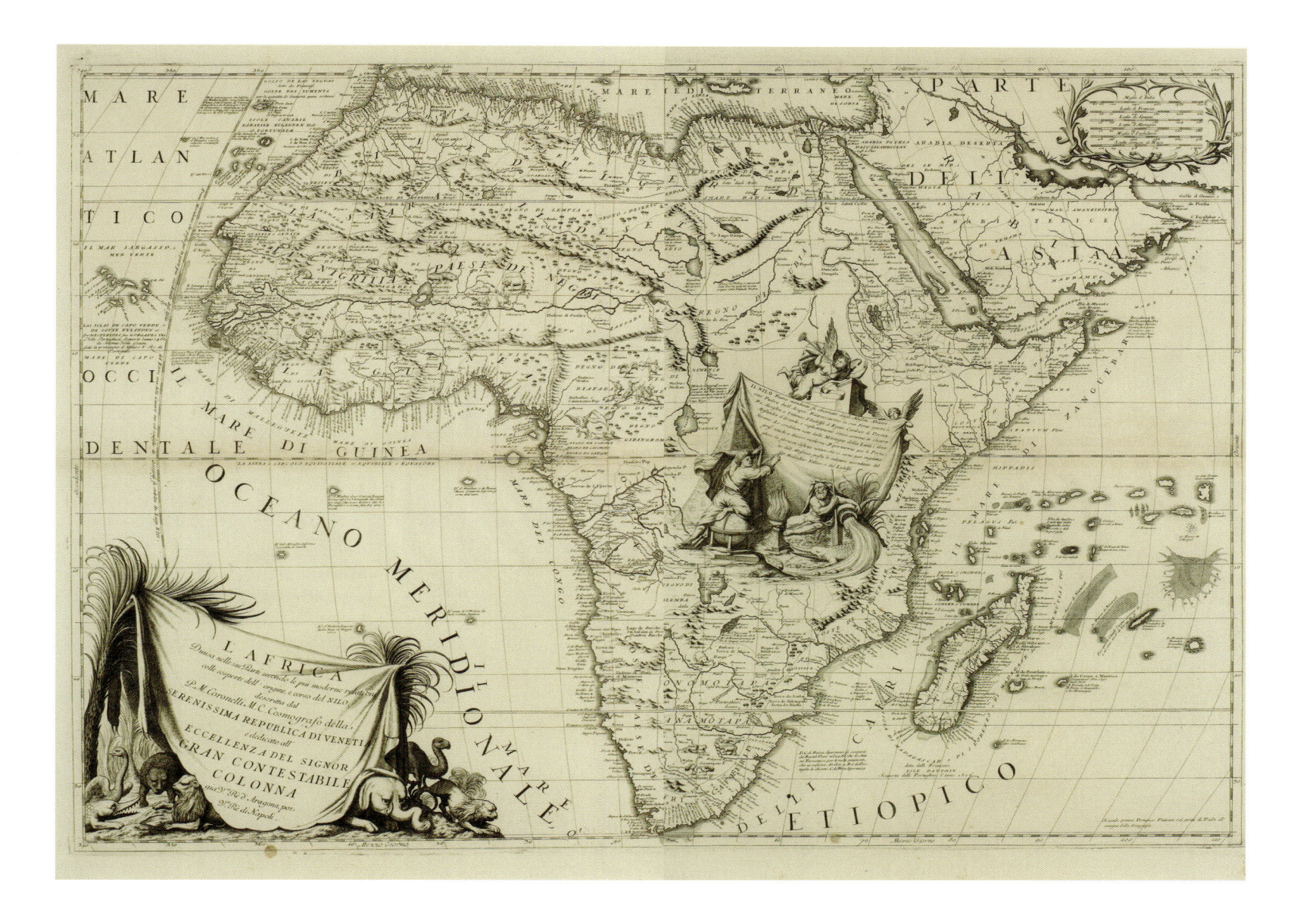

153.1 First State 1691

L'AFRICA | Diuisa nelle sue Parti secondo le piùu moderne relationi | colle scoperte dell' origine, e corso del NILO, | descritta dal | P.M. Coronelli M.C. Cosmografo della | SERENISSIMA REPUBLICA DI VENETIA | è dedicata all' | ECCELLENZA DEL SIGNOR | GRAN CONTESTABILE | COLONNA | gia V^{ce} Rè d'Aragona, poi | V^{ce} Rè di Napoli.

[at the bottom right corner of the right map sheet]:
Si uende presso Domenico Padoani sul ponte di Rialto all' insegna della Geografia.
- Scale bar is at the top of the east map sheet as follows:
  Miglia d' Italia 600 = 8.5 cm.
  Leghe di Francia 250 = 8.5 cm.
  Leghe di Spagna 180 = 8.5 cm.
  Leghe d' Alemagna 150 = 8.5 cm.
  Miglia d' Inghilterra 480 = 8.5 cm.
  Leghe comuni di Mare 120 = 8.5 cm.
- [Venice : Domenico Padouani, 1691].
- From: *Atlante Veneto, nel quale si contiene la Descrittione...* Volume One.

- 1 map : copperplate engraving ; 2 map sheets ; 60.5 x 45 cm (for each sheet).
- Latitude and longitude coordinates on all four sides of each of the two sheets. The *Primo Meridiano..* (Prime Meridian) crosses to the west of *Is del Ferro* (Ferro Island) in the Canary Islands.
- North at the top; no compass rose.
- Large, decorative title and dedication cartouche in the form of a drapery at the bottom left of the west (left) map sheet, surrounded by palm trees and various animals (some, including the elephant, are from Hiob Ludolfi's book on Abyssinia, *Historia aethiopia*).
- Large, allegorical vignette on the east (right) map sheet depicting the Nile River source with a scribe writing a legend on a drapery about the informational sources for the Nile. The text points to the source of the Blue Nile in Abyssinia and makes mention of Ptolemy with his supposition that the source of the Nile was in the Mountains of the Moon. Alexander the Great, the Portuguese Tellez, Ludolfi and their various explorations and writings are also mentioned.
- Numerous real and imaginary (e.g. phoenix) animals are set within Africa on both map sheets.
- Various informational text passages in Italian are placed throughout the map.
- No ships or sea monsters are in the oceans.

Examples: *London BL*, C.44.f.6/1 (dated 1691 on the title page of the atlas); *Author's Collection*.

153.2 Second State 1701
Date on the title cartouche of the map of 'MDCCI'.

Example: From Tooley (1969: 38).

Description
Coronelli's finely-engraved and beautifully-decorated two-sheet map contains up-to-date information on Africa. While this map bears close similarities to the Nolin-Coronelli map of 1689, both are actually derived from Coronelli's 1688 globe gores that show Africa and from his terrestrial globe of 1683. The basic models used are the maps of Africa by Sanson (1668), Jaillot (1674), and Duval (1678).

Vicenzo Coronelli (1650-1718) was a noted Italian theologian, mathematician, and cartographer. He is considered one of the more famous map and globemakers of the seventeenth century. He started as a Franciscan monk, became a Doctor of Theology in Rome in 1673, and later was appointed the Cosmographer to the Venetian Republic in 1685 where he founded a geographical society.

As a talented young monk, he was sent to Rome, and after his return to Parma, he made a large manuscript globe for the local prince. It was the beginning of his career as a globemaker. Having seen the work, Cardinal D'Estee commissioned Coronelli to build large terrestrial and celestial globes for the king of France, Louis XIV. Coronelli moved to Paris where he directed the work on the giant pair of globes which were almost four meters in diameter. He finished the work by 1683 and returned home to Venice.

Upon his return to Venice, he continued his work with globes and began work on his *Atlante Veneto*, with 65 maps including this one of Africa.

Publication Information
Coronelli's *Atlante Veneto*, Volume I, has the date of 1690 on its title page, but it was not published until 1691, the date on its dedication and colophon (*In Venetia, Appresso Girolamo Albrizzi. M.DC.XCI*). Likely, the printing of the atlas began in 1690 by Domenico Padouani, but was not completed until the following year by Albrizzi. While this map could have been printed in 1690, the author has dated the map as 1691, the date for the publication of the atlas, following Tooley's date for this map (Tooley 1969: 38).

The publisher for the map was Domenico Padouani, whose imprint and publication location is at the bottom right corner of the east (right) map sheet: 'Si uende presso Domenico Padoani sul ponte di Rialto all' insegna della Geografia'.

References
Shirley 2004: 425-426; Norwich 1997: map # 56; Tooley 1969: 38; Phillips 1909-1992: entry # 5950.

154 1691 Laurence Echard, London

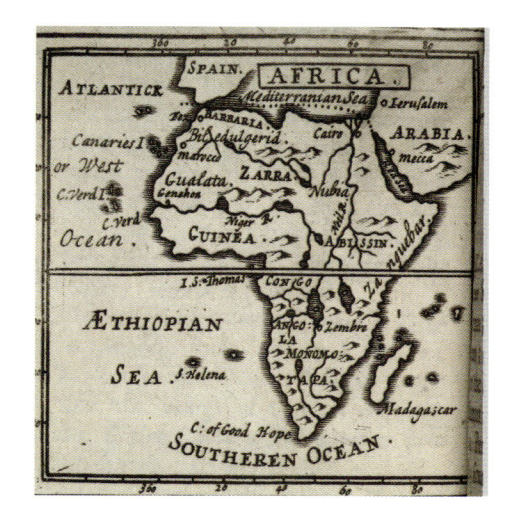

AFRICA.
- No scale bar.
- [London : Thomas Saulsbury, 1691].
- From: *A most compleat compendium of geography, general and special : describing all the empires, kingdoms and dominions in the whole world ... / collected according to the most late discoveries and agreeing with the choicest and newest maps by Laurence Eachard ...*
- 1 map : copperplate engraving ; 6.0 x 6.5 cm.
- Latitude and longitude coordinates on all four sides of the map. Equatorial Line across Africa.
- North at the top; no compass rose.
- Title set within simple rectangular box on the map above Africa.
- Devoid of decorative elements.

Examples: 1691 second edition: *London BL*, 795.a.50.(1.); 1693 edition: *Oxford CC*, a.3.237; 1697 edition: *London BL*, 1296.a.27; 1704 edition: *London BL*, 10004.a.10; 1705 edition: *London BL*, 10001.aa.2; 1713 edition: *London BL*, 571.a.26.

Description
This little map appeared in Laurence Echard's *A most compleat compendium of geography...* . Also in the book were maps of the other three continents. The Americas map has the imprint for Herman Moll as the engraver (*H. Moll fecit*). As a result, it is assumed that Moll also was responsible for engraving this map of Africa. In 1691, Moll was still employed primarily as an engraver; his major geographical works were yet to occur.

Owing in part to its small size, the map shows little geographical information besides a number of placename divisions for Africa common during this period. The three main rivers on the map are the Nubia flowing east across North Africa, the Niger flowing into the Atlantic in West Africa, and the Nile River originating in the two lakes in Central Africa. In Southern Africa, two rivers are shown; one (the Spirito Santo) beginning in the traditional lake in South-central Africa and the other (the Cuama) beginning separately in the foothills. Lake Tana is not shown in the Abyssinian Highlands.

Publication Information
The author of the book is Laurence Echard 1670?-1730, whose name sometimes appears in the literature as Eachard. Echard was born at Barsham, Suffolk, and educated at Cambridge. He took religious orders and became Archdeacon of Stow. He is primarily known as a historian and author of numerous books of history and geography. His chief works include his *History of England* covering the period from the Roman occupation to his own times, and continued to be the standard work on the subject until it was superseded by translations of Rapin's French History of England in 1725.

Echard's *A most compleat compendium of geography, general and special...* first appeared in 1691. It must have been popular as a second edition appeared in the same year. This was followed by subsequent editions in 1693, 1697, [1700], and 1704. Burden (2007: 379) mentioned a seventh edition of c.170[7] which lacks the imprint for Salusbury and Moll on the Americas map. A final edition of the book appeared in 1713.

The publisher for the book was Thomas Saulsbury who was active from *the Sign of the Temple near Temple Bar in Fleet Street* in 1691, and then from 1693 at the *Kings Arms next to Dunstan Church in Fleet Street* .

The map is known in one state. In the 1691 second edition, the map of Africa appears on a page above the map of the Americas. In the 1693 edition, the map of Africa is set within text (above the map is *150 and Oriental Islands.* and two lines of text; and below the map, *III. A- *).

References
Burden 2007: 379; Sabin 1868: entry # 21760.

155 1692 Johann Ulrich Müller, Ulm

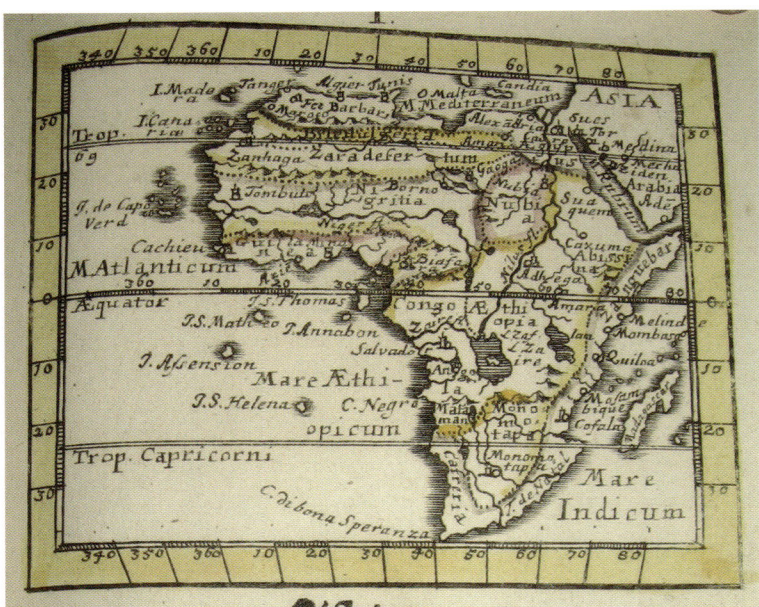

155.1 First State 1692

[Map of Africa].
- No scale bar.
- [Ulm : Georg Wilhelm Kühnen, 1692].
- From: *Kurtzbündige Abbild...Der Gantzen Welt...*
- 1 map : copperplate engraving ; 6.5 x 7.5 cm (for the map only) ; 17 x 9.5 cm (for entire page with the text).
- Latitude and longitude coordinates on all four sides of the map. Longitude coordinates across the Equatorial Line.

- North at the top; no compass rose.
- No decorative elements on the map.
- 'I.', above the map, signifying the map's position as the first map within the book. Map at top of a page of text; text below the map beginning with 'Africa... .

Examples: *London BL*, 1296.b.25 (*Geographia Totius Orbis Compendiaria* of 1692); *London, Hubbard Collection*.

155.2 Second State 1702
The addition above the map, just after the 'I' of '*b*' in italics as follows : 'I.*b*'.

Example: Müller's *Neu aussgefertigter kleiner atlas* of 1702: *London, Burden Collection*;

The map is also known with no text below the map, but with the 'I.b' above the map.

Example: Müller's undated *Atlas Minor sive Orbis Terrae*: *München SB*, MAPP 5 (with the 'I.b.' above the map, but no text below the map).

Description

Johann Ulrich Müller (1633 - 1715) was a geographer and a cartographer working in Ulm and Augsburg in the late seventeenth and early eighteenth century. This small map of Africa appears at the top of a page of text in a geography book by Müller, *Kurtzbündige Abbild...der Gantzen Welt...*, published in Ulm in 1692. This map may have been engraved by G. Karsch as his imprint as the engraver appears on other maps within the atlas in the *Neu aussgefertigter kleiner Atlas*. Gabriel C. Bodenehr engraved the book's title page as his imprint (G.C. Bodenehr Sc. Aug.) appears on that page in the *Geographia Totius Orbis Compendiaria*.

The Africa map from Johann Hoffmann's 1678 German edition of Pierre Duval's *La Geographie Universelle*, first published in c.1660, appears to be the primary source for this map.

Considering its small size, the map contains a good deal of geographic information. Besides identifying the various regions of Africa, it shows a few important places: the mine at *la Mina* in Guinea; the capital of *Monomotapa* by the same name of *Monomotapa* in Southern Africa; *Amara* in Abyssinia, etc.

Publication Information

This map of Africa is one of 103 maps that appeared in Müller's *Geographia Totius Orbis Compendiaria*, and also Müller's *Kurtzbündige Abbild...Der Gantzen Welt...*, both published by Georg Wilhelm Kühnen in Ulm in 1692. It also appeared in 1702 in an enlarged, two-volume edition of Müller's atlas entitled *Neu aussgefertigter kleiner atlas*, published by Kühnen in Ulm and by Johann Andreae in Frankfurt. This atlas contained 163 maps.

This map also appeared in Müller's *Atlas Minor sive Orbis Terrae*, an undated atlas, likely published in c.1700 in Augsburg.

References

Sotheby's October 2006: 74; Shirley 2004; King 2003; Tooley 1999-2004, K-P: 294-295; Shirley 1993: 554.

156 1692 Alexis Hubert Jaillot – Pieter Mortier, Amsterdam

156.1 First State 1692

L'AFRIQUE | divisee suivant l'estendue de ses principales parties | ou sont distingués les vns des autres. | LES EMPIRES, MONARCHIES, ROYAUMES, ESTATS, ET PEUPLES, | qui partagent aujourd huy L'AFRIQUE | sur les Relations les plus Nouvelles | Par le S^r. SANSON, Geographe Ordinaire du Roy. | PRESENTEE | A MONSEIGNEUR LE DAUPHIN | Par son tres humble tres obeissant et tres fidele Serviteur | Hubert Iaillot (above the map on two lines of text): L'AFRIQUE DISTINGUÉE EN SES PRINCIPALES PARTIES, SÇAVOIR LA BARBARIE, LE BILEDULGERID, L'EGYPTE, LE SAARA ou LE DESERT, LE PAYS DES NEGRES, LA GUINEE, LA NUBIE, L'ABISSINIE, LE ZANGUEBAR, LE CONGO | LE MONOMOTAPA, LES CAFRES, LES ISLES DE CANARIES, DU CAP-VERD, DE S^T. THOMAS, L'ISLE DAUPHINE autrement MADAGASCAR OU SONT REMARQUÉS LES EMPIRES MONARCHIES ROYAUMES, ESTATS , et PEUPLES, QUI PARTAGENT PRESENTEMENT L'AFRIQUE; Tiré sur les Relations les pl.[us] nouuelles. Par le S^r. Sanson Geogr. ord^{re}., du Roy. 1692
(at the bottom of the scale bar): A PARIS | Chez H. IAILLOT joignant les grands Augustins, aux 2 Globes | Auec Priuilege du Roy pour 20 Ans. | 1692
- Scale bar set within drapery cartouche at upper left of the map:
  Eschelle
  Milles Pas Geometriques ou Milles d'Italie: 600 = 6.9 cm.
  Lieues Communes de France: 250 = 6.9 cm.
  Lieues Communes d'Allemagne: 150 = 6.9 cm.

- Lieues Communes d'Espagne: 178 = 6.9 cm.
- Lieues d'vne Heure de Chemin: 200 = 6.9 cm.
- Dietes ou Iournees de M.P.G. Chacune: 15 = 6.9 cm.
- [Amsterdam : Pieter Mortier], 1692.
- From: separately published / *Atlas Nouveau*.
- 1 map : copperplate engraving ; 2 map sheets ; 54.5 x 87.5 cm (not including text above the map).
- Latitude and longitude coordinates on all four sides of the map. Longitude coordinates across Equatorial Line.
- North at the top; no compass rose.
- Title within cartouche at the upper right of the map.
- Dedication within the title cartouche is to the Dauphin, the son of King Louis XIV, with his coat-of-arms at the top center.
- No decorative elements on the map except for a large, decorative title cartouche at the top right and a scale bar at top left on a drapery cartouche.

Examples: 1692 edition: *Amsterdam UB*, V 9 X 14,15 (8); *Cambridge Harv MC*, 2375.1692* (1692 as a separate map); *København KB*; *London BL*, 1.TAB.1, 2 (7) + (117).

156.2 Second State 1696

Map date changed from 1692 at bottom of scale bar <u>and</u> at end of letterpress title above the map to 1696. Appeared in edition of 1696.

Examples: 1696 edition: *Amsterdam UB*; *Leiden UB*, COLLBN Atlas 15 (8); *Washington LC*, G 1015.S2 1696.

156.3 Third State after 1696

Date of 1696 removed from the map with some evidence of the erasure of 1696 remaining.

A third edition of the Pieter Mortier copy of the *Atlas Nouveau* was published in Amsterdam with no date on the title page of the atlas or on the map of Africa. A copy of this undated Mortier atlas with this map of Africa is at the Royal Geographical Society. This example has evidence of an erasure of the date of 1696. This edition was issued by Mortier some time after 1696.

Example: *London RGS*, MR 14.C.132.

156.4 Fourth State 1696-1708

At bottom of scale bar: 'A Amsterdam | chez Pierre Mortier & Compagnie | avec privilege'.

Based on Blonk's analysis in *Hollandia Comitatus* (Blonk: 2000), a fourth edition was published without a date on the title page and no date on the map of Africa. However the map of Africa in this edition does have Mortier's name and address as: 'A Amsterdam | chez Pierre Mortier & Compagnie | avec privilege' (also see Pastoureau 1984: 257: no. 4). This edition was likely published between 1696 and 1708.

Examples: *London BL*, 150.e.2.(8); *Paris BN*, Ge CC 1007-1008.

156.5 Fifth State after 1721

At bottom of scale bar: imprint of Johannes Covens & Cornelis Mortier.

There is a final, also undated, copy at the Royal Geographical Society with the imprint of Johannes Covens & Cornelis Mortier (Pieter's son). As Covens and Mortier entered into partnership in 1721 upon the marriage of Cornelis Mortier and Coven's sister, this map is after that date.

Example: *London RGS*, MR 14.C.131.

See Pastoureau (1984: 246-262) for further information on the Amsterdam 'counterfeit' copies.

Description

Prepared by Pieter (Pierre) Mortier, this map is a close copy of the Alexis Hubert Jaillot map of Africa of 1674. It is often attributed to Jaillot and is thought to be later states of Jaillot's map of 1674, although there is no indication that Jaillot was involved in this map's publication.

This Jaillot-Mortier map of Africa is distinguished from the earlier Jaillot map of Africa by the use of the dates of 1692 or 1696 at the end of the title and at the bottom of the scale bar, and by the omission of the name of Jaillot's engraver, Cordier (The Jaillot map had the imprint of 'Cordier Sculp' at the bottom right corner). There are also minor changes to the title cartouche in the Jaillot-Mortier example. Above 'PRESENTEE', there is no accent mark, whereas in Jaillot's 1672 map, there is an accent mark (PRESENTÉE). Another change includes the omission of some diacritical marks within the title above the map (See Map # 118 to closely compare the titles.) A comparison of the engraving of the two title cartouches reveals that Mortier's copyist had a more refined style which is especially apparent on the faces of the figures.

Mortier (1661-1711) was the founder of a large Amsterdam book and map publishing house which included his sons Cornelis and Pieter II. Although the atlas in which this map appears has a publication location of Paris and the imprint of Alexis Hubert Jaillot as the publisher, the atlas was actually published in Amsterdam in 1692 by Pieter Mortier and Marc Huguetan. It has no mention of Mortier or Huguetan, or the location of Amsterdam. It is unclear if this edition had Jaillot's approval but, as Koeman mentions (1967-71, III: 8), this copying of Jaillot's name and publication location of Paris was obviously done to promote the sale of Mortier and Huguetan's copy of the popular Sanson atlas, as published by Jaillot. By the late seventeenth century, the French held a dominant position in the map publishing business with Dutch and English publishers freely copying French prototypes, with and without approval.

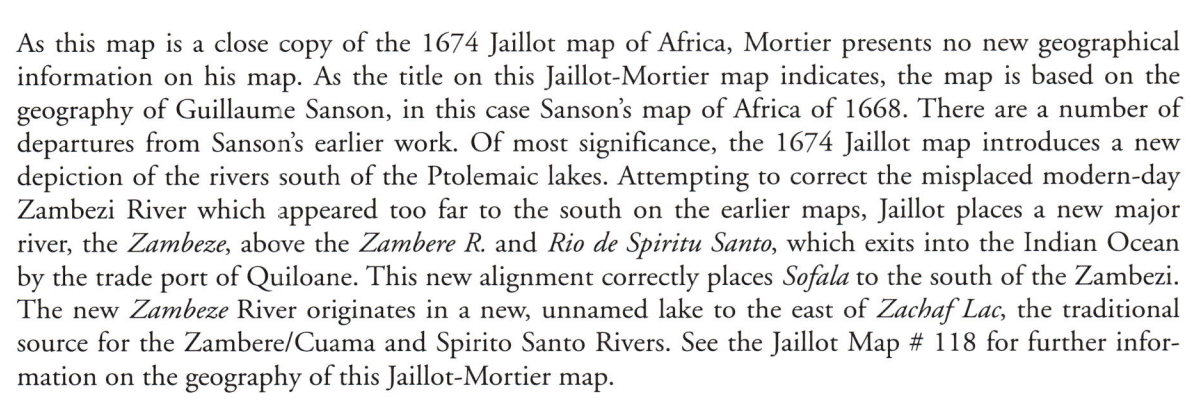

Map 156.1 Scale Bar.

As this map is a close copy of the 1674 Jaillot map of Africa, Mortier presents no new geographical information on his map. As the title on this Jaillot-Mortier map indicates, the map is based on the geography of Guillaume Sanson, in this case Sanson's map of Africa of 1668. There are a number of departures from Sanson's earlier work. Of most significance, the 1674 Jaillot map introduces a new depiction of the rivers south of the Ptolemaic lakes. Attempting to correct the misplaced modern-day Zambezi River which appeared too far to the south on the earlier maps, Jaillot places a new major river, the *Zambeze*, above the *Zambere R.* and *Rio de Spiritu Santo*, which exits into the Indian Ocean by the trade port of Quiloane. This new alignment correctly places *Sofala* to the south of the Zambezi. The new *Zambeze* River originates in a new, unnamed lake to the east of *Zachaf Lac*, the traditional source for the Zambere/Cuama and Spirito Santo Rivers. See the Jaillot Map # 118 for further information on the geography of this Jaillot-Mortier map.

Publication Information

This map appeared in Amsterdam in Pieter Mortier and Marc Huguetan's edition of Jaillot's *Atlas Nouveau* as discussed above. The map is in five states.

References

The author is grateful to Marco van Egmond, Universiteit Utrecht, who wrote his doctoral dissertation on Covens and Mortier and who kindly shared his information on Pieter Mortier.
Shirley 2004: 579-582 and 717-724 (for Mortier); Norwich 1997: 56-57; Pastoureau 1984: 229-262 (235 for map no. 4); Koeman 1967-71, III: 4-10.

157 1693 Nicolas de Fer, Paris

157.1 First State 1693

L'AFRIQUE | Dreßez Sur les dernieres | Relations par N. de Fer | Avec Privilege du Roy | 1693. | A Paris chez l'Auteur dans | l'Isle du Palais á la Sphere | Royale
(within a small box at the bottom right of the map): C. Inselin Sculp.

- Scale bar at bottom left: Echelle | Dix Degrez de l'Equateur qui font | Deux Cents Lieues d'une heures | de Chemin 10 Degrez = 2.3 cm.
- Paris : Nicolas de Fer, 1693.
- From: *Petit et Nouveau Atlas*.
- 1 map : copperplate engraving ; 21 x 27 cm.
- Latitude and longitude coordinates on all four sides of the map. Longitude coordinates across Equatorial Line. Line showing the Prime Meridian (*Ligne du Premier Meridien des Francois*) to the west of *Isle de Fer* in the Canary Islands.
- North at the top; no compass rose.
- Title cartouche at the top right corner with a simple drapery rope for decoration.
- No decorative elements on the map other than the title cartouche.

Example: Edition of *Petit et Nouveau Atlas* of 1697: *Washington LC*, G1015.F45 1697 (Phillips no. 5955).

157.2 Second State 1705
Date on title cartouche changed from 1693 to 1705.

Examples: Edition of *Petit et Nouveau Atlas* of 1705: *Washington LC*, G1015.F45 1705 (Phillips # 547); *London BL*, C.39.b.2.

Map 157.2.

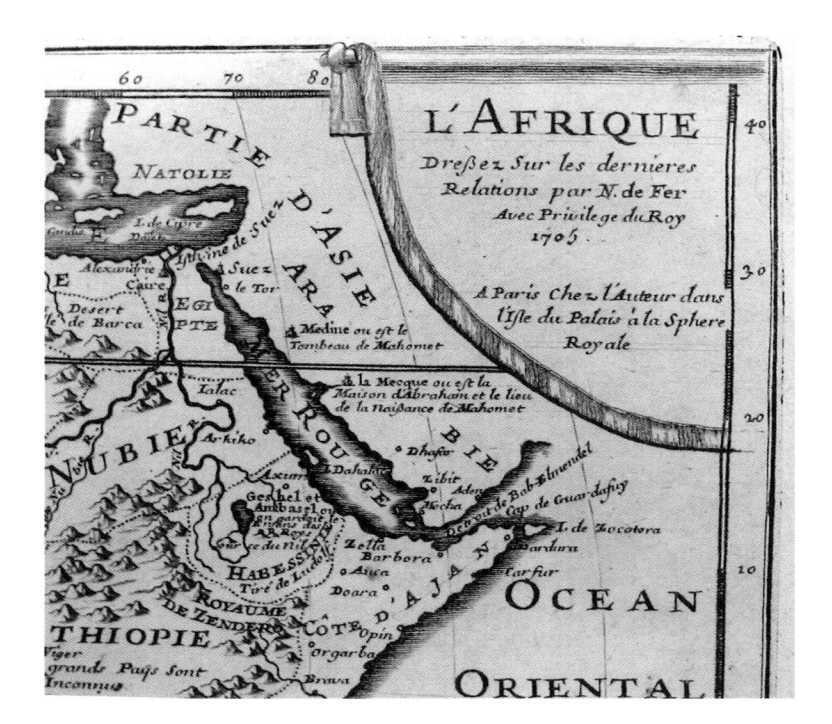

Description

Nicolas de Fer (1646-1720) is named on the title as the mapmaker and publisher of this map of Africa. It follows his earlier work of c.1684, although this map is considerably updated. While De Fer was not able to issue an atlas in c.1684 due to financial reasons and evidently sold his copperplates to Jacques Robbe to use in his *Methode Pour Apprendre Facilement la Geographie*, he always intended to publish his own quarto-size atlas. With his fortune improving in the 1690s after he was appointed 'Géographe du grand Dauphin', De Fer began work on this map of Africa and 18 other maps (the world, the continents, and regions of Europe) for his own atlas.

This map follows the Nolin-Coronelli 1689 map in its depiction of Africa. A single, detached Ptolemaic *Lake Zaire* is in Central Africa with no river connected to it. A third river, the *Zambeze*, is directly below, with its small unnamed source lake. Lake Tana as a Nile River source in Abyssinia is correctly presented. The interior of Africa is generally devoid of towns, as De Fer likely only included those locations where he felt he had sufficient information.

Publication Information

While the map of Africa was prepared with a date on it of 1693, it appears that the first edition of De Fer's *Petit et Nouveau Atlas* did not appear until 1697. Other maps in the quarto, oblong-shaped atlas have dates of 1693, 1695 and 1697. It seems that De Fer began plans for his atlas in at least 1693, but other projects or financial considerations kept him from completing his work on this atlas until 1697.

The map is known in two states. There was one further edition of the *Petit et Nouveau Atlas* (*Paris BN*, Ge DD 1232) published by Guillaume Danet, De Fer's son-in-law and successor, in 1723. This edition, according to Pastoureau (1984: 217), has a map of Africa that is identical to the second state of 1705 described above.

References

Pastoureau 1984: 216-217; Tooley 1969: 45.

158 1694 Alexis Hubert Jaillot – Pieter Mortier, Amsterdam

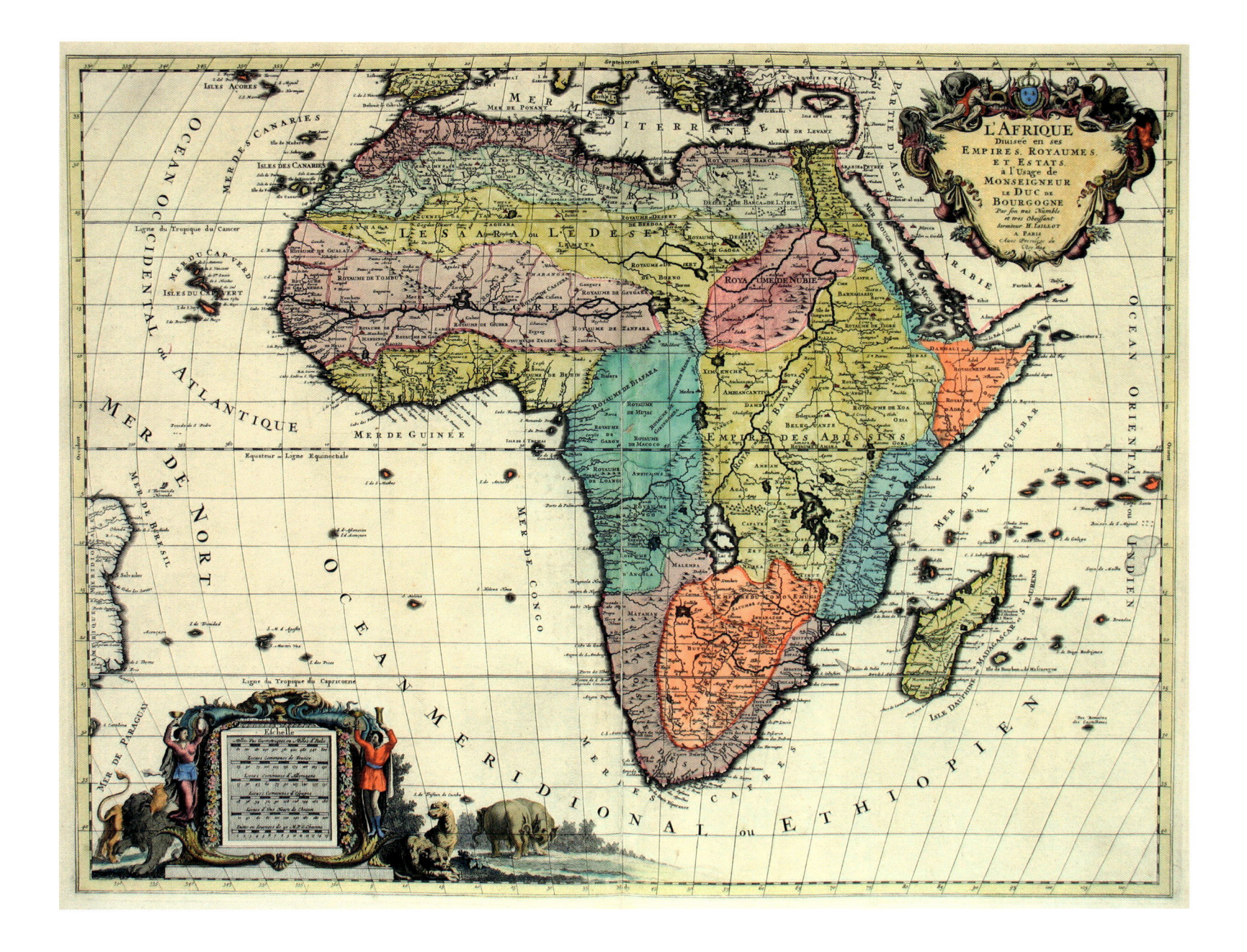

158.1 First State 1694

L' AFRIQUE | Diuiseé en ses | EMPIRES, ROYAUMES, | ET ETATS, | à l'Usage de | MONSEIGNEUR | le DUC de | BOURGOGNE | Par son tres Humble | et tres Obeissant | Serviteur H. JAILLOT | A PARIS | Avec Privilege du | Roy 1694.
[on one line above the map image (not on above illustration)]:
AFRICA ACCURATÈ IN IMPERIA, REGNA, STATUS & POPULOS DIVISA, AD USUM SERENISSIMI BURGUNDIÆ DUCIS.

- Scale bar at lower left:

Eschelle	
Milles Pas Geometriques ou Milles d'Italie	600 = 5.2 cm.
Lieues Communes de France	250 = 5.2 cm.
Lieues Communes d'Allemagne	150 = 5.2 cm.
Lieues Communes d'Espagne	180 = 5.2 cm.
Lieues d'Vne Heure de Chemin	200 = 5.2 cm.
Dietes ou Iournees de 40 M.P.G. Chacune	15 = 5.2 cm.

- [Amsterdam : Pierre Mortier], 1694.
- From: *Atlas Royal a l'Usage de Monsieur le Duc de Bourgogne...* or *Atlas François ... A Paris, Chez Nicolas de Fer. M.DC.XCV.* With a second title page of *Introduction a la Geographie... A Amsterdam, Chez Pierre Mortier...*
- 1 map : copperplate engraving ; 46 x 59 cm.
- Latitude and longitude coordinates on all four sides of the map. Longitude coordinates across Equatorial Line.
- North at the top; no compass rose.
- Elaborate and decorative title cartouche at upper right with a dedication to 'le Duc de Bourgogne'.
- Decorative scene surrounding scale bar with two figures, elephants, and lions.

Examples: 1695 edition: *London BL*, Maps C.39.f.12; 1699 edition: *Cambridge Harv MC*, MA 17.99.pf. From Pastoureau: 1697 edition: *Nantes BM*, 33896; 1699 edition [with some maps dated to 1702]: *Amsterdam UB*, 363 A 22 (85); *Washington LC*, Philipps-Le Gear entry 5956).

158.2 Second State 1721
At the bottom of the title cartouche, imprint and location of publisher changed to 'A Amsterdam | Chez Covens & Mortier'.

Example: Information from Marco van Egmond, Universiteit Utrecht.

158.3 Third State c.1730
At the bottom of the title cartouche, imprint and location of publisher changed to 'A Amsterdam | Chez R & J Ottens.'.

Example: As reported by Tooley 1969: 56 (map image # 42).

158.4 Fourth State 1792
At the bottom of the title cartouche, imprint and location of publisher with a new date changed to 'A Amsterdam | Chez I. B. Elwe 1792.'.

Example: As reported by Tooley 1969: 57.

Description
Often attributed to Alexis Hubert Jaillot, this map was actually prepared and published by Pieter (Pierre) Mortier in Amsterdam. Mortier (1661-1711) was the founder of a large Amsterdam book and map publishing house, which included his sons' Cornelis and Pieter II.

The Jaillot 1674 map of Africa is the model for this map. (See Map # 118). The title and dedicatory cartouche at the upper right has been copied in reverse with a new coat-of-arms for the Duke of Burgundy. A new cartouche appears at the bottom left with the scale bar, replacing the earlier scale bar which was at the upper left.

Even though some mapmakers in France, such as Nolin and De Fer, were beginning to question the placement of the source lakes for the Nile River, this map, not wishing to break with tradition, depicts the two Ptolemaic source lakes for the Nile River in Central Africa. In East Africa, Lake Tana, the source for the Blue Nile in the Abyssinian highlands, is not shown. The *Nubia* River above the Niger is still shown flowing eastward into the Nile. In Southern Africa, the *Zambeze* (Zambezi) River is placed correctly, though the *Zambere* River is still shown to its south.

Most of the maps in Mortier's atlas bear the imprint of Alexis Hubert Jaillot, though all of these maps, including this map of Africa, are copies of Jaillot's maps. Three maps in the atlas have the engraver's imprint of 'F. Goeree delin', though it is not certain if Goeree engraved this map of Africa as well. The British Library example of the atlas from 1695 contains an engraved frontispiece titled *Atlas Minor...Atlas François* that is signed by 'R. de Hooghe'. Romein de Hooghe was a noted Dutch engraver and artist who is known to have also worked for Mortier on other works.

Publication Information

Although the atlas in which this map appears has a publication location of Paris and the imprint of Nicolas de Fer as the publisher, the atlas was actually published in Amsterdam by Pieter Mortier in 1695 (Pastoureau 1984: 189-190). Mortier published this atlas without De Fer's approval and, in fact, had no relationship with De Fer. The use, even if improper, of the well-known De Fer name and the location of Paris were obvious incentives to sell this atlas. By this time, the French held a dominant position in the map publishing business with Dutch and English publishers freely copying French prototypes, with and without approval. Besides the 1695 publication date, there were editions of the Mortier atlas with publication dates of 1697, 1698, and 1699. There were later editions by Covens and Mortier (the son), Ottens, and finally Elwe in 1792. This map is known in four states.

Reference

The author is grateful to Marco van Egmond, Universiteit Utrecht, who wrote his doctoral dissertation on Covens and Mortier and who kindly shared his information on Pieter Mortier.

Shirley 2004: 728; Pastoureau 1984: 189-193: FER III A (1695), FER IIIB (1697), and FER III C (1699); Tooley 1969: 56-57: map # 42 (third state of this map by R & J Ottens).

159 1695 Herman Moll, London

AFRICA.
- No scale bar.
- [London : Abel Swall and Tim Child, 1695].
- From: *Thesaurus Geographicus. A new Body of Geography... Printed for Abel Swall and Tim Child...1695.*
- 1 map : copperplate engraving ; 14.5 x 19.5 cm.
- Latitude and longitude coordinates on all four sides of the map.
- North at the top; no compass rose.
- Title above the map gridline.
- No decorative elements on the map.

Examples: 1695 edition: *London BL*, 10003.e.4; 1701 edition: *London BL*, 10005.h.11. *The Compleat Geographer* 1709: *London BL*, 568.i.9; and 1723: *London BL*, 10003.f.4; *Atlas Manuele* of 1709: *London BL*, Maps.43.b.3.

Description

This is the second map of Africa attributed to Herman Moll, following the Moore-Moll map of Africa of 1681. Moll (c.1654-1732) was a geographer, mapmaker, and an engraver originally from Germany. It is thought that Moll emigrated to England in 1678 via The Netherlands. He was initially employed as an engraver in London, but, starting with this atlas, became much more involved in atlas production.

Moll conservatively follows the traditional and, by this time, quite dated Blaeu 1617 folio map as his model for this map of Africa. As Moll was still establishing himself in London in the cartographic business, it is possible that he felt the wiser course was to use the popular Blaeu map, instead of the newer, more innovative information then available through the French.

This 1695 atlas does not mention Herman Moll's name on the title page as the author of the maps. However, one map, Germany, is actually signed by Moll (as 'H. Moll Fecit'). In the edition of 1701, credit is given to Moll on the title page, clearly reflecting Moll's growing prominence as a cartographer.

Publication Information

The Africa map appears in Book III of the 1695 atlas *Thesaurus Geographicus*. It is set within a page of text with the map title engraved into the plate above the map image. This map was also used in a further edition of Moll's *A System of Geography...* in 1701, and in *The Compleat Geographer...* in 1709, 1719, and 1723. The same series of maps were also used in Moll's *Atlas Manuele* of 1709, 1713, and 1723.

The map is unchanged in all editions. There are differences in the text and pagination within the editions. Among the editions located and examined by the author, the differences are as follows:

1695 edition: On the Africa map, at top right of page in letterpress: '447'.
1701 edition: On the Africa map, at top right of page in letterpress: '99'.

References

Shirley 2004: entry T-MOLL-1a to 1e; Reinhartz 1997; Shirley 1993: 566; Tyacke 1978: 122-123. Information on Moll's publication record from Philip Burden.

160 169[5] Alexis Hubert Jaillot, Paris

160.1 First State 169[5]

L' AFRIQUE | divisée suivant l'estendüe de ses Principales Parties, | ou sont distingués les vns des autres | LES EMPIRES, MONARCHIES, ROYAUMES, ESTATS, ET PEUPLES. | qui partagent aujourd'huy L'AFRIQUE. | sur les Relations les plus Nouuelles: | Par le Sr SANSON, Geographe Ordinaire du Roy. | DEDIÉ AU ROY | Par son tres humble tres obeissant tres-fidele Sujet et Seruiteur, | HUBERT IAILLOT. | 169[5].

[at the top left of the map, set within a cartouche drapery screen, immediately below the scale bar]:
A PARIS, Chez H. IAILLOT | joignant les Grands Augustins, aux 2. Globes | Auec Priuilege du Roy.
[at the bottom right corner of the map):
L. Cordier Sculp.

- Scale Bar at upper left corner within a cartouche drapery screen:
  Eschelle
  Milles Pas Geometriques, ou Milles d'Italie 480 = 4.5 cm.
  Lieues Communes d'Allemagne 120 = 4.5 cm.
  Lieues Communés d'Espagne 144 = 4.5 cm.
  Lieues d'Vneheure de Chemin 160 = 4.5 cm.

- Dietes, ou Iourneés de 40 M.P.G. Chacuns 12 = 4.5 cm.
- Paris : Alexis Hubert Jaillot, 169[5].
- From: *Atlas François* (in the London Royal Geographical Society example, the map is in the first volume of the two volume atlas).
- 1 map : copperplate engraving ; 46 x 64.5 cm.
- Latitude and longitude coordinates on all four sides of the map. Longitude coordinates across Equatorial Line.
- North at the top; no compass rose.
- Title at top right of the map.
- No decorative elements on the map apart from the title cartouche and the scale bar.
- Manuscript date of '5' as part of 169[5] at the bottom of the title cartouche.

Examples: *Cambridge Harv MC*, MA 17.95.2. pf (with all of 1695 in manuscript; Africa is map no. 8 in the atlas); *London RGS*,1.B.4 [plate 8]; *Paris BN*, Ge D 12324.

The example of this map within the *Atlas François* at the Harvard Map Collection has all four digits of 1695 in manuscript. It is likely that this example of the Africa map is a proof copy, before the first three digits were printed.

Pastoureau (1984: 283, map no. 4) lists an edition of the *Atlas François* of 1700 (*Lyon BM*: 7006) with a manuscript '9' added in place of the '5' for a new date of 169[9].

160.2 Second State 1708
Date changed to 1708, as printed on the cartouche.

Example: *Dijon BM*, 24559 (from Pastoureau 1984: 280).

160.3 Third State 1719
Date changed to 1719 as printed on the cartouche on a line above 'DEDIÉ AU ROY'. Also, there are significant geographical changes on the map reflecting the rapid advances made by the French during this period. This is especially apparent with an absence of the two Nile source lakes in Central Africa and the sole source of the Nile as Lake Tana in the Abyssinian highlands.

Examples: *Cambridge Harv MC*, MA 18.00.5 pf* (title page for this atlas has a date of 1700); *Grenoble BM*, H.84 (from Pastoureau 1984: 278).

Map 160 Title cartouches of state 1 and 3.

Description
This is Alexis Hubert Jaillot's second map of Africa, prepared for his *Atlas François*. Jaillot likely had this map engraved to better fit into his *Atlas François* since his earlier map of 1674 was simply too large and, by that time, somewhat out-of-date. A separate two pages of 'Tables Geographiques des DIVI-

SIONS DE L'AFRIQUE Par Sr Sanson...' accompanies the map.

This map generally follows the Sanson – Jaillot model. Please refer to Jaillot, Map # 118, for a discussion of the geography of this map. The engraver for this map was Louis Cordier whose imprint also appears on Jaillot's map of 1674 as 'L. Cordier Sculp' at the bottom right corner.

As with the world map in the *Atlas François* (Shirley 1993: 565), the last digit of '5' in 169[5] is added in manuscript. It appears that since Jaillot wanted to have this map appear up-to-date and since he did not know when the atlas would be published, he printed the map first and had the final digit added in manuscript at the last moment. The title page for the atlas has a printed date of 1695.

Publication Information
After 1695, there were later editions of the *Atlas François* with this map of Africa in 1696, 1698, and 1700. It also appears in the specially assembled *Atlas François* with maps dated 1708 and 1719. The map is known in three states, with information on the second and third states partially based on Pastoureau:

References
Shirley 1993: 565, with map image no. 569; Pastoureau 1984: 262-292 (map entry # 4 on p. 264).

161 c. 1696 Gerard Valk, Amsterdam

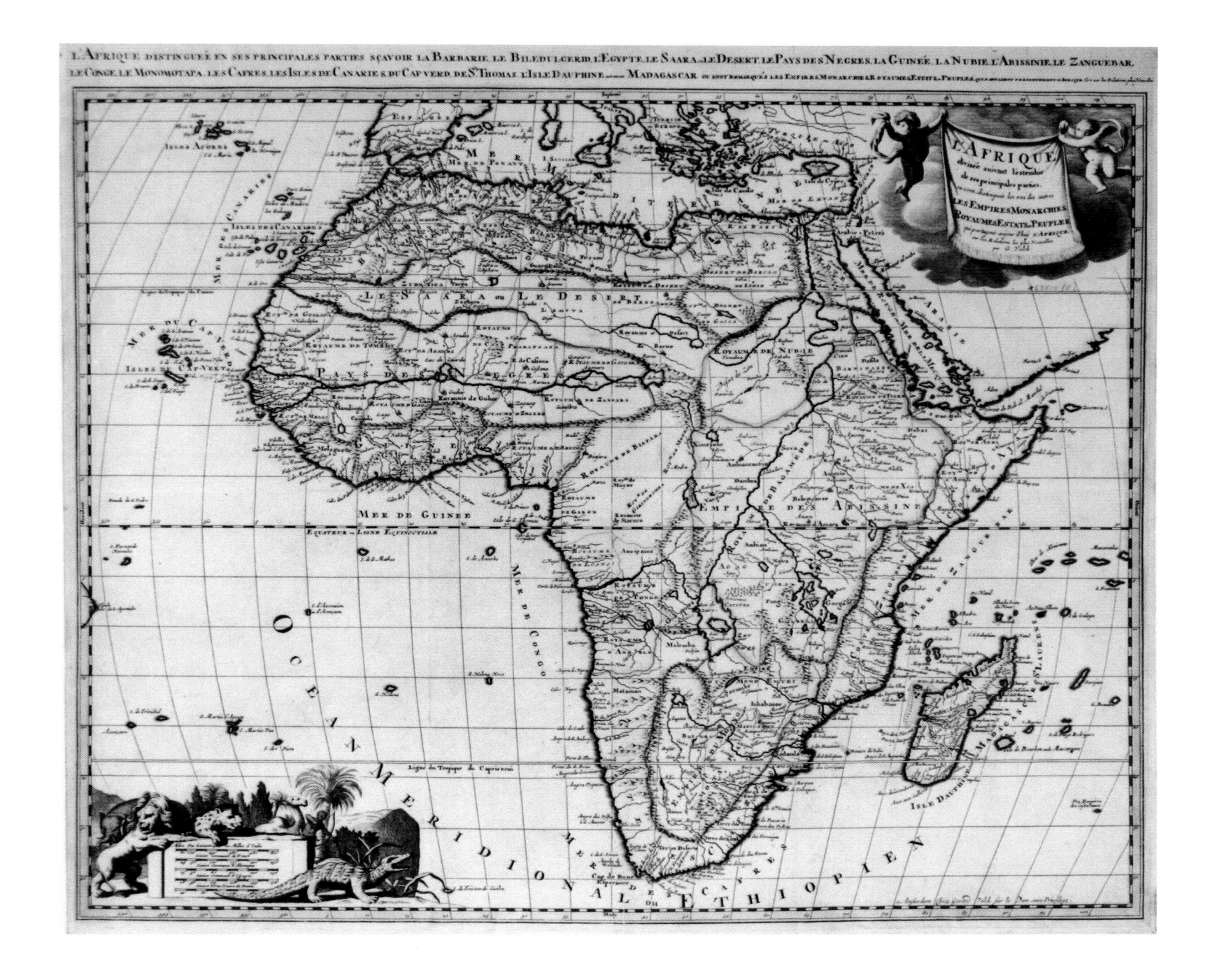

L'AFRIQUE | diviseé suivant l'estendıe | de ses principales parties. | ou sont. distingués les vns des autres | LES EMPIRES, MONARCHIES, | ROYAUMES, ESTATS, et PEUPLES, | qui partagent aujour d'huÿ L'AFRIQUE. | sur les Relations les plus Nouuelles | par G. Valck.
[at bottom right corner of the map]:
A Amsterdam Chez Gerard Valck sur le Dam, auec Privilege.
[above the map on two lines of text]:
L'AFRIQUE DISTINGUEÉ EN SES PRINCIPALES PARTIES SÇAVOIR LA BARBARIE, LE BILEDUL-GERID, L'EGYPTE, LE SAARA ou LE DESERT, LE PAYS DES NEGRES, LA GUINEÉ, LA NUBIE, L'ABISSINIE, LE ZANGUEBAR, | LE CONGE, LE MONOMOTAPA, LES CAFRES, LES ISLES DE CANARIES, DU CAP VERD, DE S^T.THOMAS, L'ISLE DAUPHINE autrement MADAGASCAR OU SONT REMARQUÉS LES EMPIRES, MONARCHIES, ROYAUMES, ESTATS , et PEUPLES, QUI PARTA-GENT PRESENTEMENT L'AFRIQUE. tire sur les Relations plus Nouuelles.

- A scale bar is set within a decorative vignette with African animals at the bottom left of the map:
  Milles Pas Geometriques ou Milles d'Italie 600 = 5.5 cm.
  Lieues Communes de France 250 = 5.5 cm.
  Lieues Communes d'Allemagne 150 = 5.5 cm.
  Lieues Communes d'Espagne 178 = 5.5 cm.
  Lieues d'vne Heure de Chemin 200 = 5.5 cm.
- Amsterdam : Gerard Valk, [c.1696].
- From: separately published.
- 1 map : copperplate engraving ; 46.5 x 57.5 cm (not including two lines of text above the map impression).
- Latitude and longitude coordinates on all four sides of the map. Longitude coordinates across Equatorial Line.
- North at the top; no compass rose.
- Title at top right, on drapery held by two cherubs.
- No decorative elements on the map other than the title cartouche and the scale bar vignette.

Examples: *London BL*, *63510 (5); *Washington LC.*

Description

Gerard Valk issued this map to appeal to the increasing demands of the French market for up-to-date maps of Africa with the text in French. The map appears to be a close copy of Alexis Hubert Jaillot's map of 1674 in terms of geographic content, and the wording and design of the map. Please refer to Jaillot, Map # 118, for a discussion of the geography of this map.

Gerard Valk (1652–1726) produced a prodigious amount of cartographic material during his professional career and was a noted Amsterdam publisher, engraver, and globemaker. See Map # 148 for further information on Valk and his relationship with Schenk.

For a time, Valk worked in London. Upon his return from London in the late 1670s, Valk collaborated with Pieter Schenk, a map engraver and seller from Germany.

It appears that Valk (and Schenk) was granted a privilege to produce maps copied from Sanson (the Jaillot maps) in late 1695 (Koeman 1967-71, III: 109). A modification to this privilege was issued in December 1696 which was intended to protect the interests of Pieter Mortier who already had authority to produce similar maps. The modification stated that Valk (and Schenk) had permission to sell (certain maps) because they had already been finished. These included the world map, continent maps (including this map of Africa), and a few other regional maps.

Publication Information

According to Koeman (1967-71, III: 136-140), Valk's *Atlantis sylloge Compendiosa*, a specially-assembled, non-standardized atlas, was not published until c.1702, but Valk was active from about 1683 through the 1690s with separately-published maps, such as this one. Although the map was initially prepared to be sold separately, it does appear in Valk's atlas along with his newer map of Africa of c.1700 (see below).

The map is undated, but as Valk was issued his privilege in 1695 for his version of the Sanson (Jaillot) maps and noting the publication location of 'sur le Dam' (on the Dam), which he moved to from a house on the Kalverstraat near the Dam sometime before 1700, the author has dated this map as c.1696. The map is only known in one state.

Around 1700 in preparation for the issuance of his atlas, *Atlantis sylloge Compendiosa*, Gerard Valk prepared a newer map of Africa with details as follows: 'Africa' with the imprint of 'Gedruckt Gerard Valk tot op Dam in den Wackeren hont. Cum prov.' (two map sheets; 57.5 x 95 cm in total size). The decorative elements include allegorical representations, scenes with inhabitants, and an inset on the calculation and construction of a geographical grid (Information from Koeman 1967-71, III: 137).

Also around 1700, Pieter Schenk published his own atlas *Atlas Contractus,* a specially-assembled, non-standardized atlas with the number and composition of maps differing in the various copies known.

For his atlas, he prepared a new map of Africa with details as follows: 'AFRICA, ELABORATISSIMA. P. Schenk ex: Amst: cum: Privil:' (47 x 55.5 cm) (Information from Koeman: 1967-71, III: 119).

Finally, there was a later map of Africa by Gerard and his son Leonard Valk in about 1720 or possibly slightly earlier with details as follows: 'AFRICA | —- Maurô | Percussa Oceanô, | Niloque admota tepenti. | Auctoribus | GERARDO ET LEONARDO VALK | Cum Privilegio Ordinum | Hollandiæ et Westfrisiæ.' (49 x 60 cm) (*København KB*, Atlas major III, 70; *Author's collection*).

References
Norwich 1997: map # 49 (for Valk's copy of the Jaillot map), and map # 63 (for Schenk's map); Shirley 1993: 529; Koeman 1967-71, III: 107-121 (for Schenk) and 136-140 (for Valk).

162 c. 1696 Carel Allard, Amsterdam

NOVISSIMA | et | PERFECTISSIMA | AFRICÆ | DESCRIPTIO | EX FORMIS | CAROLI ALLARD. | AMSTELO – BATAVI: | Cum Privilegio Potentissimerum D.D. | Ordinum Hollandiæ et Westfrisiæ.
[at the bottom of the vignette, to the left]:
Ph. Tideman del.
[at the bottom of the vignette, to the right]:
G. v. Gouwen sculp.
- No scale bar.
- Amsterdam : Carel Allard, [c.1696].
- From: separately published / composite atlas.
- 1 map : copperplate engraving ; 50 x 58.5 cm.
- Latitude and longitude coordinates on all four sides of the map. Within the latitude gridlines on the left and right sides of the map is information on climates.
- North at the top; one compass rose in the Atlantic Ocean.
- Title within monument-shaped cartouche surrounded by decorative vignette is at the bottom left of the map.
- No other decorative elements on the map besides the elaborate vignette surrounding the title cartouche, with a woman as an allegorical representation of Africa, along with several African children and a lion, an ostrich, and a crocodile.

Examples: *Amsterdam UB*, 33-17-36 *and* 33-17-37; *Leiden UB*, COLLBN Port 179 N 30 (dated 1705, based on a De Wit map of Hispania in the atlas with that date); *Washington LC*, G1015.A5

1696 (within an undated *Atlas Minor* of Carel Allard), also Phillips 3466 (Braakman, atlas minor 1706), Phillips 3472 (Visscher Atlas Minor 1710), Phillips 605 (Valck Nova Totius 1748, no Africa maps nos. 121-124), Phillips 3490 (Ottens Heirs Atlas 1740), Phillips 538 (Allard magnum theatrum belli 1702), Phillips 523 (Allard Atlas minor 1676); and Phillips 4257 (Ottens Atlas maior 1641-1729).

Description

Carel Allard (1648-1706), the son of the map publisher Hugo, was an engraver and publisher working from 'op den Dam in de Kaertwinkel' in Amsterdam. He initially worked for his father; during this time he produced maps with his own imprint. Upon his father's death in 1691, Carel assumed control of the business.

Though the Carel Allard maps are not dated, Koeman (1967-71, I: 31-48) suggests dates for these maps as from 1680 to 1700. He is known to have produced a world map dated before 1683, with a second state containing a privilege. He produced a second world map in 1696 (Shirley, 1993: 573). It is known that the States of Holland and West Friesland granted Carel Allard a privilege on October 16, 1683 (Koeman 1967-71, I: 31). This was intended to protect him from copyists for the maps that he intended to produce. In anticipation of the issuance of an atlas, which was finally published in 1697 (Koeman 1967-71, I: 39), Allard probably prepared this map of Africa in c.1696.

On first appearance, this map appears to be a copy of the c.1670 De Wit map of Africa in its later states, except for the lack of decorative animals and ships, and a new decorative vignette surrounding the title cartouche. This map even includes some of De Wit's lettering within the surrounding gridline. However, Allard's map considerably updates the geography of the interior of Africa. The Allard map shows each of the two Ptolemaic lakes in Central Africa with a river flowing to the north. However, Allard abruptly ends the flow of this river just north of the Equator, suggesting that the lakes were no longer considered to be the source for the White Nile River. Further, the source for the Blue Nile at Lake *Tana* in Abyssinia is well developed. All of this suggests French influences (Duval, De Fer and others) in the development of this map.

At the bottom of the vignette is the name of the person who drew the map, Philip Tideman, as: 'Ph. Tideman del.'. To the right, is the name of the map engraver, Gilliam van Gouwen, as: 'G. v. Gouwen sculp.'

Publication Information

Besides likely appearing as a separately issued map, this map appears in the *Atlas Minor sive tabulæ...* of Carel Allard in 1697 with later issues. It is believed that this map is only known in one state after a review of numerous examples, though the Americas map is known in four states. It is possible that Covens and Mortier acquired the copperplate for this map after Allard's death and added their own imprint. An example with this imprint has not been located.

Carel Allard was also known to reissue maps produced by the Blaeus, De Wit, and others, including a later state of a map of Africa originally prepared by Dancker Danckerts that Allard sold. Allard acquired a Dancker Danckerts copperplate shortly before c.1687 and published what was the third state of this map (Map # 99.3), with a possible fourth state.

References

Norwich 1997: map # 54; Shirley 1993: 517 and 573; Koeman 1967-71, I: 31-48, Al.1; Phillips 1909-1992: entry # 523, 538, 549, 605, 3490, 3466, 3472 & 4257.

163 1696-1698 Nicolas de Fer, Paris

163.1 First State 1698

L'AFRIQUE | Ou tous les Points Principaux | sont Placez,| SUR LES OBSERVATIONS DE | MESSIEURS DE L'ACADEMIE | ROYALE DES SIENCE | Par N: DE FER Geographe de | MONSEIGNEUR LE DAUPHIN. | A PARIS | Chez l'Autheur dans l'Isle du | du Palais à la Sphere | Royale. 1698.

[above the map on a text strip (with a date of 1696 on Reiss & Sohn example, or 1698 for other examples)]:

L'AFRIQUE, DIVISÉE SELON LETENDUE DE SES PRINCIPALES PARTIES, | ET DONT LES POINTES PRINCIPAUX SONT PLACEZ SUR LES OBSERVATIONS DE MESSIEURS DE L'ACADEMIE ROYALE DES SCIENCES. | Dressée Par N. DE FER, Geographe de MONSEIGNEUR LE DAUPHIN. | A PARIS, Chez l'Autheur dans l'Isle du Palais sur le Quay de l'Horloge a la Sphere Royale. avec Privilege du Roy. 1696.

[within a cartouche at the bottom corner of the bottom right map sheet, sheet 4]:

Dediée a | MONSEIGN^R LE DAUPHIN, | Par Son tres humble et tres Obeissant | Serviteur, et Geographe. | N. de Fer | Avec Privilege du Roy.

[at bottom right of sheet 4]:
H. Van Loon Fecit.

[on the map below the dedication]
Guerard juur fecit.

- Scale bar along the bottom of sheet 4:
  Echelle
  Lieues 200 = 10.0 cm.
  Degrez 10 = 10.0 cm.
  Dix Degrez du Latitude ou 200.Lieies d'une Heure de Chemin.
- Paris : Nicolas de Fer, 1696.
- From: separately published wall map.
- 1 wall map : copperplate engraving and etching ; 4 map sheets ; 92.5 x 118.5 cm (including border vignettes with map sheet 1 as 46 x 59.5 cm, sheet 2 as 46 x 59.5 cm, sheet 3 as 46 x 59.5 cm, and sheet 4 as 46 x 59.5 cm) ; 106 x 161 cm (including map sheets, letterpress text panels, and title strip across top).
- Latitude and longitude coordinates on all four sides of the map. Longitude coordinates across the Equatorial Line.
- North at the top; two compass roses in Indian Ocean and two compass roses in Atlantic Ocean.
- Title cartouche at the top center of the map. Title text strip above the map.
- There are 17 sailing ships in the Indian Ocean, 14 sailing ships and two large sea battles in the Atlantic Ocean, and two ships in the Mediterranean Sea. Vignettes on all four edges of the map with descriptive text panels describing these scenes and peoples of different regions of Africa. No animals within Africa.
- Outside of the map, there are letterpress text strips on the left, right, and bottom sides describing Africa in French. The text begins at the upper left with 'DESCRIPTION | DE L'AFRIQUE.' and ends at the bottom right with the imprint: 'A PARIS, | Chez l'Auteur, dans l'Isle du Palais, sur le Quay de l'Orloge, à la Sphere Royale. | M. D.C. XCVI. | AVEC PRIVILEGE DU ROY.'

Examples, with variations as noted:

Map 163.1.A. *Königstein, Reiss & Sohn Buch- und Kunstantiquariat Auktionen*, May 2003 auction catalog. Title strip at the top and the text panels at bottom right are dated 1696);

Map 163.1.B. *Paris BN*, Ge DD 2987 [7778]. Without title strip at top, and text panels on both sides and at the bottom;

Map 163.1.C. *Paris BN*, Ge CC.1423. Without title strip at the top, and text panels on both sides and at the bottom);

Map 163.1.D. *Author's collection*. Title strip at the top, and the text panels at bottom right dated 1698.

163.2 Second State 1705

Map date changed to 1705 on the title cartouche of the map.

Examples: *Karlsruhe LB*, Fer, N. de. C2 (1705 Africa wall map with title and text strips; both title

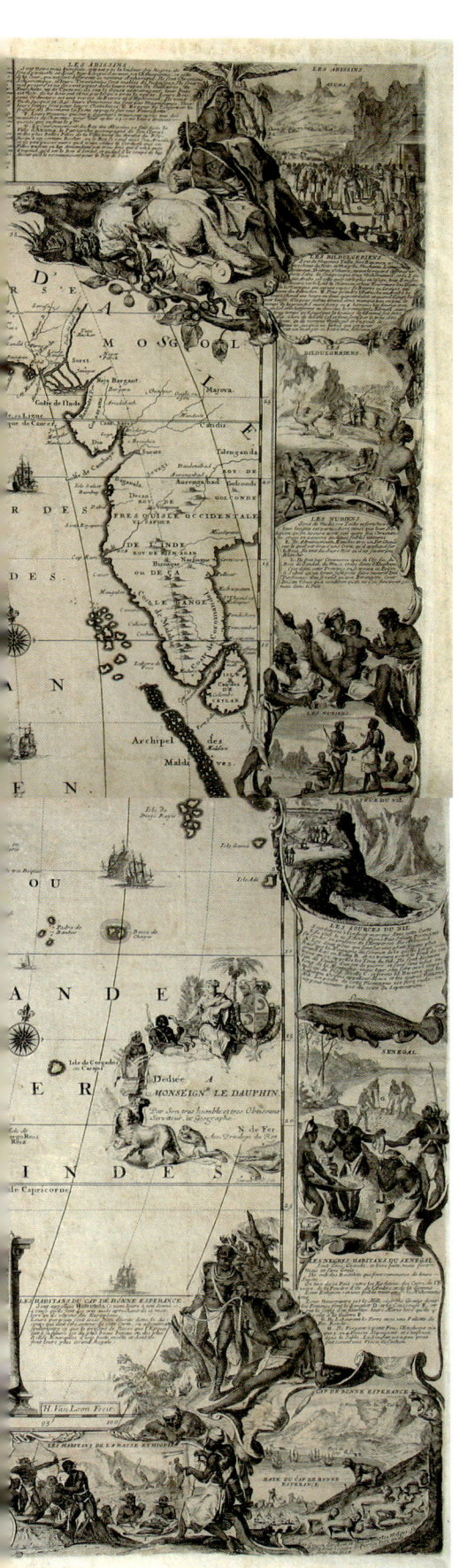

Map 163.1.C.

strip above the map and the title on the map have date of 1705); *London BL*, *Maps 63510.[267.] (for 1705 Africa wall map without the title and text strips).

163.3 Third State 1724/1728?
Map date changed to 1724/1728? on the title cartouche of the map.

Examples: *New York, Martayan Lan*, (catalog no. 26: 33: map no. 81 shows an issue of the Africa map, which they date as 1724/1728).

163.4 Fourth State 1730
Publisher's imprint and date on the map changed to: 'A Paris chez le Sr. Danet… sur le P. N. Dame a la Sphere Royale avec P. du Roy, 1730'.

Examples: *New York, Richard B. Arkway Rare Books Inc.* (catalog no. 52: map no. 6).

Map 163.1.D.

Description
This rare De Fer wall map has elaborate, highly decorative vignettes on all four edges of the map with separate, descriptive text panels describing these scenes and peoples of Africa. Shirley (1993: 645) gives the companion De Fer wall map of the world of 1694 his highest rarity rating of 'RRR'.
Nicolas de Fer (1646-1720) appears to have followed the earlier 1689 Nolin-Coronelli folio map of Africa and the 1691 Coronelli two-sheet map of Africa based on the Coronelli 1688 globe gores of Africa for his geography, supplemented with updated information from the new discoveries and, in part, based on Hiob Ludolfi's 1681 map of Abyssinia. The basic geographic models are that of Sanson, Jaillot, and Duval.

Along with the Nolin and Coronelli maps, this De Fer map is important as it represents a turning away from the traditional Ptolemaic view of the interior of Africa. This was to culminate in Delisle's landmark 1700 map. The geography in this De Fer map still shows the two Ptolemaic lakes in Central Africa, but breaks their connection with the Nile River by placing a text box to the north of the lakes with the wording 'we prefer to leave this space blank as it is unknown to the Europeans' along with further information that was known, primarily based on Portuguese discoveries. Though De Fer does question the Ptolemaic Nile source lakes, he still shows his ambivalence toward Ptolemy and the ancient sources by following the tradition of the *Montagnes de la Lune* (Mountains of the Moon) further to the south of the Nile source lakes. The Blue Nile (*Abay* River) is shown with its source in Lake

Tana in the Abyssinian highlands. The White Nile River is completely excluded and does not show its true source further to the south in East Africa.

In West Africa, the Niger River still retains its east-to-west orientation, arising in *Le Grand Lac ou Marais Niger* in Central Africa and emptying into the Atlantic Ocean. This was a relatively consistent representation on maps since Leo Africanus' book, first published by Ramusio in 1550, described the Niger River. Above the Niger, the *Nubia* River continues its west-to-east flow into the Nile River by Egypt.

In Southern Africa the *Cuama* and *St. Esprit* Rivers are shown rising jointly from *Lac ou Marais Zachaf*. A separate *Zembeze* (Zambezi) River is properly placed to the North of *Sofala* and the *Cuama* River. On this map De Fer has an odd representation of a wider southern part of the continent. Southern Africa has a distorted southwest coastline which has an inward slope to it and an incorrect placement and shape of the Cape of Good Hope.

The map itself is devoid of decorative elements and contains no animals. This was a further concession to the growing French emphasis on scientific cartography; that is, their desire was to only include elements that were proven.

On the bottom right map sheet, there is a large inset plan of the Dutch Fort at the Cape of Good Hope which was founded in 1666. Further to the right on the bottom right map sheet, there is a large dedicatory cartouche to the Dauphin of France, 'Dediée a | MONSEIGN^R LE DAUPHIN, |Par Son tres humble et tres Obeissant | Serviteur, et Geographe. | N. de Fer | Avec Privilege du Roy'.

Hendrik van Loon was the engraver for the map itself with the imprint at bottom right of the map: 'H. Van Loon Fecit'. The vignettes surrounding the map were etched by Nicolas Guerard II with the imprint on the map below the dedication: 'Guerard junr fecit'. The placenames on the map are in French.

Publication Information
De Fer is known to have issued wall maps of the world and the four continents, all engraved by Van Loon and Guerard. It appears that De Fer issued his wall map of the world first in 1694 and then followed it with wall maps of the continents.

While there may have been a map with a date of 1696 on the map itself, an example is not known. Examples of this map have been located in Europe, all dated 1698 on the map itself. At the 2003 Reiss & Sohn auction, the author has seen an example with a date of 1696 on the separate title strip, but the date on the map itself is 1698. A close examination of the 1698 date on the map title of the various examples does not show any evidence of erasure or re-engraving of the 1698 date. It is likely that De Fer began the preparation of this map of Africa shortly after the completion of his world map and this map's final completion was delayed until 1698.

This map also appeared showing dates after 1705, based on the De Fer wall map of the world. There were later issues of the De Fer 1694 wall map of the world in 1705, 1717, and possibly in 1720. The Africa map likely had a similar printing history. A Martayan Lan catalog (catalog no. 26: 33: map no. 81) shows an issue of the Africa map, which is dated 1724/1728.

Later versions of the Africa map were issued by De Fer's son-in-law, Guillaume Danet, who inherited the business on De Fer's death in 1720. The map was issued in 1730 with the imprint changed to read 'A Paris chez le Sr. Danet... sur le P. N. Dame a la Sphere Royale avec P. du Roy, 1730'. There may have been further Danet issues of this map after 1730, based on the existence of world maps in 1730 and 1737 with the Danet imprint (Shirley, 1993: 559). The map itself is known in four states.

References
Shirley 1993: 558-9 (for world map); Stopp 1974: 6: no. C3.

164 1697 Abraham Ortelius – Domenico Lovisa, Venice

AFRICAE TA | BVLA NOVA.
- No scale bar.
- [Venice : Domenico Lovisa, 1697].
- From: *Teatro del mondo di Abraamo Ortelio*.
- 1 map : copperplate engraving ; 7.5 x 10.2 cm.
- No latitude and longitude coordinates.
- North at the top; no compass rose.
- Title cartouche at bottom left in the shape of a rectangular weight with a ring handle at the top.
- One sailing ship and one sea monster in Indian Ocean. one sailing ship in Atlantic Ocean. No other decorative elements on the map.

Example: 1697 edition: *Washington LC*, G1006.T76 1697; *Various other locations*.

Description
The great success of the 1577 pocket-size version of Ortelius' *Theatrum Orbis Terrarum* spawned a number of imitators well into the late seventeenth century. This is the second issue of the pocket-size version of the *Theatrum* printed in Italy. The first was printed in Brescia under the direction of Pietro Marchetti in 1598 with subsequent editions printed in Venice. This first Italian issue had a long, apparently successful life with the last edition printed in 1667. For this second Italian issue published by Domenico Lovisa in 1697, a new series of maps were prepared.

No new geographic information was added to this map of Africa. It is copied from the 1598 Marchetti map which in turn was directly modeled after the 1577 Ortelius-Galle (1) map of Africa including the reference to Pliny's naming of Madagascar. However, this map has fewer placenames within Africa.

It appears that more than one unknown engraver was responsible for the map of Africa and the other 108 maps in the atlas. The maps lack a common engraving style.

Publication Information
The map is known in only one state. The last edition to use this map was the 1724 edition of *Teatro del Mondo* published in Venice by Domenico Lovisa (*Torino BIDB*, - a.a- e.ta ++++ [3] 1724 [R]).

The Ortelius-Marchetti map of Africa from 1598 and this Ortelius-Lovisa map of Africa from 1697 are similar, though there are the following differences:

Difference 1.:
Ortelius-Marchetti 1598 map: *OCEANVS ATLATICVS*, with insertion of an 'n' in *ATLATICVS* above the second letter 'A'.
Ortelius-Lovisa 1697 map: *OCEANVS ATLA<u>N</u>TICVS*. Generally, the Lovisa is more roughly engraved with placenames often squeezed onto map; for example, *Bagdet al Babilon* intrudes into *REGNI PE*.

Difference 2.:
Ortelius-Marchetti 1598 map: The Marchetti map has two sailing ships in the Atlantic.
Ortelius-Lovisa 1697 map: The Lovisa map only has one sailing ship in the Atlantic Ocean.

Difference 3.:
Ortelius-Marchetti 1598 map: The Marchetti map has a sea monster in the lower right with horizontal hachuring.
Ortelius-Lovisa 1697 map: The Lovisa map has a sea monster in the lower right with generally vertical hachuring on the body.

Reference
Van der Krogt 2003: atlas 33A:11-12 (Lovisa editions of 1697 and 1724); Mickwitz et al. 1979-1995: entry no. 174; Koeman 1967-71, III.: 83.

165 c. 1697 Jacob von Sandrart, Nürnberg

Map 165. Variant A.

Accuratißima | Totius | AFRICÆ | TABULA | in Lucem producta | Per Iacobum de Sandrart | Norimbergæ.
[at bottom right]:
Joann Bapt. Homann Sculpsit.
- No scale bar.
- Nürnberg : Jacob von Sandrart, [c.1697].
- From: separately published / composite German atlas.
- 1 map : copperplate engraving ; 48.5 x 57 cm.
- Latitude and longitude coordinates on all four sides of the map. Longitude coordinates across Equatorial Line.

- North at the top; one compass rose in the South Atlantic.
- Title cartouche at bottom left in the shape of a scroll.
- A large vignette with North Africans and animals is below the title cartouche.
- There are three sailing ships in the Indian Ocean and three sailing ships in the Atlantic Ocean.

Examples: *Various locations.*

165. Variant A c.1702

A variant of this map is known with the addition of two strips, pasted on to the bottom margin, that identify the locations of various religions in Africa using a color code. This variant is also not dated but may have been prepared as late as c.1702.

Example of the variant: *Author's collection.*

Description

Jacob von Sandrart (1630-1708) was most noted as a portrait painter and art dealer in Nuremberg from 1656. He is not known to have produced an atlas, but he did produce maps, notably large maps of the Rhine and Danube, and views, primarily of Central Europe.

This map is modeled after Frederick de Wit's map of Africa of c.1670 with the same geographical information, but with a completely different vignette and title cartouche. Of special interest, Von Sandrart's map has the imprint of Johann Baptist Homann as the engraver on the bottom right corner. It is one of two pre-1700 maps with Homann's imprint as the engraver; the other is the map of the Americas, *Nova tabula Americae...* . Homann went on to establish what is acknowledged to be one of the most important German map publishing firms of the eighteenth century. The Homann firm was founded in Nuremberg about 1702.

Publication Information

This attractive map of Africa does appear in some German composite atlases of the eighteenth century. The verso is blank.

The map is not dated. There is some uncertainty as to the date for it. Norwich uses a date of 1700. If this date is correct, Von Sandrart would have been 70 years old when he published this map just three years before his death. That is certainly possible, however, much of von Sandrart's work is known to have been produced from 1664. The author has used a date of c.1697, based on Heinz, and to make this map more closely correspond with von Sandrart's other published work.

References

Heinz 2002; Norwich 1997: map # 61; Moreland 1989: 84; Tooley 1969: 51.
Correspondence with Nicolaus Struck, Antiquariat Nicolaus Struck, Berlin.

166 1697 Philipp Cluver – Johann Wolters, Amsterdam

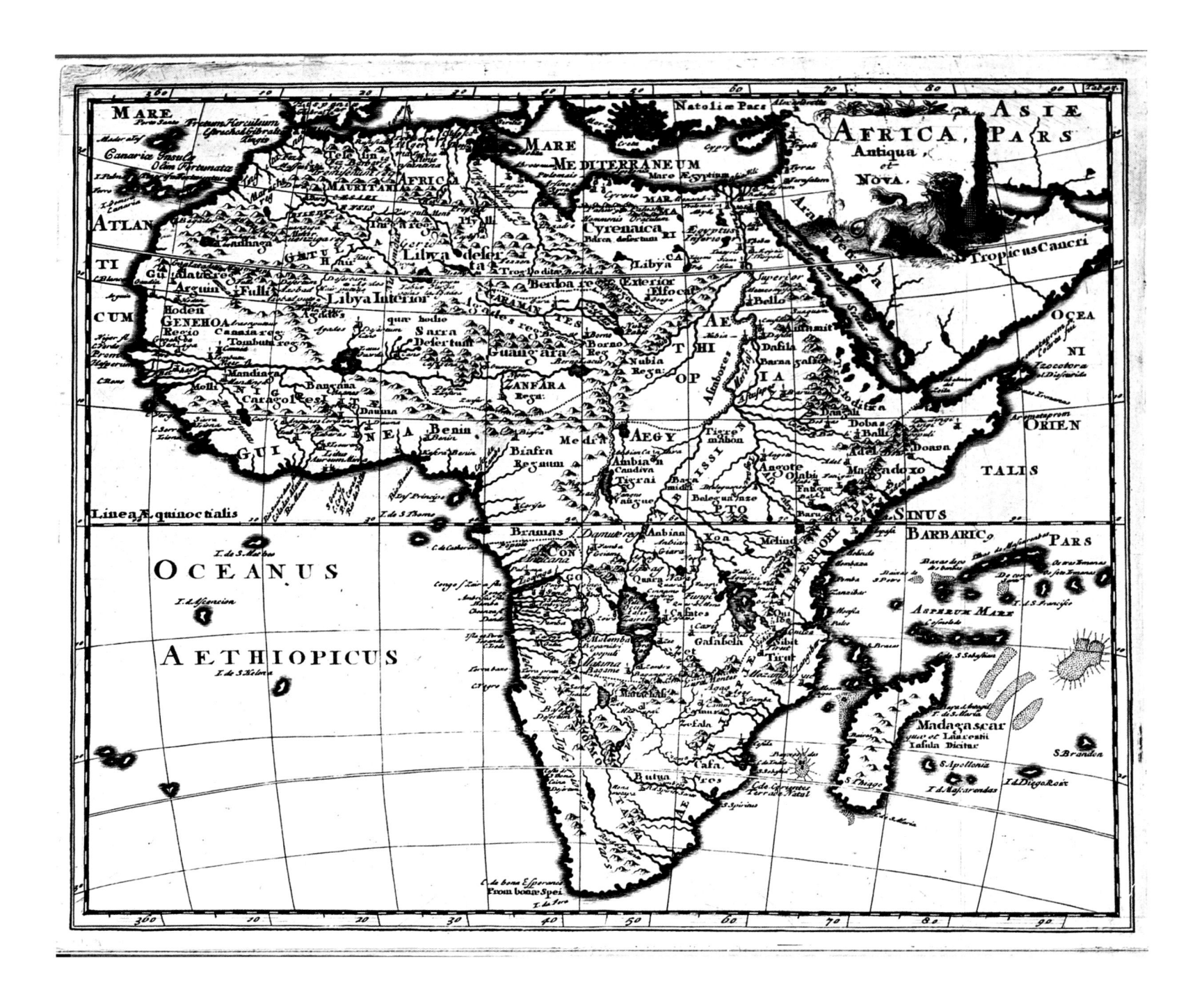

166.1 First State 1697

AFRICA | Antiqua | et | Nova.
- No scale bar.
- [Amsterdam : Johann Wolters, 1697].
- From: *Philippi Cluverii Introductio in Universam Geographicam... .*
- 1 map : copperplate engraving ; 21.5 x 25.5 cm.
- Latitude and longitude coordinates on all four sides of the map. Longitude coordinates across Equatorial Line.
- North at the top; no compass rose.
- Title cartouche in the form of a block at the top right of the map with a lion at its base.

- No other decorative elements on the map other than the title cartouche.
- The map has 'Tab 43.' engraved between the gridline and the neatline in the upper right corner indicating its placement within the book.

Example: 1697 edition: *London BL*, 568.e.3; *Utrecht UB*, THO: ALV 134-47 *and* MAG: T QU 210.

166.2 Second State 1729
The map now has an engraved 'pag. 623' at the upper right corner within the gridline, replacing the engraved 'Tab 43.'

Examples: 1729 edition: *London BL*, 215.a.23; *Cambridge Harv RB*, *GC6 C62751 1729.

Description
This edition of Philip Cluver's famous *Introductio in Universam Geographicam* was published by Johann Wolters in Amsterdam.

This map generally follows the 1661 Cluver-Johann Buno and the 1686 Cluver-Buno-Mosting maps of Africa, both of which were modeled after the 1641 Cluver-Conrad Buno map of Africa. The Cluver-Conrad Buno map in turn appears to be based on Blaeu's 1608 wall map, and, more immediately, the 1617 Blaeu map. This map shows the *Cuama* River as originating to the south of the Mountains of the Moon, whereas on the 1608 wall map, the Cuama river joins upstream with the Spirito Santo River and the Zambere River flowing from Sachaf Lacus.

Publication Information
Besides appearing in Wolters' 1697 edition, this map of Africa also appeared in a 1729 edition of Cluver published in Amsterdam by Johann Pauli. Also in 1729, an edition was published by De Coup in Amsterdam using this same Wolters map. Shirley mentions an edition of Cluver published by Wolters that is earlier than 1697 (Shirley 2004: 349), but the author has not located an earlier edition.

There were numerous other editions of Cluver's *Introductio* that were published using different maps of Africa. Shirley (2004: 345) states that over 45 editions of the *Introductio* appeared over a 100 year period. Not all editions contained maps. Those editions that did contain a map of Africa used one of the following: Cluver-Conrad Buno from 1641; Cluver Elsevier from 1659; Cluver-Johann Buno from 1661; Cluver-Buno-Mosting from 1686; Johannes Janssonius in 1661, using his 1628 Goos map; Janssonius van Waesberge in 1676 using the 1630 Cloppenburch map; and Cluver-Wolters from 1697. All are generally derived from the Cluver-Conrad Buno map of 1641, which in turn was based on the Blaeu folio map of 1617. Refer to the chart under the 1641 Cluver-Conrad Buno map, Map # 71, for a list of the different maps of Africa that were used for different editions of Cluver's *Introductio* and the distinguishing characteristics of each of these maps.

A further edition of Cluver was published in London in 1711 using new copperplates. The Africa map contains the imprint for John Senex ('John Senex sculp.', at bottom left), and page 385 ('pa.385').

References
Shirley 2004: 345-350; Burden 1996: map # 335 (1659 Americas map); Shirley 1993: 447; Phillips 1909-1992: entry # 3432; Sabin 1868: entry # 13805.

167 1697 A. Phérotée de la Croix - Johann L. Gleditsch, Leipzig

AFRICA
- No scale bar.
- [Leipzig : Johann Ludwig Gleditsch and the Heirs of Moritz Georg Weidmann, 1697].
- From: *Des Herrn de la Croix Koenigl. Majest. in Franckreich Geographi... Geographia Universalis...*
- 1 map : copperplate engraving ; 13 x 15.5 cm. (map only) 15 x 21 cm (including coats of arms along the two sides of the map).
- Latitude and longitude coordinates on all four sides of the map. Longitude coordinates across the Equatorial Line. *Premier Meridian* (Prime Meridian) is to the west of the Canary Islands.
- North at the top; no compass rose.
- Title within a cartouche at the upper right of the map.
- No decorative elements on the map. To each side of the map are coats-of-arms. From the top left are: *Barbaray*, *Maroch*, and *Egypte* and five blank coats-of-arms. From the top right are: *Abisinnia*, *Malthe*, and *Congo* and five blank coats-of-arms.

Example: *Berlin SB*, Ebd 127 (Map is in vol. 4 between p. 142 and 143); *Göttingen SUB*, 4 GEOGR 424b:4; *München SB*; *Wolfenbüttel HAB*.

Description

This is the Leipzig edition of A. Phérotée de la Croix's *Nouvelle Methode pour apprendre... La Geographie Universelle*, first published in Lyon in 1690. The map is a close copy of the De la Croix map in the 1690 edition. The notable differences are the use of Latin instead of French for place-names, and a new title cartouche at the upper right. For a discussion of the geography on this map, refer to Map # 149.

Publication Information
It appears that this map was only in the 1697 Leipzig edition of De la Croix's *Nouvelle Methode pour apprendre...La Geographie Universelle.*

Another derivative of the De la Croix 1690 map was published by Daniel de la Feuille in Amsterdam in 1702 in *Atlas Portatif*. In the De la Feuille copy, the map is distinguished by six coats of arms on each side of the map (in the 1690 map, there were only three coats of arms, with the remaining five coats of arms on each side blank).

References
Burden 2007: 399.

168 c. 1699-1700 Justus Danckerts, Amsterdam

168.1.A. First State Variant A c. 1699-1700.

NOVISSIMA | et | PERFECTISSIMA | AFRICÆ | DESCRIPTIO | Authore | I. DANCKERTS | Amstelodami. | cum Privilegio.
- No scale bar.
- Amsterdam : Justus Danckerts, [c.1699-1700].
- From: separately published / Danckerts composite atlases.
- 1 map : copperplate engraving ; 49 x 56.5 cm.
- Latitude and longitude coordinates on all four sides of the map.
- North at the top; one compass rose in South Atlantic Ocean.
- Title cartouche at the bottom left surrounded by a decorative vignette of Africans and various animals.
- No other decorative elements on the map.

Early prints have the following characteristics and comprise Variant A (*Budapest NSL*, TA 224 / 4*)*:
1. On the base part of the title cartouche panel, there are delicate lines as texture or pattern for the panel.

2. In the lakes there are sparsely etched horizontal lines (on Variant B, the lines of the lakes are reworked and in the case of the small lakes, they appear almost totally dark). For example, on Variant A, the small lakes just to the left (west) of the Nile River, north of the Tropic of Cancer and by its mouth in Egypt have finely engraved parallel horizontal lines.
3. Within the large rivers' interiors, there are vivid and intact undulating quasi-parallel lines .

168.1B First State Variant B c.1705-1707

The *Zürich ZB* example of c/.1705-1707 has the following specific characteristics, besides the copperplate being quite worn and partially reworked:

1. The cartouche appears heavily worn in Variant B; the lines on the body of the figures are also rather worn, with many original, delicate lines almost disappearing. These delicate, thinner lines of the original etching disappeared faster than the thicker, coarser ones, so the whole character of the figures around the cartouche is changed. In later printings of this state, the delicate lines are markedly worn and in some places almost absent, with the coarser lines dominating.
2. The smaller lakes are almost totally black or dark, with no or almost no visible separate horizontal parallel lines. For example, in Variant B, the small lakes just to the left (west) of the Nile River, north of the Tropic of Cancer and by its mouth in Egypt have an almost dark surface, and lack the finely engraved parallel horizontal lines of the Variant A.
3. The coastlines, i.e. the lines on the surface of the sea at the coasts are retouched. This is most easily detectable around the islands, which have differences in the length and the width of the lines and in the area covered on the surface of the water.

168.1C First State Variant C c. 1710-1715

The following are specific characteristics for prints of c. 1710-1715, besides extensive and thorough reworking of the cartouche and partly retouched geography:

1. On the base part of the title cartouche panel, the upper small vertical, parallel lines of the original etchings were retouched, and the pattern or texture on the surface of the base of the panel is also worn further.
2. In the middle of the cartouche base, there is a decorative, vertical, dark crevice in the stone panel, which was reworked at its lower and middle part only, but the upper part has not been re-worked and is worn further.
3. The cursive names along the ocean side of the sea coasts have been renewed, but inside the continent, the cursive names are worn further.
4. The large rivers' interiors are almost completely devoid of any lines.

Examples: *Budapest NSL*, TA 224 / 4 and TA 232 /9; *Glasgow UL*, Spec. Coll. e104 /4; *Pécs UL*, HH I 10 /6. *Washington LC*, Philips entry no. 540/7 and entry no. 470/4; *Wien ÖNB*, Alb 44 /4; *Zürich ZB*, TA 101 /9.

168.2 Second State 1727

The map's title is now NOVISSIMA | et | PERFECTISSIMA | AFRICÆ | DESCRIPTIO | per | FREDERICUM DE WITT. | Amstelædami | Cum Privilegio | ex Officina R. & I. OTTENS.

Example Amsterdam UB, Kaartenzl. 107.01.07.: *Washington LC*, G1015 .08 1756 / 190 [Philips no.: 522]).

Description

This is the third folio-size map of Africa prepared and issued by the Danckerts. The first map of Africa by the Danckerts is one by Dancker Danckerts dated 1661. The second one was one by Justus Danckerts in 1683.

The Danckerts family were prominent print and map publishers active in Amsterdam for almost 100 years. The founding member of the family business was Cornelis I (1603-1656) who started producing and publishing maps in the second part of the 1620s. His sons Dancker (1634-1666) and Justus (1635-1701) were also active in the business. Justus issued a number of atlases with his sons. Justus' sons, Theodorus I (1663-c.1727), Cornelis II (1664-1717), Johannes (?-1712), and Eduard (?-after 1721) and his grandson, Theodorus II (1701-1727), continued the family business into the early eighteenth century. Because of the use of the same first names by various family members, there is some confusion in the literature over authorship of some of the maps produced by the Danckerts.

The business was most active from about 1630 to 1727, when the stock of maps in the shop of Theodorus II, Cornelis I's great-grandson was sold. As Koeman states (1967-71, II: 88), 'Their cartographic work has, compared with that published by the Blaeus or Janssonius, attracted but little attention and has never received proper recognition.'

Justus appears to have been most active as a map publisher from 1664 when he was registered in the booksellers' guild of Amsterdam, until 1701 when he died. This map was prepared, using a new plate, by Theodorus I Danckerts (1663-c. 1727), who did the relief and numbers, and by Eduard Danckerts (? - after 1721) who did the lettering.

Koeman dates this new map as after 1696. In the author's correspondence with Gyuri Danku, Map Librarian of the National Széchényi Library in Budapest, Hungary, he suggests that the publishing date is closer to 1699 or 1700.

This map has a new title and also the addition of the imprint, '*cum Privilegio*'. Although the Danckerts obtained a privilege in 1684, it is interesting that they delayed taking advantage of their privilege on a map of Africa until the issue of this map. As the 1684 privilege expired in 1699, it seems logical and reasonable that a second privilege was issued around this time (1699-1700). This suggestion by Danku has been verified by the evidence that, after Justus' death in 1701, the name of Justus was still found on the Africa map along with the privilege on those maps published between 1701 and 1710 by Eduard and Johannes Danckerts. So it seems logical that the privilege granting was prolonged or a new, second one was obtained, still during the life of Justus I (before 1701).

This map is directly modeled after the earlier 1683 Danckerts map of Africa although it is slightly smaller. The primary model is the c.1670 De Wit map. On this map, the Danckerts added placenames along the coasts, particularly in West and Southern Africa, though the names seem to be used primarily to give an appearance of updated information. In Southern Africa, none of the newer Dutch information (*Tafel bay* [Table Bay], *Tafel berg* [Table Mountain], *Robben Eyl* [Robben Island], *Mossel Bay*, etc.), which had been appearing much earlier on maps, is used on this map. The map has a fine engraving style and there are numerous engraving changes from the earlier map; for example, sea names (*OCEANUS MERIDIONALIS SIVE ÆTHIOPICUS*; *MARE ZANGUEBARICUM*, etc.) were introduced in this map.

Publication Information
The map of Africa was issued separately at first. From c. 1700/1701, it appeared in Danckerts' atlases and in various Dutch composite atlases. The map is known in three variants, based on extensive and thorough reworking of the cartouche and on partly retouched geography.

Reiner and Josua Ottens acquired the copperplate for this Danckerts map of Africa, possibly in 1727, when the stock of maps in the shop of Theodorus II was sold, upon his death.

References
The author is grateful for the correspondence and discussion with Gyuri Danku, Map Librarian, Budapest, National Széchényi Library, who has conducted extensive research on the production and publication history of the Danckerts atlases and who has freely shared this information.
Koeman 1967-71, II: 88-97.

169 1699 Heinrich Scherer, Augsburg

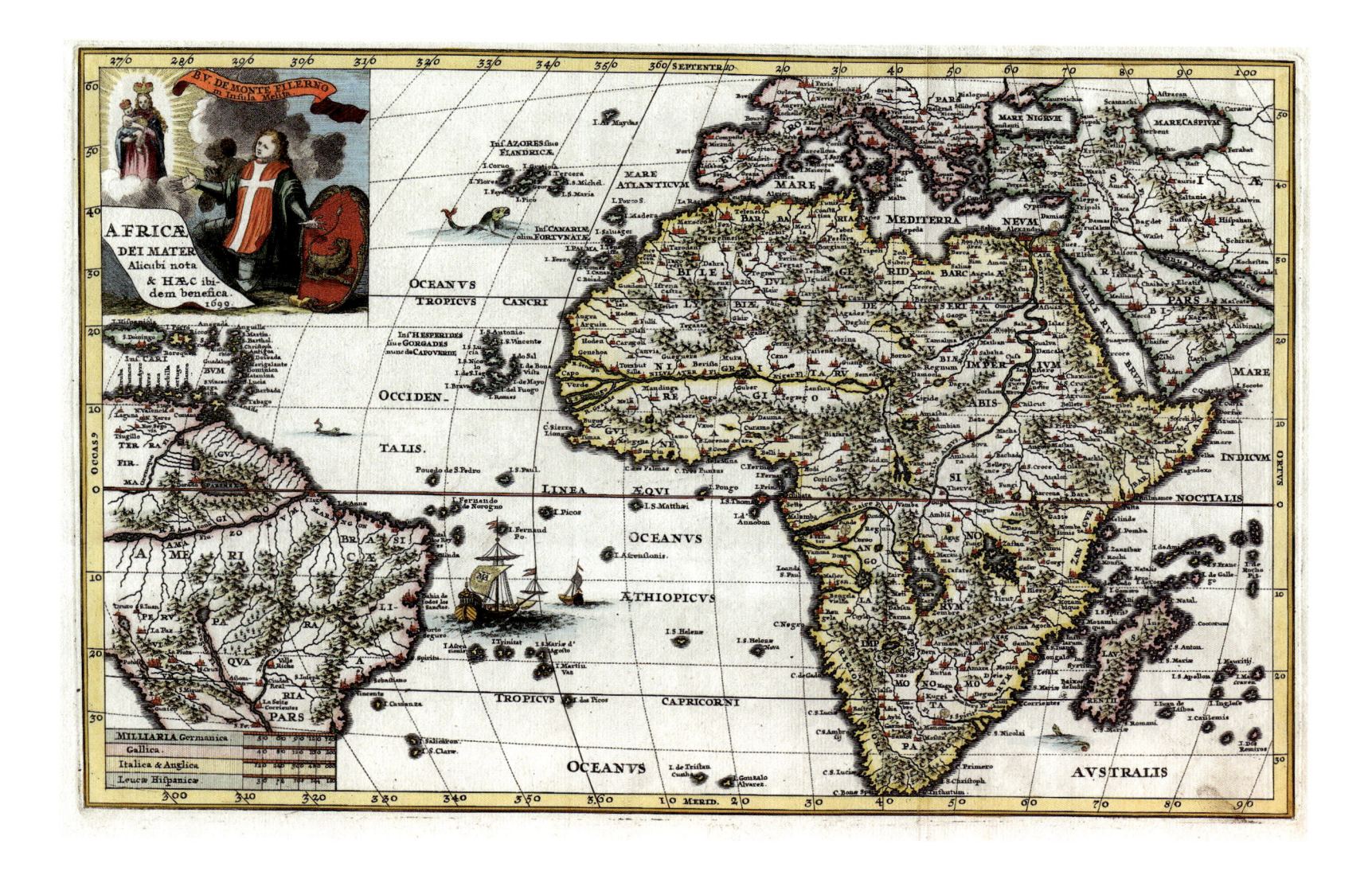

AFRICÆ | DEI MATER | Alicubi nota | & HÆC ibi- | dem benefica. | 1699.
- Scale bar at bottom left:
  MILLIARIA Germanica 150 = 2.7 cm.
  Gallica 200 = 2.7 cm.
  Italica & Anglica 600 = 2.7 cm.
  Leucæ Hispanicæ 180 = 2.7 cm.
- [Augsburg : Heinrich Scherer], 1699.
- From: *Atlas Novus Exhibens Orben Terraqueum... Anno M.DCCX* (1710).
- 1 map : copperplate engraving ; 23 x 35 cm.
- Latitude and longitude coordinates on all four sides of the map.
- North at the top; no compass rose.
- Title cartouche, as a screen, is set within a vignette showing a member of the Society of Jesuits with a large cross on his front and an African behind him, with Mary and the baby Jesus above. At the top of this vignette is: *B.V. DE MONTE FILERNO | in Insula Melita*.
- There are two sailing ships, with one flying the Jesuit's standard, and two sea monsters in the Atlantic Ocean and one sea monster in the Indian Ocean.

Examples: *London BL*, 572.k.1-8; *Author's collection* (as separate map).

Description
Heinrich Scherer (1628-1704) was a professor of mathematics in Munich and a devout Jesuit. This map was prepared for inclusion in Scherer's *Atlas Novus*. To accommodate the elaborate vignette, this map has been extended westward and depicts a portion of South America. Geographically, this map generally uses a much older representation of Africa than that of the French from this period. Much of the geography on this map is confusing, with Scherer presenting his own unique view of Africa. He shows a highly complicated series of river systems within Africa. The source lake for the Niger River, *Lacus Niger*, also is a source for the *Nilus Albus* (White Nile River). In his map, *Sachaf Lacus* feeds the *Iama* (Infante) River and also the Ptolemaic Lake Zaire. The *Spirito Santo* and the *Zambere* Rivers originate in the *Monoemvgorvm* (Mountains of the Moon). There are numerous other unique features on the map.

The engraver for this map is not known. Shirley (1993: 619) states that the Scherer maps were engraved by Leonard Hecknauer, Joseph á Montalegre, and Matthäus Wolfgang, based on the signed frontispieces and the similar style of the engraving throughout the atlas.

Publication Information
The *Atlas Novus*, which was printed in seven parts between 1702 and 1710, was one of the first atlases to have a unifying theme, in this case Catholicism. This particular map, though dated 1699, did not appear in Scherer's atlas until Part Three of his atlas of 1710. There was another edition of this atlas in 1730-37. The parts contain the various publishing locations of Augsburg, Dillingen, and Frankfurt.

Part Three of the atlas of 1710 also contained several other maps of Africa. These are:
'AFRICAE | AB AUCTORE | NATURÆ | SVIS DOTIBVS | INSTRVCTA GEOGRAPHICE | EXHIBITA | AN. MDCC', dated 1700, with a size of 22.5 x 35 cm; and, 'Africae Descriptio' undated, with a size of 12 x 18.5 cm.

References
Norwich 1997: 76; Shirley 1993: 619-625.

170 1700 Nicolas de Fer, Paris

170.1 First State 1700

L'AFRIQUE | Dressée Selon les dernieres Relat. | et Suivant les Nouvelles decouvertes | dont les Points Principaux Sont | placez Sur les Observations de | M^{rs}. de l'Academie Royale des Sciences. | Par N. de Fer. | A PARIS chez l'Auteur dans l'Isle du | Palais Sur le Quay de l'Orloge a la Sphere | Royale 1700. Avec Privil: du Roy.

[dedication at bottom left]:
Dedieé | A NOSSEIGNEURS | les Enfans de France. | Par leur tres humble et | tres obeiss. Serviteur | de Fer. | Geog^r. de Monseig^r. le Dauphin.

- Scale bar at bottom right:
  Echelles
  10 Degr.= 3.0 cm.
  200 Lieıes = 3.0 cm.
  10 degrez de l'Equat. qui sont 200. | grande Lieıes de france a 20. | au degrez.
- Paris : Nicolas de Fer, 1700.
- From: *L'Atlas Curieux ou le Monde.*

- 1 map : copperplate engraving ; 23 x 31.5 cm.
- Latitude and longitude coordinates on all four sides of the map. Longitude coordinates across Equatorial Line.
- North at the top; one compass rose in South Atlantic.
- Title cartouche at bottom right.
- Decorative element of kilns to the left of the title cartouche, and crown with coat of arms at top of dedicatory cartouche to the Dauphin at the bottom left of the map.

Examples: *Paris BN*, Ge DD 1219; *Washington LC*, G 1015.F42 1705 (with title page dated 1705, with evidence under the '5' of a '0').

170.2 Second State 1705
Dated 1705 in title cartouche as follows: 'A Paris Chez l'Auteur dans l'Isle du | Palais Sur le Quay de l'Orloge a la Sphere | Royale 1705. Avec Privil. Du Roy.' The Africa map is no. 10 in the sequence.

Examples: 1705 edition: *London BL*, Maps C.39.c.2 *and* Maps C.1.c.11.

170.3 Third State 1717
Dated changed on title cartouche to 1717.

Examples: 1717 edition: *London RGS*, 7.C.13; *Paris BMaz*, 4899, 48 A (from Pastoureau 1984: 184).

Description
This attractive map is from De Fer's *L'Atlas Curieux ou le Monde*. For this map, De Fer uses his own wall map of Africa of 1696/98 as his model. As on the wall map, he keeps both Ptolemaic lakes, but cuts off their connection with the Nile River. It seems that De Fer recognizes that the two lakes are an historic artifact, but he is still reluctant to completely eliminate them from the map. There are three lines of text above the Ptolemaic lakes indicating that the source of the Nile River, named the *Abavi* on this map, is in the province of *Tonkoua* in Abyssinia.

Following his 1696-1698 wall map, on this map De Fer has an odd representation of a wider southern part of the continent. Southern Africa has a distorted southwest coastline which has an inward slope to it and an incorrect placement and shape of the Cape of Good Hope.

The engraver is not identified on this map, however the second title page in the atlas is signed 'N. Guerard inve fecit'. It is possible that Guerard, who worked on De Fer's 1696/98 wall map of Africa, also engraved this map.

Publication Information
The publishing history of De Fer's *L'Atlas Curieux* is extremely complex. The first edition (première partie de l'atlas) of the atlas appeared in 1700 with this map of Africa. Further parts of the atlas, containing additional maps, were published from 1701 to 1705. The atlas was expanded into a *Suite de l'Atlas Curieux* of 1714-1717. Upon De Fer's death in 1720, the atlas was continued by De Fer's two sons-in-law, Guillaume Danet (an edition of the first part of the atlas in 1723) and Jacques François Bénard (an edition in 1725). See Pastoureau (1984: 170-184) for a complete history of the atlas.

References
Norwich 1997: map # 64 (second state); Pastoureau 1984: 170-184; Tooley 1969: 45.

171 1700 Nicolas de Fer, Paris

171.1 First State 1700

L'AFRIQUE, | Dressée selon les dernieres Relations et suivant | les Nouvelles decouvertes dont les Points prin= | cipaux sont placez sur les Observations de | M^{rs}. de l'Academie Royale des Sciences. | Par N. de Fer. | A Paris. | Chez l'Autheur dans l'Isle du Palais sur le Quai de | l'Orloge a la Sphere Royale 1700. | Avec Privilege du Roy.
[at bottom right between the gridline and the neatline]:
H. van Loon sculp.
[dedication at the bottom left]:
Dedieé | A NOSSEIGNEURS | Les Enfans de France. | Par leur tres humble et | tres Obeissant Serviteur | De Fer. | Geographe de Monseigneur | le Dauphin.
- Scale bar at top right:
10 Degres = 5.5 cm.
200 Lieües = 5.5 cm.
10 Degrez de l'Equateur qui font 200. grands | Lieües de France a 20 au Degrez.

- Paris : Nicolas de Fer, 1700.
- From: *Atlas ou Recüeil de cartes geographiques.*
- 1 map : copperplate engraving ; 46.5 x 60 cm.
- Latitude and longitude coordinates on all four sides. Longitude coordinates across Equatorial Line.
- North at the top; one compass rose in the Atlantic and one in the Indian Ocean
- Title cartouche at bottom right, within vignette of kilns and workers.
- Figurative cartouche at bottom left with a dedication to the Dauphin with crown and coat of arms.

Example: *Private Belgian Collection*

171.2 Second State 1705
Now dated 1705 at the bottom of the title cartouche.

Example: *Stanford SC* (as part of the Oscar Norwich Collection; See Norwich 1997, map # 64 for an image of the second state).

171.3 Third State 1722
Now dated 1722. Publisher's imprint is now 'A Paris le Sr Danet...' at the bottom of the title cartouche). This state has a large scale bar within a decorative cartouche at the bottom left of the map in place of the dedicatory cartouche. Scale bar at upper right is removed.

Example: *København KB*, Atlas Major III, 82; *Author's Collection.*

Description

This is De Fer's folio map of Africa. It bears a close resemblance to De Fer's other map of 1700 (See Map # 170). The notable differences are the larger size of this map at 46.5 x 60 cm and the imprint within the gridline at the bottom right of the engraver Hendrik van Loon (H. van Loon sculp.) on this map (the other De Fer map of 1700 does not have an imprint for the engraver). Van Loon was a skilled engraver who had previously been employed by De Fer to engrave his set of wall maps of the world and continents.

The geography on this map is similar to De Fer's other map of 1700. The common model for both maps is De Fer's wall map of Africa of 1696/98. See Map # 163 for a discussion of the geography.

Publication Information

This map was prepared for De Fer's folio atlas, *Atlas ou Recüeil de cartes geographiques.* Like the publishing history of De Fer's *L'Atlas Curieux*, the history of this atlas is also complicated. It is known that De Fer issued this folio-sized atlas in 1709, with a further edition in 1728.

Until recently, this map was only known with dates of 1705 and 1722. Wulf Bodenstein brought this map in its first state of 1700 to the attention of the author.

This map was issued in its second state in 1705. Upon De Fer's death in 1720, the copperplate for this map was acquired by his son-in-law, Guillaume Danet, who inherited De Fer's business. Danet changed the date and publishers imprint on the map to 'A Paris le Sr Danet... 1722' and issued another edition of the atlas in 1728.

References

Norwich 1997: map # 64 (second state); Pastoureau 1984: 170-184; Tooley 1969: 45.

172 1700 Paolo Petrini, Naples

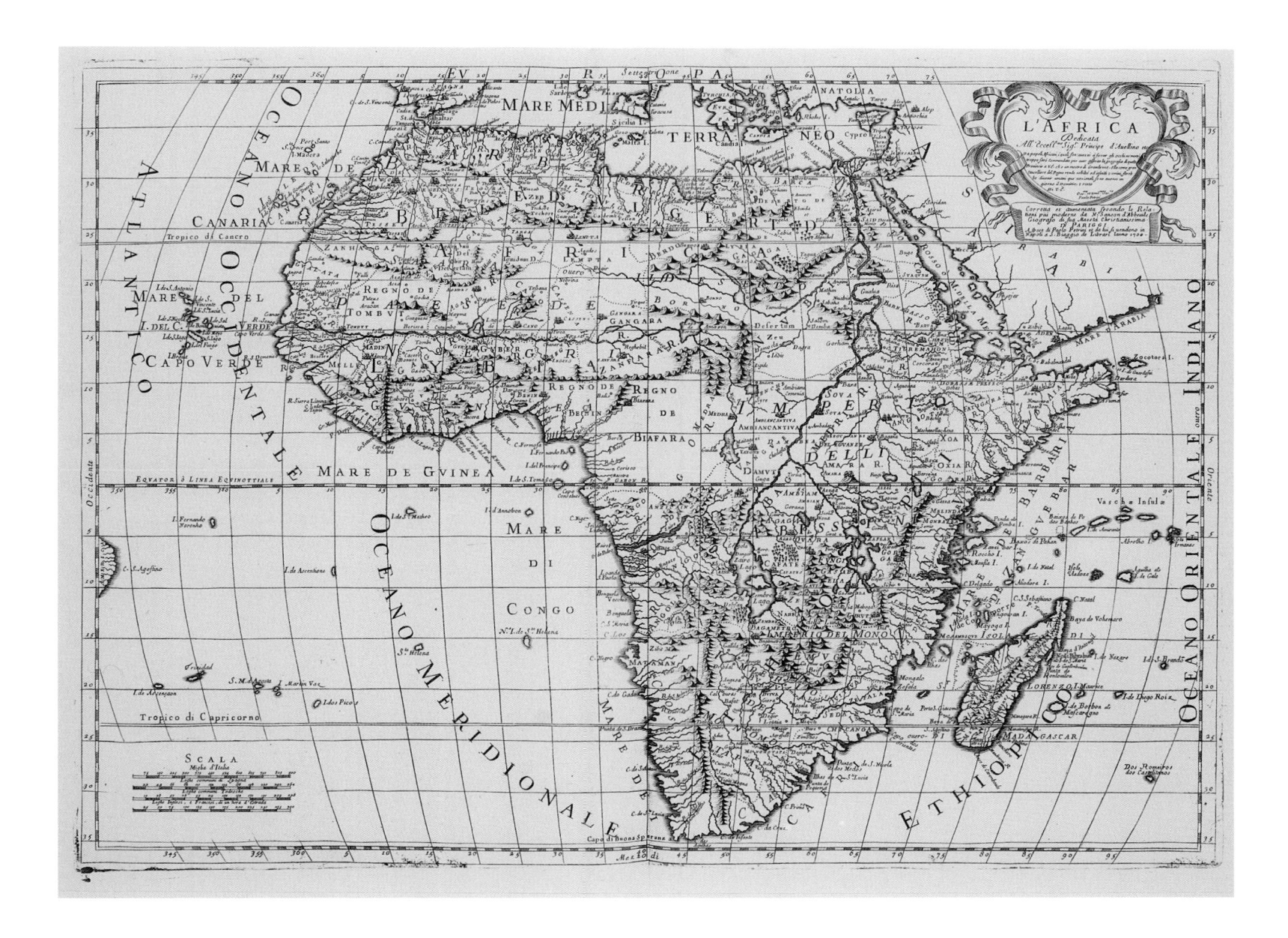

172.1 First State 1700

L'AFRICA | Dedicata | All' Eccell.mo Sig.r Principe d'Auellino etc | Da popoli Africani, i quali son' auezzi à fissar gli occhi ne'mostri | troppo farò Commendato per auer offerto la geografia di quelle | Prouincie a V.E. ch'e un mostro di Grandezza. ella come gran | cancelliere del Regno rende nobiltà ad infiniti uomini, fareb= | be diuenir umani que' nazionali se ne auesse un | giorno il Dominio, e resto | Di V.E. | Oss.mo et Umil.mo Ser.re | Paolo Petrini.
[banderole below title cartouche]:
Coretta et Aumentata secondo le Rela= | tioni più moderne da N. Sanson d'Abbeuile | Giografo di sua Maestà Christianissima. | IN PARIGGI | a spese di Paolo Petrini et da lui si uendono in | Napoli a S. Biaggio de Librari lanno 1700.
- Scale bar at the bottom left:
  SCALA
  Miglia d'Italia 900 = 7.6 cm.
  Leghe communi di Spagna 264 = 7.6 cm.
  Leghe communi Todesche 228 = 7.6 cm.

- Leghe Inglesi, e Francesi, di un' hora d'Estrada 300 = 7.6 cm.
- Naples : Paolo Petrini, 1700.
- From: *Atlante Partenopea*.
- 1 map : copperplate engraving ; 39.5 x 54.5 cm.
- Latitude and longitude coordinates on all four sides of the map. Longitude coordinates across Equatorial Line.
- North at the top; no compass rose.
- Title cartouche at top right of the map.
- No decorative elements on the map other than the simple title cartouche.

Examples: *Private Belgian collection.*

172.2 Second State 1766
Date changed from 1700 to 1766, at the end of the publisher's imprint at the bottom of the title.

Example: *Bedburg-Hau, Antiquariat Gebr. Haas.*

Description
Paolo Petrini (c.1670-1722) is known as a publisher and mapseller, working from 'S. Biaggio de Librari' in Naples. He produced a scarce atlas *Atlante Partenopea* in 1700 which contained a number of maps including this map of Africa. There was a later edition published in 1766.

This map adds little to a late seventeenth century understanding of Africa. It is a close copy of the older Guillaume Sanson model of 1668 with the toponyms changed to Italian. Petrini acknowledges Sanson in the banderole below the cartouche. It is interesting that Petrini does not use the more up-to-date geographic presentation of Africa that appears on his own wall map of c.1700. It is possible that this map was produced first, using sources available to Petrini, and that his c.1700 wall map of Africa was produced later, using the more up-to-date information available on the de Fer wall map of Africa.

References
Valerio 1999: 103-132; Valerio 1993: 147-201.

173 c. 1700 Paolo Petrini, Naples

Map 173.A.

L' A | FRICA Dedicata All' Illustriss.mo ed Exccell.mo Sig.r | Principe CARMINE NICCOLÒ CARACCIO-
LO | Principe di S. Buono etc. | L'Africa ... [with 16 lines of text] | Seruitore | Paolo Petrini.
[above the map on separate strips]:
L'AFRICA, DIVISA SECONDO L'ESTENZIONE DELLE SVE PRINCIPALI PART I | DOVE I REGNI, E
GLI STATI, SONO POSTI AL LVOCO LORO, CON TVTTE LE OSSERVAZIONI DE' SIG'. DELL'AC-
CADEMIA REALE DELLE SCIENZE DEL RE' CR. IN PARIGI | Composta da N. DE FER Geografo del
SER.' DELFINO DI FRANCIA.
- Scale bar on map below Madagascar:
  SCALA
  Miglia communi d'Italia, 60 in un Grado 600 = 10 cm.
  Leghe Spagnuole, 18 in un Grado 180 = 10 cm.

- Leghe Todesche, 15 in un Grado 150 = 10 cm.
- Leghe Inghlesi, et Francesi, 20 in un Grado 200 = 10 cm.
- Naples : Paolo Petrini, [c.1700].
- From: separately published.
- 1 wall map : copperplate engraving and etching ; 4 map sheets ; 91.5 x 118 cm, in total to outer black line (sheet 1 as 45.5 x 59 cm (est.), sheet 2 as 45.5x 59 cm, sheet 3 as 45.5 x 59.5 cm, and sheet 4 as 45.5 x 58.5 cm). A separate title strip above the map.
- Latitude and longitude coordinates on all four sides of the map. Longitude coordinates across Equatorial Line.
- North at the top; two compass roses in Indian Ocean and one in South Atlantic.
- An elaborate title cartouche is placed on the map within Arabia in the middle of sheet 2. Immediately below the title is a dedication to Carmine Niccolo Caracciolo with an accompanying 19 lines of text. Second title on separate strips above the map.
- There are six sailing ships in the Indian Ocean (sheets 2 & 4) and one sailing ship and one sea battle in the Atlantic Ocean (sheet 3).
- Vignettes on all four edges of the map show scenes and peoples of different regions of Africa (Madagascar, Ethiopia, Benin, Angola, etc.) and their buildings, customs, and religious practices. Text in Italian accompanies each vignette. No animals within Africa.

Examples of this map with variations as noted:

173.A   *USA, MacLean Collection* (this example contains the four map sheets and a title strip above the map. No letterpress text panels).

173.B.   *England, Rodney Shirley Collection* (in this example, the map is missing the first map sheet (Northwest Africa) and is without the title strip above the map).

173.C.   *Napoli BIG*, Federico II (this example has the four map sheets only).

173.D.   *London, Sotheby's* (Auction catalog no. L05403 of May 12, 2005).

Description

This map is a close copy of Nicolas de Fer's 1696-1698 wall map of Africa. Petrini acknowledges De Fer as the original maker of the map within his title strip above the map. For his own wall map, De Fer appears to have used the 1689 Nolin-Coronelli large, folio map of Africa and the earlier Coronelli globe gores of Africa. See Map # 163, Nicolas de Fer, Paris, 1696-1698, for a fuller discussion of the geography of this map.

Other than adding a new title and dedicatory cartouche, a new scale bar, and removing the inset view of the fort at Cape Town, all other elements of this map are the same as the 1696-1698 De Fer wall map, including the use of the same vignettes as De Fer. Besides Petrini's imprint on the map, it is also readily identified by the use of Italian in place of French on the map. The engraving of the map itself is somewhat rough when compared to the earlier De Fer, although the details are still vivid. The engraver is not known.

Little is known of Paolo Petrini (c.1670-1722). He was a publisher and map seller at 'S. Biaggio de Librari', Naples. In the same year that he produced his wall maps, he produced an atlas *Atlante Partenopea* containing a number of maps including a map of Africa.

Publication Information

This map is not dated. Petrini is known to have produced wall maps of the world and the four continents, which have been attributed by various writers to c.1700.

According to *Tooley's Dictionary of Mapmakers* (1999-2004: K-P: 414), there is a second state of this map, with a date of c.1722.

The two examples of this map examined by the author (Maps. 164.A, and 164.B) did not contain separate letterpress text panels on the sides and across the bottom, as was seen with examples of the De Fer 1696-98 wall map. It is possible that an example of the Petrini wall map of Africa exists with side and bottom text panels.

References

Valerio 1999: 103-132; Tooley 1999-2004, K-P: 414; Shirley 1993: 615; Valerio 1993: 147-201.

174 1700 Guillaume Delisle, Paris

174.1 First State 1700

L'AFRIQUE, | Dressée sur les Observations de M^rs. de | l'Academie Royale des Sciences, et | quelques autres, & sur les Memoi= | res les plus recens. | Par G. DE L'ISLE, Geographe | A PARIS, | Chez l'Autheur Rue des Canettes | prez de S^t. Sulpice. | Avec Privilege du Roy; pour | 20. Ans 1700.
(on the bottom right ribbon of the title cartouche): N. Guerard Inv. et Fecit
- Scale bar at the bottom left within a box:
  AVERTISSEMENT (and then six lines of text followed by):
  ECHELLE
  Lieıes marines de France et lieıes communes d'Espagne de 20 au degre 300 = 8.0 cm.
  Lieıes marines d Espagne de 17 ? au degre 270 = 8.0 cm.
  Lieıes Communes d'Allemagne de 15 au degre 230 = 8.0 cm.
  Iournées communes de 8 ou 9 lieıes Francoises 27 = 8.0 cm.
- Paris : Guilaume Delisle, 1700.
- From: separately published / *Atlas du Geographe.*
- 1 map : copperplate engraving ; 45 x 58.5 cm.
- Latitude and longitude coordinates on all four sides of the map. Longitude coordinates across Equatorial Line. The *Premier Meridien* (Prime Meridian) is identified on the map as passing to the west of *I. de Fer.* in the Canary Islands.
- North at the top; no compass rose.
- Title cartouche at top right of the map.
- No decorative elements on the map other than the circular title cartouche in upper right of the map surrounded by African animals and an African subduing a crocodile.

Examples: *Cambridge Harv MC*, 2375.1700 (as a separate map); *Paris BN*, Ge D 655; *Private French Collection*; *Author's Collection* (as a separate map).

174.2 Second State 1707
Imprint in cartouche of a new address: 'Quai de l'Horloge a la Couronne de Diamans', replacing the earlier address.

There are numerous changes to the copperplate for the second state of this map in the toponyms and in the placement of lakes and rivers. Of note, Delisle shows a much more correct Senegal River originating in West Africa and flowing due west into the Atlantic. He also removes the faint line connecting the Niger and Nile Rivers and places text in French stating that 'some pretend that the Niger is an arm of the Nile'.

Examples: *Various locations.*

174.3 Third State 1708
Imprint in cartouche of a new address: 'Quai de l'Horloge', replacing the earlier address.

Example: *London BL*: C.36.f.3.

174.4 Fourth State 1708-1718+
Immediately below the same title cartouche as the third state, the following is added: 'se Trouve a Amsterdam chez L Renard Libraire Pres de la Bourse'

Sometime after 1708 and before 1718+, the copperplate was evidently used by Louis Renard ('chez L Renard'). In this state, the following imprint is lightly etched below the title cartouche: 'se Trouve a Amsterdam chez L Renard Libraire Pres de la Bourse'. This evidently refers to an agreement to allow Renard the right to sell the map on the Amsterdam market.

Map 174.1.

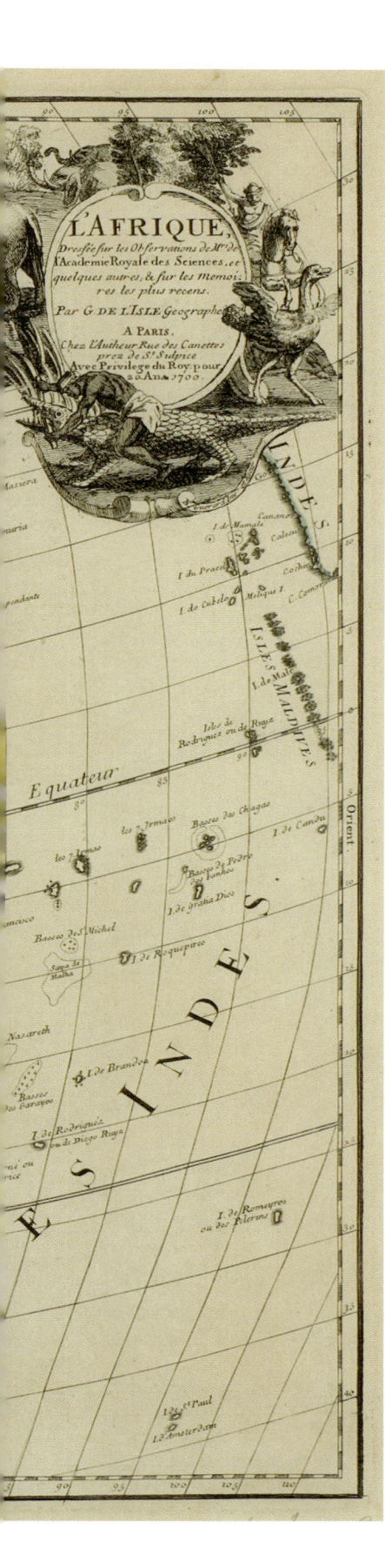

Example: *Author's collection*.

174.5 Fifth State 1718+
Imprint added before Delisle's name in cartouche of: 'Premier Geographe du Roi'.

Examples: *Various locations*.
    Description

This is a landmark map of Africa. The 1700 Delisle is the first map to show Africa without the two Ptolemaic-based, Nile River source lakes. Delisle also gives the correct longitude for the Mediterranean Sea of 42°, thus correcting the width of the northern shape of Africa.

The interior of Africa has the main regions identified with large, dark capital letters for *BARBARIE, NIGRITIE, HAUTE GUINEE, ABISSINIE,* and *PAYS DES CAFRES*. Lines divide all of Africa into numerous sub-regions labeled *Royaumes, Estats,* etc. Within the regions, considerable detail and additional placenames are shown particularly around those areas settled or explored by Europeans in West Africa, Abyssinia, the Congo, Southeast Africa on the Zambezi River, and South Africa. At the Cape of Good Hope, the Dutch settlement at *Fort des Hollandois* (the Dutch Fort at the Cape of Good Hope) and at *Hellenbok* (Stellenbosch) are shown.

In West Africa, Delisle orients the Niger River in an east-west axis, but without its source lake of Lake Niger to the south as it was presented on previous maps of Africa. The Niger seems to flow westward into *Lake de Guarde* where it ends; exiting *Lake de Guarde* in the west is the Senegal River which flows into the Atlantic. A faint line connects the Niger River with the Nile River; this was Delisle's way to show that some in the past thought that the Niger connected with the Nile, though he did not believe this to be true. The *Nil fl* (Nile River) is shown originating in Lake Tana in the Abyssinian highlands. The west-to-east Nubia River, also shown in previous maps flowing eastward into the Nile, ceases to exist on this map.

Delisle does show a vaguely engraved and unidentified lake, at about 5° S, not far inland from Zanzibar in East Africa with no rivers entering or exiting it. This lake is probably based on reports by traders of the inland lakes that, in fact, are in this part of Africa. It could represent any of the lakes in this region, or it could be an amalgam of Lakes Victoria, Malawi, and Tanganyika.

In Southern Africa, Delisle removes the middle (the Cuama) of the three traditional, major southern rivers and replaces it with an un-named smaller river, which is likely the Sabia River, in the region of *R.e de Sabia* and *Sofala*. The Infante River is also removed. He does show the *Magnice ou R. de S. Esprit* (Spirito Santo) River (the modern Limpopo). On the *Zambeze* River, both of the Portuguese interior trading towns of *Tete* and now *Sena* are shown.

This map of Africa served as the model for European mapmakers and was frequently copied throughout the eighteenth century. It is noteworthy for its attention to a scientific approach in the preparation of the map. Delisle exercised care, wherever possible, in the inclusion only of verifiable information. The map was constantly being updated; thus there were a number of later states of this map quickly following on his original 1700 publication.

Guillaume Delisle (1675-1726) is often referred to as the father of modern geography. He was elected a member of the Academie Royale des Sciences in 1702, and then was appointed Premier geographe du Roi in 1718. His total cartographic output totaled over 100 maps plus their various states.

The sources for this map were likely varied. Delisle did refer to the predecessor maps of Africa by his compatriots such as De Fer, particularly with his 1696/98 wall map of Africa, and Nolin, with the 1689 folio map. He also credited Coronelli, including his 1688 globe gores and his 1691 large, folio two-sheet map of Africa. As France was undergoing tremendous political and economic expansion in Africa and elsewhere during this period, Delisle likely had access to the numerous reports on Africa that were flowing back to Paris. This wealth of geographic information, coupled with a desire to exclude information that could not be confirmed though verifiable scientific study, lead him to create this map of Africa.

The engraver for the map was Nicolas Guerard, the Elder, (c.1648-1719). The map contains an imprint of

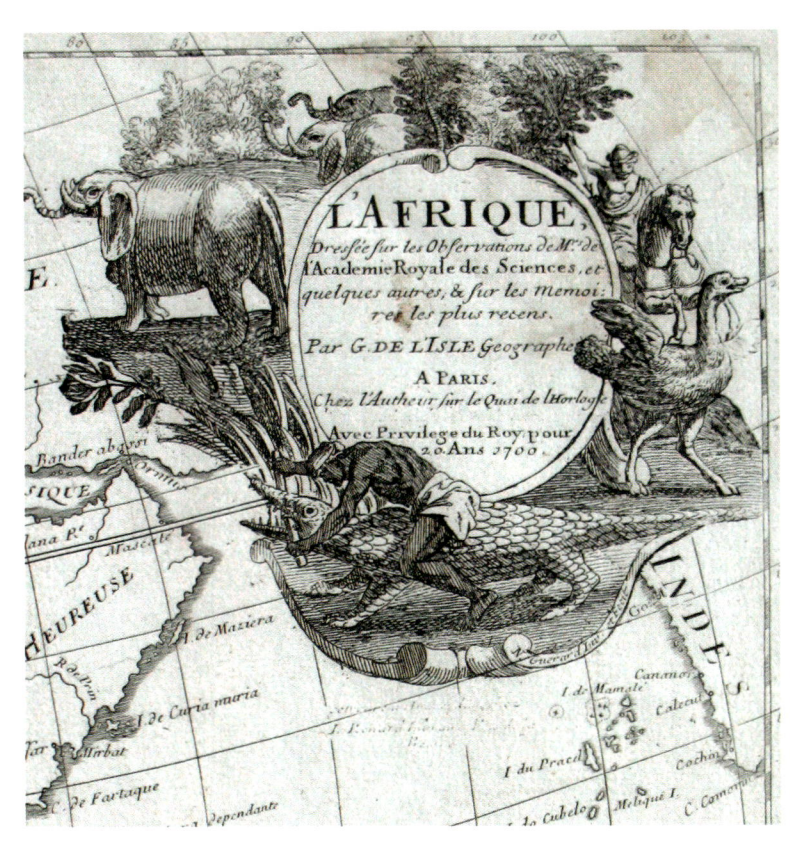

Cartouche without and with Renard's imprint (states 1 and 4).

'N. Guerard Inv. et Fecit' on the bottom right ribbon of the title cartouche.

Publication Information

Delisle's first publications were a globe in 1699 and maps of the world and the four continents in 1700. He produced an *Atlas du Geographe* in 1700-1712 with later editions, in which this map appeared.

The actual date for each of his map states is generally determined by the publication addresses that appear in the title cartouches, especially since he retained the 1700 date on the later maps. Over his career, Delisle is known to have worked at certain addresses during specific periods of time. The first address was 'Rue des Canettes pres de St. Sulpice', where he worked from 1700 to 1707, then 'Quai de l'Horloge a la Couronne de Diamans' from 1707 to 1708, and then 'Quai de l'Horloge' from 1708.

Due to the popularity of the Delisle map of Africa, it was copied by numerous other mapmakers. Pieter Mortier, in about 1706-1710, used the first state of Delisle's map as a model for his own map.

Pieter Schenk issued his copy of the Delisle map in 1708. Covens and Cornelis Mortier issued their copy of the Delisle map in c.1722. Others included: Jermiah Wolf in Augsburg in 1720 (*Paris BN*, Ge DD 2987 [7768]), Girolamo Albrizzi in Venice in 1740, Reiner and Josua Ottens in 1745, and Tobias Lotter in 1760 using the second state as a model.

Delisle produced a second map of Africa in 1722, which is easily distinguished by the date of 1722 in the title cartouche. Lake Maravi (Malawi) in Central Africa is now identified for the first time on the Delisle map. Delisle's 1722 map also separates, for the first time, the Senegal and Niger Rivers.

This 1722 map was also widely copied. Upon Delisle's death in 1726, his business was carried on by his wife Marie Angélique Delisle, the daughter of Pierre Duval. She was responsible for the posthumous publication of Delisle's *Carte de l'Afrique Francoise ou du Senegal* in 1727 (date at the bottom of the scale bar), with the map dated 18 April, 1726 (at the bottom of the title cartouche). Upon her death, the Delisle copperplates passed to her son-in-law Philippe Buache.

References
Tooley 1999-2004, A–D: 354-355; Norwich 1997: map # 59 (fourth state); Tooley 1969: 68-73.

Appendices and Indexes

List of Lost Maps that Show Africa

The following are maps that may or may not still exist. Hopefully, further information on these maps will be uncovered. There may be other "missing" maps, not known to the author.

Giovanni Battista Mazza, c.1590:
Possible second state of Giovanni Battista Mazza's c.1590 map of Africa: AFRICA | EX MAGNAE | ORBIS TER | RAE DESCRI | PTIONE.
There is a second state of Mazza's Americas map, with the removal of the imprint of Mazza ("Iouan Batista Mazza fece" on the Africa map) and with evidence of some additional hachuring to the sea.

Fausto Rughesi, 1597:
There is the possibility of an earlier proof or published state that is unrecorded. The University of Texas set of maps of the world and the four continents all bear imprint erasures. Also, as noted by Burden (1996: 133), there are some nomenclature alterations on the Americas map.

Luis Teixeira-Joannes van Doetecum, c.1600:
There is a second state of Luis Teixeira-Joannes van Doetecum c.1600 map of Africa reported by the literature: Tabula Aphricae nova sumta | ex operibus Ludouici Tercerae | cosmographi Regiae majestatis | Hispaniarum, with the addition of the following: t'Amsterdam gedruckt bij Dauit de Meijne inde weeelt cart. This state was published by David de Meijne.

Petrus Bertius, 1624:
First state of Petrus Bertius' 1624 map of Africa: *Carte de | L'AFRIQVE | Corrigee et augmentee | desus toutes les aultres | cy deuant faictes P.Bertius | L'annee 1624*. The author has found the later states of this map of 1640, 1646, and 1670.

Melchoir Tavernier (attr.), 1639:
First state of the map of Africa attributed to Melchoir Tavernier, 1639: *Carte de | L'AFRIQVE | Corrigee, et, augmentee, dessus | toutes les aultres cy deuant | faictes par | .P.Bertius. | Anno 1639*.
The author has not yet located an example of the first state of this map but is assumed to exist, based on the existence of the Americas map of 1639.

Pierre Du Val, 1660
There may be a fourth state of the Africa map of c.1687 or 1688, from Du Val's *Le Monde ou La Geographie Universelle En plusieurs Cartes* with a plate number in the lower right corner, based on the fourth state of the Americas map of c.1687 as per Burden (1996). This map might exist in an edition of 1688 by *Melle. M. Du Val* [Du Val's daughter]). The author has examined several examples of the Du Val atlas of 1682 with the second title page of 1688, but it did not contain a fourth state of the Africa map.

Dancker Danckerts, 1661:
There may be a possible fourth state of Dancker Danckerts 1661 map of Africa, with the imprint of "Apud Carolum Allard 1695".

Guillaume Sanson, 1668:
Earlier editions of Sanson's *Carte Generale de toutes les parties du Monde* contained Nicolas Sanson's map of 1650. This was replaced in subsequent editions with Guillaume Sanson's 1668 map of Africa. There is a possible third state of 1690 of Guillaume Sanson's 1668 map, based on Guillaume Sanson's Americas map (see Burden, 1996: 514). This state of the Americas map is dated 1690 and has the following after the fifth line of the title: augmentee et corrigee en cette seconde edition.

Johann Hoffmann, c. 1696-1710
Hoffmann is known to have produced a set of playing cards of the world, *Geographisches carten-spiel von Asia, Africa, und America* (Schreiber 1901: 81, G. 267).

The author has not been able to locate an example of this set of cards. The set of world cards at the British Museum has not been located and a supposed second set is no longer accessible at the Playing Card Museum in Cincinnati, Ohio, USA as that museum has closed. In appearance, it has been suggested by Shirley that the set is similar to the miniature maps prepared by Hoffmann for Beer's translation of Duval's *Le Monde* of 1678 (Map # 121).

This set of playing cards of the world has been dated as early as c.1678 by Shirley (1993: 496) and King (2003: 146). Meurer has located one copy of the booklet to accompany the cards, dated 1696. According to the text in the booklet, the general map of Africa is shown on the card "Pique 6". Meurer has found that the example in Cincinnati forms a second edition with the booklet dated 1710. Based on this evidence, the author has to conclude that the playing cards of the world including the map of Africa were first produced in 1696 and then reissued in 1710.

List of Some "Firsts" on Printed Maps of the Continent of Africa

The earliest known printed map of the continent of Africa in which the continent is represented alone as surrounded by the ocean. **Map 1. 1508 Antonio Francanzano (Fracan) Da Montalbodo.**

The earliest, readily available, printed map to show the entire continent of Africa.
Map 3. 1540 Sebastian Münster.

The first printed map of Africa in a book to show a southbound river, the *Zembere F.*, flowing out of the western Ptolemaic Lake (the unnamed Zembere/Zaire), passing southward through the *Monti de Luna* (Mountains of the Moon), and then splitting into the *Cuama F.* (Zambezi) and *Spirito Sato F.* (Limpopo) Rivers which flow to the southeast coast of Africa. **Map 4. 1554 Giovanni Battista Ramusio - Giacomo Gastaldi.**

For the first time on a printed map of the entire continent, the island of Madagascar or *S. Lorenzo Isola* as it was then known. **Map 4. 1554 Giovanni Battista Ramusio - Giacomo Gastaldi.**

The ancient city of *Tombotv* (Timbuktu) is shown in one of its earliest depictions on a printed map, though placed too far to the west. **Map 4. 1554 Giovanni Battista Ramusio - Giacomo Gastaldi.**

Evidence of the Portuguese settlements in West Africa with *Lamina* (El Mina, the Portuguese fort) on the coast of Guinea. **Map 4. 1554 Giovanni Battista Ramusio - Giacomo Gastaldi.**

Cefala (Sofala) in East Africa is identified on the map. **Map 4. 1554 Giovanni Battista Ramusio - Giacomo Gastaldi.**

The ancient city of Great Zimbabwe, depicted on the map as the twin cities of *Zimbro* and *Simbaoe*, appears in the interior of southern Africa. **Map 9. 1564 Giacomo Gastaldi – Fabio Licinio.**

The Cape of Good Hope became more pointed and the eastward extension of the continent was significantly reduced by about 1,700 kms, to 7,000 kms. **Map 12. 1570 Abraham Ortelius.**

Following Mercator, a major lake is shown to the southwest of the traditional twin Ptolemaic Nile source lakes. This third lake, the unnamed Lac Sachaf, is now shown feeding the river system to the south, notably the Cuama and Spirito Santo Rivers. This same lake is the source for the Congo River to the west and also flows into the Nile to the north. Above the Niger River in West Africa, Thevet follows Mercator by placing a river, the *Gher-Nubie*, that flows eastward into the Nile River. **Map 17. 1575 Andre Thevet.**

Several islands are shown in the South Atlantic, such as an early appearance of *Tristan de Cugna* (Tristan da Cunha), reflecting their importance to early shipping around the Cape of Good Hope. **Map 30. 1594 Giuseppe Rosaccio.**

Evidence of European advances into the interior, for example, Portuguese exploration up the Cuama River into the interior of south central Africa in the region of Monomotapa, or *Benamataxa*, as he names it on this map. *Ca. Portogal* (the Portuguese Fort) is placed on the map within the junction of the Spirito Santo and Cuama rivers. **Map 31. 1595 Gerard Mercator II.**

Vlejis baij, Vis baij, and *Mossel baij* on the south African coast. **Map 54. 1608 Willem Janszoon (Blaeu).**

R da Volta (Volta River) is more precisely placed on the map and *Accara* (Accra) appears near its mouth on the Guinea coast in present-day Ghana. **Map 78. 1650 Nicolas Sanson.**

Use of a number of Dutch names interspersed with older Portuguese names in South Africa. Among these names are *Tafel bay* (Table Bay), *Tafel berg* (Table Mountain), and *Robben Eyl* (Robben Island), and *Schorre hoek* (inland from Cape Agulhas, the southern-most point in Africa). **Map 84. c. 1655 Jan Mathisz.**

Ou de St. Laurent et Isle Daufine (Madagascar) with the placement of *Fort Dauphin* at the southern end of the island. He changes the name of the Cuama River to the *Zambere River*, though *Zefala* (Sofala) is still placed far to the north of the river. Introduction of a *Nouuelle I. de St. Helene* (St. Helena Island). **Map 104. 1666 Nicolas Sanson.**

Fort Nassau in West Africa. **Map 114. c. 1670 Frederick de Wit.**

Zambeze River, above the *Zambere R.* and *Rio de Spiritu Santo*. **Map 118. 1674 Alexis-Hubert Jaillot.**

Correct orientation of the Blue Nile in the Abyssinian highlands and its source as Lake Tana. For the first time on a map of Africa, the connection between the Nile River and the two Ptolemaic central African lakes is purposely removed, and they are not shown as a source for the Nile. French influence is depicted in West Africa with placenames for French settlements (*Petit Diepe*, etc.). **Map 122. 1678 Pierre Duval.**

Almost complete omission of the two Ptolemaic lakes in central Africa, except for the lower portion of the western Lake Zaire. *Abawi le Nil, Nilus fl.* (Abay River or the Blue Nile) with its source in *Tzana Lac* (Lake Tana) in the Abyssinian highlands. *Tete* on the Zambezi River is shown). Addition of Hout Bay in South Africa. **Map 146. 1689 Jean Baptiste Nolin – Vincenzo Coronelli.**

First map to show Africa without the two Ptolemaic-based, Nile River source lakes. Also, correction in the longitude for the Mediterranean Sea of 42°, thus correcting the width of the northern shape of Africa. **Map 174. 1700 Guillaume Delisle.**

Some Important Placenames on the Map of Africa

The following are some placenames that appear in this cartobibliography. Many of the placenames exhibit variations in spelling over time.

There were nine, principal parts of Africa commonly used in the sixteenth and seventeenth centuries, although precise geographical divisions did not exist. Some of these parts also have sub-divisions. The exact meaning of each part was far from uniform or consistent over time and the actual borders of each part were seldom depicted with accuracy. These parts were:

Egypt: extending further west and south than present day Egypt.

Barbary: the states of North Africa from Morocco to Cyrenaica.

Biledulgerid (that is, the Bilad-al-Djarid, the "Land of the Palms" to the interior of North Africa: southern Tunisia and the Algerian Sahara.

Zaara: The Sahara.

Nigritia: the western and central Sudan.

Guinea: introduced to Europe by the Portuguese, derived from the Berber word, Aguinaou, meaning Negro, identical with the Arab word Sudan (Hallett 1965: 37).

Nubia: the eastern Sudan.

Ethiopia was divided into two parts, Inferior and Superior.

Ethiopia Inferior covered the entire southern half of the continent including the west coast south of Cameroon. The region of interior southern Africa also came to be known as *Monomotapa* (from De Barros, for the interior), and the *Coast of Caffres* (for the coast of southern Africa).

Ethiopia Superior contained the Kingdom of Abyssinia, whose size was greatly exaggerated with its borders reaching well south of the Equator.

EAST AFRICA

Elephas (elefans) *Mons* (from Ptolemy) = *Ras el-Fil* (from the Arabs) = on the northeast coast of Africa.

Scotra = *Scoira* = the Island of Socotra (at the entrance to the Red Sea).

Abasce = *Abasie* (from Polo; divided into 6 kingdoms, 3 Christian Kingdoms from the days of Thomas the Apostle, 3 Moslem Kingdoms on the Aden side of this "Middle India") = Abyssinia (in northeast Africa).

Abawi = Blue Nile River or the Abay River (as it is known today).

Astapius (from Ptolemy) or River *Stapius* = *Bahr al-Azraq* = The Blue Nile or the Abay River (as it is known today.

Tzana Lac = Lake Tana.

Melido = *Mylinde* or *Mylindi* = *Malinde* = *Malindi* (present-day name), north of Mombasa on the coast of Kenya.

Maabase (Arab) = *Mombassa*.

Cefala = Sofala (present-day name, on the Indian Ocean to the south of the Zambezi River).

Xegiba (Arab) = *Zanzibar*. *Zanguebar* or *Zanzibar* (name applied to all of southern Africa by Persian & Arab authors. (Bodenstein 1998: 192) appears on the map as *Zanzibar* = southern Africa. Sanuto's map of Africa shows both a Zanguebara Terra and the Island of Zanzibara.

Zanzibar or Zenzibar = *Zanchibar* (from Marco Polo) = Zanzibar Island.

Quiloa = Kilwa, present day name, on coast south of Zanzibar.

Zanj (Arab) ("Land of the Negroes") = East Africa.

Azania (from the Greeks) (Hall 1998: 13) = East Africa.

Cast. Portugal = unknown, possibly a reference to a fort in Abyssinia from the time of Christóvão da Gama. The reference may also be to Sena or Tete on the Zambezi River, which were established as trading outposts, though hardly forts, by this time (1569). The name first appears on Gerard Mercator's world map of 1569.

Cuama (from Ramusio) = *Zuama* (from Ortelius) = Zambezi River (though its placement was to change in southern Africa).

Rio da Lagoa (renamed the *Espirito Santo* by Lourenço Marques as a result of his 1545 exploration).

Spirito Santo (from Ortelius and others) = Limpopo River (in southern Africa).

rio de Manhice (Zambezi) and the *Lorenzo Marches* (Limpopo) on the Pigafetta 1591 map.

District of *Toroa*, known as the Kingdom of *Butua*, a sometimes vassal of *Monomotapa* (location for gold mines) = In present day Zimbabwe.

Madeigascar (from Marco Polo) = *Diab* (from Fra Mauro) = *Madeisgascar* = *Magastar* (in Ramusio) = Madagascar. This was viewed as the very limit of the habitable world from which there flowed such overpowering currents that none could ever be expected to return against them (Beasley 1897-1906: Vol. III: 148).

SOUTHERN AFRICA

Ethiopia Inferior = Southern Africa.

Rio de Infante (farthest easterly point for Diaz in 1488 and named after a ship's captain, João Infante) = The Great Fish River or Keiskama River (but this is open to conjecture; see Axelson 1973: 15, for a discussion of this topic).

Mare Prasodu = the name of the ocean off southeastern Africa reached by B. Diaz (on Portuguese portolans of early 1500s).

Promentory Prasso = Likely Cape Correntes, cape on the southeast coast in present-day Mozambique.

São Bras or *Saint Blaize* (from Diaz; first landfall east of the Cape of Good Hope) = *Mossel Bay*, from the Dutch.

Porto Fragoso (from the Portuguese; on Waldseemuller's maps = Hout Bay.

Cabo de São Brandão = Cape Agulhas (southern most point in Africa).

Golfo dentro das Serras = False Bay.

R. dos Vaqueiros (based on the cattle seen there; Diaz' first sighting of land after rounding the Cape of Good Hope) = southwest of Mossel Bay.

Bafa da Lagoa = Algoa Bay.

Crucis Insule or *Ilhéu da Cruz* = Isle of St Croix (at entrance to Algoa Bay).

Angra das Voltas (from Diaz) = *Luderitz Bay*, in Namibia.

Ponta dos Farilhoes = *Walfish Bay*, in Namibia.

Monamotapa (from de Barros), the major kingdom in the interior of southern Africa, a region known for its gold mines. It was named for the ruler of the land.

Simbaoe = *Symbaoe* = *Zimbabwe* or *Great Zimbabwe* = a court where the Monomotapa (ruler) might reside.

WEST AFRICA

Niger River = derived from a Tuareg expression *N-ger-n-gereo* or "river of rivers (Hallet 1965: 47).

Cape Bojador (the southern extent of Africa on the west coast; initially this southern extent was *Cape Non* to the north) = *Cape Buzedot* (from Waldseemüller) or *Budezor* (Münster) *Buyetder* (from Catalan Atlas of c.1375) or variations on some maps = On the mainland across from the Canary Islands.

Synus Ethyopius (from Fra Mauro map) = Gulf of Guinea.

Terra dos negros = Guinea.

NORTH AFRICA

Tingis = Latinized name for Tangier.

Libya interior = Sahara, from classical writers.

Ethiopia inferior = Sudan, from classical writers.

Garamantes = a Berber tribe located in the desert of north Africa.

Babilonia, Babillonja = the medieval name for Old Cairo.

Alcair (from Blaeu and other maps of the period) = Cairo.

Niffe on portolan charts = Casablanca.

CENTRAL AFRICA

Rio poderoso or *Rio do padrom* (from Reinel portolan of 1485) = Congo River.

Capo Lobo (from Diego Cao voyage of 1485) = Cape St. Mary.

Lunae Montes = *Monte de Betsun* (Gastaldi) based on Bodenstein (1998) = *Betzum m.* (Ortelius; see Bodenstein: 1998) = Mountains of the Moon.

References:
For an extensive discussion of placenames on the southern Africa coast refer to *Diaz and his Successors* (Axelson 1998). For a discussion of the placenames in the interior of southern Africa see *The Empire of Monomotapa* (Randles 1979). For a discussion of other placenames in southeast Africa refer to *South-East Africa as Shown on Selected Maps of the Sixteenth Century* (Randles 1956).

Chronology of Historical Events

c.150	Claudius Ptolemy working at the Library in Alexandria in Egypt writes Geographia.
1st to 6th Century	Axum or Aksum, a powerful kingdom, ruled the Horn of Africa. Direct contact with the Mediterranean world. Conversion of the country to Christianity by Frumentius (early fourth century).
c. 600 to 1000	Bantu migration to southern Africa. Emergence of southeastern African societies, to become the city-states of Great Zimbabwe, Dhlo-Dhlo, Kilwa, and Sofala, which flourish through 1600.
610	Beginning of Islam.
639-641	Khalif Omar conquers Egypt with Islamic troops.
640-710	Conquest of North Africa by Arabs, beginning in Egypt and spreading westward.
700-800	Islam sweeps across North Africa; Islamic faith eventually extends into many areas of sub-Saharan African to c.1500.
740	Islamicized Africans (Moors) invade Spain, and rule it until 1492.
c. 750	Settlements on east African coast by Arabs from Saudi Arabian peninsula.
800 to 1100	Growth of trans-Sahara gold trade between North Africa and sub Saharan Africa. In West Africa a number of kingdoms emerge whose economic base lay in their control of the trans-Saharan trade routes and the trade in gold, kola nuts, and slaves being sent north in exchange for cloth, utensils, and salt.
10th Century	Height of Kingdom of Ghana, founded in 4th century. Ghana equipped its armies with iron weapons and became master of the trade in salt and gold, controlling routes extending from present-day Morocco in the north, Lake Chad and Nubia/Egypt in the east, and the coastal areas of western Africa in the south.
1054	Beginning of Islamic conquest of West Africa.
13th Century	Rise of the Mali Empire of the Mande (or Mandinka) peoples in West Africa. The Mali Empire was strategically located near gold mines and the agriculturally rich interior floodplain of the Niger River.
c. 1250	Zimbabwe (meaning stone house or building, some of which are massive), constructed in southeastern Africa by ancestors of the Shona peoples of modern Zimbabwe.
1307-1332	Height of the Mali Empire (Mandingo Empire).
1324-1325	Mali Emperor Mansa Musa's elaborate pilgrimage to Mecca spreads Mali's reputation across Sudan to Egypt and the Islamic and European worlds. Under Mansa Musa, diplomatic relations with Tunis and Egypt were established, and Islamic scholars and artisans were brought into the empire. Mali appeared on maps produced by Europeans.
1352-1353	Visit of the traveler and writer Ibn Batuta to Mali Empire (Mandingo Empire)
14th to 15th Centuries	Great Zimbabwe (with its impressive stone structures) of the Karanga ancestors of the Shona peoples of southeastern Africa controls a large part of interior southeast Africa. The Karanga peoples formed the Mwene Mutapa Empire, which derived its wealth from large-scale gold mining. At its height in the fifteenth century, its sphere of influence stretched from the Zambezi River to the Kalahari in the west, to the Indian Ocean in the east and to the Limpopo River in the south.
14th Century	Complex, advanced states located from Lakes Albert and Victoria southward were established, including kingdoms ruled by the Bachwezi, Luo, Bunyoro, Ankole, Buganda, and Karagwe, but little is known of their early history. Farther to the south, in Rwanda, a cattle-raising pastoral aristocracy founded by the Bachwezi (called Bututsi, or Bahima, in this area) ruled from the sixthteenth century onward.
After 1400	Internal disputes weakened the strength of the over-extended Mali Empire, and the northern towns and provinces revolted, eventually leading to the Empire of

	Songhai. The Songhai (or Songhay) began to spread along the Niger River. Much of Mali fell to the Songhai Empire in the western Sudan during the fifteenth century.
1405-1433	The Chinese began their series of voyages across the Indian Ocean, first reaching northeast Africa in 1417.
1415	Portuguese capture Ceuta on North African coast.
1418-c.1420	Portuguese settle in Madeira.
1426	Portuguese, Gonzalo Velho, sails beyond Cape Nun.
1427-c.1432	Portuguese discover and settle in Azores.
1433	Tuaregs from the Sahara seize Timbuktu.
1434	Portuguese, Gil Eanes, first rounded Cape Bojador after the fifteenth Portuguese attempt.
1435	Portuguese, Eanes and Baldaia, reach Angra dos Ruivos.
1436	Portuguese, Baldaia, to Rio de Ouro.
1441	Portuguese, Nuno Tristão, to Cape Blanco.
1441	Beginning of European slave trade in Africa with capture and first shipment of ten African slaves near Bojador by Antam Gonçalves.
1443	Portuguese, Nuno Tristão, to Arguin and Garças.
1444	Portuguese, Lanzarote, to Naar and Tider.
1444	Portuguese, Nuno Tristão, to Terra dos Negros.
1444	Portuguese, Dinis Dias, to Cape Verde.
1445	Portuguese, Alvaro Fernandes, to Cabo dos Mastos.
1445	The first factory or feitoria (a trading outpost) was founded at Arguim.
1446	Portuguese, Nuno Tristão, killed at Gambia(?) River.
1455	Cadamosto and Usodimare, Italians sailing for the Portuguese, reach Bissagos Islands (Guinea Bissau).
1456	Cadamosto and Diogo Gomes in separate expeditions to points further south. Cadamosto to Cape Verde Islands.
1460-1461	Portuguese, Pedro de Sintra, to Sierra Leone.
1468	Songhoy ruler re-captures Timbuktu. Songhai (or Songhay) Empire, centered at Gao, dominates the central Sudan after Sunni Ali Ber's army defeated the largely Tuareg force at Tombouctou or Timbuktu.
1471	Portuguese found the trading post of São Jorge de El Mina on the Guinea Coast. João de Santarem and Pero Escobar to Guinea Coast as far as Shama Bay, later called the Gold Coast.
1472	Portuguese to islands of Fernando Po, Sao Tome, and Principe in Gulf of Guinea.
1474	Portuguese, Rui de Sequeira, to Cape Santa Catharina.
1482	Construction begins on the Fort of São Jorge da Mina (El Mina) on the Guinea coast.
1482-1484	First voyage of Portuguese, Diego Cão, to south of Equator and the Congo River; reaches Cape Santa Maria 13° S. (reaches the Congo River in 1483).
1485	Second voyage of Portuguese, Diego Cão, to approximately 22° S at Walfis Bay in southwest Africa (Namibia).
1487	Portuguese reach Timbuktu in West Africa from the coast.
1487-1488	Portuguese, Bartlomeu Dias, around Cape of Good Hope to southeast coast of Africa. Diaz reaches the Rio Infante (possibly Great Fish River) in 1488.
1490	Portuguese ascend the Congo River in west-central Africa for about 200 miles and convert the King of the Congo to Christianity. Post at São Salvador established.
1497-1499	Portuguese, Vasco da Gama, sails around southern Africa to India and returns. (Da Gama rounds Cape of Good Hope in 1497).
1493-1529	Height of Songhoy Empire under Askia Mohammed with conquest of Mandingo Empire and expansion east beyond the Niger.
Late 1400s	Kingdom of Congo flourished on the Congo River (Zaire), a confederation of provinces under the Manikongo.
c. 1500	Benin in West Africa at the height of its power.
1500	Diogo Dias, a Portuguese ship captain and a member of Pedro Álvares Cabral's voyage to India, views Madagascar on August 10, 1500.
1505	Occupation of Sofala on the east African coast by the Portuguese. Portuguese view Seychelles Islands.

1507	Visit of Leo Africanus to Songhoy Empire.
1514	Spanish occupation of Algiers (1514-1516).
1520	Father Alvares to Abyssinia.
1549	Bareto and Homem explore Zambezi River.
1562-1576	John Hawkins initiates British slave trade on West African coast, taking slaves to the New World. Other countries involved in the European slave trade included Spain (from 1479); North America (from 1619); Holland (from 1625); France (from 1642); Sweden (from 1647); and Denmark (from 1697).
1575	Portuguese begin the settlement of Angola at São Paulo de Loanda.
1578-1584	Duarte Lopes travels through the Kingdom of the Congo.
1591	Fall of Songhai Empire to the armies of al-Mansur of Morocco resulting in a period of continual strife and economic decline.
Late 1500s	To the east of Songhai, between the Niger River and Lake Chad, the Hausa city-states and the Kanem-Bornu Empire had been established since the 10th century. After the fall of Songhai, the trans-Saharan trade moved eastward, where centers of flourishing commerce and urban life developed.
1595	First establishment of Dutch on the coast of Guinea.
1598	Dutch seize island of Mauritius to east of Madagascar.
1600	Establishment of English East India Company.
1602	Establishment of Dutch East India Company whereby the company was granted a monopoly in trade east of the Cape of Good Hope.
1612	Dutch establish a settlement on Guinea coast at Mouri (Fort Nassau).
1617	Dutch establish base on Gorée Island off Cape Verde peninsula.
1618	Frenchman, Paul Imbert, travels to Timbuktu.
1618-1619	G. Thompson ascends the Gambia River for about 400 miles.
1618	Paez visits Lake Tana in Abyssinia.
1623	Lobo explores Abyssinia.
1626	French establish a settlement, St. Louis, at mouth of the Senegal River.
1626	French establish first settlements on Madagascar.
1633	Dutch build factory at Arguim.
1638	Dutch capture the Fort of São Jorge de Mina on the Guinea coast.
1643	French found Fort Dauphin on the extreme southern coast of Madagascar.
1645	Capuchin monks ascend the Congo River, possibly as far as Stanley Falls.
1652	First Dutch settlement under Jan van Riebeeck at Cape of Good Hope.
1659	French establish the fort of Saint Louis in Senegal.
1660	Rise of Bambara Kingdom on the upper Niger River replacing the Mandingo Empire in 1670.
1662	British build a fort at James Island at the mouth of the Gambia River
1666	Dutch build a fort at Cape of Good Hope.
1673	Formation of the French trading company of Senegal.
1677	French take the island of Gorée off Cape Verde peninsula from the Dutch.
1686	King Louis XIV annexes Madagascar.
1687	French in Ivory Coast.
1697	French under André de Brue complete the conquest of Senegal and advance up the Senegal River to Mambuk by 1715.
1698	Portuguese expelled from their trading posts on the east African coast by Omani Arabs. Mombassa abandoned by Portuguese in 1739 but they retain Mozambique until the late 20th century.
1698-1699	French doctor, C. Poncet, travels from Cairo to Gondar in Ethiopia as a representative of King Louis XIV of France.
1700-1717	Asante (or Ashante) Empire of Akan peoples is unified under Osei Tutu on the Gold Coast of West Africa and dominates gold-producing areas.

References

Bibliothèque Nationale de France (web site, date consulted June 2006); Diffie and Winius 1977; Fage and Oliver 1975; Davidson 1966; Langer 1948.

Bibliography

Ade Ajayi, J.F. and Michael Crowder. 1985. *Historical Atlas of Africa*, London: Longman Group,.
Alden, John and Dennis Landis, eds. 1980-1997 *European Americana: A Chronological Guide To Works Printed* In *Europe Relating To The Americas, 1493 - 1750 [Volumes I - VI]*. John Carter Brown Library: New York.
Almagià, Roberto. 1929. *Monumenta Italiae cartographica*, Firenze: Istituto Geografico Militare,
—. 1934. *Planisfero di Arnoldo de Arnoldi, 1600*. Rome: Publicazioni dell'Instituto di Geographia della R. Universita di Roma,
—, 1944-55. *Monumenta Cartographica Vaticana*. 5 vol. Vatican City: Biblioteca Apostolica Vaticana,
—, 1951. On the cartographic work of Francesco Rosselli. *Imago Mundi* : 27-35.
Arkway, Richard B. [n.d.]. Richard Arkway Inc. catalog # 54.
Asher, Adolf. 1839 *Bibliographical essay on the collection of voyages and travels edited and published by Levinus Hulsius and his successors, at Nuremberg and Francfort from anno 1598 to 1660*. Berlin: A. Asher,
Atkinson, Geoffroy. 1927. *La littérature géographique française de la renaissance : répertoire bibliographique : description de 524 impressions d'ouvrages publiés en français avant 1610, et traitant des pays et des peuples non européens, que l'on trouve dans les principales bibliothèques de France et de l'Europe occidentale*. Paris: A. Picard,
Axelson, Eric. 1961. Portuguese Settlement in the Interior of South-East Africa in the Seventeenth Century.In *Actas do Congresso Internacional de História dos Descobrimentos, vol. V, 2 part.*. Lisboa, 1961, 1-18.
—, 1969. *Portuguese in South-East Africa 1600-1700*. Johannesburg: Witwatersrand University Press
—, 1973a. *Portuguese in South-East Africa 1488-1600*. Cape Town: C. Stuik (Pty) Ltd.,
—, 1973b. *Congo To Cape; Early Portuguese Explorers*. New York: Harper & Row,
— with **Charles Boxer, Graham Bell-Cross and Colin Martin. 1988.** *Dias and his Successors*. Cape Town: Saayman & Weber,
—, 1998. *Vasco Da Gama. The Diary of His Travels Through African Waters 1497-1499*. Somerset West: Stephan Phillips,
Bagrow, Leo. 1964 *History of Cartography*. Cambridge, Massachusetts: Harvard University Press,
Baker, JNL.1937. The Earliest Maps of H. Moll. *Imago Mundi* 2: 16
Baynton-Williams, Ashley. 1991. *Introduction to the Facsimile of the Theatre and Prospect of John Speed*. London,
Beazley, C. Raymond. 1897-1906. *The Dawn of Modern Geography. A History of Exploration and Geographical Science from the Middle of the 13th. to the Early Years of the 15th. Century.* 3 volumes. Oxford: Clarendon Press,
Beckingham, C.F. 1961. The Travels of Pero da Covilhā and their Significance. In: *Actas do Congresso Internacional de História dos Descobrimentos*, vol. III, Lisboa: 1-14.
Besterman, Theodore. 1972. *History and Geography, A Bibliography Of Bibliographies.* 4 volumes. Totowa, NJ: Rowman and Littlefield,
Biasutti, Renato. 1908. Il 'Disegno della geografia moderna' dell'Italia di Giacomo Gastaldi (1561). In: *Memorie geografiche pubblicate come supplemento alla Rivista Geografica Italiana*. vol. 2, no. 4. Firenzia,
—, 1920. La carta dell'Africa di G. Gastaldi (1545-1564) e lo sviluppo della cartografia Africana nei sec. XVI e XVII. In: *Bollettino della Società Geografica Italiana*, Ser. 5, 9: 327-346 and 387-436.
—, 1923. La carta dell'Africa del De Jode (1593) e l'influsso del Gastaldi sulla cartografia olandese. *Congresso Geografico Italiano, 8th, Firenzia, 1921*. Atti, vol. 2. Firenzia, 1923, 307-10.
Bibliothèque Nationale de France. 1971. *Catalogue De L'histoire De L'afrique*. New York: Burt Franklin, (reprint of a handwritten bibliography, dated no earlier than 1893).
—. 2006. *Background information on the history of Africa On-line*. http://gallica.bnf.fr/VoyagesEnAfrique/ (date consulted May 2006).

Birmingham, David. 1965. *The Portuguese Conquest of Angola.* Johannesburg: published under auspices of the Institute of Race Relations.
Blake, John W. 1977. *West Africa Quest for God and Gold 1454-1578.* London.
Blonk, Dirk, and Joanna Blonk-van der Wijst. 2000. *Hollandia Comitatus: Een kartobibliografie van Holland.* 't Goy-Houten: HES & De Graaf Publishers.
Bodenstein, Wulf. 1988. Ortelius' Maps of Africa. In *Abraham Ortelius and the First Atlas: Essays Commemorating the Quadricentennial of His Death 1598-1998*, edited by Marcel van den Broecke, Peter van der Krogt, and Peter Meurer. Utrecht: HES Publishers: 185-207
Bonacker, Wihelm. 1969. *Introduction to the Facsimile of Quaa's Geographisch Handtbuch 1600.* Amsterdam: Theatrum Orbis Terrarum.
Borba de Moraes, Rubens. 1958. *Bibliographia Brasiliana: A bibliographic essay on rare books about Brazil published from 1504-1900.* Amsterdam: Colibris Editora LTDA.
Borroni-Salvadori, Fabia. 1980 *Carte piante e stampe storiche delle raccolte lafreriane della Biblioteca Nazionale di Firenze.* Rome
Boxer, C.R. 1961. *Four Centuries of Portuguese Expansion, 1415-1825: a Succinct Survey.* Johannesburg: Witwatersrand University Press.
—- , 1965. *The Dutch Seaborne Empire 1600-1800.* London: Penguin Books.
—- , 1969/1978. *The Portuguese Seaborne Empire 1415-1825.* London: Hutchinson.
—- , 1981. *João de Barros: Portuguese Humanist and Historian of Asia.* New Delhi: Concept.
—- , 1984. *From Lisbon to Goa 1500-1750: Studies in Portuguese Maritime Enterprise.* London.
Brauer, R. W. 1995. *Boundaries and Frontiers in Medieval Muslim Geography.* Philadelphia: American Philosophical Society.
Bricker, Charles, R.V. Tooley, and Gerald R. Crone. 1989. *Landmarks of Mapmaking: an Illustrated Survey of Maps and Mapmaking.* Ware: Wordsworth Editions.
Briels, J.G.C.A. 1974. *Zuidnederlandse boekdrukkers en boekverkopers in de Republiek der Verenigde Nederlanden, omstreeks 1570-1630.* Nieuwkoop: B. de Graaf.
British Museum. 1963. *Exhibition of Books, Manuscripts and Antiquities from Ethiopia.* London: British Museum
Briquet, Charles M. 1997. *Filigranes, Dictionnaire Historique Des Marques Du Papier.* 4 vol. Mansfield Center, CT.: Maurizio Martino. (reprint of 1907 edition).
Broecke, Marcel P.R. van den 1996. *Ortelius Atlas Maps: An Illustrated Guide.* 't Goy-Houten: HES Publishers.
—- , 1998. Introduction to the Life and Works of Abraham Ortelius (1527-1598). In *Abraham Ortelius and the First Atlas: Essays Commemorating the Quadricentennial of his Death*, edited by Marcel van den Broecke, Peter van der Krogt, and Peter Meurer. 't Goy-Houten: HES Publishers.
—- , 2006. *Background information on Ortelius Map No. 8.* On-line http://www.orteliusmaps.com/book/ort8. (date consulted July 2006).
Bruel, Georges. 2003. *Bibliographie De L'afrique Equatoriale Française.* Mansfield Centre, CT: Martino. (reprint of 1914 edition of descriptions of 7029 books, articles, and pamphlets related to French Equatorial Africa, as compiled by Georges Bruel, Administrateur en Chef des Colonies).
Bunbury, E. H. 1878. *A History of Ancient Geography among the Greeks and Romans.* 2 vol. London.
Burden, Philip. 1996. *The Mapping of North America: A list of printed maps 1511-1670.* Rickmansworth, England: Raleigh Publications.
—- , 2007. *The Mapping of North America II: A list of printed maps 1671-1700.* Rickmansworth, England: Raleigh Publications.
Burger, C.P. 1929. Het Caert-thresoor. In *Het Boek* 18: 289-304, 321-344.
Bürmeister, Karl Heinz. 1960. *Sebastian Münster: Eine Bibliographie.* Wiesbaden: Pressler.
—- 1963. *Sebastian Münster: Versuch eines biographischen Gesamtbildes.* Basel: Helbing & Lichtenhahn.
Campbell, Tony. 1987. *The Earliest Printed Maps 1472 -1500.* London: British Library.
Caraci, Giuseppe. 1927. Avanzi di una preziosa raccolta di carte geografiche a stampa dei secoli XVI e XVII. *La Bibliofina* 29:178-192.
Cardinall, A.W. 2002. *A Bibliography of the Gold Coast.* Mansfield Centre, CT.: Martino Publishing Co. (reprint of 1932 edition published by the Gold Coast Colony).
Cartwright, Margaret Findlay. 1976. *Maps of Africa and Southern Africa in Printed Books 1550-1750.* Cape Town: University of Cape Town Libraries.
—- , 1992. *Maps of the South Western Cape of Good Hope.* Cape Town: South African Library
Caton-Thompson. G. 1931. *The Zimbabwe Culture, Ruins and Reactions.* Oxford: Clarendon Press

Chang, Kuei-sheng, 1970. Africa and the Indian Ocean in Chinese maps of the 14th and 15th centuries. *Imago Mundi* 14: 21-30.

—, 1971. The Ming maritime and China's knowledge of Africa. *Terra Incognita* 3: 33-44

—, 1979. The Han Maps: New Light on Cartography in classical China. *Imago Mundi* 31: 15.

Christie's auction catalog. 1994. *Auction Catalog.*

- - - , June 2000. *Cartography.* Auction Catalog

Chittick and Rotberg (eds.). 1975. *East Africa and the Orient: cultural syntheses in pre-colonial times.* New York ; London: Africana ; Holmes and Meier Publishers Ltd.

Church, Elihu Dwight. 1951. *A Catalogue of Books Relating to the Discovery and Early History of North and South America Forming a Part of the Library of E.D. Church.* Compiled and Annotated by George Watson Cole. 5 vol. New York: Peter Smith. (reprint of 1907 Edition).

Crone, Gerald R. 1978. *Maps and their Makers: An Introduction to the History of Cartography*, 5th edition. Dawson: Archon Books

Cortesão, Armando, and Avelino Teixeira da Mota. 1960. *Portugaliae Monumenta Cartographica volume I-VI.* Lisbon: Coimbra.

—, 1975. The Portuguese Discovery and Exploration of Africa. *Esparsos:* 2:155-180. Coimbra: BGUC, 1975 (1 ed. Lisboa: JIU, 1961).

—, 1969-71. *History of Portuguese Cartography.* 2 vols. Coimbra: Junta de Investigacoes do Ultramar-Lisboa, 1969-71.

Cortesão, Jaime. 1964-1969. Os Portugueses am Africa. *História de Portugal,* vol. XVI. Lisboa: Portugalia Editora

D'Allemagne, H.-R. 1950. *Le Noble Jeu de l'Oie de 1640 à 1950.* Paris: Grund: 218

Dames, M. L. (ed.). 1918-21. *The Book of Duarte Barbosa. An Account of the Counties Bordering the Indian Ocean and their Inhabitants…completed about the Year 1518.* London: Hakluyt Society, Second Series, XLIV & XLIX.

Davidson, Basil. 1966. *Africa, History of a Continent.* London.

—, 1960. *Old Africa Rediscovered. The Story of Africa's Forgotten Past*, London: Gollancz.

Deschamps, Pierre. 1964. *Dictionnaire de Geographie Ancienne Et Moderne, Suivi De L'Imprimerie Hors L'Europe.* Paris: G.-P. Maisonneuve & Larose. (first published in 1870 as 2nd supplement to Jacques Charles Brunet's Manuel du Libraire et de l'Amateur de Livres. Paris: Firmin-Didot, 1860-65).

Denucé, J. 1912-13. *Oud-Nederlandsche kaartmakers in betrekking met Plantijn.* 2 vols. Antwerp: De Nederlandsche Boekhandel (reprint in Amsterdam: Meridian Publishing, 1964).

—, 1937. Afrika in XVIe eeuw en de handel van Antwerpen. Met een reproductie van de wandkaart van Blaeu-Verbist van 1644 in 9 foliobladen. *Dokumenten voor de geschiedenis van den handel* 2. Antwerp: De Sikkel.

Dias, Jill 1989. As Primeiras Penetrações Portuguesas em África. In *Portugal no Mundo* vol. I, edited by Luis de Albuquerque. Lisboa: Alfa, 281-299.

Dias, Manuel Nunes. 1990. A Penetração no Continente e a Tentativa de Cristianização do Monomotapa. In *Portugal no Mundo*, vol. III, edited by Luis de Albuquerque. Lisboa: Alfa, 66-87.

Diffie, Bailey W., and George D. Winius. 1977. *Foundations of the Portuguese Empire 1415-1580 – Europe and the World in the Age of Expansion Vol. I.* Minneapolis: University of Minnesota Press.

Douwma. 1979. *Catalogue 22.* London. Maps # 2 and # 104.

Fage, F G., and R. Oliver. 1975. *The Cambridge History of Africa.* Cambridge, London, and New York: Cambridge University Press.

Fall, Yoro K. 1982. *L'Afrique a la naissance de la cartographie: Les cartes majorquines, XIVe-Xve siecles.* Paris: Centre de Recherches Africaines.

Filesi, Teobaldi. 1972. *China and Africa in the Middle Ages.* TRANSLATION David L. Morison. London: Frank Cass.

Fischer, Joseph, and Franz von Wieser. 1903. *The Oldest Map with the name America of the year 1507 and the Carta Marina of the year 1516 by M. Waldseemuller (Ilacomilus).* Innsbruck: Wager'sche Universitatis. (reprinted Amsterdam, 1968)

— **and Franz von Wieser.** 1907. *Cosmographiae Introductio Of Martin Waldseemuller In Facsimile Followed By The Four Voyages Of Amerigo Vespucci, With Their Translation Into English; To Which Are Added Waldseemuller's Two World Maps Of 1507.* New York: United States Catholic Historical Society.

Fite, Emerson D., and Archibald Freeman. 1926. *A book of old maps delineating American history from the earliest days down to the close of the Revolutionary War.* Cambridge. Mass.

Frabetti, Pietro. 1959. *La Collezione delle Antiche Carte Geografiche Museo Delle Navi.*, Bologna,.

Fracan, Montalbodo. 1992. *Itinerarium Portugallensium e Lusitania in Indiam & indem occidentem & demum ad aquilonem* (Facsimile reprint of Milan, 1508 edition). Lisboa: Fundação Calouste Gulbenkian, Servico de educão.

Fuchs, Walter. 1946. The Mongol Atlas of China by Chu Ssu-pen and the Kuang-yü-t'u. *Monumenta Serica* monograph Vlll: 9-11.

—- , 1953. Was South Africa Already Known in the 13th Century?. *Imago Mundi* 10: 50.

Gallo, Rodolfo. 1950. Gioan Francesco Camocio and his Large Map of Europe. *Imago Mundi* 7: 93-102.

—- , 1954. *Carte geografiche cinquecentesche a stampa della Biblioteca Marciana e della Biblioteca del Museo Correr di Venezia.* Venice. Instituto Veneto Di Scienze Letlere Ed Arti.

Gay, Jean. [1998]. *Bibliographie Des Ouvrages Relatifs A L'afrique Et A L'arabie : Catalogue Methodique de tous les ouvrages français & des principaux en langues étrangères traitant de la géographie, de l'histoire, du commerce, des lettres & des arts de l'Afrique & de l'Arabie* . Mansfield Centre, CT.: Martino Fine Books,(n.d. but 1998). (reprint of the 1875 first edition).

Grandidier, Guillaume. [2001]. *Bibliographie De Madagascar 1500-1903.* Mansfield Centre, CT: Martino Publishing (n.d. but 2001). (reprint of the first volume only that contains works from 1500-1905 and was originally printed in Paris by the Comite de Madagascar in 1905-1906).

Grässe, Johann Georg Theodor. 1972.. *Orbis Latinus: Lexikon Lateinischer Geographischer Namen Des Mittelalters Und Der Neuzeit.* 3 vol. Braunschweig: Klinkhardt & Biermann.

Gravier, Gabriel. 2002. *La Cartographie De Madagascar.* Mansfield Centre, CT.: Martino Publishing. (reprint from the edition originally published by E. Gagniard, Rouen, 1896).

Haley, K.H.D. 1972. *The Dutch in the Seventeenth Century.* London: Thames and Hudson Ltd.

Hall, Richard. 1998. *Empires of the Monsoon, A History of the Indian Ocean and its Invaders.* London: Harper Collins.

Hallett, Robin. 1965. *The Penetration of Africa; European Enterprise and Exploration Principally in Northern and Western Africa up to 1830*, Volume 1 to 1815. London: Routledge and Kegan Paul.

—- , 1970. *Africa to 1875: A Modern History.* Ann Arbor: The University of Michigan Press.

Hamilton, Genesta. 1951. *In the Wake of Da Gama - the Story of Portuguese Pioneers in East Africa 1497-1729.* Skeffington.

Hargrave, Catherine Perry. 1930. *A History of Playing Cards and a Bibliography of Cards and Gaming.* Boston: Houghton Mifflin Co.

Harley, J.B., and David Woodward (editors). 1987. *The History of Cartography, Volume I: Cartography in Prehistoric, Ancient and Medieval Europe and the Mediterranean.* Chicago and London: University of Chicago Press.

—- and —- , 1992. *The History of Cartography, Volume 2, Book 1: Cartography in the Traditional Islamic and South Asian Societies.* Chicago and London: University of Chicago Press.

—- and —- , 1994. *The History of Cartography, Volume 2, Book 2: Cartography in the Traditional East and Southeast Asian Societies.* Chicago and London: University of Chicago Press.

Harrell, J. A. 2007. *Turin Papyrus Map from Ancient Egypt.* On-line http://www.eeescience.utoledo.edu/Faculty/Harrell/Egypt/Turin%20Papyrus/Harrell_Papyrus_Map_text.htm date consulted March 2007.

Harrell, J. A., and V. M. Brown. 1992. The oldest surviving topographical map from ancient Egypt (Turin Papyri 1879, 1899 and 1969). *Journal of the American Research Center in Egypt* 29 81-105.

Harris, Elizabeth. 1985.. The Waldseemüller world map: A typographic appraisal. *Imago Mundi* 37: 30-53.

Hart, Henry. 1950. *Sea Road to the Indies: An account of the voyage and exploits of the Portuguese navigators, together with the life and times of Dom Vasco da Gama, Capitâo-Mor, viceroy of India and Count of Vidigueira.* New York: The MacMillian Company.

Heawood, Edward. 1923. A Hitherto Unknown Map of the World. *The Geographical Journal London*.

—- , 1940. Florentine world-maps of Francesco Rosselli. *The Geographical Journal London* 95 (1940): 452-454

Heijden, H.A.M. van der. 1992. *De Oudste Gedrukte Kaarten Van Europa.* Alphen aan den Rijn: Canaletto.

—- , 1998. Heinrich Bunting's Itinerarum Sacrae Scripturae,1581. *Geography of the Bible in Quarendo* vol 28 no. 1: 49-71.

Hein, Jeanne. 1993. Portuguese Communication with Africans on the Sea Route to India. *Terrae Incognitae: The Journal for the History of Discoveries* 25: 41-51.

Heinz, Markus, and Ruth Bach-Damaskinos. 2002. *Auserlesene Und Allerneueste Landkarten' Der Verlag Homann In Nurnberg 1702-1848*. Nurnberg: Tummels.

Hellwig, Fritz. 1994.. On Giov. Lorenzo Anania and his Fabrica del Mondo: The Hitherto Unknown Maps of the First Edition (Naples, 1573). In *Liber Amicorum: Essays on Art, History, Cartography and Bibliography in Honour of Dr. Albert Ganado*, ed. Joseph Schirò, et al. Malta: Msida, 105-120.

Hoppen, Stephanie. 1975. Fifty Small and Miniature Maps of Africa. *Map Collector Circle* no 108.

Hourani, Georg Fadl. 1922. *Arab Seafaring in the Indian Ocean in Ancient and Early Medieval Times*. New Jersey-Princeton.

Huan, Ma. 1997. *Ying-yai Sheng-lan: The Overall Survey of the Ocean's Shores*. Bangkok: White Lotus Press

Hubbard, Jason. 2004. Nicolas De Fer's Untitled World Atlas (1684). *Map Forum Magazine*. 16-24.

—- , 2005. Nicolas De Fer's Untitled World Atlas (1684) Part II: Appendix and addenda. *Map Forum Magazine*. 46-63

Jacobson, William R. 2004. *The Rediscovery of Africa 1400-1900, Antique Maps & Rare Images. A narrative history and catalogue for an exhibition of antique African maps and rare books including the Oscar I. Norwich Collection at the Stanford University Libraries commencing*. Stanford, CA.: Stanford University Libraries.

Jäger, Eckard. 1982. *Prussia Karten : 1542 - 1810 ; Geschichte d. kartograph. Darst. Ostpreussens vom 16. bis zum 19. Jh.*. Weissenhorn: Konrad.

Joucla, Edmond. 2000. *Bibliographie De L'afrique Occidentale Francaise*. Mansfield Centre, CT.: Martino Publishing. (reprint of the 1912 first edition published by E. Sansot et Cie of Paris).

Kamal, Yusuf. 1926 – 1951. *Monumenta Cartographica Africae et Aegypti*. Volumes in 16 parts. Cairo: Privately Published.

Kammerer, Albert. 1947. *La Mer Rouge, l'Abyssinie et l'Arabie aux XVI et XVII Siécles et la Cartographie des Portulans du Monde Oriental. Premiére Partie, XVI Siècle. Abyssins et Portuguais Devant l'Islam*. Le Caire: Societé Royale de Geographie d'Egipte.

—- , 1949. *La Mer Rouge, l'Abyssinie et l'Arabie aux XVI et XVII Siécles et la Cartographie des Portulans du Monde Oriental. Seconde Partie, XVII Siècle. Les Jesuites Portugais et l'Ephemere Triomphe du Catolicisme en Abyssinie (1603-1632)*. Le Caire: Societé Royale de Geographie d'Egipte.

Karrow, Robert W. 1993. *Mapmakers of the Sixteenth Century and Their Maps*. Chicago: Newberry Library by Speculum Orbis Press.

Kayser, Gabriel. 2003.. *Bibliographie D'ouvrages Ayant Trait A L'afrique En General Dans Ses Rapports, Avec L'exploration & La Civilisation De Ces Contrees*. Mansfield Centre, CT: Martino Publishing Co. (reprint from the original edition published in Bruxelles, Belgium, 1887).

Keuning, Jonannes. 1948. Jodocus Hondius Jr. *Imago Mundi* 5: 63-71

—- and Marijke Donkersloot-de Vrij. 1973. *Willem Jansz Blaeu: A Bibliography And History Of His Work As A Cartographer And Publisher*. Amsterdam: Theatrum Orbis Terrarum.

—- , 1960. Pieter van den Keere (Petrus Kaerius), 1571-1646(?). *Imago Mundi*: 66-72

Kimble, George H.T. 1935. The Laurentian world map with special reference to its portrayal of Africa. *Imago Mundi* I: 29-33.

—- , 1938. *Geography in the Middle Ages*. London: Methuen & Co. Ltd.

King, Geoffrey L. 2003. *Miniature Antique Maps, second edition*. Oxfordshire, England: Tooley Adams & Co.

Klaus, Wolfram. 1990. Pläne Und Grundrisse Afrikanischer Städte (1550-1945). *Kartographische Bestandsverzeichnisse* 5. Berlin: Deutsche Staatsbibliothek.

Klemp, Egon, 1968. *Africa on Maps Dating from the Twelfth to the Eighteenth Century*. Leipzig: Ed. Leipzig.

—- , 1971. *Commentary on the Atlas of the Great Elector*. Stuttgart, Berlin, and Zurich: Belser

Koeman, Cornelis. 1967-1971. *Atlantes Neerlandici. Bibliography of terrestrial, maritime, and celestial atlases and pilot books, published in the Netherlands up to 1880: Vol. I Van der Aa-Blaeu, Vol. II Blussé-Mercator, Vol. III Merula-Zeegers, Vol. IV: Celestial and maritime atlases and pilot books, Vol. V.: Indexes*. Amsterdam: Theatrum Orbis Terrarum Ltd.

Kraus, H. P. 1956. *The Eightieth Catalogue, Remarkable Manuscripts, Books, Maps*.

—- , [1970]. *Monumenta Cartographica* (Catalogue 124), undated.

Krogt, Peter van der. 1985. *Advertenties voor kaarten, atlassen, globes e.d. in Amsterdamse kranten 1621-1811*. Utrecht: HES Uitgevers.

—- with Marc Hameleers, and Paul van den Brink. 1993. *Bibliografie van de Geschiedenis van de Kartografie van de Nederlanden/Bibliography of the History of Cartography of the Netherlands.* Utrecht: HES.

—- , 1994. Commercial Cartography in the Netherlands with particular reference to atlas production (16th-18th centuries). In : *La Cartografia dels Paises Baixos.* Barcelona: Institut Cartografic de Catalunya. p. 70-140

—- , 1997. *Koeman's Atlantes Neerlandici, Vol. I: The Folio Atlases published by Gerard Mercator, Jodocus Hondius, Henricus Hondius, Johannes Janssonius and Their Successors.* 't Goy-Houten: HES.

—- , 2000. *Koeman's Atlantes Neerlandici, Vol. ll: The Folio Atlases published by Willem Jansz. Blaeu and Joan Blaeu.* 't Goy-Houten: HES & De Graaf.

—- , 2002. The Map Image of Africa. Dutch atlases of the sixteenth and seventeenth centuries in Dutch Geography and Africa. In *Dutch Geography and Africa*, International Geographical Union, Section The Netherlands, 118-158. Nederlandse Geografische Studies, 300. Utrecht: Koninklijk Nederlands Aardrijkskundig Genootschap; Faculteit Ruimtelijke Wetenschappen, Universiteit Utrecht

—- , 2003. *Koeman's Atlantes Neerlandici, Vol. III: Ortelius' Theatrum Orbis Terrarum, De Jode's Speculum Orbis Terrarum, The Epitome, Caert-Thresoor and Atlas Minor, The Atlases of the XVII Provinces, and some other Atlases Published in the Low Countries up to c.1650.* 't Goy-Houten: HES & De Graaf.

Lach, Donald F. 1965. *Asia in the Making of Europe. Vol. I: The Century of Discovery.* Chicago, University of Chicago Press.

Lacroix, W. F. G. 1998. *Africa in Antiquity. A linguistic and toponymic analysis of Ptolemy's map of Africa, together with a discussion of Ophir, Punt and Hanno's voyage.* Saabrucken: Nijmegen Studies in Development and Cultural Change.

Lane-Poole, E.H. 1950. *The Discovery of Africa: A history of the exploration of Africa reflected in the maps in the collection of the Rhodes-Livingstone Museum.* Occasional Paper. Livingstone North Rhodesia: The Rhodes-Livingstone Museum.

Langer, William L. 1948. *An Encyclopedia of World History.* Boston: Houghton Mifflin Co.

Langlands, B.W. 1961. *Maps of Africa, 1540-1850: An Introduction and catalog to an exhibition of maps displayed in the Uganda Museum, September 1961.* Kampala: Uganda Museum.

Lana, Gabriella et al. 1967. *Glossary of Geographical Names in Six Languages: English, French, Italian, Spanish, German, and Dutch, Glossaria interpretum.* Amsterdam: Elsevier.

La Roncière, Charles de. 1924. *La Découverte de l'Afrique au Moyen Âge* vol.1. Cairo: Mémoires de la Société Royale de Géographie d'Egypte.

—- . 1925. *La Découverte de l'Afrique au Moyen Âge* vol.2. Cairo: Mémoires de la Société Royale de Géographie d'Egypte.

—- . 1927. *La Découverte de l'Afrique au Moyen Âge* vol.3. Cairo: Mémoires de la Société Royale de Géographie d'Egypte.

Lawrence, A.W. 1963. *Trade Forts and Castles of West Africa.* London: Jonathan Cape.

Lenox, James. 1877. *Voyages of Hulsius, etc.* New York: Printed for the Trustees, NY Public Library.

Lexikon *zur Geschichte der Kartographie von den Anfängen bis zum Ersten Weltkrieg.* Hrsg. von Ingrid Kretschmer et al. Wien: Franz Deuticke, 1986.

Loeb-Larocque, Louis. 1989. Ces Hollandaises habillées à Paris ou L'exploitation de la cartographie hollandaise par les éditeurs parisiens au XVIIe siècle. In *Theatrum Orbis Librorum. Liber Amicorum presented to Nico Israel on the occasion of his seventieth birthday*, edited by Tom Croiset van Uchelen, Koert van der Horst, and Günter Schilder. Utrecht: Hes Publishers: 15-30.

Löwenhardt, Werner, and Rodney Shirley. 1983. *Seek and ye shall Find... an unidentified world map in a Dutch Bible.* The Map Collector 25: 10-12.

Maggs Brothers. 1929. *Bibliotheca Asiatica et Africana Part IV.* London: Maggs Brothers Catalogue.

Magna Gallery catalogue 1984, Oxford, UK.

Martayan Lan. *Martayan Lan Fine Antique Maps, Atlases & Globes catalogue # 17*, New York City, undated.

Marques, Alfredo Pinheiro. 1988. *Novos Elementos Sobre a Cartografia Portuguesa Vinte e Sete Anos Depois da Primeira Publicação dos 'Portugaliae Monumenta Cartographica'. New Materials Relating to Portuguese Cartography Twenty-Seven Years After the First Publication of 'Portugaliae Monumenta Cartographica. vol VI.* Lisbon: INCM, [7]-[114].

—- , 1989. The Dating of the Oldest Portuguese Charts. *Imago Mundi* 41: 87-97.

Mayhew, Robert. 2003. *Cosmographie / Peter Heylyn*, a facsimile edition as part of historical cultures and geography series, 1600-1750. Bristol, U.K.: Thoemmes.

McEvedy, Colin. 1980. *Atlas of African History, from 175 million years ago to the present, concisely explained in 59 clear maps and a lucid text.* New York: Facts of File.

McIlwaine, John. 1997. *Maps and the Mapping of Africa: A Resource Guide.* Wayne, N.J.: James Currey Publishers.

McLaughlin, Glen, and Nancy Mayo. 1995. *The Mapping of California as an Island. An illustrated checklist.* California Map Society: Occasional Paper no. 8.

Mendelssohn, Sidney. 1993. *South African Bibliography.* 2 volumes. Cambridge: Maurizio Martino Publisher. (reprint of the 1910 first edition).

Meurer, Peter. 1988. *Atlantes Colonienses Die Kolner Schule der Atlas Kartographie 1570-1610.* Bad Neustadt: Verlag Dietrich Pfaehler.

—- , **2001.** *Corpus der älteren Germania-Karten: Ein annotierten Katalog der gedruckten Gesamtkarten des deutschen Raumes von den Anfängen bis um 1650.* Alphen aan den Rijn: Canaletto / Repro-Holland.

—- , **2004.** *The Strabo Illustratus Atlas : A unique sixteenth century composite atlas from the House of Bertelli in Venice.* Bedburg-Hau: Antiquariat Gebr. Haas; Paris: Librairie D. Le Bail & Weissert.

—- , **2007.** Sanson-Jaillot Nachstiche des Nürnberger Verlages Johann Hoffmann. *Cartographica Helvetia*, 35: 9-19.

Mickwitz, Ann-Mari, Leena Miekkavaara, and Tuula Rantanen (compiled). 1979-1995. *The A.E. Nordenskiöld Collection in the Helsinki University Library, Annotated Catalogue of Maps made up to 1800.* 5 volumes in 6 books. Helsinki: Helsinki University Library.

Mollat du Jourdin, Michel, and Monique de la Ronciere. 1984. *Sea Charts of the Early Explorers.* Translated from the French by L. le R. Dethan. New York: Thames and Hudson.

Moorehead, Alan. 1960. *The White Nile.* New York: Vintage Books

—- , **1962.** *The Blue Nile.* New York: Vintage Books.

Moreland, Carl, and David Bannister. 1989.. *Christie's Collectors Guides: Antique Maps.* 3rd ed. Oxford: Phaidon.

Morris, John Gottlieb. 1955. *Martin Behaim, the German Astronomer and Cosmographer.* Baltimore: Maryland Historical Society.

National Maritime Museum, Greenwich 1975. *Catalogue of the Library. Vol. 1 Voyages & Travel, 1968. Vol. 2 Biography, Parts One and Two, 1969. Vol. 3 Atlases & Cartography, Parts One and Two, 1971. Vol. 4 Piracy and Privateering, 1972. Vol. 5, Naval History, Part One: The Middle Ages to 1815.* Greenwich.

Nebenzahl, Kenneth. *Compass*, issue no. 6, item 29 (State 4 of Americas map).

Needham, J. 1975. *Science and Civilization in China,* vol. 3. London: Cambridge University Press, 551-555.

Newton, A.P. ed. 1926. *Travel and Travelers of the Middle Ages,* London: Kegan Paul.

Nicolardi, Silvia Curi. 1984. *Una Societa Tipografico-Editoriale A Venezia Nel Secolo XVI Melchiorre Sessa E Pietro Di Ravani (1516-1525).* Firenzia: Leo S. Olschki Editore.

Nordenskiöld, A.E. 1973. *Facsimile Atlas to the Early History of Cartography with Reproductions of the Most Important Maps Printed in the XV and XVI Centuries.* New York: Dover Publications, Inc.

Norwich, Oscar I. 1976. *Italian Copperplate Map Engravings - Rare Examples of Separate Ones of Africa. Africana Notes and News* Vol 22, No. 2. Johannesburg: Africana Society of Africana Museum: 76-83.

—- , **1993.** *Maps of Southern Africa.* Johannesburg: AD Donker Publisher.

—- , **1997.** *Norwich's Maps of Africa: An Illustrated And Annotated Carto-Bibliography.* revised and edited by Jeffery C. Stone. Norwich, VT: Terra Nova Press.

O'Donoghue, Freeman M. 1901. *Catalogue of the collection of playing cards bequeathed to the trustees of the British Museum by the late Lady Charlotte Schreiber.* London: Longmans & Co., B. Quartitch, and Asher & Co.

Oehme, R. 1968. *Cosmographei: Basel 1550 / Sebastian Munster* Amsterdam: Theatrum Orbis Terrarum.

Oxford Atlas of the World. 2002. New York: Oxford University Press.

Palma, Maria Terea di. 1991. *Four Wall maps of the continents made by J. Blaeu/de Rossi (1666).* Paper presented at the 14th International Conference on the History of Cartography in Sweden, 14-19 June 1991.

Parker, John. 1956. *From Lisbon to Calicut.* Transl.: Alvin E. Prottengeier. Minneapolis: University of Minnesota Press for James Ford Bell Collection.

Pastoureau, Mireille 1980. *Les Atlas Imprimés en France avant 1700*. *Imago Mundi* 32: 45-72.

—, 1984. *Les Atlas Français XVIe-XVIIe Siècles*. Paris: Bibliotheque Nationale.

—, 1987. Maps at the Bibliothèque Nationale: a collection of collections. *The Map Collector* 40: 8-12.

Peirce, Benjamin. 1931. *A Catalogue of the Maps and Charts in the Library of Harvard University in Cambridge, Massachusetts*. Cambridge: E.W. Metcalf and Company, Printers to the University.

Penfold, P.A. 1982. *Maps and Plans in the Public Record Office: Vol. 3. Africa*. London: H.M.S.O.

Penney, Clara Louisa. 1982. *Printed Books 1468-1700 in the Hispanic Society of America*. New York: Hispanic Society of America.

Pennington, Richard. 1982. *A Descriptive Catalog of the Etched Works of Wenceslaus Hollar 1607-1677*. Cambridge: Cambridge University Press.

Peeters-Fontainas, Jean Felix. 1933. *Bibliographic des impressions espagnoles des Pays - Bac avec preface de Maurice Sabbe*. Louvain, Antwerpen: J. Peeters-Fontainas, Musée Plantin-Moretus (with reprint in 1965 by B. de Graf).

Phillips, Philip Lee, and Clara Egli LeGear. 1909-1992. *A List of Geographical Atlases in the Library of Congress, with Bibliographical Notes*. 9 volumes. Washington: Government Printing Office/Library of Congress.

Plak, Adrian. 1989. The editions of the [Hondius] Atlas Minor until 1628. In: *Theatrum Orbis Librorum. Liber Amicorum presented to Nico Israel*. edited by Tom Croiset van Uchelen, Koert van der Horst, and Günter Schilder. Utrecht: Hes Publishers

Randles, W.G.L. 1956. Southeast Africa as Shown on Selected Printed Maps of the Sixteenth Century. *Imago Mundi* 13: 69-88.

—, 1958. *Southeast Africa and the Empire of Monomotapa as Shown on Selected Printed Maps of the 16th Century*. Lisbon: Centro de Estudoas Historicos Ultramarinos.

—, 1979. *The Empire of Monomotapa from the Fifteenth to the Nineteenth Centuries*. Salisbury (Harare): Mambo Press.

—, 1988. *Bartolomeu Dias and the Discovery of the South-East Passage Linking the Atlantic to the Indian Ocean. RUC, Coimbra BGUC*: 24: 19-28

—, 1993. The Alleged Nautical School Founded in the Fifteenth Century at Sagres by Prince Henry of Portugal, Called the 'Navigator'. *Imago Mundi* 45: 20-28.

Rauchenberger, D. 1999. *Johannes Leo der Afrikaner: Seine Beschreibung des Raumes zwichen Nil und Niger nach dem Urtext*. Harrassowitz Publishing House, Orientalia Biblica et Christina 13.

Ravenstein, Ernest G. 1890. The Voyages of Diogo Cão and Bartholomeu Dias 1482-88. *The Geographical Journal London*: 16 (6).

—, 1891. The Lakes Region of Central Africa. A contribution to the cartography of Africa. *The Scottish Geographical Magazine* 7: 299-310.

—, 1908. *Martin Behaim, his life and his Globe*. London: George Philip and Son.

Reinhartz, Dennis. 1997. *The Complete Geographer and the Literati – Herman Moll and his Intellectual Circle*. Lewiston, NY: Texas A & M University Press.

Reiss & Auvermann Buch und Kunstantiquariat Auction Catalog. 1989. Travel and Exploration – Portugal and Spain. Auction 40. Königstein, Germany.

Reiss & Sohn 1999. Reiss auction catalog of 1999. Königstein, Germany.

—, 2001. Reiss auction catalog. of October 2001.

—, 2005. Reiss auction catalog. of April 2005.

Relaño, Francesc. 1995. Against Ptolemy: The Significance of the Lopes-Pigafetta Map of Africa. *Imago Mundi* 47: 49-66.

—, 2002. *The Shaping of Africa: Cosmographic Discourse and Cartographic Science in Late Medieval and Early Modern Europe*. Aldershot, England: Ashgate Publishing Company.

Rey, Charles F. 1929. *The Romance of the Portuguese in Abyssinia (1490-1633)*. London: Witherly.

Ristow, Walter. 1967. America and Africa: Two Seventeenth Century Wallmaps. *Quarterly Journal of the Library of Congress*. Washington, DC: Library of Congress.

—, 1972. *Jaillot wall map of Africa in A la Carte: Selected Papers on Maps and Atlases*. Washington, DC: Library of Congress.

Rizzo, Gerald. (2006) The Patterns and Meaning of a Great Lake in West Africa. *Imago Mundi* 58 I: 80-89.

Ruland, Harold. 1962. A Survey of Double-page Maps in the Thirty-Five Editions of the Cosmographia Universalis 1544-1628 of Sebastian Munster and in his Editions of Ptolemy's Geographia 1540-1552. Imago Mundi 16: 84-96.

Sabin, Joseph. 1998. *Bibliotheca Americana. Dictionary of Books Relating to America From its Discovery to the Present Time.* New York: 2 vols. Mansfield Centre, CT: Maurizio Martino (reprint from 1868).

Sanceau, Elaine. 1944. *The Land of Prester John, A Chronicle of Portuguese Exploration.* New York: Alfred Knopf.

—- , 1968. *Good Hope: The Voyage of Vasco Da Gama.* Lisboa: Academia Internacional Da Cultura Portuguesa.

Schilder, Günter. 1981. The Cartographical Relationships between Italy and the Low Countries in the Sixteenth Century (with an illustration of the sole surviving example of the second state of De Jode's map of Africa). *Map Collector* 17: 2-8.

—- , 1987. *Monumenta Cartographica Neerlandica, II.* Alphen aan den Rijn: Canaletto.

—- , 1990. *Monumenta Cartographica Neerlandica, Volume III (covering the two states of the Blaeu world map, 1619 & 1645-6).* Alphen aan den Rijn: Canaletto.

—- , 1992. An unrecorded set of thematic maps by Hondius. *The Map Collector* 59: 44-46.

—- , 1993. Monumenta Cartographica Neerlandica, IV: Single sheet maps and topographical prints published by Willem Jansz. Blaeu. Alphen aan den Rijn: Canaletto.

—- , 1996. *Monumenta Cartographica Neerlandica*, V: *Ten Wall Maps by Blaeu and Visscher.* Alphen aan den Rijn: Canaletto.

—- , 2000. *Monumenta Cartographica Neerlandica, VI: Dutch folio-sized single sheet maps with decorative borders.* Alphen aan den Rijn: Canaletto.

—- , 2003. *Monumenta Cartographica Neerlandica, VII: Cornelis Claesz (c.1551-1609) Stimulator and Driving Force of Dutch Cartography.* Alphen aan den Rijn: Canaletto.

—- , 2007. *Monumenta Cartographica Neerlandica, VIII: Jodocus Hondius (1563-1612) and Petrus Kaerius (1571- c.1646).* Alphen aan den Rijn: Canaletto.

Seltzer, Leon E. (ed), 1952.) *Columbia Lippincott Gazetteer of the World.* Morningside Heights, N.Y.: Columbia University Press, by arrangement with J.B. Lippincott Co.

Servicio Geografico del Ejercito 1946. *Exposición de Cartografía Africana. Organizada por el Servicio Geográfico del Ejército.* Patrocinada por la Dirección General de Marruecos y Colonias. Madrid.

—- , 1975. *Seccion de Documentacion. Cartoteca Historica- Indice de Mapas y Planos historicos de Africa* (list of maps and plans related to Africa). Madrid.

Seville, Adrian. 2005. *Le Jeu de France-Pierre Duval's Map Game. Brussels International Map Collectors' Circle (BIMCC)* Newsletter No.: 24-27.

Sheehy, Eugene P. 1986. *Gazetteers and Geographical Names and Terms,.* In *Guide to Reference Books* 10th ed. p. 944-957. Chicago: American Library Association.

Shefrin, Jill, 1999a. *Make it a pleasure and not a task.* Princeton University Library Chronicle, no. 2, 1999. 5: 60.

—- , 1999b. *Neatly Dissected for the Instruction of Young Ladies and Gentlemen in the Knowledge of Geography: John Spilsbury and Early Dissected Puzzles.* Los Angeles: The Cotsen Occasional Press.

Shirley, Rodney. 1993. *The Mapping of the World: Early Printed World Maps 1472-1700.* London: New Holland Publishers Ltd.

—- , 2004. *Maps in the Atlases of The British Library: a descriptive catalogue c. AD 850 - 1800.* London: The British Library.

Short, John Rennie. 2003. *The World through Maps: A History of Cartography.* Buffalo, New York: Firefly Books, Inc.

Silverberg, Robert. 1972. *The Realm of Prester John.* Athens, Ohio: Ohio University Press.

Sims, Douglas. [2003]. *Giacomo Gastaldi and The Four Continents* (unpublished work dated 2003).

Skelton, Raleigh A. 1961a. *English* Knowledge of the Portuguese Discoveries in the 15th Century: a New Document. *Actas do Congresso Internacional de História dos Descobrimentos,* Lisboa, 2: 365-374.

—- , 1961b. Gastaldi's Map of 1564. In *The Prester John of the Indies: A True Relation of the Lands of the Prester John Being the Narrative of the Portuguese Embassy to Ethiopia in 1520 Written by Father Francisco Alvares.* (Hakluyt Society publications, Series 2, vols. 114-115). Cambridge, Vol. 2, 562-67.

—- , 1964. *Introduction to the facsimile of Abraham Ortelius: Theatrum Orbis Terrarum. Antwerp 1570.* Amsterdam: N. Israel, Publisher; Meridian Publishing Co.

—- , 1965a. *Bibliographical Note.* in *Geografia dell'Africa, Venice 1588,* etc., by Livio Sanuto. Amsterdam: Theatrum Orbis Terrarum.

—, **1965b.** *Introduction to the facsimile edition of G. de Jode's Speculum Orbis Terrarum 1578.* Amsterdam: Theatrum Orbis Terrarum Ltd.

—, **1965c.** *Introduction to the facsimile edition of Livio Sanuto's Geografia Dell' Africa. Venice 1588.* Amsterdam: Theatrum Orbis Terrarum Ltd.

—, **1966a.** Ptolemaeus, Claudius (edited by Munster, Sebastian). *Geographia*, Basel 1540. Amsterdam: Theatrum Orbis Terrarum Ltd.

—, **1966b.** *Bibliographical Note to the Facsimile of Speed's Prospect.* Amsterdam: Theatrum Orbis Terrarum Ltd.

—, **1967-1970.** *Introduction and an analysis of the contents by George B. Parks. Navigationi et viaggi* Venice, 1563-1606, Gian Battista Ramusio, in 3 vol. Amsterdam: Theatrum Orbis Terrarum Ltd.

—, **1968a.** *Introduction to the facsimile edition of the English edition of the Mercator Hondius Atlas, 1636.* Amsterdam: Theatrum Orbis Terrarum Ltd.

—, **1968b.** *Bibliographical Notes to the facsimile of Abraham Ortelius, The Theatre of the Whole World.* Amsterdam: Theatrum Orbis Terrarum Ltd.

Smith, A.H. 1952.. *Exhibition of decorative maps of Africa up to 1800, 4-16 August 1952; descriptive catalog.* Johannesburg: Johannesburg Public Library & Africana Museum.

Snow, Philip. 1988. *The Star Raft: China's Encounter with Africa.* New York: Weidenfeld & Nicolson.

Soulsby, B. H. 1902. *The First Map Containing the name America.* The Geographical Journal London

Sotheby's Catalog, 1985. Sotheby's Catalog Nov. 7, 1985, lot # 392.

—, **1987.** Sotheby's London auction catalog of 25 June 1987, lot 339

—, **1990.** Sotheby's auction catalogue of February 6, 1990.

—, **1992.** Sotheby's auction catalogue of 25 June 1992, lot 364.

—, **1993.** Sotheby's auction of 7 December 1993

—, **2000.** Sotheby's December 2000 catalog.

—, **2005 October.** The Wardington Library, Important Atlases & Geographies Part One: A-K.

—, **2006 October.** The Wardington Library, Important Atlases & Geographies Part Two: L-Z.

Sprent, F. P. 1924. *A map of the world designed by G. M. Contarini, engraved by F. Rosselli, 1506.* London: British Museum.

Stevens, Henry N. 1972. *Ptolemy's Geography. A Brief Account of All the Printed Editions down to 1730,* Amsterdam: Meridian Theatrum Orbis Terrarum Ltd.

Stevenson, Edward Luther, (trans.and ed). 1991. *Claudius Ptolemy: The Geography*, with an introduction by Joseph Fischer. New York: Dover Publications, Inc.

Stone, Jeffrey C. 1995. A Short History of the Cartography of Africa. *African Studies, vol. 39.* Lewiston, N.Y.: The Edwin Mellon Press.

Stopp, Klaus, and Herbert Langel. 1974. *Katalog der alten Landkarten in der Badischen Landesbibliothek Karlsruhe.* Karlsruhe: G. Brown.

Strauss, Gerald. 1959. *Sixteenth-Century Germany: Its Topography and Topographers.* Madison: The University of Wisconsin Press.

Strandes, Justus. 1961. *The Portuguese Period in East Africa.* Nairobi: East African Literature Bureau, (translation of 1899 first edition by Jean F. Wallwork; edited with topographical notes by J. S. Kirkman).

Teixeira da Mota, A. 1964. *A Cartografia Antiga da Africa Central e a Travessia entre Angola e Mozambique (1500-1860).* Lorenzo Marques: Sociedade de Estudos de Moçambique.

Terea di Palma, Maria. 1991. *Four Wallmaps of the continents made by J. Blaeu/de Rossi (1666).* Paper presented by Maria Terea di Palma at the 14th International Conference on the History of Cartography in Sweden, 14-19 June 1991.

Thrower, Norman J.W. 1978. *The Compleat Plattmaker. Essays on Chart, Map, and Globe Making in England in the Seventeenth and Eighteen Centuries.* Berkeley: University of California Press.

Tibbetts, G.R. 1971. *Arab Navigation in the Indian Ocean Before the Coming of the Portuguese.* London: Royal Asiatic Society.

—, **1973.** Comparisons Between Arab and Chinese Navigational Techniques. London: *Bulletin of the School of Oriental and African Studies*, 36 part 1: 97-108.

Tiele, P.A., and Frederik Muller. 1867. *Mémoire bibliographique sur les journaux des navigateurs néerlandais. Réimprimés dans les collections de Bry et de Hulsius, et dans les collections hollandaises du XVIIe siècle, et sur les anciennes éditions hollandaises des journaux de navigateurs étrangers; la plupart en la possession de Frederik Muller à Amsterdam.* Amsterdam: Nico Israel, 1969. (reprint from 1867).

Tooley, Ronald V. 1939. Maps in Italian Atlases of the Sixteenth Century. *Imago Mundi* 3: 12-47.

—, 1966. *The Printed Maps of the Continent of Africa and Regional Maps South of the Tropic of Cancer, 1500-1600, Part I and II.* London: Map Collectors' Circle (Map Collectors' Series no. 29 & 30).

—, 1968. *Maps of Africa: A Selection of Printed Maps from the Sixteenth to the Nineteenth Century, Part I and II.* London: Map Collectors' Circle (Map Collectors' Series no. 47 & 48).

—, 1969. *Collectors' Guide to Maps of the African Continent and Southern Africa.* London: Carta Press.

—, 1972. *A Sequence of Maps of Africa.* London: Map Collectors' Circle (Map Collectors' Series no. 82).

—, 1978. *Maps and Map-makers.* London: B. T. Batsford Ltd.

—, 1980. *The Mapping of America.* London: The Holland Press Ltd.

—, 1999-2004. *Tooley's Dictionary of Mapmakers, Revised Edition, A - D, E – J, K – P, and Q – Z.* Edited by Josephine French and Valerie Scott. Riverside CT: Early World Press.

Tyacke, Sarah. 1978. *London Map-sellers 1660-1720.* Tring, Hertfordshire: Map Collector Publications Ltd.

Valerio, Vladimiro. 1993. *Atlanti italiani dall'invenzione della stampa all'affermazione della litografia.* In *La Cartografia Italiana, Cicle de conferences sobre Historia de la Cartografia.* Barcelona: Istitut Cartografic de Catalunya, 1993, 147-201.

—, 1999. *Atlanti italiani dal XV al XVII secolo, L'Universo.* LXXIX. p. 103-132.

Vasconcellos, Ernesto de. 1903-1904. *Exposicao De Cartographia Nacional* Lisbon: Sociedade De Geographia De Lisboa.

Vogt, John. 1979. *Portuguese Rule on the Gold Coast, 1469-1682.* Athens: University of Georgia Press.

Wallis, Helen. 1982. 'So Geographers in Afric-Maps'. *The Map Collector* 35: 30-34.

- - and Dorothy Middleton. [n.d.]. *The Mapping and Exploration of Africa and Joseph Banks in Maps and Mapping of Africa.* SCOLMA.

Warnsinck, J.C.M. 1939. *De wetenschappelijke voorbereiding van eerste schipvaart naar Oost-Indie.* –'s-Gavenhage: Nijhoff.

Welch, Sidney R. 1935. *Europe's Discovery of South Africa.* Cape Town: Juta & Co.

—, 1952. *Portuguese and Dutch in South Africa, 1641-1806.* Cape Town & Johannesburg.

Werner, Jan W.H. 1994. *Inde Witte Pascaert: Kaarten en atlassen van Frederick de Wit.* Amsterdam: Universiteitsbibliotheek Amsterdam.

—, 1998. *Caert-Thresoor, inhoudende de tafelen des gantsche werelts landen, met beschryvingen verlicht, tot lust vanden leser, nu alles van nieus met groote costen ende arbeyt toegereet.* Weesp: Robas.

Wieder, F.C. 1925-1933. *Monumenta Cartographica: reproductions of unique and rare maps, plans and views in the actual size of the originals; accompanied by cartographical monographs,* 5 volumes, The Hague: Martinus Nijhoff.

—, 1925-1933. Descriptive Catalogue of Maps published separately by Blaeu. In *Monumenta Cartographica vol III.* The Hague: Martinus Nijhoff.

Willoughby, Francis. 2003. *Francis Willoughby's book of games: a seventeenth-century treatise on sports, games, and pastimes,* edited and introduced by David Cram, Jeffrey L. Forgeng and Dorothy Johnston. Aldershot, Hants, England & Burlington, VT: Ashgate.

Winius, George D., and Bailey W. Diffie. 1977. *Foundations of the Portuguese Empire 1415-1580.* Minneapolis: University of Minnesota Press.

—, 1995. *Portugal: the Pathfinder. Journeys from the Medieval toward the Modern World. 1300-ca.1600.* Madison: Luso-Brazilian Review-University of Wisconsin: 363-372.

Wing, Donald. 1945-1951. *Short-title Catalogue of Books Printed in England, Scotland, Ireland, Wales, and British North America and of English Books Printed in Other Countries 1641-1700.* New York: Columbia University Press.

Witte, Guy De. 1994. Preservatie, conservatie en restauratie van de IATO-atlas van de Koninklijke Oudheidkundige Kring van het Land van Waas. *Annalen van de Koninklijke Oudheidkundige Kring van het Land van Waas* 97: 195-232.

Wittkower, Rudolf. 1942. Marvels of the East: A Study in the History of Monsters. *Journal of the Warburg and Courtauld Institutes* 5: 194-195.

Wolff, Hans. 1992. Martin Waldseemüller, The Most Important Cartographer in a Period of Dramatic Scientific Change, p. 111-126. In *America, Early Maps of the New World, edited by* Hans Wolff. Munich: Prestel.: 111-126

Woodward, David (ed.). **1975.** The Woodcut Technique. In: *Five Centuries of Map Printing.* Chicago: 25-50

—, **1985.** Reality, symbolism, time, and space in medieval world maps. *Annals of the Association of American Geographers* 75: 510-21.

—, **1990.** *The Maps and Prints of Paolo Forlani: A Descriptive Bibliography.* Chicago: Hermon Dunlap Smith Center for the History of Cartography, Occasional Publication 4.

—, **1997.** *The Four Parts of the World: Giovanni Francesco Camocio's Wall Maps.* Minneapolis: James Ford Bell Lectures 34.

— **and Lewis, G. Malcolm,** ed. **1998.** *The History of Cartography, Volume 2, Book 3: Cartography in the Traditional African, American, Arctic, Australian, and Pacific Societies.* Chicago-London: University of Chicago Press.

Zisska, F. & R. Kistner. 2001. *Auction Catalog No. 37 of 8-11 May 2001.* Munich.

Libraries cited in this Cartobibliography

Where the names for libraries are abbreviated within the cartobibliography, the full name is listed below along with the international code for the country.

Amsterdam HM = Amsterdam Historisch Museum (NL)
Amsterdam UB = Universiteitsbibliotheek Amsterdam (NL)
Amsterdam SMA = Nederlands Scheepvaartmuseum, Amsterdam (NL)
Antwerpen PM = Plantin-Moretus Museum, Antwerp (NL)
Ann Arbor Clem = William L. Clements Library (USA)
Ann Arbor UML = University Of Michigan Libraries (USA)
Austin HR = Harry Ransom Humanities Research Center of the University of Texas at Austin (USA)
Basel UB = Universitätsbibliothek Basel (CH)
Berkeley UC = University of California, Berkeley (USA)
Berlin SB = Staatsbibliothek zu Berlin (D)
Bern LB = Schweizerische Landesbibliothek (CH)
Bern SUB = Stadt – und Universitätsbibliothek (City and University Library) (CH)
Besançon BM = Bibliothèque Municipale d'étude Besançon (F)
Birmingham CL = Birmingham Central Library (GB)
Bologna BU = Biblioteca Universitaria (I)
Bologna BUN = Museo delle Navi, Biblioteca Universitaria (I)
Boston Afriterra = Boston Afriterra Collection (USA).
Boston PL = Boston Public Library (USA)
Bordeaux BM = Bibliothèque Municipale Bordeaux (F)
Bourges BM = Bibliothèque Municipale Bourges (F)
Breslau SB = Stadtbibliothek, Breslau (D)
Brussels AB = Afrika Bibliotheek (B)
Budapest NSL = National Széchényi Library (H)
Burgdorf R = Rittersaalverein, Burgdorf (CH)
Cambridge Harv MC = Harvard University Map Collection, Pusey (USA)
Cambridge Harv RB = Harvard University Houghton Rare Book Collection (USA)
Charleston CL = College of Charleston (USA)
Chicago NL = Newberry Library (USA)
Copenhagen KB = Det Kongelige Bibliothek (DK)
Den Haag KB = Koninklijke Bibliotheek Den Haag (NL)
Dijon BM = Bibliothèque Municipale de Dijon (F)
Dresden LB = Sächsische Landesbibliothek – Staats und Universitätsbibliothek Dresden (D)
Dresden SH = Sächsisches Hauptstaatarchiv, Dresden (D)
Edinburgh NLS = National Library of Scotland (GB)
Enkhuisen MB = Municipale Bibliotheek (NL)
Erlangen UB = Universitätsbibliothek Erlangen-Nürnberg (D)
Evanston NWU = Northwestern University Library, Herskovits Library, Evanston, IL, (USA)
Firenze BML = Biblioteca Medicea Laurenziana (I)
Firenze IGM = Instituto Geographico Militare (I)
Firenze BN = Biblioteca Nazionale (I)
Fulda LB = Hessische Landesbibliothek (D)
Glasgow UL = Glasgow University Library, Special Collection (GB)
Göttingen SUB = Niedersächsische Staats- u. Universitätsbibliothek (D)
Graz LB = Steiermärkische Landesbibliothek (A)
Greenwich NMM = National Maritime Museum (GB)
Grenoble BM = Bibliothèque Municipales de Grenoble (F)
Helsinki AEN = University Library, A.E. Nordenskiöld Collection (FIN)
Johannesburg PL = Harold Strange Library of African Studies, Johannesburg Public Library (RSA)
Karlsruhe LB = Badische Landesbibliothek Karlsruhe (D)

Kyoto, RU = Ryukoku University Library (JP)
Leeuwarden Tr = Tresoar (NL)
Leiden UB = Universiteitsbibliotheek Lieden (NL)
London BL = British Library (GB)
London BM = The British Museum (GB)
London RGS = Royal Geographical Society (GB)
Lüneburg RB = Lüneburg Ratsbücherei (D)
Madrid BM = Biblioteca Nacional Madrid (E)
Madrid NM = Museo Naval (E)
Milwaukee AGS = American Geographical Society Collection (USA).
Minneapolis Bell = James Ford Bell Library, University of Minnesota (USA)
Modena Estense = Biblioteca Estense Universitaria, Modena (I)
München SB = Bayerische Staatsbibliothek, München (D)
Nantes BM = Bibliothèque Municipale (F)
Napoli BIG = Biblioteca dell'Istituto di Geografia, Universitaria (I)
New Haven Yale = Beinecke Rare Book & Manuscript Library, Yale University (USA)
New York HSA = Hispanic Society of America, New York (USA)
New York PL = New York Public Library (USA)
Newport News MM = Mariner's Museum of Virginia (USA)
Nürnberg SA = Staatsarchiv, Nürnberg (D)
Nürnberg GNM = Germanisches National Museum, Nürnberg (D)
Oxford Bodl = Oxford University Bodleian Library (GB)
Oxford Magd = Oxford University Magdalen College Library (GB)
Oxford CC = Oxford University Christ Church Library (GB)
Paris BArs = Bibliothèque de 'lArsenal (F)
Paris BMaz = Bibliothèque Mazarine (F)
Paris BN = Bibliothèque Nationale de France (F)
Pécs UL = Pécs University Library (H)
Perugia BU = Perugia Biblioteca Universitaria (I)
Pisa BU = Biblioteca Universitaria (I)
Portland USM = University of Southern Maine, Osher Map Library (USA)
Princeton UL = Princeton University Library (USA)
Providence JCB = John Carter Brown Library, Brown University (USA)
Quimper BM = Municipal Bibliothèque (F)
Regensburg SB = Staatliche Bibliothek, Regensburg (D)
Rennes BM = Bibliothèque Municipale de Rennes (F)
Rotterdam MM = Maritiem Museum, Rotterdam (NL)
Roma SGI = Societa Geografica Italiana, Rome (I)
Roma BVat = Biblioteca Apostolica Vaticana (V)
Rostock UB = Universitätsbibliothek Rostock (D)
Sint-Niklaas KOKW = Koninklijke Oudheidkundige Kring van het Land van Waas (B)
Stanford SC = Stanford University Libraries Special Collection (USA)
Stockholm KB = Kungliga Biblioteket (S)
Stuttgart WL = Würtembergische Landesbibliothek (D)
Tervuren RMCA = Royal Museum for Central Africa, Tervuren (B)
Toulouse BM = Toulouse Municipal Bibliothèque (F)
Torino BIDB = Biblioteca dell'Istituto Internazionale Don Bosco, Torino (I)
Utrecht UB = Universiteitsbibliotheek (NL)
Venezia BNM = Biblioteca Nazionale Marciana, Venezia (I)
Venezia Correr = Biblioteca del Museo Correr (I)
Washington FSL = Folger Shakespeare Library (USA)
Washington LC = Library of Congress (USA)
Weimar HAAB = Herzogin Anna Amalia Bibliothek, Stiftung Weimarer Klassik (D)
Wien ÖBN = Österreichische Nationalbibliothek (A)
Wien UB = Universitätsbibliothek Wien (A)
Wesleyan UL = Wesleyan University Library, CT. (USA)
Wolfenbüttel HAB = Herzog August Bibliothek (D)
Zürich ZB = Zentralbibliothek (CH)

Alphabetical Index of Map Titles

The following is an alphabetical list of map titles. The number after the title is the number of the map in the cartobibliography. The wording of [Map of Africa] signifies that the map has no title on it.

[Map of Africa] / [Antonio Francanzano (Fracan) Da Montalbodo]. 1.
[Map of Africa] / [Caius Plinus Secondus]. 2.
[Map of Africa] / [Leo Africanus – Jean Temporal] 5.
[Map of Africa] / [Paulo Forlani] 6.
[Map of Africa] / [Giovanni Francesco Camocio - Paulo Forlani] 8.
[Map of Africa] / [Giacomo Gastaldi – Fabio Licinio] 9.
[Map of Africa] / [Giacomo Gastaldi – Giovanni Franceso Camocio] 11.
[Map of Africa] / [Claudio Duchetti]. 19.
[Map of Africa] / [Giuseppe Rosaccio] 29.
[Map of Africa] / [Giuseppe Rosaccio] 30.
[Map of Africa] / [Levinus Hulsius]. 48.
[Map of Africa] / [G]iuseppe Rosaccio – Giuseppe Moretti]. 144.
[Map of Africa] / [Johann Ulrich Müller]. 155.

A New and Exact Map | of | AFRICA | and the Islands thereun to | belonging | Anno 1666 103.
A | new and most Exact map | of | AFRICA | Described by N:I: Vischer and don | into English Enlarged and Corrected acording | to I Bleau and Others With the Habits of y^e | people & ye manner of ye Cheife sitties: ye like neuer before | LONDON | Printed Colloured and are to be sould by Iohn Ouerton at ye White | horse in Little Brittaine neare the Hospitall | 1668 109.
A New Mapp of | AFRICA | Designed by Mounsi^r Sanson, Geograph^r | to the French King. Rendered into | English and Ilustrated with Figurs | By Richard Blome By the Kings | Especiall Command | 1669 111.
A New Mapp of | AFRICA | Divided into Kingdoms and | Provinces Sold by J Overton | at y^e White Horse with out New= | gate and by Philip Lea at y^e | Atlas and Hercules in | Cheapside | LONDON 141.
Accuratißima | Totius | AFRICÆ | TABULA | in Lucem producta | Per Iacobum de Sandrart | Norimbergæ. 165.
Affrica... 26.
AFRICA. 20.
AFRICA. 28.
AFRICA. 32.
AFRICA. 34
AFRICA. 35.
AFRICA. 36.
AFRICA. 37.
AFRICA. 38.
AFRICA. 40.
AFRICA. 42.
AFRICA. 43.
AFRICA. 44.
AFRICA. 47.
AFRICA. 51.
AFRICA. 56.
AFRICA. 49.
AFRICA. 69.
AFRICA. 71.
AFRICA. 76.
AFRICA. 84.

AFRICA. 93.
AFRICA 102.
AFRICA 114.
AFRICA 117.
AFRICA 119.
AFRICA. 121.
AFRICA 126.
AFRICA 128.
AFRICA 139.
AFRICA. 134.
AFRICA 147.
AFRICA 148.
AFRICA 152.
AFRICA 154.
AFRICA 159.
AFRICA. 167.
AFFRI= | CAE TA= | BVLA | NOVA 21.
AFRIC | A 15.
AFRICA | Ex magna orbis ter- | r´ descriptione Gerardi | Mercatoris de sumpta, | Studio & industria | G.M. | Iunioris. 31.
AFRICA | pars orbis 3. ad Meridiem exposita magnisq[ue] | solis ardoribus obnoxia est, terminatur ab ori: | ente Mari Rubro, a meridie Oceano Aethiopico, ab | occidente mari Atlantico, a Septentrione undiq[ue] | mari Mediterraneo: Regiones multas insignes ha: | bet, quarū interiora loca adhuc non satis sunt | cognita, multa tamen inculta & arenis sterilib | obducta, aut ob cæli situm humanæ habita | tioni incommoda, multo deniq[ue], venefico ani: | maliū genere infestata, sed ubi inco: | liture eximie fertilis est. 41.
AFRICA | Petrus Kærius Cælavit 75.
AFRICA | VETVS | Autore N. Sanson Abbavilleio | Christianiss. Galliar. Regis Geographe | Parisiis | Apud Autorem | Et apud Petrum Mariette via Iacoboea | sub signo spei |1650 | Cum Privilegio | Annorum Viginti 79.
AFRICA | IOANNE BAPTISTA NICOLOSIO S.T.D. | Sic Describente 94.
AFRICA | Antiqua | et | Nova 98.
AFRICA | Antiqua | et | Nova 140.
AFRICA | Antiqua | et | Nova 166.
AFRICA | by | R. Morden 125.
AFRICA | nova Tabula | Auct Jud Hondio 73.
AFRICÆ NOVA DELINEATIO. 33.
AFRICA | Divided according to the extent of its Principall Parts | in which are distinguished one from the other | the EMPIRES, MONARCHIES, KINGDOMS, STATES, and PEOPLES, | which at this time inhabite AFRICA. | To the Most Serene and Most Sacred Majesty of | CHARLES II. | by the Grace of God KING of Great Brittain, France, and Ireland. | This Map of AFRICA, is humbly Dedicated, and Presented, By | your Majesties Loyal Subject, and Servant | William Berry. 124.
AFRICA | divisa in suas principales partes, | nempe: Imperia | MONARCHIAS, REGNA, PRINCIPA- TUS, ET INSULAS | Per .S.r Sansonium, Geographum | Regis Galliae Ordinarium. | Noribergæ, | ap. Johannem Hoffmannum. 127.
AFRICA | EX MAGNÆ | ORBIS TER= | RÆ DESCRI= | PTIONE. 25.
AFRICA TERTIA PARS TERRÆ. 24.
AFRICA XVIII NOVA TABVLA 3.
AFRICÆ 151.
AFRICÆ ACCURATA TABULA | ex officina | Nic. Visscher. 87.
AFRICÆ | ACCURATA TABULA | ex officina | IACOBUM MEURSIUM. 108.
AFRICÆ | ACCURATA TABULA | ex officina | GERARDI A SCHAGEN. 115.
AFRICÆ | ACCURATA | TABULA | EX OFFICINA | JOAN: DE RAM. 130.
AFRICÆ | DEI MATER | Alicubi nota | & H Æ C ibi- | dem benefica. | 1699. 169.
AFRICÆ | DESCRIPTIO 64.
AFRICÆ | DESCRIPTIO. 53.
AFRICÆ | Descriptio. 90.
AFRICÆ | Descriptio Nova | Impensis | Henrici Seile | 1652 83.

AFRICÆ, described, | the manners of their Ha | bits, and their buildinge: newly | done into English by I. S. | and published at the cha: | rges of G. Humble Ano 1626 62.
AFRICÆ | nova descriptio. | Auct. Guil: Janssonio.   57.
AFRICAE | NOVA DESCRIPTIO | Impensis | Annæ Seile | 1663.   100.
AFRICAE | NOVA DESCR. | Auctore Petro Kærio | Excusum in edibus Amsterodami | Anno Domini 1614   55.
AFRICÆ | nova | discriptio.   99.
AFRICÆ | nova Tabula. | Auct. Jud: Hondio. 58.
AFRICÆ | nova Tabula | Auct. Jud. Hondio 60.
AFRICÆ | nova Tabula | Auct J. Hondio. 65.
AFRICÆ | nova Tabula. | Auct: Jud: Hondio 66.
AFRICÆ | noua | TABVLA 80.
AFRICAE | NOVISS:CATERES | TABULA | AUCT: I. BORMEESTER 150.
Africae | ta bula nou | a 18.
AFRICAE TABVLA XII. 22.
AFRICÆ | TABULA 23.
AFRICAE TA: | BVLA NOVA. 39
AFRICAE VERA FOR= | MA, ET SITVS. 27.
AFRI= | QUE 144.
AFRI= | QUE | Par | N. De Fer 135.
AFRI= | QUE | Par | N. De Fer 142.
AFRI= | CAE TA= | BVLA | NOVA. 12.
AFRI= | CAE TA= | BVLA | NOVA. 16.
AFRICAE TA | BVLA NOVA 164.
Africæ ut terra mariq[ue], lustrata est, proprijßima | ac verè genuina descriptio, obseruatis ad unguem gra= | dibus longitudinis et latitudinis. Autore M. Iacobo | Castaldo . 14.
AFRICA | VETVS | NICOLAI SANSON Christianiss. Galliar. Regis | Geographi. | Recognita Emendata et | Multis in locis Mutata, | Conatibus Geographicis | GULIELMI SANSON N. FILII. | Lutetiae Parisiorum | Apud Petrum Mariette Via Iacobaea | sub Signo Spei. | CIƆ IƆ C LXVII | Cum Privilegio ad Viginti Annos. 106.
AFRIQVE | ANCIENNE 133.
AFRIQVE | ANCIENNE | das Alte | Africa 137.
AFRIQVE | MODERNE 132.
AFRIQVE | MODERNE | Newes | Africa 136.
AFRIQVE | Par N. Sanson d'Abbevile, Geog. du Roy | A Paris | Chez l'Auteur | Et chez Pierre Mariette, rue S. Iacques a l'Esperance | 1650 | Auec privilege du Roy pour vingt ans 78.
AFRIQVE. | Par le Sr. Sanson d' Abbeville, Geographe du Roy. | Avec Privilege pour vingt ans. | A Paris | chez l'Auteur. | 1656. 86.
AFRIQVE | Par P. DV VAL | Geogr. du Roy. 95.
AFRIQUE, | Par N. SANSON Geographe ordinaire du-Roy | corrigeé et changeé en plusieurs endroits | suivant les Relations les plus recentes; | Par le Sr SANSON le Fils. | A PARIS; | Chez PIERRE MARIETTE rue S. Iacques a l'Esperance | Avec privilege de S. Mai. te po. 20 ans. | 1668. 107.
AFRIQVE | Par P. DV VAL | Geogr. du Roy. 143.
AFRIQVE | selon les Relations les plus Nouvelles | Dresseé et Dedieé | Par le P. Coronelli Cosmographe de la Sere- | nissime Republique de VENISE. | A Monseigneur le Duc de BRISSAC | Pair de France. | A PARIS | Chez. I.B. Nolin. sur le Quay de l'Horloge du Palais, | proche la Rue de Harlay, a l'Enseigne de la | Place des Victoires. | Avec Privilege du Roy | 1689. 146.
A= | PHRI= | CA 46.
Carte de | L'AFRIQVE | Corrigeé et augmenteé | desus toutes les aultres | cy deuant faictes P.Bertius | L'anneé 1624 61.
Carte de | L'AFRIQVE | Corrigeé, et augmenteé, deßus toutes | les aultres cy deuant faictes par | P. Bertius. 63.
Carte . de | L'AFRIQVE | Corrigeé, et, augmenteé, deßus toutes | les aultres cy deuant faictes par | .P.Bertius . | Anno · 1639 68.
Carte de | L'AFRIQVE | Corrigeé et augmenteé, deßus toutes | les aultres cy deuant faictes 89.
CARTE | DE | L'AFRIQVE | Nouuellement | Dresseé sur les Memoires des | Meilleurs Geographes de nostre | temps et distinguée suiuant les | Royaumes, souuerainetés et | principales parties, qui | se trouuent iusques | apresent 1669. 112.

L'AFRICA | Dedicata | All' Eccell.mo Sig.r Principe d'Auellino etc | Da popoli Africani, i quali son' auezzi à fissar gli occhi ne'mostri | troppo saro' Commendato per auer offerto la geografia di quelle | Prouincie a V.E. ch'e un mostro di Grandezza. ella come gran | cancelliere del Regno rende nobiltà ad infiniti vomini, fareb= | be diuenir umani que' nazionali se ne auesse un | giorno il Domino, e resto | Di V.E. | Oss.mo et Umil.mo Ser.re | Paolo Petrini. 172.

L' A | FRICA Dedicata All' Illustriss.mo ed Exccell.mo Sig.r | Principe CARMINE NICCOLÒ CARACCIOLO | Principe di S. Buono etc. | L'Africa... and at the end of 17 lines of text, the imprint of Paolo Petrini. 173.

L'AFRICA | Diuisa nelle sue Parti secondo le piùu moderne relationi | colle scoperte dell' origine, e corso del NILO, | descritta dal | P.M. Coronelli M.C. Cosmografo della | SERENISSIMA REPUBLICA DI VENETIA | è dedicata all' | ECCELLENZA DEL SIGNOR | GRAN CONTESTABILE | COLONNA | gia V^ce Rè d'Aragona, poi | V^ce Rè di Napoli. 153.

L'AFRICQUE, | divisée en ses principales parties, | et en ses ESTATS, ROYAUMES, &c. | Tirée de Livio Sanuto, de Marmol, et autres ·› | Par le Sr SANSON d'Abbeville Geogr ord^ie de Sa Majesté | A PARIS. | Chez Pierre Mariette, Rue S Iaques a l'Esperance | Avec Privilege du Roy pour Vingt Ans . | 1666. 104.

L'AFRIQUE 110.

L'AFRIQUE | divisée suivant l'estendie de ses principales parties | ou sont distinqués les vns des autres | LES EMPIRES, MONARCHIES, ROYAUMES, ESTATS, ET PEUPLES, | qui partagent aujourd'huy L'AFRIQUE. | sur les Relations les plus Nouvelles. | Par le S^r. SANSON, Geographe Ordinaire du Roy. | PRESENTÉE | A MONSEIGNEUR LE DAUPHIN | Par son tres-humble, tres-obeissant, et tres-fidele Serviteur | Hubert Iaillot 118.

L'AFRIQUE | divisee suivant l'estendue de ses principales parties | ou sont distinqués les vns des autres. | LES EMPIRES, MONARCHIES, ROYAUMES, ESTATS, ET PEUPLES, | qui partagent aujourd huy L'AFRIQUE | sur les Relations les plus Nouvelles | Par le S^r. SANSON, Geographe Ordinaire du Roy. | PRESENTEE | A MONSEIGNEUR LE DAUPHIN | Par son tres humble tres obeissant et tres fidele Serviteur | Hubert Iaillot 156.

L' AFRIQUE | Diuiseé en ses | EMPIRES, ROYAUMES, | ET ETATS, | à l'Usage de | MONSEIGNEUR | le DUC de | BOURGOGNE | Par son tres Humble | et tres Obeissant | Serviteur H. JAILLOT | A PARIS | Avec Privilege du | Roy 1694 158.

L' AFRIQUE | divisée suivant l'estendüe de ses Principles Parties, | ou sont distingués les vns des autres | LES EMPIRES, MONARCHIES, ROYAUMES, ESTATS, ET PEUPLES. | qui partagent aujourd'huy L'AFRIQUE. | sur les Relations les plus Nouuelles: | Par le Sr SANSON, Geographe Ordinaire du Roy. | DEDIÉ AU ROY | Par son tres humble tres obeissant tres-fidele Sujet et Seruiteur, | HUBERT IAILLOT. | 169[5] 160.

L'AFRIQUE | diviseé suivant l'estendie | de ses principales parties. | ou sont. distingués les vns des autres | LES EMPIRES, MONARCHIES, | ROYAUMES, ESTATS, et PEUPLES, | qui partagent aujour d'huÿ L'AFRIQUE. | sur les Relations les plus Nouuelles | par G. Valck 161.

L'AFRIQUE | Dreßez Sur les dernieres | Relations par N. de Fer | Avec Privilege du Roy | 1693. | A Paris chez l'Auteur dans | l'Isle du Palais á la Sphere | Royale 157.

L'AFRIQUE | Dressée Selon les dernieres Relat. | et Suivant les Nouvelles decouvertes | dont les Points Principaux Sont | placez Sur les Observations de | M^rs. de l'Academie Royale des Sciences. | Par N. de Fer. | A PARIS chez l'Auteur dans l'Isle du | Palais Sur le Quay de l'Orloge a la Sphere | Royale 1700. Avec Privil: du Roy. 170.

L'AFRIQUE, | Dressée selon les dernieres Relations et suivant | les Nouvelles decouvertes dont les Points prin= | cipaux sont placez sur les Observations de | M.rs de l'Academie Royale des Sciences. | Par N. de Fer. | A Paris. | Chez l'Autheur dans l'Isle du Palais sur le Quai de | l'Orloge a la Sphere Royale. 171.

L'AFRIQUE | Dresseé sur les Observations de M^rs. de | l'Academie Royale des Sciences, et | quelques autres, & sur les Memoi= | res les plus recens. | Par G. DE L'ISLE, Geographe | A PARIS. | Chez l'Autheur Rue des Canettes | prez de S^t. Sulpice. | Avec Privilege du Roy; pour | 20. Ans 1700. 174.

L'AFRICA | Nuouamente corretta, et | accresciuta secondo le relationi piu | moderne da GVGLIELMO SANSONE, | Geografo di S.M. Christianissima. | E data in luce da GIO · GIACOMO DE | ROSSI, in Roma, nella sua Stamperia | alla Pace l'Anno 1677. 120.

L' AFRIQUE | Où sont exactement decrites | toutes les Costes de la Mer, | suivant les Routiers et Portolans de divers Pilotes | et le Dedans du Pais | Selon les Relations les plus nouvelles. | Par P. DU-VAL Geographe du Roy | A PARIS | Chez l'Auteur, en l'Isle du Palais, Sur le quay de l'Orloge, | proche le coin de la rüe de Harlay | Avec Priuilege de sa Majesté | 1678 122.

L'AFRIQUE | Ou tous les Points Principaux | sont Placez,| SUR LES OBSERVATIONS DE | MESSIEURS DE L'ACADEMIE | ROYALE DES SIENCE | Par N: DE FER Geographe de | MONSEIGNEUR LE DAUPHIN. | A PARIS | Chez l'Autheur dans l'Isle du | du Palais à la Sphere | Royale. 1698. 163.

L'AFRIQVE | Par le Sieur Du Val Geographe Ord. du Roy | A PARIS 85.

L'AFRIQVE | Par P. Du Val | Geographe Ordinaire | du Roy. | A PARIS. | Chez l'Auteur prés le Palais. | Avec privilege du Roy | pour Vingt Ans. | 1664. 101.

L'AFRIQVE | Selon les | Autheurs les plus | Modernes 149.

LE IEV | DV | MONDE dedié | A Monsieur | Monsieur le Comte de Viuone | Premier Gentilhome de la Chambre | du Roy, par son tres humble et tres | obeissant serviteur | Du Val 74.

New, Plaine, & Exact Mapp of | AFRICA, described by: N:I: Vischer | and done into English, enlarged and | corrected, according to: I:Blaeu: with | the habits of the countries and | manner of the cheife Citties; | the like never before. 1658.| Printed, colered and Are to be | Sould by Robert Walton at the | Globe and Compass in St. Paules Church yard between | the two north doores. 88.

NOVA | AFRICA | DESCRIPTIO | Auct. F. de Wit. 96.

NOVA | AFRICA | Hugo Allardt | Excudit. 97.

NOVA | AFRICAE | Per | Nicolaum Io Visseher | Iacques Honervogt | Excudit 70.

NOVA AFRICAE GEOGRAPHICA ET HYDROGRAPHICA NOVA AFRICAE DESCRIPTIO, auct: G: Ianss. 54.

NOVA | AFRICÆ| TABULA | AUCTORE | Jodoco Hondio 52.

nova descriptio | AFRICÆ. 67.

NOVA TOTIVS AFRICAE (at the bottom of sheet one and two) DESCRIPTIO (at top of sheet five and six). 13.

NOVA | TOTIVS | AFRICAE | TABVLA | Auctore Guillel. Iansonio | Gottifredus de Scaicki | excudebat. 59.

NOVA | TOTIVS | AFRICAE | TABVLA | AVCTORE IOANNES BLAEV | AMSTELÆDAMI | Jo. Jacobus de Rubeis | Formis Romæ. 105.

NOVISSIMA | AFRICÆ | DESCRIPTIO. | AMSTELAEDAMI, Apud I. BLAEV. 1659. 91.

NOVISSIMA et | PERFECTISSIMA | AFRICÆ | DESCRIPTIO | EX FORMIS | CAROLI ALLARD. | AMSTELO – BATAVI: | Cum Privilegio Potentissimerum D.D. | Ordinum Hollandiæ et Westfrisiæ 162.

NOVISSIMA | et | PERFECTISSIMA | AFRICÆ | DESCRIPTIO | Authore | I. DANCKERTS | Amstelodami. | cum Privilegio. 168.

Nouuelle description | DAFRIQVE 72.

Nouuelle description | DAFRIQVE 81.

Nouuelle description | DAFRIQVE 82.

Nouuelle description | DAFRIQVE 92.

PRIMA TAVOLA 4., 7., and 10.

TABLE D'AFRIQVE. 17.

Tabula Aphricæ nova sumta | ex operibus Ludouici Terceræ | cosmographi Regiæ majestatis | Hispaniarum 45.

TOTIUS | AFRICÆ | Accuratissima | TABULA | Authore | FREDERICO DE WIT | Amstelodami. 114.

TOTIUS | AFRICÆ | Accuratissima | TABULA | Authore | FRIDERICO DE WIT | Amstelodami. 114A.

TOTIUS | AFRICÆ | Accuratissima | TABULA | Authore | I. DANCKERTS | Amstelodami. 131.

TOTIVS AFRICÆ NOVA ET EXACTA, ... SOVVERAINETES, ET PRINCIPALES PARTIES QVY SY TROVVENT A PRESENT 77.

TOTIVS AFRICAE NOVA ET EXACTA TABVLA EX OPTIMIS TVM GEOGRAPHORVM TVM ALIORVM | SCRIPTIS COLLECTA ET AD HODIERNAM REGNORVM PRINCIPATVVM ET MAIORVM PARTIVM DISTINCTIONEM ACCOMMODATA PER GVLIELMV BLAEW AMSTELODAMI M D C LXX 113.

Names of Persons

Anonymous **365-366** (map # 114A)

Aa, Pieter van der 248, 322
Abreu, Jorge de 31
Aefferden (or Van Afferden), Franciscus van 443-444
Afonso V, King of Portugal 26, 27
Aguiar, Jorge de 49
Al-Bakri 20
Al-Biruni, Abu Rayhan 18
Albrizzi, Giambatista 448, 498
Albuquerque, Afonso de 32, 54
Al-Idrisi, Mohammed 18, 19, 20
Allard, Carel 67, 329, 330, **468-469** (map # 162), 500
Allard, Hugo 66, 292, **324-325** (map # 97), 397, 399, 400
Almeida, Francisco de 31
Almeida, Manuel de 31, 32
Alvares, Francisco 31, 32, 46
Amaulry, Thomas 71, 414, 426, **430** (map # 145)
Andreae, Johann 451
Andreas, Lambert 179-180
Angelieri, Agostino 206-207
Aristotle 16
Arnoldi, Arnoldo di 59, **191-192** (map # 44)
Azambuja, Diogo de 29
Balbus, Cornelius 16
Baldini, Vittorio 158
Barbier, Jean Baptiste 439-440
Barbosa, Duarte (Odoardo) 112, 146
Barreto, Francisco 33, 34
Barros, João de 29, 33, 60, 63, 64, 96, 112, 122, 135, 146, 190, 197 209, 218
Bassett, Thomas & Richard Chiswell 242, 243, 273, 290
Battuta, Ibn 20
Beer, Johan Christoph 320, 382
Behaim, Martin 43-4
Belleforest, François de 59, **132** (map # 16)
Bellere, Jean 100
Bello, Sultan Mohammed 15
Berey, Nicolas 66, 67, 232, **285-286** (map # 81), 288, 312, **304-305** (map # 89), 356-358
Berey II, Nicolas 305
Berlinghieri 37, 38
Berry, William 71, **389-390** (map # 124)
Bertelli, Donato 59, 113, 115, 117, **126-128** (map # 13)
Bertelli, Ferando 57, 102, **114-115** (map # 10), 128
Bertius, Petrus 62, 66, 178, **223-224** (map # 56), 232, **238-240** (map # 61), **244-245** (map # 63),
    256-257 (map # 68), 305, 314
Bianco, Andrea 25, 26, 27
Billaine, Louis & Thomas Jolly 297
Bishop, George 100, 189-190
Blaeu, Joan 65, 66, 224, 227, **307-309** (map # 91)

Blaeu, Willem Jansz. 62-68, 186, 209-210, **214-219 (map # 54)**, 221, 224, **225-228 (map # 57)**, 231-232, 234, 237, 240, 242, 245, 248, 250, 252, 255, 257, 263, 265-266, 273, 275, 278, 280, 292-293, 300, 302, 307, 314, 322-323, 327, 340, 342, 351, 356-358, 360, 363, 370, 372-373, 380, 399-400, 405, 408-409, 422, 437, 460-461, 469, 479, 484, 502, 505

Blome, Richard 69, **354-355 (map # 111)**

Boisseau, Jean 66, 67, 232, 232, 248, **251-253 (map # 66)**, **265-266 (map # 72)**, 269, 285, 286, 288

Boissevin, Louis 251

Bormeester, Joachim 67, **441-442 (map # 150)**

Botero, Giovanni 59, 140-142, **166-167 (map # 32)**, 171, 173, 179-180, 182, 188, 190, 195-196, 206-207, **258-259 (map # 69)**

Buache, Philippe 498

Bucelin, Gabriel 59, **306 (map # 90)**

Bulderen, Henry van 414, 425-426

Buno, Conrad 67, **262-264 (map # 71)**, 313-314, **326-327 (map # 98)**, 421-422, 479

Buno, Johann 67, 263-264, 314, 326-327, **421-422 (map # 140)**, 479

Bünting, Heinrich 57, **149-151 (map # 24)**, **153 (map # 26)**, **204-205 (map # 50)**

Bussemacher, Johann 195-196

Cabral, Pedro Alvares 32, 43

Cadamosto (da Cà da Mosto), Alvise 29, 54, 79-80, 85, 96, 99, 146

Camocio, Giovanni Francesco 59, 102, **105-107 (map # 8)**, **116-117 (map # 11)**, 138, 146

Cantino, Alberto 45-46

Cão, Diogo 29, 30, 34, 41, 45, 46

Carvajal, Luis del Marmol 297

Castanhoso, Miguel de 31

Castro, João 47

Caveri, Nicolo de 47, 48, 49, 50, 51

Certe, Jean 69, 382, **427-428 (map # 143)**, 444

Chamereau (Chemereau), N. 403

Chaudiere, Guillaume, & Pierre L'Huilier 133-134

Cherso, Francesco de 27

Chesneau, Nicolas 132

Chetwind, Phillip 289-290

Child, Tim, & Abel Swall 460

Chiswell, Richard, & Thomas Bassett 242, 243, 273, 290

Chrysoloras, Emanuel 37

Chuan Chin (or Kwon Kun) 22

Claesz., Cornelis 20, 35, 59, 62, 63, 125, 164, **176-178 (map # 37)**, 184, 190, 193-194, **202-203 (map # 49)**, 208-209, 211-212, 218, 223, 265, 325, 418

Clapperton, Hugh 15

Cloppenburch, Jan Evertsz. 67, 190, 224, **249-250 (map # 65)**, 264

Cluver, Philipp 67, 248, 250, **262-264 (map # 71)**, 313-314, **326-327 (map # 98)**, **421-422 (map # 140)**, **478-479 (map # 166)**

Coignet, Michael 197

Columbus, Christopher 29, 30, 31, 44, 54

Compagnia Bresciana 181-182

Contarini, Giovanni 45, 48, 49, 50

Cordier, Louis 352, 375-376, 396, 454, 462-464

Cordova (Heirs), Diego Fernandez de 187-188

Coronelli, Vincenzo Maria 70, 71, **431-433 (map # 146)**, **447-448 (map # 153)**, 456, 472, 494, 497, 503

Cosa, Juan de la 44-47, 50

Covens, Johannes & Cornelis Mortier 345, 364, 453, 458-459, 498

Covilhã, Pero da 30, 31, 41

Cresques, Abraham 25

Croix, A. Phérotée de la 71, 414, 426, 430, **439-440 (map # 149)**, **480-481 (map # 167)**

Cunha, Tristan da 54

Curio, Coelius Augustin 94

Curti, Stefano 171
D'Ailly 23
D'Anania, Giovanni Lorenzo 57, 59, **131 (map # 15)**, **140-142 (map # 20)**, 166-167, 170-173, 190, 206-207, 258-259
D'Angiolo, (Jacobus Angelus), Jacopo 37
D'Avity, Pierre 239, 240, 255, 257, 268
Danckerts, Cornelis I 66, 245, **277-278 (map # 77)**, 360
Danckerts, Cornelis II 292-293
Danckerts, Dancker 66, **328-330 (map # 99)**
Danckerts, Justus 66, 67, 364, **406-408 (map # 131)**, **419-420 (map # 139)**, **482-484 (map # 168)**
Danckerts, Theodorus I, 408, 442, 484
Danckerts, Theodorus II 408, 484
Danet, Guillaume 456, 472-473, 488, 490
De Coup, 479
Delisle (Duval), Marie Angélique 498
Delisle, Guillaume 72, 472, **495-498 (map # 174)**
Dell'Aquila, Giuseppe Cacchii 131
Dias, Bartolomeu 30, 41-45, 47
Dias, Dinis 26, 29
Dias, Diogo 32, 33, 46, 96
Diest, Gielis Coppens van 118, 123
Diogenes 17
Doetecum, Joannes van 62, 129-130, 169
Doetecum II, Joannes van **193-194 (map # 45)**
Doetecum, Lucas 129-130, 169
Duchetti, Claudio 59, 102, **137-139 (map # 19)**
Dufresnoy, Abbé Nicolas Lenglet 298, 388, 403
Dulcert, Angelino 25
Duren, Jan (Juan) 444
Duval, Pierre 67-72, **270-271 (map # 74)**, 278, 288, **294-295 (map # 85)**, 311-312, **318-320 (map # 95)**, **333-334 (map # 101)**, **352-353 (map # 110)**, 362, **381-386 (maps # 121, 122)**, 393, 405, 413-414, **427-428 (map # 143)**, 432, 437, 440, 448, 451, 469, 472, 498, 501, 503
Eanes, Gil 25, 29
Echard, Laurence **449 (map # 154)**
Elwe, I. B. 458-459
Elzevier 67, 313-314
Ende, Josua van den 215-219, 234, 275-276, 308, 373
Eratosthenes 16, 17
Fei Xin 23
Fer, Antoine de 285, 288, 295, 413
Fer, Nicolas de 71, 363, **412-414 (map # 135)**, **425-426 (map # 142)**, **430 (map # 145)**, **455-456 (map # 157)**, 458-459, **470-473 (map # 163)**, **487-490 (maps # 170, 171)**, 492, 493-494
Fernandes, Alvaro 26
Fernandes, Antonio 33
Ferrer, Jaume 28
Ferrera, Duke (of Ercole d'Este) 45
Feuille, Daniel de la 405, 440, 481
Florian, Johannes 100
Florimi, Matteo 191-192
Forlani, Paulo 59, **101-102 (map # 6)**, **105-107 (map # 8)**, 122, 138, 146
Fra Mauro 25-27
Fries, Laurent 37, 55, 56, 96
Galle, Filips 59, 123, **135-136 (map # 18)**, **156-157 (map # 28)**, 181, 182, 183, 184, 197
Gama, Cristóvão da 31, 34
Gama, Vasco da 23, 30, 38, 45-47, 49, 54
Gastaldi, Giacomo 55, 57, 58-59, 84, **95-97 (map # 4)**, 99, 100, 102, **103-104 (map # 7)**, **108-113 (map # 9)**, 115, **116-117 (map # 11)**, 122, **126-128 (map # 13)**, 130, 131, 146, 152, 155, 163, 169, 177, 186, 190

Germanus, Nicolaus 41
Gerritsz, Hessel 217
Giunti, Lucatonio 69, 141-142, 258-259
Giunti, Thomaso 103-104
Gleditsch, Johann L. 71, **480-481 (map # 167)**
Glockengiesser 44
Goeree, F. 459
Gomes, Fernão 29
Goos, Abraham 241-242
Görlin, Johannes 305
Gottfried, Johann Ludwig 255
Gouwen, Giliam van der 468
Grassi, Bartolomeo 35
Greco, Giorgio 166-167
Grüniger, Johann (Joannes) 54, 55
Guerard I, Nicolas 495-498
Guerard II, Nicolas 471-473, 488
Hakluyt, Richard 100
Halma, François 401-402, 403, 414, 425, 426
Hamy, King 47
Harrewijn, Jacob (Jacques) 444
Hawqal, Ibn 20
Henry, Prince of Portugal 29
Herodotus 16, 20, 57, 60
Heylin (Heylyn), Peter 290, 332
Heyns, Pieter 136, 183
Heyns, Zacharias 59, **183-184 (map # 40)**
Hoffmann, Johann 69, 320, 349, 376, **381-382 (map # 121)**, 393, **395-396 (map # 127)**, 451
Hogenberg, Frans 123
Holbein, Hans 86
Hollar, Wenceslaus 66, **337-338 (maps # 103, 103A)**, 351
Homann, Johann Baptist 476-477
Homem, Vasco Fernandes 33, 34
Hondius, Henricus 139, 209-210, 230-231, 338
Hondius, Jodocus 59, 62, 65, 67, **147-148 (map # 23)**, 164, 177, **185-186 (map # 41)**, 194, 203, **208-213 (maps # 52, 53)**, 221, 223-224, 232, 237, 240, 248, 250, 300, 309
Hondius Jr., Jodocus 62, 65, 209, 223-224, **229-232 (map # 58)**, 235-237 (map # 60), 240, 242, 245, 250, 252-253, 257, 268-269, 284, 293, 305, 322, 323, 325, 338, 351, 363, 405, 422, 437
Honervogt, Jacques 62, **260-261 (map # 70)**
Hong Wu (Ming Emperor) 22
Honter, Johannes 55
Hooghe, Romein de 458
Houtman, Cornelis de 177
Huguetan, Marc 454
Hulsius, Levinus 62, **198-201 (map # 48)**
Humble, George 241-243
Humble, William 67, **272-273 (map # 75)**
Huot 311
Jaillot, Alexis-Hubert 65-66, 68, 70-72, 219, 347, **356-358 (map # 112)**, **374-376 (map # 118)**, 385, **395-396 (map # 127)**, 433, 448, **452-454 (map # 156)**, **457-459 (map # 158)**, **462-464 (map # 160)**
Janssonius, Johannes, 62, 65, 67, 178, **202-203 (map # 49)**, 205, 209, 210, 229-231, 235-237, **246-248 (map # 64)**, 250, 263-266, 285, 288, 300, 301, 312, 347, 356-358, 374-376, 385, 390, 392, 395-396, 411, 433, 437, 440, 448, 452, 454, 457-459, 462-464, 466-467, 472, 503
Janssonius van Waesberge, Johannes 250, 264, 479
Jansz., Jan 151, 173, 178, 184, **204-205 (map # 50)**, 211-212
Jenner, Thomas 66, **337-338 (map # 103)**
João, Infante 30

João II, King of Portugal 29, 30, 31, 44
Jode, Cornelis de 58, 59, 63, 113, 130, **154-155 (map # 27)**, **168-169 (map # 33)**, 186, 218
Jode, Gerard de 58, 59, 112-113, **129-130 (map # 14)**, 155, 169, 186
John, Prester 27-28, 30-31, 47, 64, 84, 127, 138, 150, 163-164, 169, 194, 208, 209, 218, 220, 220, 222, 303, 310, 351, 440
Jollain, Gerard 261, 269, 310, 311, 312
Jolly, Thomas, & Louis Billaine 297
Jonghe, Clement de 329-330
Joosten, Jacob 204
Juba (Berber King) 16
Kalperger 41
Karera (Heirs), Simonis Galignani de 170-171
Keerbergen, Johannes van 59, **197-198 (map # 47)**
Keere, Pieter van den 62, **220-222 (map # 55)**, 235, 237, 250, 261, 267, 269, 272
Keschedt, Petrus 59, 142, **172-173 (map # 35)**
Khalduna, Ibn 20
Kirchner, Ambrosius 150-151
Kühnen, Georg Wilhelm 450-451
Küsell, Melchior 306
L'Huilier, Pierre & Guillaume Chaudiere 133-134
Lafreri, Antonio 102, 115, 138
Lagniet, Jacques 294
Langenes, Barent 59, **176-178 (map # 37)**, 184, 190
Lasor a Varea (Raphael Savonarola) 171
Laurensz., Hendrick 178
Lawrence, John 69, **418 (map # 138)**
Lea, Philip 67, **423-424 (map # 141)**
Lebna Dengel (or King David) 31
Leclerc, Jean 59, **147-148 (map # 23)**
Leo Africanus 15, 56-57, 59-60, 63, 85, 96, **98-100 (map # 5)**, 112, 122, 142, 146, 163, 167, 171, 173, **189-190 (map # 43)**, 207, 209, 218, 250, 259, 340, 433, 473, 509, 524
Leo X, Pope (Johannes Leo de Medici) 99
Leonard, Juan (Jean) 443-444
Licinio, Fabio 59, **108-113 (map # 9)**, 117, 127
Lima, Rodrigo de 31
Linschoten, Jan Huygen van 177, 186, 218
Lo Hung-hsien (or Luo Hongxian) 20-21
Lobo, Jerome 31-32
Lochom, Michael van 239
Loon, Hendrik van 431, 432, 433, 471, 473, 489, 490
Lopes (Lopez), Duarte 34, 63-64, 186, 200
Lotter, Tobias Conrad 498
Lovisa, Domenico 59, **474-475 (map # 164)**
Lucena, Vasco Fernandes de 30
Lucini, Antonio Francesco 57, **335-336 (map # 102)**
Ludolf, Hiob 32, 432, 448, 472
Ma Huan 23
Macrobius 23
Madjid, Ahmad ibn 31
Madrignano, Archangelo 79
Maggiolo 47
Magini, Giovanni Antonio 59, 142, **170-173 (map # 34)**
Mallet, Alain Manesson 67, **409-410 (maps # 132, 133)**, **415-417 (maps # 136, 137)**
Malocello, Lanceroto or Lanzarotto 38
Mandeville, Sir John 24
Manesson Mallet, Alain 67, **409-410 (maps # 132, 133)**, **415-417 (maps # 136, 137)**
Manuel, King of Portugal 34, 45
Marchetti, Pietro Maria 59, **181-182 (map # 39)**

Mariette, Pierre 277-282
Mariette II, Pierre 66, 279-282, **283-284 (map # 80)**, 339, 344-345
Marinus of Tyre 17
Martellus, Henricus 30, 40-46
Mascardi, V. 315-317
Massaio, Pietro 51-52
Maternus, Julius 16
Mathisz., Jan 66, **291-293 (map # 84)**, 300, 307, 325, 371
Mattiolo, Pietro 96
Mazza, Giovanni Battista 59, **152 (map # 25)**
Medrano, Sebastián Fernández de 67, **443-445 (map # 151)**
Mela, Pomponius 16
Mercator, Gerard 34, 60-61, 63, 64, 66
Mercator II, Gerard 61-62, **162-165 (map # 31)**, 194, 209, 502
Merian, Matthäus 66, **254-255 (map # 67)**
Metellus, Johannes Matalius 62, **179-180 (map # 38)**
Meteren, Emanuel van 226
Meufves, Alexandre de 253
Meurs, Jacob van 67, 301, **348-349 (map # 108)**
Moll, Herman 67, **393-394 (map # 126)**, 449, **460-461 (map # 159)**
Montalbodo, Antonio Francanzano (Fracan) da 54, **78-79 (map # 1)**
Moore, Jonas 69, **393-394 (map # 126)**
Morden, Robert 71, **391-392 (map # 125)**
Moresini, Francesco 152
Moretti, Giuseppe 59, 161, **429 (map # 144)**
Moretus Brothers 123
Mortier, Cornelis & Johannes Covens 345, 364, 453, 458-459, 498
Mortier, Pieter 71, **452-454 (map # 156)**, **457-459 (map # 158)**, 466, 469, 498
Mosting, Herman 67, **421-422 (map # 140)**
Moxon, James 57, 424, **446 (map # 152)**
Müller, Gottfried 262-263
Müller, Johann Ulrich 67, **450-451 (map # 155)**
Münster, Sebastian 57, **83-94 (map # 3)**, 112, 132, **143-144 (map # 21)**, 150
Muschio, Andrea 140-141
Mutapa (paramount chief of the Makalanga Confederation of Monomotapa) 33
Nelli, Niccolò 114-115
Guerard (the elder), Nicolas 495-498
Guerard Jr., Nicolas 471-473, 488
Nicolosi, Giovanni Battista 69, **315-317 (map # 94)**
Nobili, Pietro de 138-139
Noli, António 33
Nolin, Jean Baptiste 71, **431-433 (map # 146)**, 448, 456, 458, 472, 494, 497
Nolin II, Jean Baptiste 433
Nzinga Nvemba (or Dom Afonso I) 34
Ortelius, Abraham 31, 35, 58-64, 66, 85, 112, **118-125 (map # 12)**, 130-132, **135-136 (map # 18)**,
    141, 144, 148, 151-152, 155, **156-157 (map # 28)**, **181-182 (map # 39)**, **197-198 (map # 47)**,
    212, 218, 265, 306, 434, **474-475 (map # 164)**, 502, 504-505
Ottens, Reiner & Josua 458, 459, 483, 484, 498
Overton, John 67, **350-351 (map # 109)**, 392, **423-424 (map # 141)**
Padouani, Domenico 447-448
Paez, Pero 31, 32
Paiva, Afonso de 30
Pauli, Johann 479
Peeters, Jacques 67, **443-445 (map # 151)**
Petri, Heinrich 83
Petri, Sebastian 59, 94, **143-144 (map # 21)**
Petrini, Paolo 69, 71, **491-494 (maps # 172, 173)**
Phérotée de la Croix, A. 71, 414, 426, 430, **439-440 (map # 149)**, **480-481 (map # 167)**

Phillip II, King of Portugal 34
Picart, Huges 285-286
Picart, Nicolas 67, 248, **267-269 (map # 73)**, **287-288 (map # 82)**, **310-312 (map # 92)**
Pigafetta, Filippo 34-36, 63-64, 186, 200, 203, 209, 218
Pisarri, Antonio 429
Plantin, Christoffel 100
Plato 16
Pliny the Elder 16, **81-82 (map # 2)**
Polo, Marco 24, 33, 43-44, 57
Polybius 16
Ponton, Pedro 444-445
Porro, Girolamo 166-167, 171
Pory, John 59, **189-190 (map # 43)**
Prester John 27-28, 30-31, 47, 64, 84, 127, 138, 150, 163-164, 169, 194, 208, 209, 218, 220, 220, 222, 303, 310, 351, 440
Ptolemy, Claudius 16-17, 19-20, 27, 35, 37-38, 40, 43-44, 51, 64, 72, 80, 84, 85, 96, 102, 122, 131, 158, 170, 172, 218, 282, 345, 432, 448, 472, 497
Quad, Matthias 62, 173, 180, **195-196 (map # 46)**
Ram, Johannes de 67, **404-405 (map # 130)**
Ramusio, Giovanni Battista 57, 60, 63, **95-97 (map # 4)**, 99-100, 102, **103-104 (map # 7)**, 115, 122, 131, 163, 190, 209, 218
Ravani, Pietro de 82
Reinel, Jorge 25, 47
Reinel, Pedro 40
Renard, Louis 495
René, Duke of Lorraine 48, 50
Ribbius, Joannes 297, 388, 401-403
Ribeiro, Diogo 47
Riebeeck, Jan de 35
Ringman, Matthias 31
Robbe, Jacques 71, **412-414 (map # 135)**, **425-426 (map # 142)**, **430 (map # 145)**
Robert, Lewes 290
Roger II, Norman King of Sicily 20
Rosaccio, Giuseppe 59, 142, **158 (map # 29)**, **159-161 (map # 30)**, **206-207 (map # 51)**, **429 (map # 144)**
Rosselli, Francesco 42-43
Rossi, Giovanni Giacomo de 66, **341-343 (map # 105)**, **359-360 (map # 113)**, **379-380 (map # 120)**
Rughesi, Fausto 59, **174-175 (map # 36)**
Ruscelli, Girolamo 258
Ruysch, Johannes 39
Sandrart, Jacob von 67, **476-477 (map # 165)**
Sanson, Guillaume 69, **344-347 (maps # 106, 107)**, 355, 360, 363, 375-376, 380, 390
Sanson, Nicolas 68, 71, **279-282 (maps # 78, 79)**, **296-298 (map # 86)**, **339-340 (map # 104)**, **387-388 (map # 123)**, **401-403 (map # 129)**
Sanson II, Nicolas 345
Santos, João dos 34
Sanuto, Livio 59, **145-146 (map # 22)**
Saulsbury, Thomas 449
Sauzet, Henri du 250
Scaicki (van Schayck), Godefridus de 66, **233-234 (map # 59)**
Schaep, Gerard 299-301
Schagen, Gerrit Lucasz. van 67, 301, **367-368 (map # 115)**
Schenk, Pieter 437-438, 466, 467, 498
Scherer, Heinrich **485-486 (map # 169)**
Scolari, Stefano Mozzi 66, 127, **274-276 (map # 76)**
Sebastião, King of Portugal 33, 35
Seile, Anne 66, **331-332 (map # 100)**

Seile, Henry 66, **289-290 (map # 83)**
Seller, John 57, 71, **377-378 (map # 119)**, **411 (map # 134)**, 418
Senex, John, 479
Septimus, Flaccus 16
Sernigi, Girolamo 96
Sessa, Melchior 57, **81-82 (map # 2)**
Sintra, Pero de 29, 54, 80, 96, 99
Soler, Guillelmus 25
Soligo, Christoforo 30
Solinus 84
Solis, Hernando de **187-188 (map # 42)**
Solomon, King 57
Sonnius, Michael 132
Sparke, Michael 212
Speed, John 66, **241-243 (map # 62)**, 273, 290
Strabo 16, 17, 122
Stridbeck, Jr., Johann 57, **434-435 (map # 147)**
Swall, Abel & Tim Child 460
Sylvanus 37
Tassin, Christophe (Nicolas) 245, 285
Tavernier, Melchior 66, 232, **244-245 (map # 63)**, **256-257 (map # 68, attr.)**, 278, 280
Teixeira, Luis 62, **193-194 (map # 45)**
Tellez, Baltazar 32
Temporal, Jean 57, **98-100 (map # 5)**
Thevet, André 57, **133-134 (map # 17)**
Tillemon (Du Tralage), Jean Nicolas 432
Todeschi, Pietro 66, **372-373 (map # 117)**
Tosi, Francesco 159-160
Trevethen, William 289-290
Tristão, Nuno 29
Valk, Gerard 66, 71, **436-438 (map # 148)**, **465-467 (map # 161)**
Valk, Leonard 437, 467
Vaughan, Robert 331-332
Velde, Jan van de 324-325
Verbist, Pieter 216
Verdussen, Henricus (Henry) Cornelius 444
Vesconte 25
Vespucci, Amerigo 54
Visscher, Claes Jansz. 178, 216-217, 221, 261, 300, 322-323
Visscher I, Nicolaas 64-65, 67, 177, 216-217, 221-222, **299-301 (map # 87)**, 303, 307, 322-323, 349-351, 367
Visscher II, Nicolaas 66, 293, 300, **397-400 (map # 128)**, 442, 469
Vivaldi, Ugolino and Vadino 28
Vrients, Joan Baptista 123, 125, 155, 157, 169, 198
Waesberge, Johannes Janssonius van 250, 264, 479
Waldseemüller, Martin 35, 37, 39, 47-48, 50-52, 54-56, 80, 96, 163
Walton, Robert 62, **302-303 (map # 88)**
Weidmann (Heirs), Moritz Georg 480
Weleslavina, Daniel Adam z **153 (map # 26)**
Wesel, Jacques van 297
Winter, Antoine de 69, 297, 388, **401-403 (map # 129)**, 414, **425-426 (map # 142)**
Wit, Frederick de 66, 67, **321-323 (map # 96)**, **361-366 (maps # 114, 114A)**, **369-371 (map # 116)**
Wolf, Jermiah 498
Wolters, Johann 67, **478-479 (map # 166)**
Yi Hewi (or Yi Hoe) 22
Zanni, Agostino de 82
Zenaro, Damiano 145-146

Zheng He 18, 23
Zhu Siben (or Chu ssu-Pen) 20
Zunner, Johann David 67, 69, 297, **387-388 (map # 123)**, 415-417 **(maps # 136, 137)**

Photo Credits

Cartobibliography only, reference to map numbers.

Amsterdam UB: 12, 14.2, 40, 45, 64.1, 87, 114e, 116, 118, 131, 158, 162
Austin HR: 36
Author's collection: all other maps
Bedburg-Hau, Antiquariat Gebr. Haas: 9, 55.4, 56, 85.2, 85.3, 101.2, 101.4
Berlin SB: 167
Berlin, Nicolas Struck Book and Map Dealer: 113
Bern SUB: 127
Besançon BM: 143
Bonn, Hellwig Collection: 15
Boston Afriterra: 16, 17, 22, 79, 93, 100, 109, 125
Budapest NSL: 24, 168
Cambridge Harvard: 3 (1542), 156, 160
Chicago, Baskes Collection: 81, 82, 86, 92.1
Evanston NWU: 6, 114, 124,
Gent, Antiquariaat Sanderus: 3 (1540 and details), 60.2
Greenwich NMM: 76C, 76D, 139
Grenoble BM: 51, 149,
Hertfordshire, Burden Collection: 92.3, 92.4, 102, 152, 155
Johannesburg PL: 19
København KB: 122 (wall map)
Königstein Reiss & Sohn: 114.4
La Jolla, Barry Lawrence Ruderman Antique Maps: 138
Leeuwarden Tr 129.3
Leiden UB: 18, 28, 60.1, 99.3, 114
London BL: 1.2, 5.3, 20, 21, 26, 30, 35, 39, 43, 44, 47, 74, 84, 91, 94/2, 99.3, 103, 111, 114.2, 115, 119, 129, 144, 145, 148, 154, 159, 161
London BM: 147
London RGS: 160,
London, Altea Gallery: 66.2
London, Ashley Baynton-Williams: 85.1
London, Bernard Shapero Rare Books: 10, 141
London, Hubbard Collection: 129, 135, 142, 151, 157
Minneapolis Bell: 11
München SB: 123, 151
New Jersey, Private Collection: 25
New York PL: 2, 8.2, 29, 90, 95, 121, 126
New York, Graham Arader Galleries: 117
New York, Martayan Lan, Fine Antique Maps: 112, 122
New York, Richard B. Arkway Rare Books Inc.: 110, 122.2
Newport News MM: 32
Nürnberg Staatsarchiv: 33
Orvieto, Museo dell'Opera del Duomo: 105.1
Paris BN: 49, 54, 70, 77, 99.1, 104, 128B
Paris Loeb-Larocque and Pierre Joppen/Paulus Swaen: 23, 61, 63, 73, 89, 101.1
Private Belgian collection: 150.2, 171
Private Collection: 75
Private European collection: 50
Rotterdam MM: 8.4, 14, 37, 41, 105
Stanford SC: 130

Stockholm KB: 128A
Stopp collection: 57.1
Toronto, University of Toronto Libraries: 103A
United States MacLean Collection: 59, 173
Utrecht UB: 57.3, 150, 166
Venezia, Biblioteca del Museo Correr: 13
Venezia BNM: 8
Washington LC: 1, 54.5, 57.1, 66.2, 69, 70.2, 72, 99.2, 120, 129.4, 134, 136, 137, 157.2, 164, 170
Washington, Private Collection: 7
Wien, Österreichische Akademie der Wissenschaften: 114A